Plant & Soil Water Relationships

McGRAW-HILL SERIES IN ORGANISMIC BIOLOGY

Kramer: Plant and Soil Water Relationships: A Modern Synthesis

Plant & Soil Water Relationships: A Modern Synthesis

Paul J. Kramer, Ph.D.

James B. Duke Professor of Botany
Duke University

McGraw-Hill Book Company

New York, St. Louis, San Francisco, London, Sydney, Toronto, Mexico, Panama

Plant and Soil Water Relationships:
A Modern Synthesis

Library of Congress Catalog Card Number 77–77958

35348

1 2 3 4 5 6 7 8 9 0 HDMM 7 6 5 4 3 2 1 0 6 9

To Edith Vance Kramer

Botany

Preface

This book attempts to cover the entire field of soil-plant-atmosphere water relationships in an integrated manner, using modern terminology and concepts. It contains discussions of the properties of water and solutions, cell water relations, soil water, root systems, water and salt absorption and movement through plants, transpiration, the effects of water deficits on plants, and the measurement of plant water stress.

The interrelationships among the processes involved in water movement from soil to roots, through plants, and into the air has been emphasized by use of the Gradman-van den Honert concept. This concept treats water movement through the soil-plant-air continuum as analogous to flow of electricity through a conductor, the rate being determined by the differences in potentials and resistances in various parts of the system. Although an oversimplification, this approach is useful both in teaching and in research because it permits identification and evaluation of the importance of the various factors affecting water movement at each stage in its progress through the plant.

This treatment emphasizes the fact that no step in the movement of water through plants can be treated entirely independently of other steps: all must be considered in relation to one another. For example, the rate of water absorption is closely linked to the rate of transpiration through the continuous sap stream of the xylem, but it also is affected by the resistance to water flow through roots and by various soil factors which affect the availability of water to the roots. The rate of transpiration is controlled primarily by the energy supply, but it also is affected by stomatal opening and the supply of water to the leaves. Finally, the internal water balance of plants is controlled by the relative rates of absorption and transpiration.

The terminology has been modernized by the use of the term "water potential" in place of diffusion pressure deficit to describe the free energy or chemical potential of water in various parts of the soil-plant system. The term potential seems preferable to older terms such as suction force or diffusion pressure deficit because it permits separation and evaluation of the various components (osmotic, matric, and pressure) which affect the free energy of water in soils and plants. Furthermore, potential is widely used in soil science and the physical sciences and its meaning is clear to workers in many fields.

This book is intended for teachers, investigators, and students in both basic and applied plant sciences. It should be useful to botanists, agronomists, foresters, horticulturists, soil scientists, and others interested in plant water relations. An attempt has been made to present the basic facts underlying the various phenomena in relatively simple terms, intelligible to readers from all fields of plant science. If the treatment of some areas seems inadequate to specialists in these areas, they are reminded that this book was not written for specialists in plant and soil water relationships, but for those plant scientists who need an introduction to the general field. It is believed that it will aid plant scientists in developing a better understanding of how soil and plant water relationships affect the growth of plants. Such information is not only basic to an understanding of plant physiology and ecology, but is also important to workers in the various fields of applied plant science.

The need for a book which summarizes the entire field of plant water relations has been increased by the tremendous amount of publication during the past two decades. In 1949 there were only two other books in the field of plant water relations. Today there are a dozen books and thousands of papers in this rapidly expanding field. The great volume of publication makes it difficult to produce a balanced summary of the literature. An attempt has been made to provide examples of the important literature in various fields and from various parts of the world, but many good papers are omitted for lack of space to discuss them. However, enough have been cited on each topic to introduce the reader to the literature of that field.

In addition to summarizing the literature, the author has attempted to evaluate it and to draw what seem to be logical conclusions about various processes. Since these conclusions naturally reflect the personal viewpoint of the author, it is inevitable that some readers will disagree with certain of them. Furthermore, everyone should understand that in expanding areas such as plant physiology our information is often too limited to permit final conclusions, and as more is learned, it may become necessary to modify or even to abandon some of our present beliefs. Many so-called scientific facts are true only in the sense that they represent the most logical conclusions that can be drawn from existing data. When the data are incomplete, different workers often come to differing conclusions, and acquisition of further information may necessitate revision of existing views.

The author is indebted to numerous colleagues, friends, and graduate students for many valuable suggestions. He is especially grateful to R. O. Slatyer of the Australian National University, Canberra, whose extensive contributions to this book would have justified his name appearing as coauthor, were he not already the author of an important book in this field. Readers should not be surprised if they find similarities between the treatments of certain topics in the two books, because both were started while Dr. Slatyer was spending a year at Duke University, and he made numerous contributions to this work. The entire manuscript was read by M. R. Kaufmann and D. W. Lawlor, and certain chapters were read by H. D. Barrs, K. R. Knoerr, C. W. Ralston, and C. P. Weatherspoon. It has undergone considerable revision since it was reviewed by these persons, hence the author assumes full responsibility for any errors which are found in the text.

The author gratefully acknowledges the support provided by a series of grants from the National Science Foundation.

These grants greatly aided the author's research program on plant water relations and have also contributed both directly and indirectly to the writing of this book.

Paul J. Kramer

Contents

one

Water & Its Role in Plants

This chapter deals with the ecological and physiological importance of water, its various roles in plants, its peculiar properties, how it occurs in cells and tissues, and the forces involved in its movement. The terminology of cell water relations also will be discussed.

The importance of water

The importance of water will be treated in terms of its ecological and physiological roles. Much additional material can be found in the first three volumes of Ruhland's "Encyclopedia of Plant Physiology," also in Crafts et al. (1949), Hagan et al. (1967), Kozlowski (1964), Rutter and Whitehead (1963), Slatyer (1967), and Slavik (1965).

Ecological importance of water

The distribution of vegetation over the surface of the earth is controlled more by the availability of water than by any other single factor. Regions where rainfall is abundant and fairly evenly distributed over the growing season have lush vegetation. Examples are the rain forests of the tropics, the vegetation of the Olympic Peninsula of the Northwestern United States, and the luxuriant cove forests of the Southern Appalachians. Where summer droughts are frequent and severe, forests are replaced by grasslands, as in the Steppes of Asia and the prairies of North America. Further decrease in rainfall results in semidesert with scattered shrubs, and finally deserts.

Even the effects of temperature are partly exerted through water relations because decreasing temperature is accompanied by decreasing rates of evaporation and transpiration, and increasing temperature is accompanied by increasing rates. Thus, an amount of rainfall adequate only for grasslands

HEREDITARY POTENTIALITIES

Depth and extent of root systems

Size, shape and total area of leaves, and ratio of internal to external surface

Number, location, and behavior of stomata

ENVIRONMENTAL FACTORS

SOIL. Texture, structure, depth, chemical composition and pH, aeration, temperature, waterholding capacity, and water conductivity

ATMOSPHERIC. Amount and distribution of precipitation

Ratio of precipitation to evaporation

Radiant energy, wind, vapor pressure and other factors affecting evaporation and transpiration

↓

PLANT PROCESSES AND CONDITIONS

Water absorption

Ascent of sap

Transpiration

Internal water balance as reflected in water potential, turgidity, stomatal opening, and cell enlargement

Effects on photosynthesis, carbohydrate and nitrogen metabolism, and other metabolic processes

↓

QUANTITY AND QUALITY OF GROWTH

Size of cells, organs, and plants

Dry weight, succulence, kinds and amounts of various compounds produced and accumulated

Root/shoot ratio

Vegetative versus reproductive growth

FIG. 1.1 **Diagram showing how the quantity and quality of plant growth is controlled by hereditary and environmental factors operating through the internal processes and conditions of the plant, with special reference to factors affecting water relations.**

in a hot climate can support forests in a cooler climate where the rate of evapotranspiration is much lower. This fact was responsible for development of the concept of the rainfall-evaporation ratio by Transeau (1905) and for the inclusion of data for potential evapotranspiration along with temperature in formulas from which life zones are predicted from climatic data (Holdridge, 1962). Furthermore, temperature may affect life processes as much by its effects on the properties of water as by its effects on rates of chemical reactions (Tanford, 1963).

Physiological importance of water

The ecological importance of water is the result of its physiological importance. The only way in which an environmental factor such as water can affect plant growth is by affecting internal physiological processes and conditions. The interrelationships between plants and their environment with respect to water are summarized in Fig. 1.1.

Almost every plant process is affected directly or indirectly by the water supply. Many of these effects will be discussed later, but it can be emphasized here that within limits metabolic activity of cells and plants is closely related to their water content. For example, the respiration of young, maturing seeds is quite high, but it decreases steadily during maturation as water content decreases (see Fig. 1.2). Incidentally, the decrease in water content during maturation of seeds encased in fleshy fruits, such as tomato, is diffi-

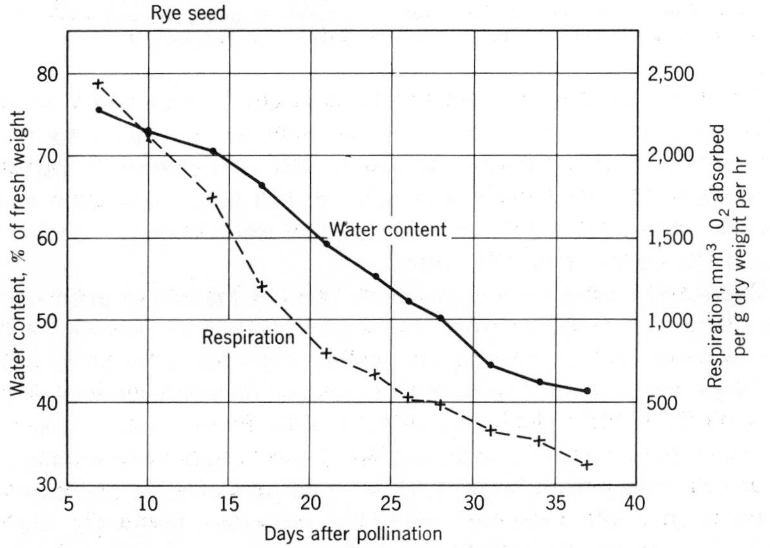

FIG. 1.2 Decrease in water content and respiration during maturation of rye seed. This shows the relationship between water content and rate of respiration often found in plant tissues. (*From Shirk,* 1942.)

cult to explain (McIlrath et al., 1963). The respiration rate of air-dry seeds is very low and increases slowly with increasing water content up to a critical point, where there is a rapid increase in respiration with further increase in water content (Fig. 1.3). The practical importance of the relationship between water content and respiration is shown in connection with storage of grain, because high water contents result in "heating" and spoilage.

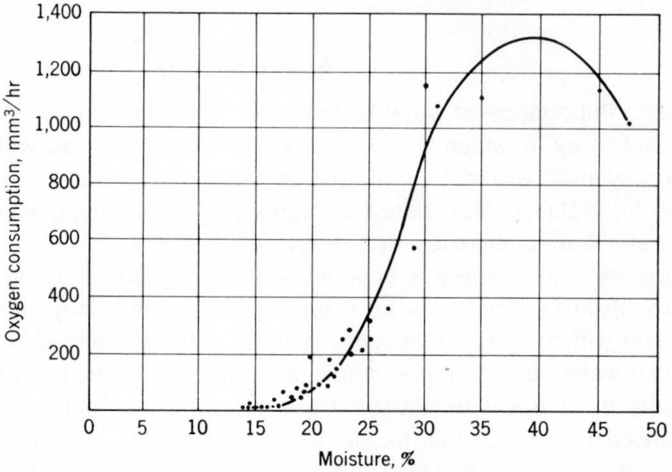

FIG. 1.3 **Relationship between water content of oat seeds and rate of respiration. Note the rapid increase in respiration with increasing water content above approximately 16 percent. Presumably most of the water is so firmly bound in air-dry seeds that it is unavailable for physiological processes. (*From Bakke and Noecker,* 1933.)**

Growth of plants is controlled by rates of cell division and enlargement and by the supply of organic and inorganic compounds required for the synthesis of new protoplasm and cell walls. Cell enlargement is particularly dependent on at least a minimum degree of cell turgor, and stem and leaf elongation are quickly checked or stopped by water deficits (Loomis, 1934; Miller, 1965; Thut and Loomis, 1944).

Decrease in water content invariably reduces the rate of photosynthesis (Brix, 1962) and usually reduces the rate of respiration. However, with decreasing water content, drying pine needles and a few other kinds of tissue sometimes show a temporary increase in respiration before the final decrease and death (Brix, 1962; Montfort and Hohn, 1950; Parker, 1952). In summary, decreasing water content is accompanied by loss of turgor and wilting, cessation of cell enlargement, closure of stomata, reduction in photosynthesis, and interference with many basic metabolic processes. Eventually, continued dehydration causes disorganization of the protoplasm and death of most organisms. The effects of water deficits on physiological processes are discussed in more detail in Chap. 10.

Uses of water in plants

The importance of water can be summarized by mentioning its most important functions, which will be grouped under four general headings.

A CONSTITUENT OF PROTOPLASM. Water is as important quantitatively as qualitatively, constituting 80 to 90 percent of the fresh weight of most herbaceous plant parts and over 50 percent of the fresh weight of woody plants. Some data on water content of various plant structures are shown in

Table 1.1 Water Content of Various Plant Tissues Expressed as Percentage of Fresh Weight *

Plant parts		Water content %	Authority
Roots	Barley, apical portion	93.0	Kramer and Wiebe, 1952
	Pinus taeda, apical portion	90.2	Hodgson, 1953
	Pinus taeda, mycorrhizal roots	74.8	Hodgson, 1953
	Carrot, edible portion	88.2	Chatfield and Adams, 1940
	Sunflower, av. of entire root system	71.0	Wilson et al., 1953
Stems	Asparagus stem tips	88.3	Daughters and Glenn, 1946
	Sunflower, av. of entire stems on 7-week-old plant	87.5	Wilson et al., 1953
	Pinus banksiana	48–61	Raber, 1937
	Pinus echinata, phloem	66.0	Huckenpahler, 1936
	Pinus echinata, wood	50–60	Huckenpahler, 1936
	Pinus taeda, twigs	55–57	McDermott, 1941
Leaves	Lettuce, inner leaves	94.8	Chatfield and Adams, 1940
	Sunflower, av. of all leaves on 7-week-old plant	81.0	Wilson et al., 1953
	Cabbage, mature	86.0	Miller, 1938
	Corn, mature	77.0	Miller, 1938
Fruits	Tomato	94.1	Chatfield and Adams, 1940
	Watermelon	92.1	Chatfield and Adams, 1940
	Strawberry	89.1	Daughters and Glenn, 1946
	Apple	84.0	Daughters and Glenn, 1946
Seeds	Sweet corn, edible	84.8	Daughters and Glenn, 1946
	Field corn, dry	11.0	Chatfield and Adams, 1940
	Barley, hull-less	10.2	Chatfield and Adams, 1940
	Peanut, raw	5.1	Chatfield and Adams, 1940

* *From Kramer (1955a)*

Table 1.1. Water is just as important a part of the protoplasm as the protein molecules which constitute the protoplasmic framework, and reduction of water content below some level is accompanied by changes in structure and ultimately in death. It is true that a few plants and plant organs can be dehydrated to the air-dry condition, or even to the oven-dry condition in the case of some kinds of seeds and spores, without loss of viability; but a marked decrease in physiological activity always accompanies decrease in water content.

SOLVENT. A second essential function of water in plants is as the solvent in which gases, minerals, and other solutes enter plant cells and move from cell to cell and organ to organ. The permeability of most cell walls and membranes to water results in a continuous liquid phase extending throughout the plant in which translocation of solutes of all kinds occurs.

REAGENT. Water is a reactant or reagent in many important processes, including photosynthesis and hydrolytic processes such as the hydrolysis of starch to sugar. It is just as essential in this role as carbon dioxide or nitrate.

MAINTENANCE OF TURGIDITY. Another essential role of water is in the maintenance of the turgidity so essential for cell enlargement and growth, and for maintaining the form of herbaceous plants. Turgor also is important in the opening of stomata and the movements of leaves, flower petals, and various specialized plant structures. Inadequate water to maintain turgor results in an immediate reduction of vegetative growth.

Properties of water

The importance of water in living organisms results from its unique physical and chemical properties. Bernal (1965) comments that nearly all of its properties are anomalous. The importance of these properties in life processes has been recognized for several decades and discussed by many writers from Henderson (1913), Bayliss (1924), and Gortner (1938) to Hutchinson (1957, pp. 195–200). Important collections of papers were published in *Symposium no.* 19 *of the Society for Experimental Biology* and in vol. 125, art. 2 of the *Annals of the New York Academy of Sciences.* The properties of water and solutions also are discussed in some detail in chap. 1 of Slatyer (1967).

Water plays such a special role in living organisms that we can scarcely imagine life without it. In fact, the absence of water on the moon and planets would be regarded as sufficient evidence that life as we know it cannot exist there.

Isotopes of water

The three isotopes of hydrogen having atomic weights of 1, 2, and 3 make it possible to differentiate "tracer water" from ordinary water. In the 1930s "heavy water," water containing deuterium (hydrogen of atomic weight 2), became available and was used widely in biochemical studies. It was also used extensively in studies of permeability of animal and plant membranes (Ordin and Kramer, 1956; Ussing, 1953, for example). However, in recent years deuterium has been largely supplanted as a tracer by water containing tritium (hydrogen of atomic weight 3), for example, in the experiments of Raney and Vaadia (1965). Tritium is radioactive and therefore more convenient as a label, being easier to detect than deuterium, which requires use of a mass spectrometer.

A stable isotope of oxygen with an atomic weight of 18 makes it possible to study the role of the oxygen in water. An example is the series of experiments with $H_2{}^{18}O$ which demonstrated that the oxygen released during photosynthesis comes from water rather than from carbon dioxide, as was previously supposed (Ruben et al., 1941).

Unique properties of water

Data on the chemical and physical properties of water were collected recently by Kavanau (1964) and Kohn (1965). A few of its more unique properties will be discussed. Water has the highest specific heat of any substance except liquid ammonia, which is about 13 percent higher. The standard unit for measuring heat, the calorie, is the amount of energy required to warm one gram of water one degree. The heat of vaporization is the highest known, 540 cal/g at 100°C, and the heat of melting, 80 cal/g, is also unusually high. The high specific heat of water tends to stabilize temperatures and is reflected in the uniform temperature of islands and land near large bodies of water. The high heat of vaporization means that evaporation of water has a pronounced cooling effect and condensation has a warming effect. Water likewise is an extremely good conductor of heat compared with other liquids and nonmetallic solids, although it is poor compared with metals.

Water is fairly transparent to visible radiation. It allows light to penetrate and makes it possible for algae to carry on photosynthesis and grow to considerable depths in bodies of water. It is opaque to long wavelengths, so that water filters are fairly good heat absorbers (see Fig. 1.4).

Water has a much higher surface tension and viscosity than most other liquids because of the high internal cohesive forces between molecules. This also provides the tensile strength required by the cohesion theory of the ascent of sap. It has a density which is exceeded among liquids only by

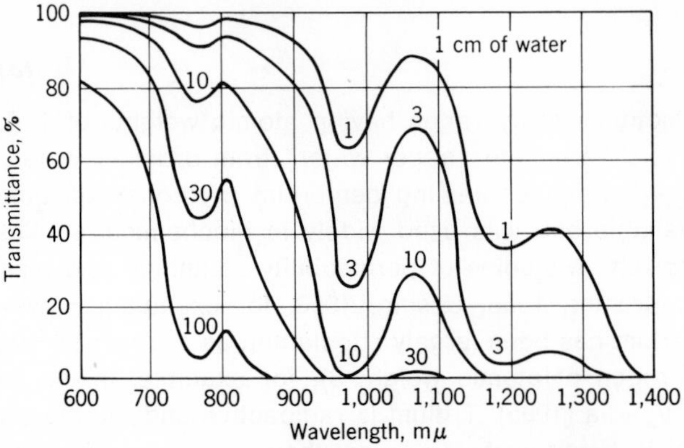

FIG. 1.4 **Transmission of radiation of various wavelengths through layers of water of different thicknesses. The numbers on the curves refer to the thickness of the layers. Note that transmission is high in shorter wavelengths and decreases for longer wavelengths. (*After Hollaender*, 1956, vol. 3, p. 195.)**

molten metals such as mercury and is remarkable in having its maximum density at 4°C instead of at freezing. Even more remarkable is the fact that water expands on freezing, so that ice has a volume about 10 percent greater than the liquid water from which it was formed (see Fig. 1.5). This explains the fact that ice floats and that pipes and radiators burst when the water in them freezes. Incidentally, if ice sank, bodies of water in the cooler parts of the world would all be filled permanently with ice.

Water is very slightly ionized; only one molecule in 5.5×10^8 is disso-

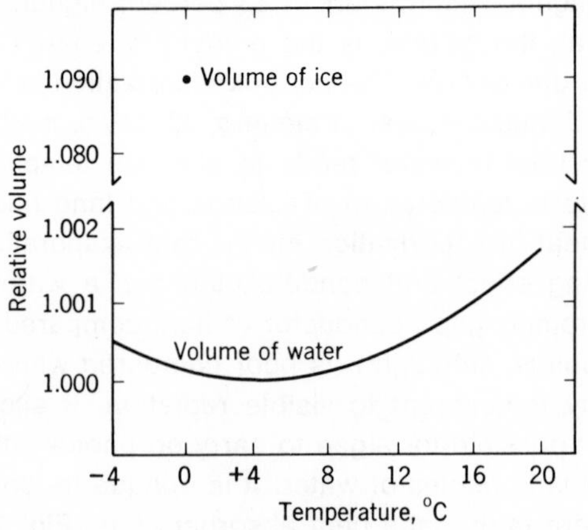

FIG. 1.5 **Change in volume of water with change in temperature. Note that the minimum volume is at 4°C and that below that temperature there is a small increase in volume because more molecules are incorporated into the lattice structure. Above 4°C there is an increase in volume due to increasing thermal agitation of the molecules. The volume of ice is much greater than that of liquid water at the same temperature because all molecules are incorporated into the widely spaced lattice.**

ciated. Since a hydrogen ion is a bare proton and cannot exist alone, it becomes associated with a water molecule to form a hydronium ion. The dissociation of water can therefore be described as $2H_2O \rightarrow H_3O^+ + OH^-$. Because it is so little ionized, water has a high dielectric constant which contributes to its behavior as an almost perfect solvent. It is a good solvent for electrolytes because the attraction of ions to the positive and negative charges on water molecules forms dipole bonds. As a result, each ion is surrounded by a shell of water molecules which keeps ions of opposite charge separated. It is a good solvent for most nonelectrolytes because it forms hydrogen bonds with amino and carboxyl groups. It tends to be adsorbed or bound strongly to the surfaces of cellulose, clay micelles, protein molecules, and many other substances. This characteristic is of great importance in soil and plant water relations.

Explanation of unique properties of water

The unusual combination of properties which makes water so important ecologically and physiologically has been recognized for several decades (Bayliss, 1924; Gortner, 1938; Henderson, 1913). Although the idea was proposed by Latimer and Rodebush (1920), this unique combination of properties has only recently been explained by assuming that water molecules are associated in an orderly structure by hydrogen bonding (Bernal, 1965; Crafts et al., 1949; Kavanau, 1964; Pauling, 1960; Tanford, 1963).

To make this explanation clearer we shall review briefly the kinds of forces which operate among atoms and molecules. First are the ordinary chemical bonds such as the covalent bonds between oxygen and hydrogen in water or carbon and hydrogen in methane. These are broken during chemical reactions, as when starch is hydrolyzed to sugar. If covalent bonds were the only type, there would be no liquids or solids, because covalent bonding does not hold molecules together. However, there are various physical or electrical forces, such as the effects of dipoles, van der Waals' forces, and hydrogen bonds, which operate between adjacent molecules. Some molecules are polar or electrically unsymmetrical because they have positively and negatively charged areas which attract one molecule to another. Water molecules with a structure $\begin{matrix} H^+ \\ \rangle O^= \\ H^+ \end{matrix}$ show this dipole effect rather strongly, while substances such as methane or carbon tetrachloride do not because they have no unsymmetrical distribution of electrical charges.

There is a tendency of the positively charged nucleus of a molecule to attract the negatively charged electrons of neighboring molecules. These attractions, called van der Waals' forces, are rather weak and are effective only if the molecules are close together. Liquefaction of gases depends on van der Waals' forces holding the molecules together, and the boiling point of most liquids therefore depends on the energy required to break the van der Waals' forces. Liquefied gases evaporate rapidly because the molecules are so weakly bound together.

Physical properties such as boiling point, heat of vaporization, viscosity, and surface tension depend on the strength of intermolecular bonding. Obviously, the peculiar properties of water depend on intermolecular forces much stronger than those already mentioned. These attractive forces are supplied by hydrogen bonds, which result from attraction of the hydrogen atoms of one water molecule to the oxygen atoms of adjacent molecules. The forces thus produced by the peculiar distribution of charges on water molecules (Bernal, 1965; Tanford, 1963) bind water molecules in the symmetrical lattice structure shown diagrammatically in Fig. 1.6. In ice, all water molecules are arranged in a lattice with unusually wide spacing. This arrangement results in a density less than that of liquid water.

As ice melts, about 15 percent of the bonds break, and perhaps 8 percent of the molecules escape from the lattice. This results in a partial collapse of the lattice and an increase to maximum density at 4°C. As the temperature rises above 4°, increase in thermal agitation of the molecules produces an

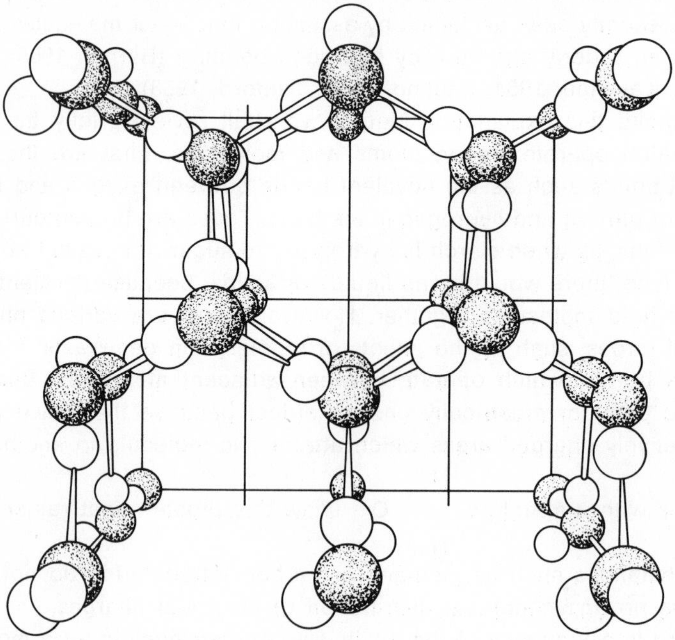

FIG. 1.6 **Diagram showing how water molecules are bound together in a lattice structure by hydrogen bonds in ice. The dark spheres are hydrogen atoms; the light spheres, oxygen atoms. In liquid water the molecules would be more loosely organized. (*After Buswell and Rodebush,* 1956.)**

increase in volume (see Fig. 1.5). There is some uncertainty about the exact structure of liquid water, and some investigators propose that it consists of "flickering clusters" or "icebergs" of structured molecules separated by regions of randomly arranged molecules (Kavanau, 1964). In any event, it seems that the structure is only partly destroyed by increase in temperature and at least 70 percent of the hydrogen bond energy found in ice remains at 100°C.

The structure is somewhat modified by the pH, which affects the distance of the hydrogen from the oxygen atoms, and by ions, because of their attraction for water molecules. Ions form dipole bonds with water molecules. The result is that the ions become surrounded by shells of water molecules which are firmly bound (Fig. 1.7). In fact, Bernal (1965) describes ions, protein

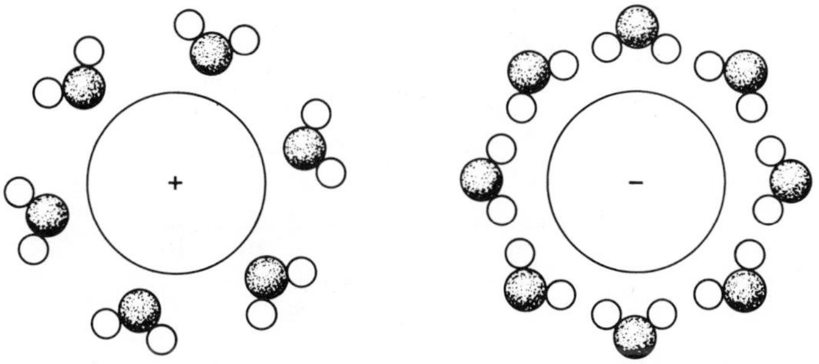

FIG. 1.7 **Arrangement of water molecules in oriented layers or shells around ions with the hydrogen ions nearest to cations, the oxygen ions nearest to anions. These shells tend to separate ions of opposite charge and enable them to exist in solution. They also disrupt the normal structure of water and slightly increase the volume.** (*After Buswell and Rodebush,* 1956.)

molecules, and cell surfaces in general as being coated with "ice," i.e., with layers of structured water molecules. Solution of alcohols, amides, and other polar organic liquids in water results in a more strongly structured system than occurs in the separate substances. This is shown in the high viscosity of such solutions. For example, the viscosity of an alcohol-water mixture at 0°C is four times that of water or alcohol alone. However, this structure is easily broken by high temperatures (see Fig. 1.8). The addition of nonpolar substances such as benzene or other hydrocarbons to water breaks hydrogen bonds and produces "holes" or disorganized areas in the structure which are surrounded by areas with a tighter structure. The water bound on large molecules such as proteins has an important effect on their structure; and Tanford (1963) cites evidence suggesting that the relative stability of the structure of viruses, DNA, and globular proteins is determined by the stability of the water associated with them.

It was mentioned earlier that the changes in volume of water during freezing and thawing are caused by changes in the proportion of water

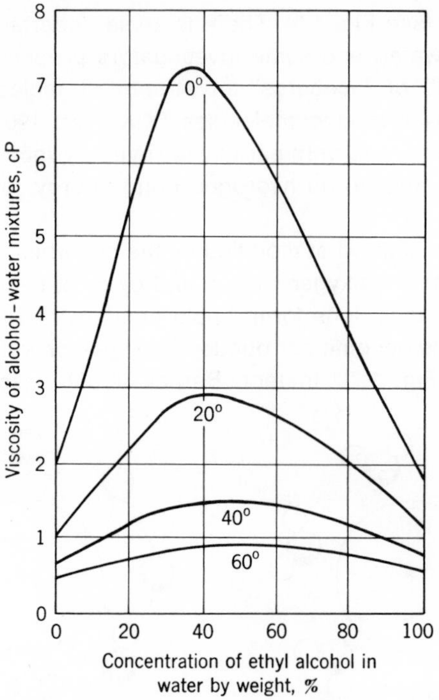

FIG. 1.8 **The effect of ethanol on the viscosity of water. Mixtures of water with polar organic liquids show large increases in viscosity at low temperatures because they have a more orderly and compact structure than the component liquids. (*After Bingham and Jackson,* 1918.)**

molecules bound in an organized lattice by hydrogen bonds. The high boiling point results from the large amount of energy required to break hydrogen bonds, and two must be broken for each molecule evaporated. Methane (CH_4) has nearly the same molecular weight as water, but it boils at $-161°C$ because no hydrogen bonds occur and only a small amount of energy is required to break the van der Waals' forces holding the molecules together in the liquid.

The unusually high viscosity and surface tension also result from the fact that hydrogen bonds between water molecules resist rearrangement. Water wets and adheres to glass, clay micelles, cellulose, and other substances which have exposed oxygen atoms at the surface with which hydrogen bonds can be formed. It does not wet paraffin and other hydrocarbons because it cannot form hydrogen bonds with them. Water wets cotton, because it forms numerous hydrogen bonds with oxygen atoms of the cellulose molecules, but not nylon, which has few atoms with which hydrogen bonds can be formed.

It has been suggested by Drost-Hansen (1965) and others that there are discontinuities in the properties of water which are of biological importance. Drost-Hansen claims that there are anomalies in the properties of water at 15, 30, 45, and 60°C which are reflected in the behavior of living organisms. This claim has been received rather sceptically and Falk and Kell (1966) attribute the alleged discontinuities to errors in measurements.

Properties of aqueous solutions

In plant physiology we seldom deal with pure water because the water in plants and in their root environment contains a wide range of solutes. Therefore, it is necessary to understand how the properties of water in solution differ from those of pure water. Only a brief discussion is possible here, and readers are referred to physical chemistry texts for a full development of these ideas.

The characteristics of water in solution can be shown concisely by tabulation of its colligative properties, i.e., the properties associated with the concentration of solutes dissolved in it. These are shown in Table 1.2 and include the effects on vapor pressure, boiling and freezing points, and osmotic pressure.

Table 1.2 ***The Colligative Properties of a Molal Solution of a Nonelectrolyte Compared with Water***

	Pure water	Molal solution
Vapor pressure	4.58 at 0°C 760.0 at 100°C	Decreased according to Raoult's law
Boiling point	100°C	100.518°C
Freezing point	0°C	−1.86°C
Osmotic pressure	0	22.69 bars
Chemical potential μ_w	Set at zero	Decreased

VAPOR PRESSURE. The decrease in vapor pressure of water in solution is essentially the result of its dilution by the addition of solutes. This is shown by Raoult's law, which states that the vapor pressure of solvent vapor in equilibrium with a dilute solution is proportional to the mole fraction of solvent in the solution:

$$e = e^{\circ} \frac{n_w}{n_w + n_s} \qquad (1.1)$$

where e is the vapor pressure of the solution, e° the vapor pressure of pure solvent, n_w the number of moles of solvent, and n_s the number of moles of solute. This is strictly applicable only to dilute molal solutions, i.e., those prepared with moles of solute per 1,000 g of water.

BOILING AND FREEZING POINTS. The effects of solutes on the boiling and freezing points are exerted through their effects on the vapor pressure of water (see Fig. 1.9). The addition of solute lowers the freezing point be-

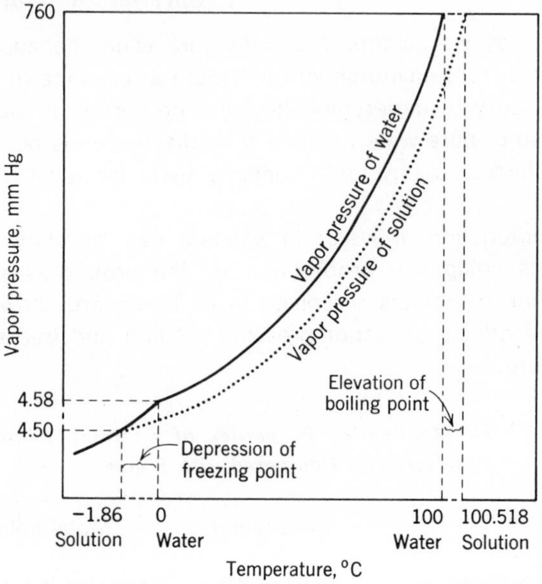

FIG. 1.9 **Effects of 1 mole of nonelectrolyte per 1,000 g of water (a one-molal solution) on the freezing and boiling points and vapor pressure of the solution. Note that this diagram is not drawn to scale.**

cause it lowers the vapor pressure and thus decreases the temperature at which the vapor, liquid, and solid phases are in equilibrium. It can be calculated that the vapor pressure at freezing of a molal solution of a nonelectrolyte in water is decreased from 4.58 to 4.50 mm and that a reduction in temperature of 1.86°C is required to bring about freezing (Daniels and Alberty, 1963).

Water boils when its vapor pressure is raised to that of the atmosphere. When the vapor pressure has been lowered by the addition of solute, the water in a solution must be heated to a higher temperature than pure water to produce the required increase in vapor pressure.

OSMOTIC PRESSURE OR OSMOTIC POTENTIAL. Raoult's law shows that the vapor pressure of water in a solution is lowered in proportion to the extent to which the mole fraction of water in the solution is decreased by adding solute. Therefore, if water is separated from a solution by a membrane permeable to water but impermeable to the solute, water will move across the membrane along a gradient of decreasing vapor pressure or chemical potential into the solution until the vapor pressures of solution and pure water become equal. The pressure which must be applied to the solution to prevent movement in a system, such as that shown in Fig. 1.10, is termed the osmotic pressure. It is often denoted by the symbol π.

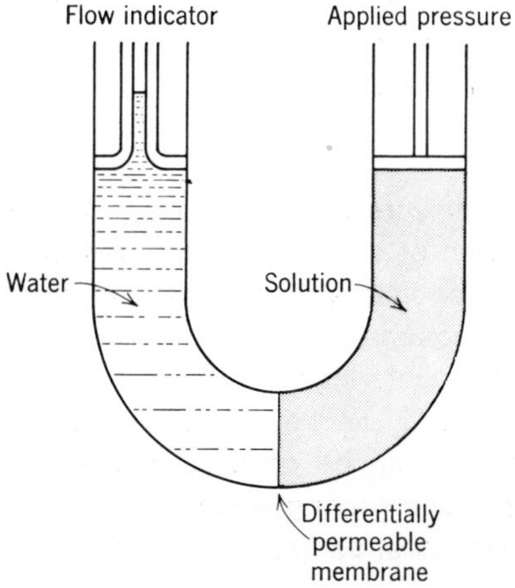

FIG. 1.10 **Diagram of an osmometer in which a membrane permeable to water but impermeable to the solute separates water from a solution. The osmotic pressure of the solution is equal to the pressure which must be applied to prevent movement of water into it. Movement is observed by change in level of water in the capillary tube indicator on the left.**

Van't Hoff developed an equation which relates the osmotic pressure to the concentration of the solution. Mathematically expressed,

$$\pi V = n_s RT \tag{1.2}$$

where π is osmotic pressure in bars or atmospheres, V the volume of solvent in liters, n_s the moles of solute, R the gas constant (0.0820 liter atm/degree mole), and T the absolute temperature. For 1 mole of solute in 1 liter of solvent at 273°K, this equation gives a value for π of 22.4 atm of 22.7 bars.

Direct measurements have shown that this relationship is approximately correct for dilute solutions of nondissociating substances. However, there are large deviations from the theoretical value for electrolytes which ionize in solution and release more particles than nondissociating substances. Thus, the osmotic pressure of a molal solution of NaCl is approximately 43.2 bars instead of the theoretical 22.7 bars. Some nondissociating molecules become hydrated or bind water molecules. This binding of water reduces the effective concentration of water and increases the observed osmotic pressure. An example is a sucrose solution where each sucrose molecule apparently binds six molecules of water, and the osmotic pressure of a molal solution of sucrose is approximately 25.1 bars instead of the expected 22.7.

The relationships between concentration, vapor pressure, freezing point, and osmotic pressure make it possible to calculate the osmotic pressure of a solution from the freezing point depression or from the vapor pressure. Since the theoretical depression of the freezing point of an ideal molal solution is 1.86°C and the osmotic pressure is 22.7 bars at 0°C, the osmotic pressure of a solution can be calculated from the depression of the freezing point (T) by the following equation:

$$\pi:22.7 = T:1.86 \tag{1.3}$$

where π is the osmotic pressure of the solution in bars and T is the observed depression. The derivation of this relationship can be found in Crafts et al. (1949), or in a physical chemistry text. The cryoscopic method is widely used; and if suitable corrections are made for undercooling, it gives accurate results. The problems encountered in using it on plant sap are discussed in Chap. 7 of Crafts et al. (1949) and by Barrs in Kozlowski (1968).

The osmotic pressure can be calculated from the vapor pressure, or more readily from the relative vapor pressure or relative humidity, $e/e^\circ \times 100$, because according to Raoult's law [see Eq. (1.1)],

$$e/e^\circ = \frac{n_w}{n_w + n_s} \tag{1.4}$$

In recent years thermocouple psychrometers have been used extensively to measure e/e°. Their operation will be discussed in Chap. 10 in connection with the measurement of water stress in plants.

There is an increasing tendency in the field of plant water relations to substitute potential for pressure and to use the term osmotic potential in place of osmotic pressure. The two terms are numerically equal, but osmotic potential carries a negative sign. The basis for the use of the term potential is discussed in the following section.

CHEMICAL POTENTIAL OF WATER. The chemical potential of a substance in a system is a measure of the capacity of that substance to do work. It is generally considered to be equal to the partial molal Gibbs free energy (Spanner, 1964). In a simple solution of a nonelectrolyte in water, the chemical potential of the water depends on the mean free energy per molecule and the concentration of water molecules, i.e., on the mole fraction of the water. The degree to which the presence of solute reduces the chemical potential of the water in the solution below that of pure free water can be expressed as

$$\mu_w - \mu_w^\circ = RT \ln N_w \tag{1.5}$$

where μ_w is the chemical potential of water in the solution, μ_w° is that of pure water (in units such as ergs per mole), R and T have the usual meaning, and N_w is the mole fraction of water. For use with ionic solutions the mole fraction is replaced by the activity of water a_w, and for general use, where water may not be in a simple solution, by the relative vapor pressure, e/e°. Equation (1.5) is then written:

$$\mu_w - \mu_w^\circ = RT \ln \frac{e}{e^\circ} \tag{1.6}$$

When the vapor pressure of the water in the system under consideration is the same as that of pure free water, ln e/e° is zero, and the potential difference is also zero. Thus, pure free water is defined as having a potential of zero. When the vapor pressure of the system is less than that of pure water, ln e/e° is a negative number; hence, the potential of the system is less than that of pure free water and is expressed as a negative number.

The expression of chemical potential in units of ergs per mole is inconvenient in discussions of cell water relations. It is more convenient to use units of energy per unit of volume. The measurements are compatible with pressure units which can be obtained by dividing both sides of Eq. (1.6) by the partial molal volume of water, $\overline{V}_w$ (cm^3/mole). The resultant term is called the water potential, Ψ_w:

$$\Psi_w = \frac{\mu_w - \mu_w^\circ}{\overline{V}_w} = \frac{RT \ln e/e^\circ}{\overline{V}_w} \tag{1.7}$$

The units of ergs per cubic centimeter, equivalent to pressure units of dynes per square centimeter, are usually expressed as bars or atmospheres. They are related as follows: 1 bar $=$ 0.987 atm $= 10^6$ dynes/cm^2 or ergs/cm^3.

The water potential in any system is decreased by those factors which reduce the relative vapor pressure:

(1) Addition of solutes which dilute the water and reduce its activity by hydration of solute molecules or ions

(2) Matric forces, which consist of surface forces, and microcapillary forces found in soils, cell walls, protoplasm, and other substances which adsorb or bind water

(3) Negative pressures or tensions such as those in the xylem of transpiring plants

(4) Reduction in temperature T

The water potential in any system is increased by those factors which increase the relative vapor pressure:

(1) Pressure, as that of the expanded cell wall on the cell contents

(2) Increase of temperature T

Cell water terminology

Study of cell water relations began in the second quarter of the nineteenth century, when Dutrochet introduced the concept of osmosis. Later Traube (1867) and Pfeffer (1877) developed the concept of differential permeability in membranes, and de Vries (1884) studied plasmolysis and explained the conditions necessary for development of turgidity in cells.

During the early part of the twentieth century it began to be realized that the movement of water from cell to cell cannot be explained in terms of gradients of osmotic pressure but rather in terms of gradients of what today is termed water potential. This led to the publication of a variety of terms intended to describe the ability of cells and tissues to absorb water. Among these terms were Saugkraft or suction force (Ursprung and Blum, 1916), water-absorbing power (Thoday, 1918), Hydratur (Walter, 1931, 1965), net osmotic pressure (Shull, 1930), diffusion pressure deficit (Meyer, 1938, 1945), water potential (Owen, 1952; Slatyer and Taylor, 1960), and others.

Diffusion pressure deficit

The term diffusion pressure deficit (usually abbreviated as DPD), introduced by Meyer and spread by the widely used textbook of Meyer and Anderson (1952), has been employed more extensively than any other. The DPD of water in a cell or solution was defined as the amount by which its diffusion pressure is lower than that of pure water at the same temperature and under atmospheric pressure. The DPD of a cell could be regarded as a measure of the pressure with which water would diffuse into the cell when immersed in pure water. The equation for the DPD of a cell is:

$$\mathrm{DPD} = \mathrm{OP} - \mathrm{TP} \tag{1.8}$$

where OP is the osmotic pressure of the cell contents and TP is the turgor pressure within the cell.

The diffusion pressure deficit terminology has been very useful, but plant science has developed to a level of scientific sophistication where it should use the more basic terminology of physical chemistry and thermodynamics. The term diffusion pressure is used rarely today in physical chemistry, and the term DPD puzzles workers in the physical sciences. In contrast, the term chemical potential is widely used in the physical sciences and in soil science. Its meaning is clear to investigators in many fields. Furthermore, use of the more basic terminology of thermodynamics makes it easier to separate and evaluate the various components: osmotic, matric, and pressure, which affect the total water potential of cells and tissues (Kramer et al., 1966; Slatyer and Taylor, 1960). Therefore, the term water potential will be used in this book instead of DPD.

The use of the terminology of thermodynamics is not new. Buckingham (1907) used the term capillary potential, and Edlefsen (1941) proposed the term specific free energy to describe what we now term the soil water potential. The net influx free energy of Broyer (1947) is equivalent to the term water potential. The terminologies proposed by Broyer and Edlefsen were never widely used, probably because plant scientists were not ready for them.

Cell water potential

The concept of water potential as a measure of the free energy status of water was developed earlier in this chapter and now will be applied to plant cells. Under isothermal equilibrium conditions the various factors involved in cell water relations can be summarized by the equation:

$$\Psi_{\text{cell}} = \Psi_s + \Psi_p + \Psi_m \tag{1.9}$$

where Ψ_{cell} is the potential of the water in the cell and the other terms express the contributions to Ψ_{cell} by solutes (Ψ_s), pressure (Ψ_p), and matric forces (Ψ_m). Ψ_s and Ψ_m are negative, Ψ_s expresses the effect of solutes in the cell solution, and Ψ_m expresses the effect of water-binding colloids and surfaces in the cell. Ψ_p is positive unless there is a negative wall pressure, a rare occurrence. In the xylem, however, Ψ_p may be negative during transpiration or positive in guttating plants as a result of root pressure. The sum of the three terms is a negative number, except in fully turgid cells when it becomes zero. In this case, the positive pressure potential balances the sum of the negative osmotic and matric potentials, as shown later.

Except in very dry tissue or in cells with small vacuoles, Ψ_m is very small relative to Ψ_p and Ψ_s. If we disregard Ψ_m, Eq. (1.9) becomes

$$\Psi_{\text{cell}} = \Psi_s + \Psi_p \tag{1.10}$$

Assuming no change in cell volume or Ψ_s, the relationships for cells of various turgidities can be shown as follows:

	$\Psi_{\text{cell}} = \Psi_s + \Psi_p$
Turgid	$0 = -20 + (+20)$ bars
Partly turgid	$-10 = -20 + (+10)$ bars
Flaccid	$-20 = -20 + 0$ bars

Another form of Eq. (1.10) has been used by some writers in which hydrostatic pressure P is substituted for Ψ_p and osmotic pressure π for Ψ_s:

$$\Psi_{\text{cell}} = P - \pi \tag{1.11}$$

From Eqs. (1.8) and (1.9) it can be seen that DPD is numerically equal to Ψ_{cell} but opposite in sign, that is:

$$\Psi_{\text{cell}} = -\text{DPD} \tag{1.12}$$

As pointed out previously, the potential of water in a cell is less than that of pure water, i.e., is negative, whereas DPD is positive because it is defined as a deficit.

Since Ψ_w is a negative number, it must be thought of as becoming lower as water stress increases. The same situation occurs in reading temperatures below zero on a thermometer; negative temperatures increase in magnitude but decrease in absolute value.

The interrelationships between Ψ_{cell} and the factors which control it can be shown by a diagram such as that in Fig. 1.11 for a cell which undergoes

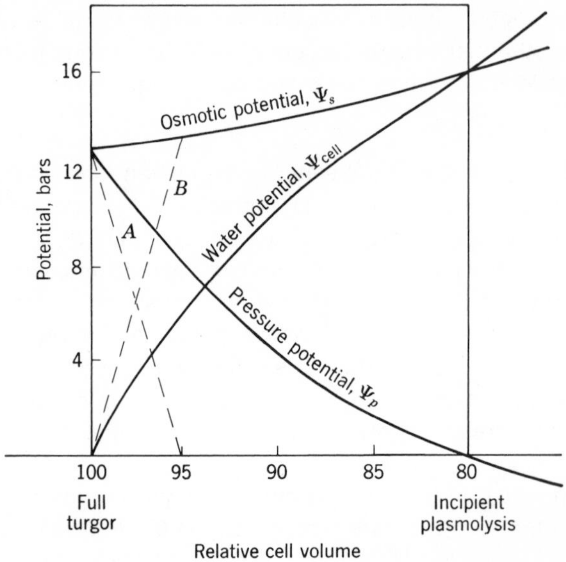

FIG. 1.11 **Interrelationships among cell volume, osmotic potential, turgor pressure, and cell water potential. The solid lines are for a highly extensible cell. The dashed lines are for a slightly extensible cell, line *A* representing pressure potential; line *B*, cell water potential. Water potential and osmotic potential are negative; pressure potential, positive. (*Modified from Höfler*, 1920 *and Bennet-Clark*, 1959.) Gardner and Ehlig (1965) show the pressure potential as a straight line, changing slope abruptly at a pressure potential of 2 or 3 bars. If the walls collapse during plasmolysis, the turgor pressure would become negative and Ψ_{cell} would be less than Ψ_s.**

considerable change in volume with change in turgor. This shows the changes in osmotic potential Ψ_s and in Ψ_{cell} as the volume and the turgor pressure Ψ_p change. When $\Psi_p = \Psi_s$, the water potential of the cell Ψ_{cell} reaches zero; and conversely when Ψ_p decreases to zero at incipient plasmolysis, $\Psi_{cell} = \Psi_s$. The line representing Ψ_p is often drawn as though turgor pressure is linearly proportional to volume, but this is not necessarily correct. Thick-walled cells show little change in volume (perhaps only 2 to 5 percent), but some thin-walled parenchyma cells show changes in volume up to 40 percent, and the lines for pressure and osmotic potential are assumed to be curvilinear in such cells (Bennet-Clark, 1959, pp. 171–174; Crafts et al., 1949, chap. 7; Kamiya et al., 1963). However, Gardner and Ehlig (1965) show an abrupt change in the shape of the line representing turgor pressure at a pressure potential of 2 or 3 bars. Evidently cell volume does not change linearly with cell turgor. According to Kamiya et al. (1963) the curves are somewhat different for enlarging and shrinking cells of Nitella.

When cells are plasmolyzed, the protoplast usually breaks away from the cell wall. Occasionally, however, it adheres so firmly to the cell wall that the wall is pulled inward. This develops a tension on the water in the cells (Crafts et al., 1949, pp. 83–85). Plasmolysis probably seldom occurs in nature, but the occurrence of negative wall pressures is inferred from measurements which indicate that Ψ_{cell} is sometimes lower than Ψ_s in plants under severe water stress. For example, Slatyer (1957*a*) found Ψ_{cell} was 5 to 10 bars lower than Ψ_s in wilted cotton and tomato, and even larger differences were found in acacia (Slatyer, 1960*a*). Still larger differences were reported in cells of the annulus of fern sporangia and elators of liverworts (see Bennet-Clark, 1959, p. 174).

MATRIC AND SOLUTE POTENTIALS IN RELATION TO WATER CONTENT. There is some uncertainty concerning the effect of change in water content on the relative importance of the osmotic and matric components of cell water potential. The two components are additive in the sense that if a quantity of water-binding material is added to a solution, the water potential will be decreased. However, as Bolt and Frissel (1960) show, there will be a discrepancy between the total water potential and the sum of Ψ_s and Ψ_m measured separately because of interaction between solutes and the matric component. Slavik (1963*a*) found that Ψ_s decreased more rapidly than expected with decreasing leaf water content and suggested that an increasing fraction of the total water content is bound as the water content decreases and therefore is unavailable as a solvent. Wilson (1967) also found that Ψ_s varied more with change in water content than would be expected from change in solute concentration. He likewise attributed this to an increasing proportion of the water being bound as the water content decreases. On the other hand, Gardner and Ehlig (1965) found that Ψ_s changed in proportion to change in water content in the four species studied by them. They did find a change in modulus of elasticity of cell walls at a turgor pressure below 2 bars, which results in wilting occurring slightly above zero turgor. Ψ_s may have considerable effect on movement of vapor through intercellular spaces in plant tissue (Curtis, 1937) and on vapor movement in soils (see Chap. 2).

TEMPERATURE AND WATER POTENTIAL. It was stated earlier that increasing temperature increases the water potential, and decreasing temperature decreases it. Spanner (1964, p. 207) states that an increase of 1° results in an increase of over 80 bars in the sense that water will distill from an open solution with an osmotic potential of −80 bars to pure water one degree cooler. One therefore might conclude that since temperature differences of several degrees often occur among various parts of a plant, the effects of temperature should override the pressure, matric, and osmotic components of water potential. It is self-evident that they do not, because water moves from the cooler roots to the warmer leaves at midday. In fact, temperature differences affect movement of liquid water only under very special conditions. Temperature affects water transfer across membranes only when the heat conductance of the membrane is very low and the pore size is very small. It is very difficult to maintain a temperature gradient across a single cell membrane of sufficient magnitude to cause thermoosmosis, and it could not operate at all in a system such as the xylem. Even where there is a tendency for liquid water to move by diffusion from warmer to cooler regions, the movement is counterbalanced by mass flow in the reverse direction caused by gradients in hydrostatic pressure. It therefore appears that temperature can have no significant direct effect on movement of water as a liquid in plants (Briggs, 1967). Temperature gradients can cause measurable movement of water from the warm to the cool side of fruits and vegetables (Curtis, 1937; Veto, 1963), but this movement probably occurs as vapor.

Hydratur

Any discussion of cell water terminology must take into account the term "hydratur," or hydrature, proposed by Walter in 1931 and used in a number of subsequent papers (Walter, 1963, 1965; Kreeb, 1967). There has been considerable uncertainty concerning the precise meaning of this term (Renner, 1933; Shmueli and Cohen, 1964; Stocker, 1960). Walter (1965) states that hydrature is the relative water vapor pressure, i.e., the vapor pressure of the water in vacuoles or cytoplasm relative to that of pure water at the same temperature and pressure. Thus, in a fully turgid cell, although the actual vapor pressure of the cell sap is equal to that of pure free water, the relative vapor pressure is lower, because, according to the definition of hydrature, the vapor pressure of the cell sap is expressed as a percentage of that of free water at the same pressure. Thus, different parts of a cell have different hydratures. As Slatyer (1967) pointed out, the hydrature of the cell as a whole is equal to its water potential, while the hydrature of the vacuolar sap and cytoplasm is equal to the osmotic potential. This situation suggests that hydrature is a confusing and unsatisfactory term. Readers are referred to Weatherley (1965) and Kreeb (1966) for further discussion of this term.

The confusion over the meaning of hydrature has obscured Walter's argument that the degree of hydration of the cytoplasm is determined by osmotic potential of the vacuolar sap rather than by the water potential of the cell as a whole. However, he agrees that water exchange between cells is controlled by the water potential.

Bound water

The term bound water is often encountered in the older literature of plant physiology, especially that dealing with cold and drought resistance. Gortner (1938) discussed the origin of this term and methods of measuring bound water. The subject was also reviewed by Kramer (1955*b*). The concept of bound water is based on the fact that a variable fraction of the water in both living and nonliving materials behaves differently from free water. It has a low vapor pressure, remains unfrozen at temperatures far below zero, does not function as a solvent, and seems to be unavailable for physiological processes. In practice, bound water has most often been defined as that water which remains unfrozen at some low temperature, usually −20 or −25°C. The binding forces constitute the matric potential Ψ_m of Eq. (1.9).

Readers should understand that there is no sharp line between unbound and bound water but a gradual transition, with decreasing water content from free water, easily frozen or removed by drying, to water so firmly bound to surfaces that it cannot be removed by temperatures far below zero or even by a long period in an oven at 100°. Water bound this firmly doubtless plays an important role as a cell constituent in the resistance to drying of some seeds, spores, microorganisms, and a few higher plants. However, efforts to explain differences in cold or drought resistance of flowering plants by

***Table 1.3** Effect of Temperature at Which Tissue Is Frozen on the Bound Water Content* *

Material	Total water as % of fresh wt	Freezing temperature	Bound H_2O as % of total H_2O	Bound H_2O as % of dry wt
Pine needles	56.8	−12	35.0	46.0
		−22	28.2	37.0
Unhardened clover roots	82.0	−10	14.5	72.2
		−20	7.4	38.7
		−30	2.0	11.8
		−40	2.0	11.5
		−50	1.4	10.6
Hardened clover roots	82.3	−15	15.4	78.0
		−25	13.6	69.1

* *From Kramer (1955b)*

differences in amount of bound water have been disappointing. The amount of bound water varies widely among different kinds of tissues, with age and past history, and according to the method used to measure it (see Table 1.3). Levitt (1951) reviewed the literature on bound water in relation to stress.

Cell water relations

The water relations of plants are dominated by the water relations of individual cells because nearly all the water occurs in cells, chiefly in their vacuoles. Thus, an understanding of plant water relations requires an understanding of cell structure and cell water relations. Plant cells vary greatly in size, shape, water content, permeability, and other characteristics.

Cell structure

A typical parenchyma cell which forms most of the water storage tissue of plants is shown in Fig. 1.12. In contrast to the young cell in the figure, it con-

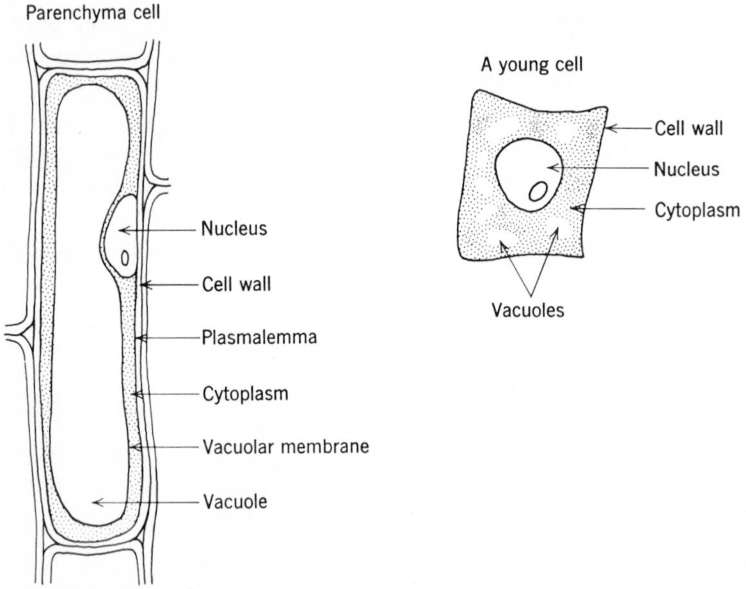

FIG. 1.12 **Diagrams of a meristematic cell and a mature, vacuolated parenchyma cell. The layer of cytoplasm is usually much thinner than shown in this diagram.**

sists of a relatively thin wall enclosing a layer of cytoplasm which in turn encloses a large central vacuole. Only a brief discussion of cell structure is given here. Readers are referred to cytology texts for more details. The development of electron microscopy is opening up new avenues for the study of the relationship between structure and function in cells. Unfortunately, it is somewhat too early to evaluate and apply the information being obtained by this method to the present discussion of salt and water uptake. Frey-Wyssling and Mühlethaler (1965) provide a detailed discussion of the ultrastructure of cells, and an example of an electron micrograph of a plant cell is shown in Fig. 1.13.

THE WALL. Plant cells are characterized by their relatively tough, rigid walls which limit expansion and cause pressure on the cell contents. The primary walls are produced by the deposition of carbohydrates in the form of cellulose, hemicellulose, and pectic compounds. According to Setterfield and Bayley (1961), the dry primary wall contains up to 50 percent hemicellulose, 30 to 50 percent cellulose, 10 to 20 percent pectic substances, and small amounts of proteins and lipids. Recent work suggests that proteins are a normal constituent of cell walls (Lamport, 1965; Thompson and Preston, 1967). In living cells, over half the volume of the primary cell wall is water. In young cells, the walls are plastic, and cell enlargement occurs with only a small amount of turgor pressure. Deposition of additional cellulose and infiltration of the cellulose framework with lignin results in thickening, loss of plasticity, and cessation of enlargement. Many cell walls remain somewhat elastic, and the protoplasts enclosed by them undergo measurable changes in volume with changes in turgidity. The deposition of lipids and related substances modifies the permeability of cell walls. Everyone is familiar with the suberized walls of cork cells in bark and the cutin layer on leaves, but it is not so well known that cutin and suberin occur in walls of cells in the interior of plants. Suberin occurs internally in the walls of specialized cells such as the endodermis; and according to Scott (1950, 1964), all cell walls exposed to the air are coated with a layer of lipid material. This includes the surfaces of leaf mesophyll cells, root cortical cells bordering intercellular spaces, and root epidermal cells including root hairs. However, a mucilaginous deposit of pectic substances often occurs external to the lipid layer on the outer surfaces of root epidermal cells.

CYTOPLASM. Within the wall, there is a layer of cytoplasm which ranges in consistency from sol to gel and varies widely in viscosity. It consists of a matrix or groundplasm bounded at the outer surface by the plasmalemma and at the inner surface by the vacuolar membrane or tonoplast. Studies with the electron microscope indicate that cytoplasmic structure is considerably more complex than was supposed from observations made with the light microscope. Not only does it contain a nucleus and plastids but a variety of other organelles including mitochrondria, ribosomes, Golgi apparatus, and endoplasmic reticulum (see Fig. 1.13).

Most important to us is the fact that the cytoplasm normally contains

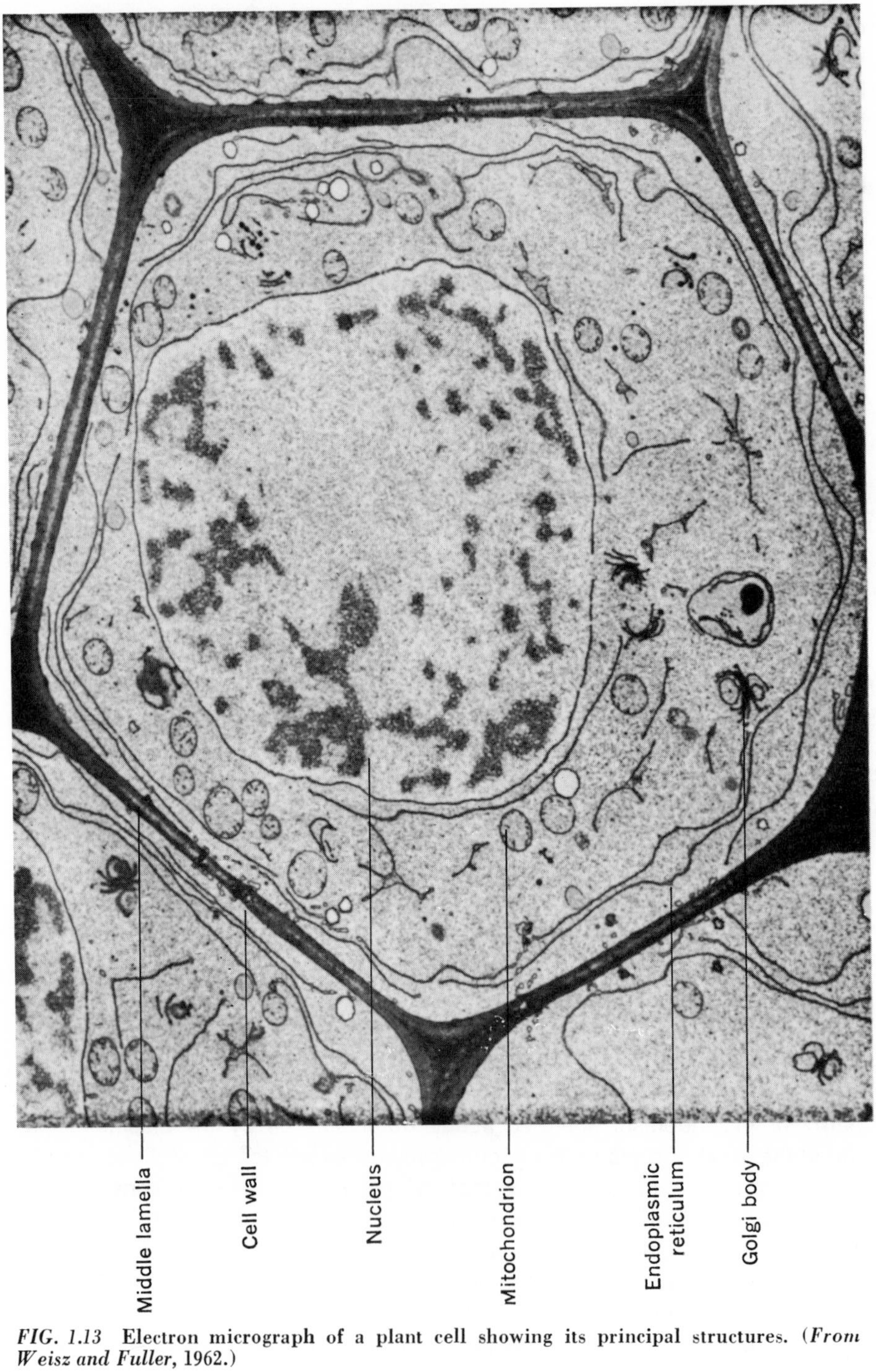

FIG. 1.13 Electron micrograph of a plant cell showing its principal structures. (*From Weisz and Fuller*, 1962.)

amounts of water bound to its protein framework and between the fibrils, also that its surface membranes are differentially permeable and control the entrance and exit of solutes. The vacuolar membrane appears to be somewhat stronger mechanically than the plasmalemma and probably is higher in lipid content. It retains its differential permeability for some time after being separated from the cytoplasm, while the differential permeability of the plasmalemma is lost immediately upon separation (Frey-Wyssling and Mühlethaler, 1965). The role of the Golgi apparatus and the endoplasmic reticulum in salt and water uptake are too uncertain to be discussed at this time. The role of the Golgi apparatus in secretion was reviewed by Mollenhauer and Morré (1966), and some workers implicate it in vacuole formation (Marinos, 1963). It has been suggested that pinocytosis, the invagination of plasmalemma vesicles into the cytoplasm, where they are said to dissolve, might be involved in the uptake of water and solutes by plant cells (Frey-Wyssling and Mühlethaler, 1965; Weiling, 1962), but this also seems very uncertain at present.

The role of the plasmodesmata is more certain. They are thin strands of cytoplasm about 0.2μ in diameter (Livingston, 1964) which extend through cell walls and connect the protoplasts of adjacent cells (see Fig. 1.14). Thus,

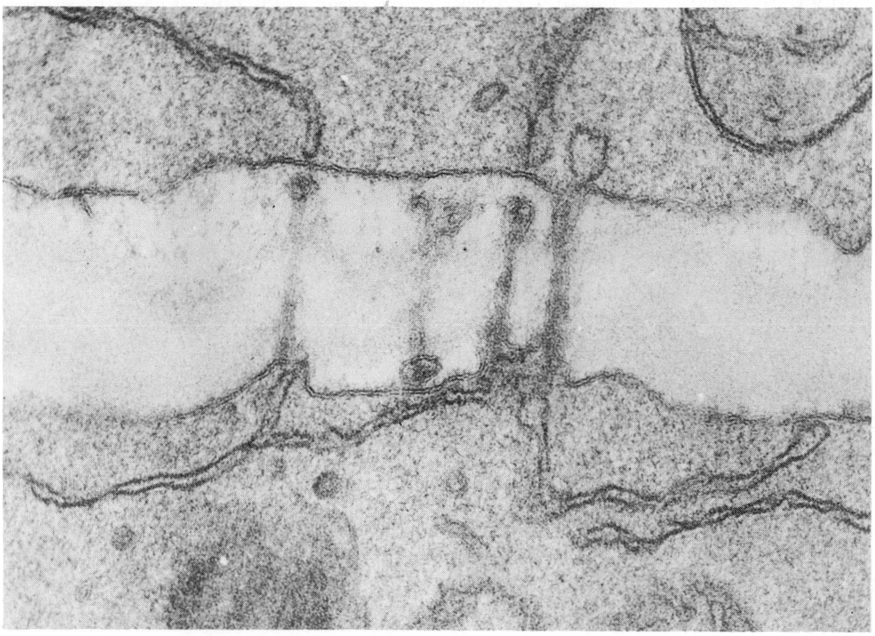

FIG. 1.14 **Electron micrograph ($\times$70,000) of the wall of an onion root cell showing plasmodesmata, the plasmalemma on each side of the wall, and the endoplasmic reticulum.** *(From Laboratorium für Elektronenmikroskopie, ETH, courtesy of Professor K. Mühlethaler.)*

they convert a collection of cells into an organized tissue, the symplast of Münch (1930) and others, in which solutes can move considerable distances without crossing differentially permeable membranes (Arisz, 1956; Crafts and Broyer, 1938).

THE VACUOLE. Vacuoles range from tiny spherical or rod-shaped structures characteristic of meristematic tissue to the large central vacuoles of parenchyma cells occupying more than 50 percent of the cell volume. Their size and shape can change; small vacuoles coalesce to form large ones, and large ones break up to form smaller ones. For example, during seed maturation the vacuoles lose water and shrink, but they enlarge again during germination. Vacuoles contain considerable amounts of sugar and salts which largely account for the osmotic potential of the cell sap. They also contain a wide variety of other substances including amino acids, amides, proteins, lipids, gums, tannins, anthocyanins and other pigments, organic acids, and crystals of minerals such as calcium oxalate.

Frey-Wyssling (1953) regarded the vacuoles as regions in which substances physically incompatible with cytoplasm accumulate by the action of physical factors such as surface tension and differential solubility. Frey-Wyssling and Mühlethaler (1965) suggested that vacuoles develop because plants cannot produce enough protein to fill their enlarging cells with protoplasm. It formerly was supposed that vacuoles develop from preexisting vacuoles, but it is now suggested that vacuoles develop from invagination of the plasmalemma or from the endoplasmic reticulum (Buvat, 1963; Frey-Wyssling and Mühlethaler, 1965). Ions and other substances are accumulated in vacuoles by active transport, and these substances can be regarded as outside the mainstream of metabolism. For example, ions once accumulated in the vacuoles are released very slowly. However, these ions have large effects on the osmotic pressure of cells.

Distribution of water in cells

The water in plants forms a continuous system through the water-saturated cell walls and water-permeable cytoplasm and organelles. It is therefore free to move from place to place along gradients of water potential. At equilibrium, the water in cells is distributed among the various structures: wall, cytoplasm, organelles, and vacuole, according to their relative volumes and their capacities to hold water. The equilibrium is very unsteady, and a change in concentration of solutes or water-binding substances, or loss of water from a plant by transpiration, is followed by movement of water until a new equilibrium in water potential is established. Because water is free to move, it occurs in different quantities and is held by different forces in the various parts of cells. This situation will be discussed briefly.

WATER IN CELL WALLS. Water is held in walls by matric or imbibitional forces which bind water molecules to the surfaces of the fibrils by hydrogen bonds, and by capillary forces in the submicroscopic spaces between the fibrils. These may be 10 to 100 mμ in width and are interconnected. They provide a considerable volume for water storage and movement. Although water in the layers next to fibrils is bound rather firmly, that in the larger spaces is able to move freely. The role of cell walls as a pathway for salt and water movement will be discussed in later chapters (also see Strugger, 1949; Ziegenspeck, 1945).

As much as half the volume of cell walls may consist of water, and some types of walls shrink up to 50 percent in volume when dehydrated (Crafts, 1931). For this reason, study of sections of plant tissue which have been killed and dehydrated often gives an erroneous impression of the volume of the cell walls available for water storage and movement. However, during maturation, deposition of pectic compounds, lignin, suberin, and other substances reduces the volume available for water storage and movement.

WATER IN THE CYTOPLASM. The water content of the cytoplasm of active tissue may exceed 90 percent; hence, even in cells containing large vacuoles, considerable water is associated with the cytoplasm. In meristematic regions and other tissue where the vacuolar volume is small and the walls are thin, most of the water may occur in the cytoplasm. Electron microscope studies are changing the classical views concerning the structure of cytoplasm. However, it seems safe to assume that much water is hydrogen-bonded to the side chains of the polypeptide strands which form the solid structure of cytoplasm, and a larger fraction occurs in the spaces among the protein strands where it is held very loosely. Sponsler et al. (1940) concluded that there is a large increase in spacing and increased freedom of movement of water molecules in gelatin as the water content increases to 30 or 35 percent, but that there is little change with further increase because all hydration centers have been satisfied. The situation probably is similar in cytoplasm. The presence of a large volume of very loosely bound water contributes to the fluidity and low viscosity characteristic of active cytoplasm. Reduction in water content leaves only the more firmly bound fraction of the water. This causes increased viscosity, and the final result is the gel condition found in dormant seeds and other dehydrated and physiologically inactive structures.

The nucleus, plastids, mitochondria, and other organelles are enclosed in differentially permeable membranes and form distinct osmotic entities within the cytoplasm. However, they contain a relatively small fraction of the water and will not be discussed separately.

The hydrophilic properties of the proteins constituting the framework of cytoplasm are modified by the amounts and kinds of ions present. There are numerous binding sites on the framework and many free ions in the liquid phase. In general, an excess of divalent ions decreases hydration, while excess of monovalent ions increases it, and ions can be arranged in the so-called lyotropic series according to their effects on hydration and other properties of colloidal systems and membranes. The hydrogen ion concentration also has marked effects on the hydration of amphoteric substances such as protein; minimum hydration occurs at the isoelectric point.

WATER IN VACUOLES. In most plant cells, the major fraction of the water occurs in vacuoles, and the water relations of plants are usually dominated by the amount and potential of the vacuolar water. Guilliermond (1941) and Frey-Wyssling (1953) emphasized the role of vacuoles as regions in which various products of metabolism and even foreign substances such as dyes are collected. Some of these substances, notably the sugars and salts, occur in solution; but others such as proteins, tannins, mucilages, and dextrins occur in the colloidal condition. As a result, the viscosity of the vacuolar sap is ordinarily about twice that of water (Frey-Wyssling, 1953), and a rigid gel is sometimes formed (Guilliermond, 1941). The characteristics of vacuoles were discussed in more detail by Kramer (1955*c*). The importance of vacuolar water results from its relatively large volume and high concentration of solutes so that it more or less dominates and controls the water potential of most plants.

WATER IN THE VASCULAR SYSTEM. Although it is essential for survival, the water in the xylem elements of a typical herbaceous plant constitutes a negligible percentage of the total volume. Emmert (1961) estimated from uptake of ^{32}P that functional xylem constitutes less than 2 percent of the volume of young bean plants. In woody plants, the percentage is much larger because most of the stem consists of xylem. Kramer and Kozlowski (1960) summarized some of the data on the water content of tree trunks, which seems to range from 50 to 250 percent of the dry weight. However, we know of no data separating the water in the xylem elements from that in the cell walls or that in the functional xylem elements from that in the nonfunctional region of the xylem.

In cell walls of woody tissues the volume accessible to water is materially reduced by lignification. The water in the lumina of xylem vessels and tracheids usually is a dilute solution of salt and organic substances. As will be discussed later, in transpiring plants this water may be subjected to tensions of over 100 bars. Under these conditions some of the water columns break, and the cavities of vessels and tracheids become occupied by air or water vapor. The entry of air occurs only if the tension is great enough to displace water from the micropores in the walls.

In contrast to the xylem, where water occurs in the walls and cavities of dead cells and vessels, water in the conducting elements of phloem occurs in living cells, the sieve tubes. These contain sap, which ordinarily has a high concentration of organic solutes. The phloem plays an important role in translocation of organic solutes, and its capacity to function in translocation is materially affected by water stress (Roberts, 1964). There is also considerable exchange of water and solutes between the phloem and adjacent xylem, which results in recirculation of some solutes (see Chap. 7).

Cell permeability

Any discussion of water and solute movement in plants requires consideration of the permeability of the various membranes through which these substances move. One of the most important characteristics of living cells is their ability to maintain combinations and concentrations of solutes quite different from those existing in their environment. For example, the concentration of several kinds of ions is tens or even hundreds of times higher in plants than in the soil in which they are rooted. Such accumulation requires the presence of membranes relatively impermeable to the ions and an active transport mechanism to move them across the cell membranes.

Definition of permeability

Permeability is a property of membranes rather than of the substances moving across them. It was defined by Brooks and Brooks (1941) as the rate of movement across a membrane under a given driving force. According to Ussing (1954), a membrane is a boundary that is less permeable than the regions separated by it. Any quantitative expression of the permeability of a membrane to a particular substance therefore requires knowledge of the magnitude of the driving force as well as of the volume of material moved per unit of time. These concepts are clearly expressed in Fick's law [Eq. (1.13)], which is applicable to passive diffusion through membranes. However, the protoplasmic membranes of living cells are relatively impermeable to polar compounds, such as salts and sugars, which play essential roles in plant metabolism. Most movement of these essential substances into and out of cells occurs by some kind of active transport mechanism powered with energy released by respiration. Thus, both the permeability of cell membranes and the driving forces causing movement across them are dependent on metabolic energy. This fact increases the difficulty of studying permeability because treatment with respiration inhibitors intended to reduce the driving force may also significantly affect the passive permeability of the membranes (Burström, 1962).

Membrane structure

Plant membranes vary greatly in thickness, structure, and permeability. They range from layers of cells, such as the epidermis or endodermis, through layers of cutin or suberin deposited on or in cell walls which are easily visible with a light microscope, to the submicroscopic structure of protoplasmic membranes, such as the plasmalemma and vacuolar membrane. Not even the electron microscope reveals the complete structural details of protoplasmic membranes, and current ideas concerning structure are based on their permeability, surface tension, and other physical properties. The most widely used model is that of Davson and Danielli (1952), who suggested that the typical membrane consists of a central bimolecular layer of oriented lipid molecules covered on each surface by films of protein molecules, somewhat as shown in Fig. 1.15. This structure is compatible with observations

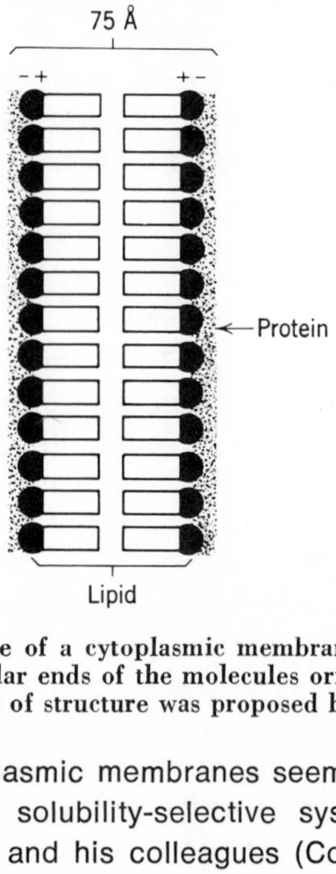

FIG. 1.15 **Diagram of structure of a cytoplasmic membrane. This diagram shows a central region of lipid with the polar ends of the molecules oriented outward toward the surface layers of protein. This type of structure was proposed by Davson and Danielli (1952).**

which indicate that protoplasmic membranes seem to act simultaneously as molecular sieves and as solubility-selective systems. This is based on observations by Collander and his colleagues (Collander, 1959) and others that permeability of some kinds of cells to nonelectrolytes of similar molecular size is related to their solubility in oil, and permeability to nonelectrolytes of equal solubility in oil is inversely proportional to their molecular size. An exception is that very small molecules penetrate more rapidly than would be expected on the basis of size. The term "pore" sometimes used in connection with the molecular sieve theory of permeability should not be taken too literally because it merely refers to hydrophilic regions in the lipoid layer, possibly formed by protein molecules extending into it, through which polar compounds can pass. It is doubtful if there are large, discrete, water-filled pores in protoplasmic membranes (Dainty, 1963*a*; Gutknecht, 1967).

It is claimed by some writers that all protoplasmic membranes have a similar structure. This resulted in the concept of a "unit membrane" advanced by Robertson (1959). The concept is regarded as an oversimplification by Korn (1966), who points out that biological membranes differ in respect to their chemical composition, enzymatic composition, permeability, and even their electron microscope image. Frey-Wyssling and Mühlethaler (1965, p. 149) point out that static structures such as that shown in Fig. 1.15 could not carry on the diverse functions of protoplasmic membranes. In addition to serving as barriers to diffusion of solutes and functioning in active transport of solutes, membranes contain enzymes and are involved in metabolic processes leading to the deposition of cell walls. They must be considerably more complex in structure than would be inferred from the usual models (Kavanau, 1964). Electron micrographs of the surface of the plasmalemma, such as Fig.

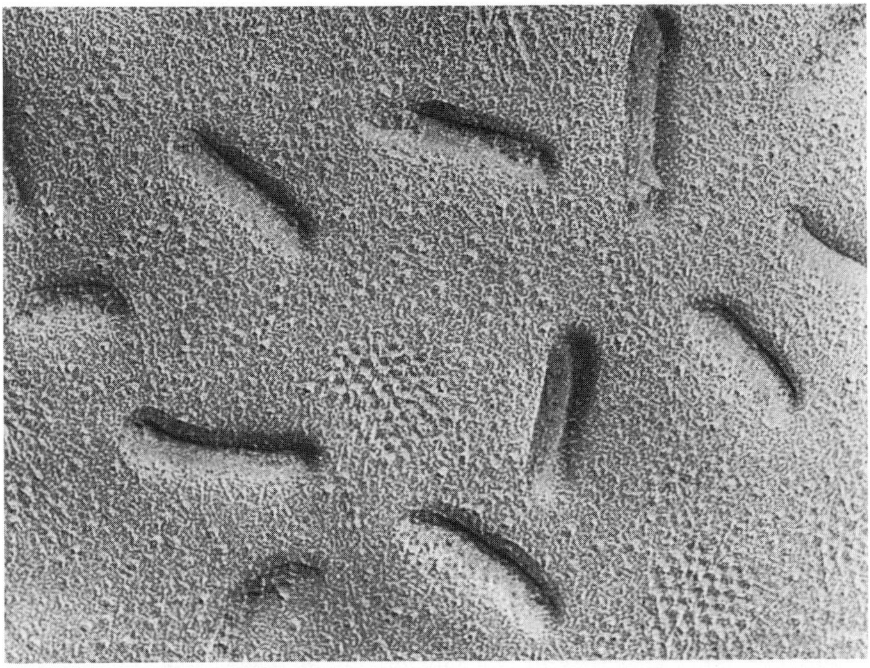

FIG. 1.16 **Surface view of the plasmalemma of a yeast cell ×150,000 showing the mosaic structure and large wrinkles or folds. The material for this study was prepared by the "freeze-etching" method.** (*From Laboratorium für Elektronenmikroskopie, ETH, courtesy of Professor K. Mühlethaler.*)

1.16, indicate that the surface is not completely uniform but appears to be a mosaic of various components.

Factors affecting permeability of protoplasmic membranes

As might be expected, the permeability of living membranes is materially affected by a variety of substances and treatments which either act indirectly by affecting metabolism or directly by modifying membrane structure. This subject was reviewed by Kramer (1955*a*, pp. 205–209) and by several other authors in vol. 2 of Ruhland's "Encyclopedia of Plant Physiology." Only a few examples will be discussed here. In general, permeability of cells and tissues to water is reduced by low concentrations of various toxic substances and increased by higher concentrations, which often cause irreversible injury to protoplasm. For example, Glinka and Reinhold (1964) observed this difference in effect between chloroform at 2×10^{-2} *M* and 5×10^{-2} *M* (also see Fig. 1.17). Moderate concentrations of fat solvents such as benzene, chloro-

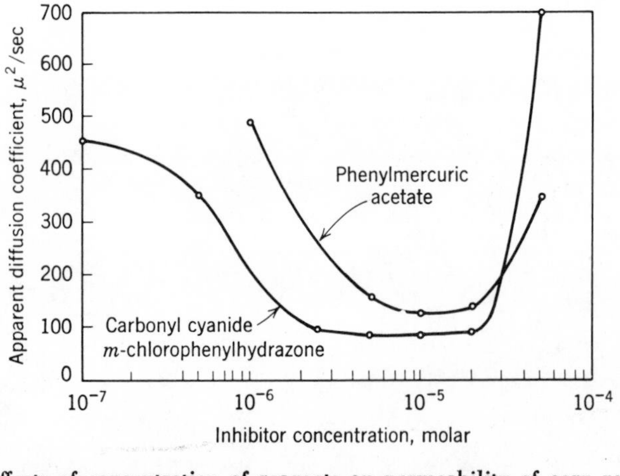

FIG. 1.17 **Effects of concentration of reagents on permeability of corn roots to tritiated water. The first effect is to reduce permeability, but since the increasing concentration causes injury, permeability suddenly begins to increase. (*Woolley*, 1965.)**

form, and ether usually increase permeability, presumably because they affect the lipid fraction of membranes. Kuiper (1964*a*) reported that a 10^{-3} *M* solution of decenyl succinic acid increased permeability of roots to water eightfold, while the same concentration of decylsuccinic acid decreased it 55 percent. He proposed that incorporation of unsaturated hydrocarbon chains into the lipid layers of cell membranes increases their permeability to water. However, his conclusions are questionable because further investigation revealed that decenyl succinic acid causes permanent injury to roots (Newman and Kramer, 1966).

Treatments which reduce metabolism, such as respiration inhibitors, oxygen deficiency, and low temperature, usually reduce permeability. Examples are the studies of Bogen and his coworkers (Bogen, 1953, 1956; Bogen and Prell, 1953) and Burström (1962). Burström found that 10^{-4} *M* NaN_3 decreased permeability of cells of *Rhoeo discolor* to glycerol and made them completely impermeable to sugar. Saturating the water containing sections of sunflower hypocotyl with carbon dioxide reduced their permeability (Glinka and Reinhold, 1964) much as it reduces the permeability of entire root systems.

Low concentrations of various ions are said to have different effects on permeability, probably depending on how they affect hydration of the protein fraction of membranes. The classical work of Osterhout, Stiles, and others is summarized in Miller (1938, pp. 97–108), in Brooks and Brooks (1941), and in vol. 2 of Ruhland's "Encyclopedia of Plant Physiology." It will suffice to mention that the presence of calcium is necessary to maintain normal structure and functioning of cell membranes, and its absence disturbs ion uptake (Epstein, 1961; Rains et al., 1964). In general, calcium decreases permeability to water, and sodium and potassium increase it (see references in Kramer, 1955*a*, p. 207). Kuiper (1963) reported that a low concentration of $NaNO_3$ increases permeability much more than the same concentration of KNO_3. Decreasing temperature markedly decreases permeability to water as well as to solutes, and sudden cooling seems to cause a greater reduction than slow cooling. Slight dehydration sometimes increases permeability (Bogen, 1940; Myers, 1951), possibly because, as Huber and Höfler (1930) claim, pressure of protoplasts against cell walls in turgid cells decreases permeability. However, decreasing the osmotic potential of the ambient solution materially reduces the permeability of *Nitella* cells to water and to urea (Dainty and Ginzburg, 1964*a*), probably because the membrane structure contracts when it is dehydrated. Effects of temperature, aeration, and dehydration on permeability of roots to water are discussed further in Chap. 6, and their effects on salt uptake are discussed in Chap. 7.

Relative permeability of various cell membranes

Water moving into a cell passes through the wall, plasmalemma, body of the cytoplasm, and vacuolar membrane before reaching the vacuole. In general, the permeability of parenchyma cell walls is very high, and they offer negligible resistance to water movement (Levitt et al., 1936; Russell and Woolley, 1961). Of course, walls which are suberized or cutinized have a high resistance, and heavily lignified walls are less permeable than cellulose walls. It is difficult to evaluate the relative resistances offered by the various parts of the protoplast, but the total effect is a relatively high resistance to water movement. This is indicated by the fact that the half time for diffusion of water into *Vicia faba* roots was reduced from 0.6 to 0.2 or 0.1 min by killing the roots (Ordin and Kramer, 1956). Also, killing root systems always results in a large increase in mass flow of water through them (Brouwer, 1954*a*; Kramer, 1933) which indicates that death of the protoplasm greatly reduces the resistance to water movement.

In studies at the cellular level, Dainty and Ginzburg (1964*b*) found the plasmalemma to be 30 times as permeable to urea as the vacuolar membrane, and they infer from their data that it also is more permeable to water. MacRobbie and Dainty (1958) found it more permeable to ions in *Nitellopsis* than the vacuolar membrane; but in *Chara* and *Nitella,* the plasmalemma is less permeable than the vacuolar membrane (MacRobbie, 1962).

Measurement of permeability

Slatyer (1967, chap. 6) has discussed the measurement of cell and tissue permeability in detail, so only a few general observations will be made about methods. Few reliable data are available because of various difficulties inherent in the two principal methods used. The hydraulic conductivity or permeability is often estimated from change in tissue or protoplast volume with a known difference in water potential between the cells and the surrounding solution (Huber and Höfler, 1930; Levitt et al., 1936; Stadelmann, 1963). Errors result from the difficulty of measuring volume changes and estimating the water potential of the cell sap. Another method is to estimate the permeability by measuring the diffusion rate of an isotope of water out of cells or pieces of tissue which first have been equilibrated with solutions of deuterium or tritium. However, it is difficult to measure the internal concentration of the isotope. The third method is transcellular osmosis (see Fig. 1.18), in which water movement is caused by immersing the opposite ends of large cells in solutions of different concentrations (Kamiya and Tazawa,

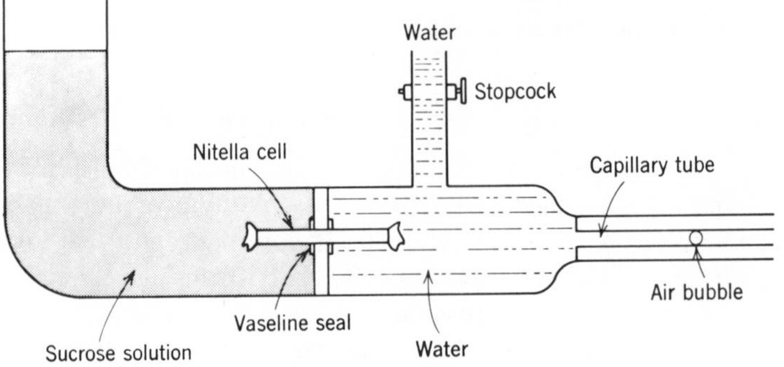

FIG. 1.18 **Apparatus for measuring the permeability or hydraulic conductivity of *Nitella* cells by transcellular osmosis. The rate of water flow can be varied by changing the concentration of sucrose in the chamber on the left. (*After Dainty and Ginzburg,* 1964.)**

1956; Dainty and Ginzburg, 1964*a*). The hydraulic conductivity measured by this method on large algal cells is 1 to 2 × 10^{-5} cm/(sec) (bar), compared to values by the plasmometric method of 5 × 10^{-7} cm/(sec) (bar).

It is generally assumed in permeability measurements that the solutions on both sides of the membrane are so well stirred that the concentration is uniform at various distances from the membrane. However, next to the membrane there is usually an unstirred layer 20 to 500 μ or even 1 mm in thickness where the concentration often differs markedly from that in the bulk liquid (see Fig. 1.19). Thus, adjacent boundary layers rather than the perme-

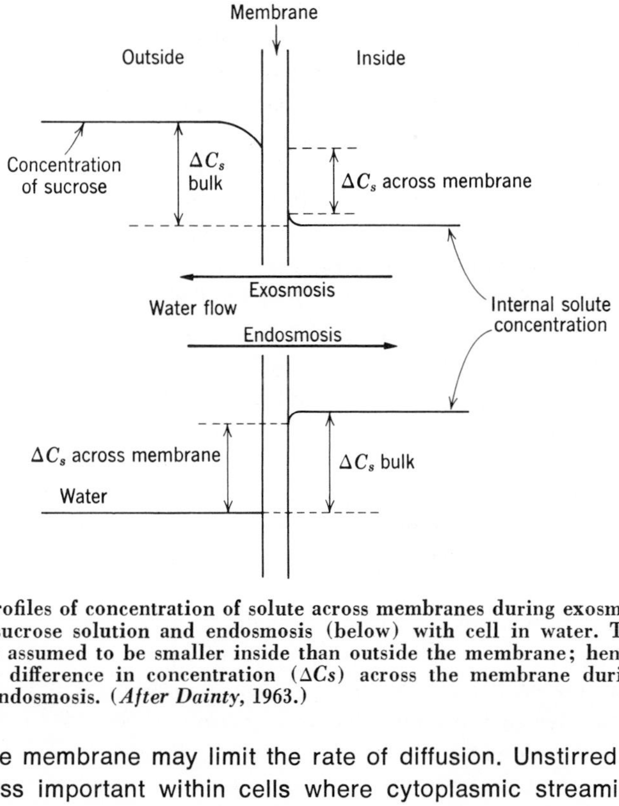

FIG. 1.19 **Profiles of concentration of solute across membranes during exosmosis (above) with cell in sucrose solution and endosmosis (below) with cell in water. The unstirred layer effect is assumed to be smaller inside than outside the membrane; hence there is a much greater difference in concentration (ΔCs) across the membrane during exomosis than during endosmosis. (*After Dainty*, 1963.)**

ability of the membrane may limit the rate of diffusion. Unstirred layers are probably less important within cells where cytoplasmic streaming occurs, but such layers often have important effects on permeability measurements (Dainty and Hope, 1959; Dainty, 1963*a*).

The movement of water and solutes in plants

The water and solutes in cells are in continual motion, moving within cells, from cell to cell, and from tissue to tissue. The distances involved vary from those across cell membranes measured in millimicrons to those from roots to shoots which may be measured in meters or even tens of meters. Obviously, different mechanisms and factors are involved in movement over such widely different distances.

Several forces are involved in the movement of water and solutes. These can be classified in general terms as (1) passive movement by diffusion and mass flow and (2) active transport dependent on expenditure of metabolic energy.

Passive movement

Passive movement of materials obeys physical laws in the sense that movement occurs along gradients of decreasing free energy or potential. If gradients of pressure potential or hydrostatic pressure constitute the driving force, the movement is generally termed mass flow. If gradients of total potential are involved, the movement is generally regarded as diffusional.

MASS FLOW. Movement of materials by mass flow occurs when force is exerted on the moving substance by some outside agent so that the molecules tend to move in the same direction in mass, whereas diffusion results from random movement of individual molecules or ions (Spanner, 1956). Water flows in streams because of the hydraulic head produced by gravity and in the plumbing system of a building because of the unidirectional force applied to it by a pump. Water and solutes move through the xylem of plants by mass flow caused by a hydrostatic pressure or pressure potential gradient extending from the roots to the shoots. The pressure flow hypothesis of phloem transport assumes that mass flow occurs in the sieve tubes because of pressure developed in the receiving solutes. Cyclosis or the streaming of the cytoplasm in cells also can be regarded as mass flow.

DIFFUSION. Movement by diffusion results from the random movement of molecules, ions, or colloidal particles caused by their own kinetic energy. While mass flow is a macroscopic process operating on material in mass, diffusion operates at the molecular level. However, Spanner (1956) reminds us that the differences between mass flow and diffusion are largely statistical, and the distinction between them becomes less important as the numbers of molecules under study decrease.

Familiar examples of diffusion are the evaporation of liquids, osmosis, and imbibition. The first mathematical treatment of diffusion seems to have been by Fick in 1855. Fick's law can be written as

$$\frac{dm}{dt} = -DA\,\frac{dc}{dx} \qquad (1.13)$$

where dm refers to the amount of substance moved per unit of time dt; D is the diffusion coefficient, a constant which varies with the substance; A is the area over which diffusion is occurring; and the minus sign indicates that diffusion occurs "downhill," or from higher to lower concentration. The term *dc* refers to the difference in concentration which is the driving force, and *dx* to the distance over which diffusion occurs. The equation indicates that, for a given substance and area, the rate of diffusion is proportional to the concentration gradient and inversely proportional to the distance over which it occurs.

Diffusion over long distances is very slow. De Vries calculated that 319 days would be required to move 1 mg NaCl from a 10 percent solution 1 m through a tube with a cross section of 1 cm^2, and 940 days would be required to move 1 mg sucrose under the same conditions. However, diffusion over short distances is very rapid because the rate is inversely proportional to the square of the distance. Diffusion over a distance of 1 μ is 10^8 times more rapid than diffusion over a distance of 1 cm. Thus movement by diffusion in and out of cells and within cells is relatively rapid, but diffusion could not possibly account for movement of water or solutes over distances measured in meters.

Diffusion is much less affected by the area of the openings through which it occurs than is mass flow, which follows Poiseuille's law and varies with the fourth power of the radius or the square of the area. Thus, subdividing an area into smaller parts has little effect on diffusion but a large effect on mass flow. This explains the fact that gelatin and water-saturated cell walls have a low resistance to movement by diffusion but are definite barriers to mass flow through the small spaces between the micelles or fibrils.

OSMOTIC MOVEMENT OF WATER. The movement of water from cell to cell is usually described as occurring by osmosis. Osmosis is generally defined as diffusion across a membrane caused by a difference in chemical potential. However, in recent years it has been claimed that osmosis really involves mass flow of water through the pores of differentially permeable membranes (Dainty, 1963*a*, 1965; Mauro, 1960; Ray, 1960).

This claim is based on both experimental and theoretical considerations. Ray (1960) reviewed a number of experiments which indicated that osmotic movement of water is much more rapid than movement of tracers such as deuterium and tritium, which must move by diffusion. However, Dainty (1963*a*, 1965) attributes these apparent differences largely to the effects of unstirred layers and other experimental errors which make reliable comparisons of osmotic movement and tracer diffusion very difficult. The idea of mass flow is also supported by mathematical analyses, which indicate that a difference in osmotic potential across a membrane caused by solutes on one side causes the same kind of water flow in the pores as a difference in hydrostatic pressure (Dainty, 1963*a*, 1965; Slatyer, 1967, chap. 6).

Mass flow through differentially permeable membranes separating water from solution can be explained only if it is assumed that they contain pores filled with water. This results in a sharp concentration gradient at the inner surface. Water molecules would then tend to pass out of the pores, thus producing a pressure drop which causes water to flow through the pores along a gradient of hydrostatic pressure (see Fig. 1.20).

As mentioned earlier, evidence for the existence of water-filled pores in plant cell membranes is not convincing. Perhaps the problem is largely semantic anyway, except to those deeply interested in detailed studies of membrane structure and function. The overall driving force causing water movement is the difference in water potential across the membrane. At least one stage of osmotic water movement seems to involve diffusion, and this is probably the rate-limiting stage.

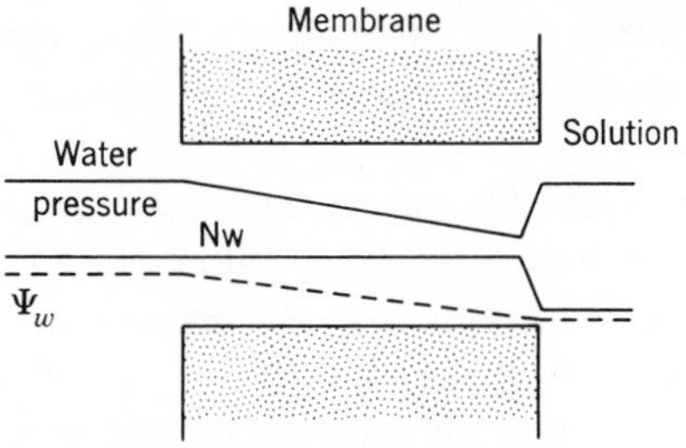

FIG. 1.20 **Drop in pressure across a membrane. The pores of the membrane are assumed to be filled with water, as shown by line N_w, indicating the mole fraction of water. Pressure is assumed equal on both sides, but it decreases toward the inner end of the pore because of resistance to water movement through the pore. The water potential Ψ_w also decreases across the membrane. (*After Dainty*, 1963.)**

REFLECTION COEFFICIENT. Discussion of osmotic phenomena requires mention of the reflection coefficient σ, first defined by Staverman (1951). It is an expression of the extent to which a membrane is impermeable to a given solute and can be defined as the ratio of the observed osmotic pressure difference across a membrane to the theoretical osmotic pressure. If σ equals 1, the membrane is completely impermeable; but if σ is less than 1, the membrane is permeable to the solute.

Disregarding the fact that plant membranes are often somewhat permeable to solutes used in plasmolytic experiments may lead to various errors. For example, if cell membranes are somewhat permeable to such solutes (σ less than 1), the osmotic potential of the cell sap may appear to be higher than that of the solution required to cause plasmolysis. Likewise, if roots have σ less than 1, the osmotic potential required to stop exudation would be lower than the osmotic potential of the xylem sap, as reported by van Overbeek (1942). Observations of these situations have been interpreted as evidence for active uptake of water but may, in fact, have resulted from the leakiness of the cell membranes to solutes. Recent measurements of the reflection coefficient by Slatyer (1966) indicate that it is 0.6 to 0.7 for sucrose and 0.8 to 0.9 for mannitol. This may explain many of the discrepancies between osmotic potentials measured cryoscopically and plasmolytically.

Active or nonosmotic uptake of water

The discussion of cell water relations has proceeded thus far on the assumption that all exchange of water between cells and tissues and their surroundings can be explained in terms of movement along gradients of decreasing water potential. However, a number of writers have claimed that active transport against gradients of water potential also occurs by a nonosmotic mechanism dependent on the expenditure of metabolic energy. Most of these proposals were reviewed by Kramer (1955*a,* 1956*d*), and the possibility of nonosmotic uptake of water by entire root systems is discussed in Chap. 5.

Evidence for the active or nonosmotic transport of water is based on observations that differences exist in osmotic potential measurements made on expressed sap and by plasmolytic methods on intact cells of the same tissue (Bennet-Clark et al., 1936; Currier, 1944; Roberts and Styles, 1939; and others), that respiration inhibitors and anaerobiosis reduce water uptake (Bogen, 1953; Bonner et al., 1953; Hackett and Thimann, 1952; Kelly, 1947), and that auxin increases water uptake (Reinders, 1938, 1942; Bonner et al., 1953; Brauner and Hasman, 1952; Hackett and Thimann, 1952; and others). However, these observations can be explained without invoking active transport of water.

Discrepancies between cryoscopic and plasmolytic measurements result chiefly from errors inherent in the methods, especially failure to take into account penetration of the plasmolyzing solute into cells (Mercer, 1955; Slatyer, 1966). There is also uncertainty concerning the amount of change which occurs in composition and osmotic potential of sap during expression from cells (Crafts et al., 1949, p. 122). Respiration inhibitors and anaerobiosis appear to operate chiefly through decreased permeability of cell membranes, reduced salt uptake, and changes in the kinds and amounts of cell solutes. All these changes affect water movement indirectly, rather than through nonosmotic processes. It has been shown that auxin greatly increases the permeability of cells to water (Guttenberg and Meinl, 1952; and others), and it also increases the extensibility of cell walls. It therefore appears that auxin-induced water uptake results from increased permeability during short-term experiments and from increased cell enlargement or growth during long-term experiments.

ACTIVE TRANSPORT IN THE XYLEM. Several investigators have reported that the osmotic potential of the xylem sap is lower in stems than in roots because of movement of ions out of the xylem sap into surrounding tissues (Barrs, 1966; Eaton, 1943; Klepper and Kaufmann, 1966; Oertli, 1966). As a result water flows upward through the xylem from a region of lower to a region of higher water potential. Oertli (1966) described this as an example of active transport of water. However, movement of water through the xylem is caused by the pressure developed in the roots, where the water moves inward by diffusion from a region of higher to a region of lower potential. The cause of water movement is a passive diffusion process occurring in the roots, not the change in potential of the xylem sap from roots to shoots. Such passive movement should not be treated as an example of active transport (Levitt, 1967).

ELECTROOSMOSIS. Another mechanism sometimes proposed to explain anomalous water movement is electroosmosis (Fensom, 1957; Spanner, 1958; and others). It is well known that water can move across the negatively charged cell membranes and even through plant tissue from positive to negative poles under the influence of a difference in electrical potential. However, Dainty (1963*a*) points out that while the ion pumps operating in plant cell membranes might produce appreciable electroosmotic flow, cell membranes are so permeable to passive water movement that the action of an electroosmotic pump would be short-circuited by leakage outward along normal pathways. This view is supported by results of experiments such as those of Blinks and Airth (1951) who reported that applied potentials of up to 1,500 millivolts from surface to interior of *Nitella* cells produced no significant movement of water. Brauner and Hasman (1946) attributed 10 percent of the water movement in potato tissue to electroosmosis. This is probably a generous estimate. The role of electroosmosis in water absorption through intact roots is discussed in Chap. 5, where it is dismissed as unimportant.

POLAR MOVEMENT OF WATER. It has been reported that seed coats of certain species are more permeable to inward movement of water than to outward movement (Denny, 1917; Brauner, 1930, 1956). Polar water movement is also said to occur through insect cuticle (Beament, 1965) and various other animal membranes. Polar water movement was discussed by Bennet-Clark (1959, pp. 164–167) who followed Brauner in attributing it to electroosmosis caused by electrical asymmetry of the membranes under study. Dainty (1963*b*) suggests that differences in rate of inward and outward flow might result from differences in hydration on the two sides of a complex membrane. However, he thinks many examples of supposed polar flow of water result from the effect of the unstirred layers shown in Fig. 1.19.

Kamiya and Tazawa (1956) carried out some interesting experiments on *Nitella* cells mounted unsymmetrically in a double chamber (see Fig. 1.18).

The water in one side was replaced by a sugar solution, which caused flow from the chamber containing water through the cell to the chamber containing sucrose. Observations of differences in water movement, when first the short end and then the long end of the cell were immersed in sucrose solution, led Kamiya and Tazawa (1956) to conclude that the permeability to outward movement of water is less than the permeability to inward movement. Similar results were obtained by Dainty and Hope (1959), who attributed their findings to the effect of unstirred layers inside and outside the cells. These layers reduce the driving force more when exomosis is through the short end than when it is through the long end of the cell. Further research indicated that unstirred layers are less important than the fact that the cell membrane is dehydrated more by contact with the sucrose solution than with water, and that consequently its permeability is reduced (Dainty and Ginzburg, 1964*a*). These explanations seem to take the mystery out of polar water movement.

WATER EXCRETION. Any discussion of nonosmotic movement of water must include mention of examples of apparent excretion or secretion of water from various kinds of glandular structures. In his review of excretion, Stocking (1956) classified plant glands into four types: (1) glands secreting oils and resins, (2) nectaries, or glands secreting a sugary liquid, (3) glands characteristic of carnivorous plants which secrete nectar, mucilages, or digestive enzymes, and (4) water glands such as hydathodes. Hydathodes are dealt with in connection with guttation, and we are not concerned with oil- and resin-secreting glands in this book. The secretion of sugar solution from nectaries of flowers is well known, and significant amounts of sugar are also secreted from nectaries on leaves of some species. It seems probable that the solutes are excreted by an active transport process and the water then diffuses out along the resulting gradient in water potential. However, according to Stocking, some writers claim that water and salt excretion are independent processes. More puzzling is the removal of water from the bladders of *Utricularia* and the secretion of liquid into unopened "pitchers" of various insectivorous plants. The mechanism bringing about secretion obviously deserves more investigation. Salt glands will be discussed in Chap. 7.

NONOSMOTIC WATER UPTAKE IN ANIMAL CELLS. It might seem profitable to investigate the movement of water across animal membranes to learn if any clues are provided concerning water movement in plants. This problem was discussed recently by several writers in Symposium 19 of the Society for Experimental Biology. Beament (1965) argued that there must be an active uptake mechanism in insects because they can absorb water from air, which is far below saturation water vapor pressure. Robinson's (1965) discussion of water regulation in animal cells treats them as somewhat imperfect osmometers but concludes with the statement that we do not know exactly how the water content of mammalian cells is regulated. Smyth (1965) concluded that water movement across the gut probably is by what he calls endogenous osmosis, meaning that active transport of salt across the membrane is followed by water transport. Durbin and Moody (1965) are inclined to attribute water movement to a double-membrane effect, in which ions are moved across one membrane into a space between it and another membrane (see Diamond, 1965, p. 338).

Seven theories were proposed by Diamond (1965) to explain water transport across animal membranes. These are classical osmosis, filtration, electroosmosis, pinocytosis, local osmosis, the double-membrane effect, and codiffusion. Codiffusion refers to movement of water molecules accompanying solute molecules. Diamond started by denying that absorption and secretion can be explained passively by the laws of diffusion and osmosis, but he finally concluded that local osmosis (the endogenous osmosis of Smyth) is the probable cause of water transport. Thus, when semantic differences are eliminated, animal physiologists seem to believe that water movement in animal tissue occurs as it does in plant tissue, by osmosis (Dick, 1966, chap. 4). The situations they describe are very similar to those in plant cells and roots, where salt is actively transported across membranes and the water moves by diffusion along the water potential gradient produced by salt movement.

The preceding discussion indicates that there is no positive evidence for nonosmotic movement of water in plants and that its occurrence in animals is questionable. Most of the alleged examples of nonosmotic movement of water involve active transport of salt, which produces gradients in water potential along which water moves passively. It is doubtful if active transport of water can be demonstrated even if it occurs, because the membranes surrounding the vacuoles are so permeable to water that it would probably leak out as fast as it was transported inward. Therefore, even if nonosmotic uptake occurs, it can play no significant role in plant water relations.

Summary

Water plays essential roles in plants as a constituent, a solvent, and a reagent in various chemical reactions, also in the maintenance of turgidity. The physiological importance of water is reflected in its ecological importance, plant distribution on the earth's surface being controlled by the availability of water wherever temperature permits growth. Its importance is a result of its numerous unique properties, many of which arise from the fact that water molecules are organized into a definite structure held together by hydrogen bonds. Furthermore, the water bound to proteins, cell walls, and other hydrophilic surfaces has important effects on their physiological activity.

There are few reliable estimates of the relative proportions of the water occurring in cell walls, cytoplasm, and vacuoles; but plant water relations are usually dominated by the potential of the water occurring in the vacuoles. The water in the walls is important as a pathway for translocation of solutes from cell to cell; and the water in walls, cytoplasm, and vacuoles forms a continuous system through the differentially permeable membranes which isolate the solutes in the cytoplasm and vacuoles from the general hydrodynamic system.

Water tends to move from regions of higher to regions of lower water potential. It is sometimes claimed that active transport of water occurs against gradients of water potential, but the evidence for such movement is questionable. It appears likely that water movement can be explained in terms of osmosis and mass flow, and if any active transport of water occurs, it is of negligible importance in the overall water relations of plants.

two
Water in the Soil

From the viewpoint of this book, interest in soil is centered chiefly on its role as a storage place for water and a medium for root growth, but it is also a reservoir of mineral nutrients and provides anchorage for plants. Furthermore, it contains an active microbiological population, and many larger organisms such as earthworms which have important effects on its chemical and physical characteristics and on root growth. This chapter deals with the important characteristics of soils and with factors which especially affect water retention and availability to plants. For additional information on soil characteristics readers are referred to Baver (1956), Hagan et al. (1967), Luthin (1957), Marshall (1959), Rose (1966), Russell (1961), and Slatyer (1967).

Important characteristics of soil

Composition

Soil is a complex system consisting of varying proportions of four principal components. These are the mineral or rock particles and nonliving organic matter which form the solid matrix, and the soil solution and air which occupy the pore space within the matrix. In addition to these four components, soil usually contains numerous living organisms such as bacteria, fungi, algae, protozoa, insects, and small animals which directly or indirectly affect soil structure and plant growth.

The mineral particles are the chief components of most soils on a volumetric basis, except in an organic soil such as peat, and also are the most stable. They consist of rock particles developed in situ by weathering or deposited in bulk by wind or water. The nonliving organic matter usually constitutes less than 5 percent of the volume, except in the surface layer and in peat soils, but it can vary considerably in a given soil with cultural practices. Organic residues consist of a wide range of materials grading in size from large root fragments through rootlets and litter to colloidal products of decomposition.

The most conspicuous property of the solid matrix is its particulate nature. In contrast, the pore space forms a continuous but geometrically complex system which usually constitutes 30 to 60 percent of the total volume (see Fig. 2.1). It may be filled entirely with water, as in saturated

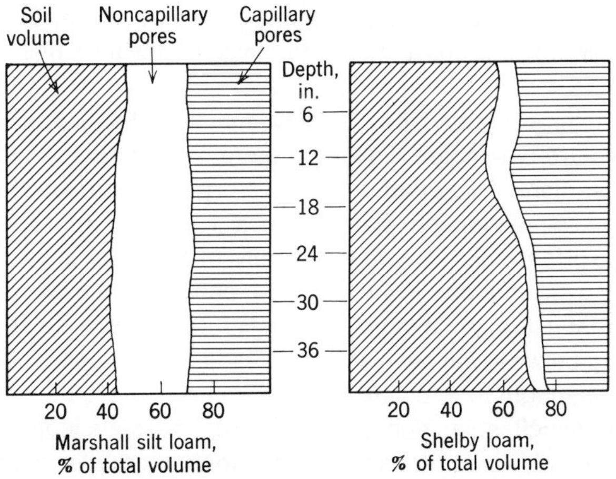

FIG. 2.1 **Differences in amount of capillary and noncapillary pore space in two dissimilar soils. A large proportion of noncapillary pore space is desirable because it promotes drainage and improves aeration.** *(Reproduced by permission from L. D. Baver, "Soil Physics," 2d ed., John Wiley & Sons, Inc., New York,* **1948.)**

soils, or largely with air, as in dry soils. In agricultural soils at field capacity the water fraction usually ranges from 40 to 60 percent of the pore space. The high degree of continuity in the water phase is of great importance in respect to water and salt movement in the soil and to roots.

In addition to these nonliving components, the soil is permeated by living roots and contains large populations of microorganisms, especially in the surface layers and in the vicinity of roots, the rhizosphere. Soil organisms play an important role in the decay of organic matter and the release of nitrogen and mineral nutrients which then become available for reabsorption. They also tend to deplete the oxygen content and increase the carbon dioxide content of soil, thus modifying the soil atmosphere in which roots grow. Organic products of decomposition are said to play a significant role in cementing soil granules together, thereby improving soil structure. The roots, together with small animals such as earthworms (see Barley, 1962), insects, and tunneling vertebrates, exert important effects on soil water relationships, particularly infiltration and distribution of water in the soil. The role of insects and the soil microflora was discussed in detail by Russell (1961), and the role of soil organisms in plant disease is covered in a symposium edited by Baker et al. (1965).

The physical characteristics of a soil depend chiefly on the texture or the size distributions of mineral particles, on the structure or the manner in which these particles are arranged, on the kind of clay minerals present and the kind and amount of exchangeable ions adsorbed upon them, and on the amount of organic matter incorporated with the mineral matter.

Characteristics of the clay fraction

The clay fraction provides most of the internal soil surface, therefore it controls the important soil properties; for this reason, it will be discussed in more detail.

There are three major types of clay minerals: kaolinite, which is most common in mature weathered soils, and montmorillonite and illite, which are the chief constituents of young soils. Unit crystals or micelles of kaolinite consist of silica and alumina platelets in a 1:1 ratio. These micelles form rigid lattices, so that soils composed chiefly of kaolinite show little swelling and shrinking with changes in hydration. Montmorillonite and illite micelles are composed of silica and alumina platelets in a 2:1 ratio. In illite, potassium ions occurring between silica platelets of adjacent micelles form chemical bonds strong enough to prevent separation and swelling. No such bonds exist between micelles in montmorillonite; hence soils containing a large proportion of montmorillonite swell and shrink markedly with changes in

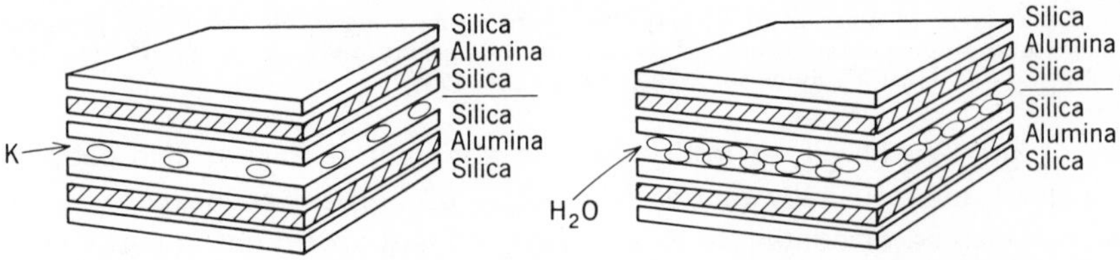

FIG. 2.2 Diagram showing the arrangement of silica and alumina sheets in illite crystals (left) and montmorillonite crystals (right). Entrance of water between silica layers causes the swelling characteristic of soils containing a large proportion of montmorillonite. (*From Thompson*, 1952.)

hydration. Such soils often develop broad, deep cracks during prolonged droughts. Two types of clay minerals are shown in Fig. 2.2.

Clay micelles are negatively charged, chiefly by replacement of silicon and aluminum ions with other cations within the crystal lattice. Another source of charge is the incomplete compensation of charge where bonds are broken at lattice crystal edges. The intensity of the negative charge determines the cation exchange capacity or ability to hold cations. The amount of water and ions bound on soil colloids has important effects on soil properties. Organic matter also has a large ionic exchange capacity (see Table 2.1), and consequently ionic exchange reactions are influenced by organic matter content as well as the amount and kind of clay. The small cation exchange capacity of soils low in clay resides in the silt and sand fractions.

Soil texture

Textural classifications of soil are based on the relative amounts of sand, silt, and clay predominating in the solid fraction. Soils are classified as sand, loam, silt, or clay, with various intermediate classes such as sandy loam, silt loam, or clay loam. Table 2.2 gives examples of soils classified according

***Table 2.1** Cation Exchange Capacity of Humus and Clay Minerals in Milliequivalents per 100 g Dry Soil* *

Vermiculite	160
Humus	100–300
Montmorillonite	100
Illite	30
Kaolinite	10

* *From data of Thompson (1952)*

***Table 2.2** Classification of Soil Particles According to System of International Society of Soil Science, and Mechanical Analysis of Three Soils* *

Fraction	Diameter, mm	Sandy loam, percent	Loam, percent	Heavy clay, percent
Coarse sand	2.00–0.20	66.6	27.1	0.9
Fine sand	0.20–0.02	17.8	30.3	7.1
Silt	0.02–0.002	5.6	20.2	21.4
Clay	below 0.002	8.5	19.3	65.8

* *From Lyon and Buckman, 4th ed., p. 43*

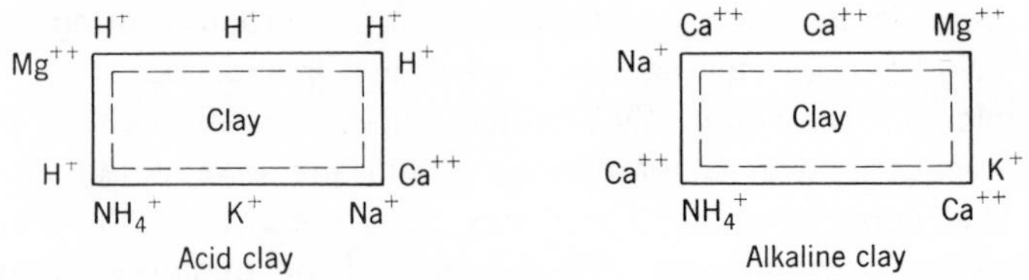

FIG. 2.3 **Attraction of cations to the surface of negatively charged clay micelles. If the clay holds a high proportion of hydrogen ions, the soil is acid; if most of the exchange positions are held by basic ions such as Ca^{++}, K^{+}, and Na^{+}, it is alkaline. (*From Thompson,* 1952.)**

to particle size, and Fig. 2.4 shows the principal textural classes graphically. The least complex soil is a sand, which by definition contains less than 15 percent of silt and clay. Such soils form relatively simple capillary systems with a large volume of noncapillary pore spaces, which ensures good drainage and aeration. Sandy soils are relatively inert chemically, are loose and noncohesive, and have a low water-holding capacity and usually a low cation exchange capacity.

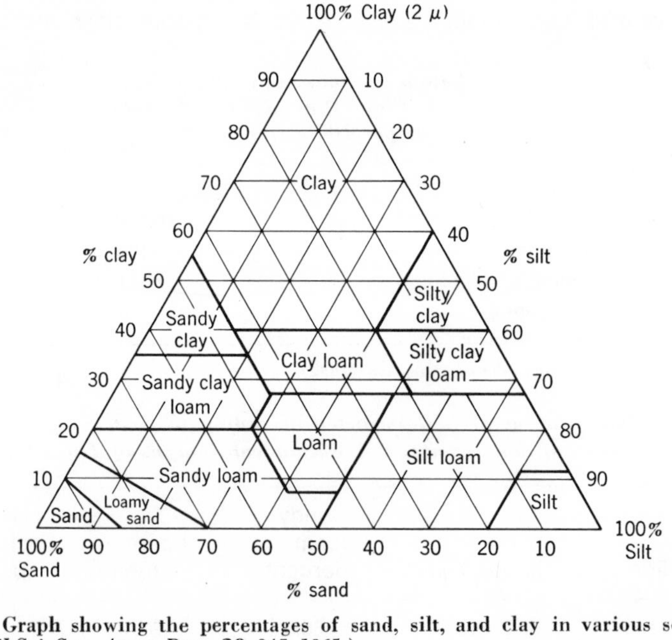

FIG. 2.4 **Graph showing the percentages of sand, silt, and clay in various soil classes. (*From Soil Sci. Soc. Amer. Proc.* 29:347, 1965.)**

Clay soils are at the other extreme with reference to size of particles and complexity because they contain more than 40 percent of clay particles and less than 45 percent sand or silt. The clay particles are usually aggregated together in complex granules. Because of their plate-like shape, clay particles have a much greater surface area than cubes or spheres of similar volume. Their extensive surface enables clay particles to hold more water and minerals than sandy soils. The surface possessed by even a small volume of particles of colloidal dimensions is tremendous. A cubical sand grain 1 mm on the edge has a surface of only 6 mm^2; but if it is divided into particles of colloidal size, 0.1 μ on the edge, the total surface resulting would be 60,000 mm^2. Day et al. (1967) state that the surface available to bind water ranges from less than 1,000 cm^2/g in coarse sands to over 1,000,000 cm^2/g in clays. The size of the clay fraction largely controls the chemical and physical properties of mineral soils.

Loam soils contain more or less equal amounts of sand, silt, and clay and therefore have properties which are intermediate between those of clay and those of sand. Such soils are considered most favorable for plant growth because they hold more available water and cations than sand and because they are better aerated and easier to work than clay.

Textural classification has only an approximate relationship to the behavior of a soil as a medium for plant growth because textural properties may be modified appreciably by organic matter content, the kinds of clay minerals present, and the amounts and kinds of ions associated with them. Aggregation effects of organic matter tend to give a fine-textured soil high in clay some of the pore space properties of a coarser textured soil, and colloidal effects of organic additions to a coarse-textured sandy soil give it some of the moisture and cation retention characteristics of a finer-textured soil. A clay soil composed chiefly of swelling clay minerals of the montmorillonite type has the properties of a finer-textured soil than one composed of kaolinitic clay minerals. In addition, when the exchange complex is dominated by sodium ions, clay micelles are dispersed and the soil appears to have a finer texture than when calcium or hydrogen are the dominant ions, because sodium ions cause dispersion of aggregates and decrease in pore space.

Structure and porosity

Combination of the ultimate soil particles into aggregates (crumbs, granules, or even clods) produces soil structure. The degree of structure existing in a soil affects the amount and size of pores and thereby greatly affects water movement and soil aeration. The pore space is that fraction of the soil volume occupied by air and water. It usually amounts to about half of the soil volume, but the total pore space is not particularly indicative of moisture and aeration properties of soils. Baver (1956) designates two major size classes of pores: (1) "noncapillary" or large pores which do not hold water tightly by capillarity, and (2) small "capillary" pores, which do. The noncapillary pores drain freely after rain or irrigation and are normally assumed to be filled with air. The capillary pores contain the water which remains after most free drainage is completed, that is, the water in the soil at field capacity. As will be seen later, this type of subdivision is very empirical since pore space, water retention, and drainage characteristics are linked dynamically, and there is no sharp break in the water retention curve at any one porosity value. However, this concept has been very useful as an initial subdivision of total porosity and has considerable significance with respect to plant growth; it is often valuable in ecological and agricultural field studies. The pore space relationships of two dissimilar soils are shown in Fig. 2.1. A soil such as the Shelby loam is an unfavorable medium for root growth because the small volume of noncapillary pore space results in poor aeration. Another example of differences in capillary and noncapillary pore space in two soils is shown in Fig. 2.5.

The large noncapillary porosity of sandy soils results in better drainage and aeration, but it also results in a lower water-holding capacity than that of clay soils, which have a larger proportion of small noncapillary pores. According to Baver (1956), an ideal soil has its pore space about equally divided between large and small, or noncapillary and capillary, pores. Such a soil has enough large pores to permit adequate drainage and aeration and enough small pores to give adequate water-holding capacity.

The development of soil structure is a complex series of processes leading to the arrangement of the primary particles into fairly stable aggregations which provide the soil with a certain matrix character and pore size distribution. Aggregation depends upon the presence of small primary particles that can be aggregated, the flocculation of these particles, and the cementation of the flocculated material into stable aggregates. It is often assumed (Russell, 1961; Baver, 1956) that sands and gravels which do not contain small primary particles are structureless. However, the single-grained arrangement of the large particles of which they are composed often appears to be stable, as do the pore size distribution and permeability characteristics.

The nature of the flocculating and cementing mechanisms is still not fully understood, even though several aspects have been established (Baver, 1956). Flocculation of clay particles appears to occur when there are sufficient cations in the soil solution to neutralize the negative charge of the clay micelles so that adjacent particle surfaces no longer repel one another and may come together, forming loose aggregations. Subsequent stabilization and cementation are thought to require the cementing action of irreversible or slowly reversible colloidal reactions of clay particles themselves, of inorganic colloids such as iron and aluminum oxides, or of organic colloids produced during decay of organic matter. The aggregation processes are hastened by mechanical deformation of the soil such as occurs by alternate wetting and drying, or freezing and thawing. Roots also appear to exert substantial pressures on the soil particles they adjoin, and fine roots, root hairs, and their organic exudates help to bind particles together.

Because all these factors operate most intensively in the surface soil, it generally tends to be better aggregated than that deeper in the profile.

However, aggregates can readily be broken down by repeated wetting and by tillage, particularly if the soil is not covered by vegetation and the organic matter content is low, as for example in a fine-textured soil after a bare fallow. Under such conditions the tendency of rain to cause dispersion of granules and puddling of the surface soil results in the formation of a relatively impermeable surface crust, which often hinders infiltration of water, gas exchange, and seedling emergence. Conditions conducive to downward drainage and accumulation of fine particles and salts at a certain depth in the soil often cause development of dense and relatively impermeable claypans or hardpans. Mixtures of sand particles of various sizes sometimes produce a soil structure containing only 4 or 5 percent of pore space (Lutz, 1952).

Preservation of good structure is essential for the infiltration of water and maintenance of conditions in the soil mass favorable for root growth. The dense mat of roots found beneath grass is particularly effective in improving and maintaining structure, while cultivation, especially if the soil is wet,

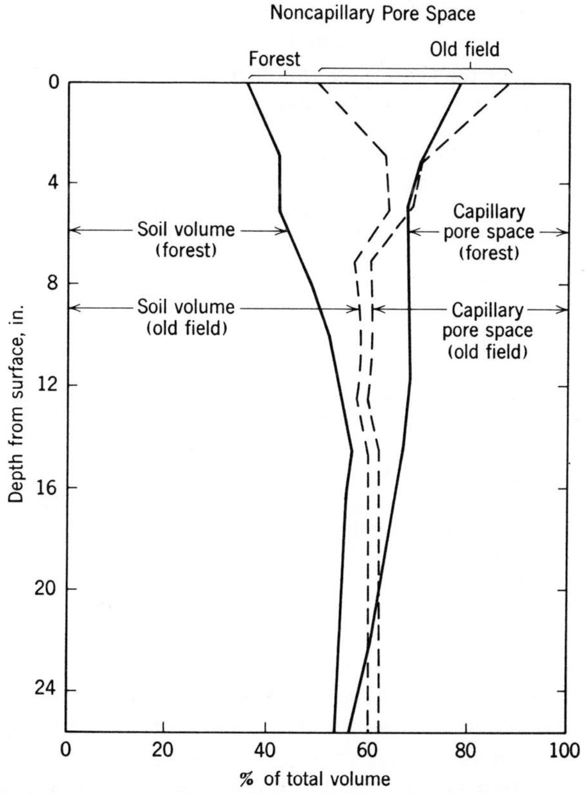

FIG. 2.5 **Differences in amount of capillary pore space in an old field and in an adjacent forest. The large percentage of capillary pore space increases the rate of infiltration (see Fig. 2.11) and decreases surface runoff during heavy rains. (*From Hoover,* 1949.)**

destroys it. The direct effect of root penetration, followed by death and decay, is to open up numerous channels. Earthworm activity is also claimed to improve soil structure. The difference in noncapillary pore space between the soil in an open field and in a nearby forest is shown in Fig. 2.5.

Soil structure is also affected by the kind of ions which predominate. For example, Richards and Fireman (1943) found that the permeability of a soil saturated with sodium was much less than when it was saturated with calcium. Sodium causes dispersion of the clay aggregates into their constituent micelles, reducing the number of large pores through which water can move rapidly.

Soil profile

Although some soils, such as recently deposited alluvium, may have a uniform texture to a depth of several feet, there are generally important changes in texture and even in structure at different depths in the soil. This is particularly true in older soils, where much downward leaching has occurred. As a result undisturbed soils usually possess a characteristic "profile" consisting of definite horizons or layers of soil which differ in their properties. A hypothetical soil profile is shown in Fig. 2.6. The upper or "A" horizon usually

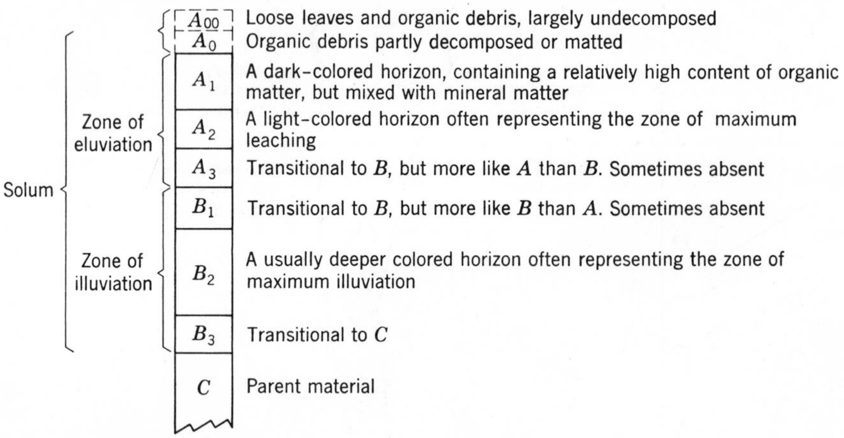

FIG. 2.6 **Hypothetical soil profile, showing the various possible horizons which might be found in a highly developed soil. (*From Thompson*, 1952.)**

differs appreciably in texture and hence in water retention characteristics (such as field capacity and permanent wilting percentage) from the "B" horizon which lies below it. It is therefore necessary to sample at least as deeply as roots penetrate and to determine the water-retaining properties of all horizons penetrated by roots in order to understand the water supply of plants. Collection of quantitative data on water retention therefore constitutes

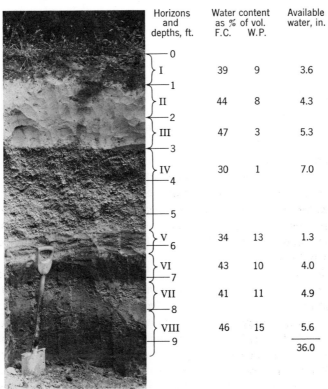

Horizons and depths, ft.	Water content as % of vol. F.C.	Water content as % of vol. W.P.	Available water, in.
0			
I	39	9	3.6
1			
II	44	8	4.3
2			
III	47	3	5.3
3			
IV	30	1	7.0
4			
5			
V	34	13	1.3
6			
VI	43	10	4.0
7			
VII	41	11	4.9
8			
VIII	46	15	5.6
9			
			36.0

FIG. 2.6a **Profile of a deep pumice soil in New Zealand formed by deposits of volcanic origin. Measurements of water content at field capacity (FC) and a permanent wilting (WP), expressed as percentages by volume, are shown for each horizon, also the readily available water in each horizon and the total water to a depth of 9 ft. This soil has an extraordinarily high storage capacity for readily available water. The photograph of the soil profile is a New Zealand Forest Service photograph by H. G. Hemming. The data are from Will and Stone (1967). The exact depths of the various horizons in the photograph vary slightly from those given by Will and Stone. Root distribution of a Monterey pine tree growing in this soil is shown in Fig. 4.13.**

a major problem for deep-rooted plants. For example, Patric et al. (1965) reported absorption of water to a depth of 6 m under forest trees.

Retention of water by soil

The water content of a sample of soil is usually defined as the amount of water lost when dried at 105°C, expressed either as the weight of water per unit weight of dry soil or as the volume of water per unit volume of bulk soil. Although useful, such information tells little about the availability of water for plant growth, because a sand can be saturated at a water content at which a clay is too dry for plant growth. This difference exists because of the different ways in which water is retained by different soils. If water is withdrawn by tension or pressure from a soil which does not shrink during drying, air must replace the water in the pore space. As a result air-water interfaces develop and form numerous curved water surfaces between adjacent particles of soil. The surface tension acting in these curved interfaces balances the tension or pressure exerted on the water and is one mechanism by which water is retained in the soil. If, on the other hand, the soil shrinks progressively as water is removed so no air enters the pore space, the soil particles are brought closer and closer together. As these particles carry a negative surface charge, they repel one another, and as they are brought closer together, the repulsive force becomes greater. The shrinkage and associated development of repulsive forces balances the pressure or tension applied to remove water and constitutes the second principal mechanism by which water is retained in the soil. Both mechanisms can be regarded as a part of the matric forces mentioned in Chap. 1 which cause lowering of the free energy or potential of water.

A third mechanism which lowers the free energy and increases the retention of water is the presence of osmotically active solutes, chiefly salts, in the soil solution. Solutes lower the vapor pressure of the soil water and decrease its free energy or water potential, as described in Chap. 1. However, solutes do not directly influence the amount of water retained against pressure or tension, unless it is applied across a membrane impermeable to solutes. For this reason the role of the osmotic potential of the soil solution in plant water relations will be postponed to Chap. 6, although it has important effects on the availability of water to plants.

Other mechanisms of water retention exist, but they are of limited importance in the free energy range within which plants grow (see Slatyer, 1967, pp. 67–72). For example, when dry soil is exposed to water vapor, several layers of molecules are bound by the hydratable cations, and water is adsorbed by polar forces on the surfaces of the negatively charged soil particles. However, this water is too firmly bound to be of any use to plants.

Classification of soil water

In Chap. 1, the importance of the chemical potential of water in plant cells and tissues was discussed in terms of the water status of plants and the tendency for water to move into or out of different cells, tissues, or organs. The chemical potential of soil water, termed the soil water potential, Ψ_{soil}, is of equal importance in soil water relations. As in the case of plant water potential, soil water potential can be related to the relative vapor pressure by Eq. (1.6). Several measurement procedures for soil water (see Chap. 3) make use of this relationship since they involve the determination of the relative vapor pressure of air in equilibrium with the soil sample being studied.

Thermodynamic terminology

The principal forces which contribute to the soil water potential are those associated with the soil matrix, those associated with the osmotic characteristics of the soil solution, and those which affect the total pressure on the soil water. The two chief mechanisms, described earlier, by which water is retained in shrinking and nonshrinking soils, are clearly associated with the structure and characteristics of the soil matrix and are termed matric forces. Together these constitute the matric potential Ψ_m. The osmotic forces associated with the soil solution comprise the solute, or osmotic potential Ψ_s, and any pressure component contributes the pressure potential Ψ_p. Thus, the soil water potential may comprise three main component potentials

$$\Psi_{\text{soil}} = \Psi_m + \Psi_p + \Psi_s \tag{2.1}$$

The soil water potential may also be defined to include external force fields such as gravity, which constitutes a gravitational potential Ψ_g. The equation can then be written

$$\Psi_{\text{soil}} = \Psi_m + \Psi_p + \Psi_s + \Psi_g \tag{2.2}$$

More details about the thermodynamics of soil water and soil water terminology can be found in Rose (1966) and Slatyer (1967). Detailed definitions of these terms can be found in Aslyng (1963) and in the glossary of soil science terms in vol. 29 of the *Proceedings of the Soil Science Society of America.*

Apart from these thermodynamic terms, several other terms, which have been in use for many years, describe soil water characteristics of particular significance to plant growth. The two most important of these are *field capacity* and the *permanent wilting percentage.*

Field capacity

The field capacity of a soil is the water content after drainage of gravitational water has become very slow and water content has become relatively stable. This situation usually exists 1 to 3 days after the soil has been thoroughly wetted by rain or irrigation. The field capacity has also been termed the field-carrying capacity, normal moisture capacity, and the capillary capacity. It is not a true equilibrium value, but only a condition of such slow movement of water that the moisture content does not change appreciably between measurements. While deep soils reach field capacity rather quickly, the presence of a water table near the surface will prolong the time required for drainage. Also, if a deep soil is initially saturated to a depth of several meters, drainage of the surface layer to field capacity will be much slower than if only the surface meter is wetted. Lack of homogeneity in the soil will also affect the water content at field capacity. For example, a fine-textured soil overlying a coarse-textured soil will have a higher water content than a uniformly fine-textured soil. Thus the field capacity of a soil is related to the conditions under which it is measured as well as to the characteristics of the soil itself. For example, the water content of soil allowed to drain in the field might be quite different from the water content of a cylinder of the same soil drained in the greenhouse on a layer of sand. Such factors should be considered in the interpretation of field capacity data.

Field capacity is determined most simply by ponding water on the soil surface and permitting it to drain for 1 to 3 days (depending on soil type) with surface evaporation prevented. Soil samples are then collected by auger for gravimetric measurement, or soil water content is measured by one of the techniques discussed in Chap. 3 and the results expressed on a gravimetric or, preferably for water storage purposes, a volumetric basis. Field capacity is a relatively reproducible value when determined in this manner, as long as precautions are taken to avoid sampling in the transition zone or

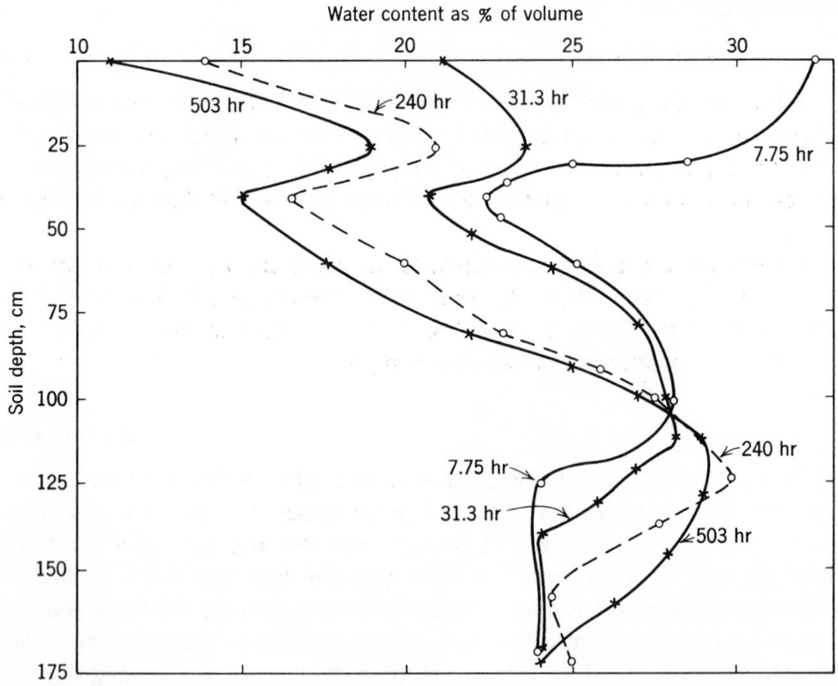

FIG. 2.7 **Profiles of water content of a nonuniform soil after drainage for various periods of time following irrigation. The surface horizon is a sandy loam, changing to a fine sandy loam at about 25 cm. Transition to a clay which holds much more water at −0.3 bar Ψ_m occurs at 75 to 100 cm. The original water content at a depth of 125 to 175 cm was about 24 percent. The profiles show progressive decrease in water content with time near the surface and increase in the 100 to 175 cm horizon after 240 to 503 hr. Slow readjustment in water content occurs for 10 to 20 days after irrigation. (*After Rose et al.*, 1965.)**

at the wetting front. It is a more significant and better-defined value in coarse- than in fine-textured soils because the larger pores of coarse soils soon empty, and the resulting rapid decline in permeability tends to cause a sharper transition from wet to dry soil than in fine-textured soils. Even so, it can vary significantly, because in any one profile at field capacity, the upper drainage zone is drying, while the lower zone is wetting, so that these two parts of the profile are on different sides of the hysteresis loop. This type of behavior is shown in Fig. 2.7.

Because field capacity is affected by soil profile and soil structure, laboratory determinations are not reliable indicators of the value in the field. Nevertheless, it is often desirable to make laboratory determinations. Most laboratory measurements are made by simulating the tension which develops during drainage in the field by use of pressure membranes or tension tables. There is some difference of opinion about the proper tension to apply. Marshall (1959) proposed a tension of 100 cm of water on undisturbed samples of soil, while Colman (1947) and Richards and Weaver (1944) recommended 0.3 bar or 330 cm on samples which have been dried, ground, and sieved. As field capacity has no fixed relationship to soil water potential, it cannot be regarded as a soil moisture constant. The amount of water retained at field capacity decreases as the soil temperature increases (Richards and Weaver, 1944).

The moisture equivalent has sometimes been used as a means of estimating the field capacity of soil. This is the water content of a wetted sample of pulverized and sieved soil drained in a centrifuge with a gravitational acceleration of 1,000 g. Richards and Weaver (1944) found that on the average the soil moisture content of 71 different soils exposed on a pressure plate to a 0.3-bar pressure differential approximated the moisture equivalent. However, there are large differences between field capacity and moisture equivalent in some soils, and since pressure cell equipment is more convenient to use and far less expensive than a centrifuge, moisture equivalent determinations are seldom cited in recent literature.

Permanent wilting percentage

Field capacity has been widely used to refer to the upper limit of soil water storage for plant growth, while the permanent wilting percentage has been just as widely used to refer to the lower limit. It is the soil water content at which plants remain permanently wilted (assuming that the leaves exhibit visible wilting), unless water is added to the soil. Briggs and Shantz (1912) first emphasized the importance of this soil water value and termed it the "wilting coefficient." They determined it by growing seedlings in small containers under conditions of adequate water supply until several leaves were developed. The soil surface was then sealed, and the plants were permitted to deplete the soil water until wilting occurred. The containers were then placed in a humid atmosphere and, if recovery occurred overnight, were again exposed to normal atmospheric conditions. This procedure was repeated until no recovery was observed, at which point the soil water content was determined.

Briggs and Shantz (1911, 1912) conducted a large number of measurements on a wide variety of plants and found little variation in the soil water content at which wilting occurred. Subsequently, a number of other workers (see, for example, Richards and Wadleigh, 1952) have determined that the soil water potential at wilting approximates −10 to −20 bars, with a mean value of about −15 bars. In consequence the percentage of water at −15 bars has become identified with the permanent wilting percentage and is frequently used as an index of it. This value is usually determined on soil samples in pressure membrane equipment (see Chap. 3). Furr and Reeve (1945) modified the definition to introduce a "first" permanent wilting percentage when the first (usually the lower) leaves wilted and an "ultimate" value when all leaves are wilted, which usually occurs at a much lower soil water potential (Slatyer, 1957*b*).

Although Briggs and Shantz were careful to state that the permanent wilting percentage represented the lower limit of soil water available for plant growth, and that water extraction may occur to lower water contents, Veihmeyer and Hendrickson (1927, 1949, 1950) and Hendrickson and Veihmeyer (1929, 1945) considered that it also marked the lower limit for water absorption. Their experiments were conducted chiefly with deep-rooted orchard crops, and they considered that at the permanent wilting percentage measurable reductions in soil water were no longer observed in the bulk of the root zone. These observations may have been complicated by the absence of knowledge about deeper roots and water table accessibility; they differ from those of other workers, who have observed soil water depletion to much lower soil water contents (Slatyer, 1957*b,* 1962).

There seems to be no physical reason why continued extraction of water may not occur after growth ceases or even after the death of the plant, although it can be expected to be much reduced by stomatal closure (see Chaps. 9 and 10). Also the appearance of visible wilting in a field crop may be due to a transient inability of the water supply system in the plant to meet the evaporative demand, rather than to conditions associated with permanent wilting. Philip (1957*a*) has shown how the appearance of wilting in the field depends on meteorological conditions, root density, and volume of soil occupied, as well as on the osmotic conditions in the plant which are the direct cause of wilting. Soil water conductivity might also affect onset of wilting.

It should be mentioned that Slatyer (1957*b*) strongly criticized the concept of the permanent wilting percentage as a soil constant. He pointed out that basically wilting occurs because of loss of turgor in the leaves and that the point of zero turgor pressure associated with wilting is dependent on the osmotic characteristics of the leaf tissue sap. Consequently, wilting occurs when there is a dynamic balance between the water potential in the soil and the water potential in the plant. Therefore, the soil water potential at wilting can be expected to vary as widely as the variation in osmotic potential in plants, which can range from −5 to −200 bars. However, an important point to remember here is that, because of the shape of the water potential/water content curve of soils (see Fig. 2.8) marked changes in water potential often accompany small changes in water content, so that for many practical purposes the permanent wilting percentage, or the percentage at −15 bars, can still be regarded as an important soil value. This approximation is particularly true for most crop plants, as the osmotic potentials of many species range from about −10 to −20 bars. However, it should always be borne in mind that in the equilibrium measurement as described by Briggs and Shantz the permanent wilting percentage is a function of the index plant for any given soil.

Readily available soil water

This expression, generally referring to the availability of soil water for plant growth, is taken as the amount of water retained in a soil between field capacity and the permanent wilting percentage. Since field capacity represents the upper limit of soil water availability and the permanent wilting percentage a lower limit, this range has considerable significance in deter-

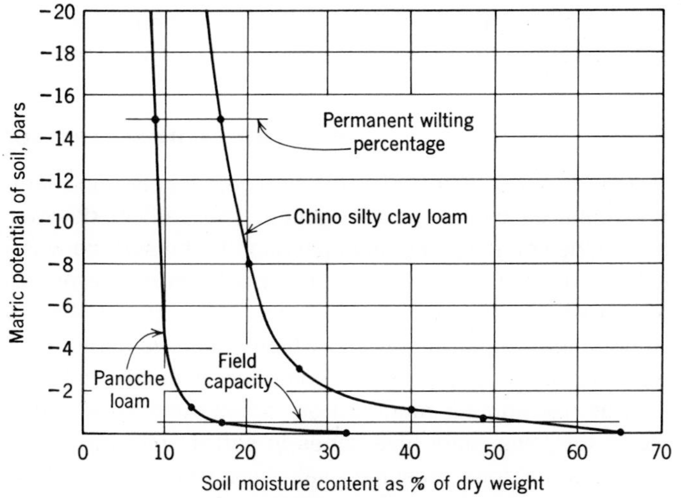

FIG. 2.8 **Matric potentials of a sandy loam and a clay loam soil plotted over water content. The curve for Panoche loam is from Wadleigh et al., 1946, and that for Chino loam is from data of Richards and Weaver, 1944.**

mining the agricultural value of soils. The available water range in different soils varies widely, as shown in Fig. 2.8 and Table 3.1.

In general, finer-textured soils have a wider range of water between field capacity and permanent wilting than coarse-textured soils. Also, the slope of the curve for water potential over water content indicates a more gradual release of the water with decreasing soil water potential. In contrast, sandy soils with their larger proportion of noncapillary pore space release most of their water within a narrow range of potential because of the predominance of large pores.

Data on readily available water must be used cautiously because the availability of water depends on several variables. For example, in any given soil, increased rooting depth in the profile as a whole can compensate for a narrow range of available water in one or more horizons. Conversely, restricted root distribution combined with a narrow range of available water results in considerable hazards for plant growth from an inadequate water supply, especially in climates where summer droughts are frequent. Also, it should be remembered that in many soils the range of water available for survival is substantially greater than that available for good growth. Furthermore, within the range of available water the degree of availability usually tends to decline as soil water content and Ψ_{soil} decline (Richards and Wadleigh, 1952; Hagan et al., 1961). It should be clear that there is no sharp limit between available and unavailable water and that the permanent wilting percentage is only a convenient point on a curve of decreasing water potential and decreasing availability. A more detailed discussion of soil water availability is given in Chap. 6. For the present it need only be noted that the range of soil water between field capacity and the permanent wilting percentage constitutes an important field characteristic of soil when interpreted properly.

Movement of water into soils

The rate of infiltration of water into soil is an extremely important factor in soil moisture recharge by rain and irrigation. The path of downward movement of water in the soil following its application to the surface was described in detail by Bodman and Colman (1944) for a uniform profile and by Colman and Bodman (1945) for a nonuniform profile. They found that the wetted portion of a column of uniform soil into which water was entering at the top and moving downwards appeared to comprise a stable gradient through which water was transmitted, ranging from a saturated zone at the top to a wetting zone at the lower end. Five zones in series were described as (1) a saturation zone, i.e., a zone presumed saturated which reached a maximum depth of 1.5 cm; (2) a transition zone, a region of rapid decrease of water content extending to a depth of about 5 cm from the surface; (3) the main transmission zone, a region in which only small changes in water content occurred; (4) a wetting zone, a region of fairly rapid change in water content; and (5) the wetting front, a region of very steep gradient in water content which represents the visible limit of water penetration. These zones are shown diagrammatically in Fig. 2.9.

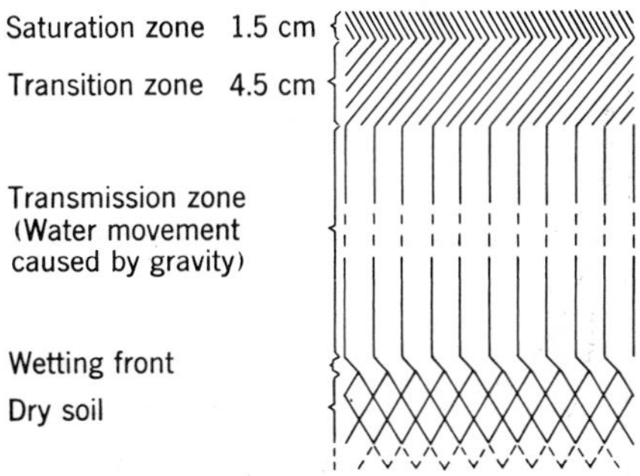

FIG. 2.9 **Diagram of zones in a uniform soil wetted from the top. There is little change in water content from top to bottom of the transmission zone, through which water moves by gravity, but a very steep decrease in water content and in matric potential at the wetting front. (*After Bodman and Colman,* 1944.)**

Philip (1957*d*) made a theoretical analysis of infiltration which accounts for all these zones except the transition zone. This discrepancy is probably caused by air entrapment in the thin surface layer. It can be seen that the transmission zone is a continually lengthening unsaturated zone of fairly uniform water content and potential. According to Marshall (1959), in this zone the matric potential Ψ_m is probably higher (closer to zero) than −0.025 bars, and the degree of pore space saturation is about 80 percent. The gradient of Ψ_m in the transmission zone is usually very small after the water has penetrated to a reasonable depth (Taylor and Heuser, 1953), so that movement within that zone is caused primarily by gravity. The rate of advance of the wetting front depends on the rate at which water is supplied to it through the transmission zone and is little affected by the size of the potential gradient between it and the dry soil beyond it. The infiltration rate usually decreases with time following application of surface water, so that a fairly stable minimum value is frequently observed in infiltration experiments. This is illustrated in the three curves of Fig. 2.10. For practical purposes infiltration data are often presented in units such as centimeters per hour, but they really represent the total amount of water entering in a given time, so the type of presentation in Fig. 2.10 provides much more information.

Factors affecting infiltration

The factors affecting infiltration of water into the soil are important in agriculture and watershed management because if infiltration is impeded,

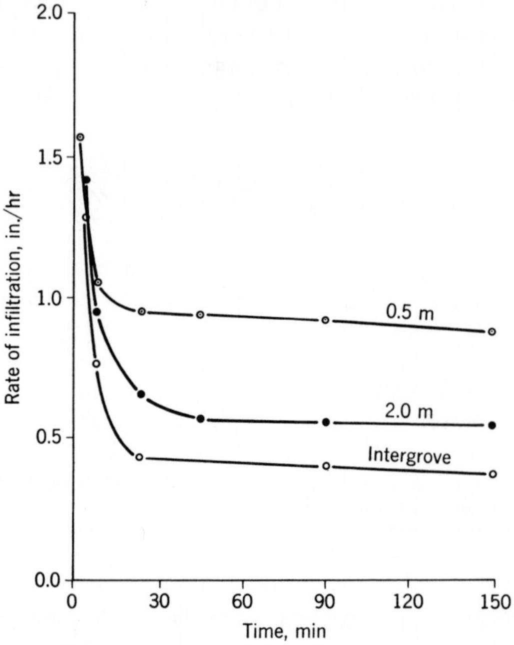

FIG. 2.10 **Rate of infiltration of water into a clay soil at various distances from the trunk of an isolated acacia tree. The effect of roots is shown by the more rapid infiltration near the tree. (*From Slatyer*, 1962.)**

recharge of soil water is hindered and rapid runoff occurs, often accompanied by erosion and flooding. According to Musgrave (1955) the major factors affecting the infiltration of water into soil are the initial water content, surface permeability, internal characteristics of the soil (such as pore space), degree of swelling of soil colloids and organic matter content, duration of rainfall, and temperature of the soil and water.

A number of investigators have reported that infiltration is reduced as initial soil water content increases (Tisdall, 1951; Ayers and Wikramanayake, 1958). These results are to be expected because the difference in potential across the wetting front is reduced, and hydration and swelling of clay particles reduces the cross-sectional area available for the entrance of water. Philip (1957*c*) states that as the initial water content increases, the initial infiltration rate is reduced, but the effect is less pronounced with increasing time, and all curves eventually tend to approach the value for initially saturated soils.

Infiltration is greatly decreased by zones of low soil permeability such as surface crusts (McIntyre, 1958), claypan and hardpan zones (Fishback and Duley, 1950), surface compaction caused by farm implements (Diebold, 1954) and human and animal traffic, and by dispersed clay structure caused by an excess of alkali (Richards, 1954). Permeability characteristics of soils often change during infiltration, not only because of increasing water content, but also because of "puddling" of the surface caused by reorientation of surface particles and washing of finer materials into soil. Musgrave (1955) proposed four main groups of soils in terms of their expected minimum infiltration rates. These groups are presented in Table 2.3, where it can be seen that at least a tenfold range in permeability can be expected.

Table 2.3 Minimum Infiltration Rates for Wet Soil *

Type of soil	Rate in cm/hr
Plastic, swelling clays, and certain saline soils	0.0–0.125
Shallow sandy and silty soils, soils low in organic matter, and soils high in clay	0.125–0.375
Sandy loams and shallow loess soil	0.375–0.75
Deep sand and loess	0.75–1.125

** From Musgrave (1955)*

Protective mulches are often used to improve surface permeability and increase infiltration (Fishback and Duley, 1950), and organic matter is incorporated into the surface soil (Pillsbury and Richards, 1954). The data in Fig. 2.10 taken at increasing distances from the trunk of an isolated tree demonstrate the effect of organic matter and root activity on both initial and final filtration rates. Differences in rate of infiltration into a forest soil and the soil of an adjacent eroded old field are shown in Fig. 2.11. In some areas the primary purpose of cultivation is to improve infiltration, and some procedures appear to be effective to considerable depths and for extended periods (Diebold, 1954; van Duin, 1955).

Occasionally it is reported that rainfall and irrigation water fail to penetrate properly because of difficulty in wetting the surface soil. Jamison (1946) reported very uneven infiltration of water under citrus trees on light, sandy soils in central Florida because the sand particles were difficult to wet. Wander (1949) attributed this to formation of a coating of water-repellent metallic soap on the soil particles. Some success has been reported in improving infiltration of water into badly trampled soils and soils difficult to wet by addition of wetting agents. Attempts have also been made to improve infiltration on burned slopes and hydrophobic soils of watersheds by applying wetting agents (Letey et al., 1962). However, more research is needed on their use before any general recommendations can be made.

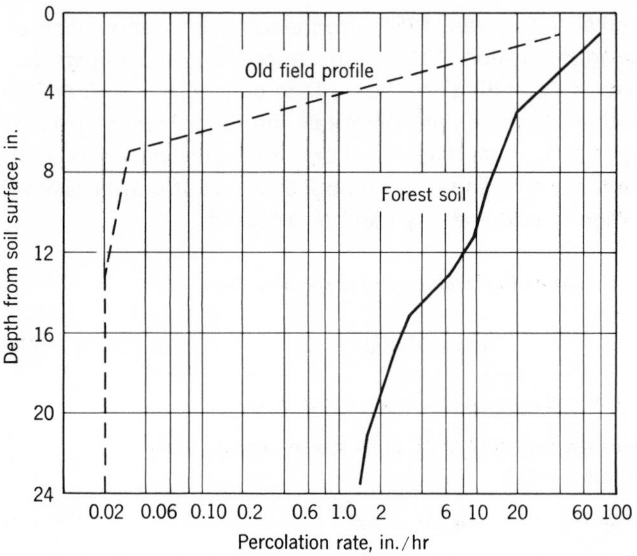

FIG. 2.11 Comparison of rates of infiltration of water into a forest soil and an adjacent old field. Figure 2.5 shows differences in capillary pore space in these two soils. (*From Hoover*, 1949.)

Movement of water within soils

The movement of water within soil controls not only the rate of infiltration but also the rate of supply to roots and the rate of underground flow to springs and streams. In conventional terms liquid water flows through the water-filled pore space under the influence of gravity, and in the films surrounding soil particles under the influence of surface tension forces. Water also diffuses as vapor through the air-filled pore spaces along gradients of decreasing vapor pressure. In all instances movement is along gradients of decreasing water potential, but the gradients are produced by different components of the total soil water potential. Miller and Klute (1967) and Cary and Taylor (1967) present detailed discussions of water flow in soil.

Movement of liquid water

It has been customary to differentiate between saturated flow or saturated conductivity in saturated soils and unsaturated flow or capillary conductivity in soils which are unsaturated. However, the term hydraulic conductivity, formerly used for water flow in saturated soils, is now being used for both saturated and unsaturated flow (Slatyer, 1967, p. 95). The chief difference is that in saturated soils gravity controls the water potential gradient, while in drained soils it is controlled by the matric potential, and water moves in films surrounding the soil particles rather than by gravity flow through the pores.

SATURATED FLOW. The theory of movement of liquid water is based on a generalized form of Darcy's law which states that the quantity of water passing a unit cross section of soil is proportional to the difference in hydraulic head. If the hydraulic head is replaced by the difference in total potential $\Delta\Phi$ and a proportionality coefficient is introduced, the resulting equation is

$$V = -K\frac{\Delta\Phi}{\Delta z}$$

If V is the velocity of flow in centimeters per second in the z direction, Φ is the total water potential in centimeters of water, and Δz refers to depth, the coefficient K is called the hydraulic conductivity and is expressed in centimeters per second.

UNSATURATED FLOW. As drainage proceeds and the larger pores are emptied of water, the contribution of the hydraulic head or gravitational component to total potential becomes progressively less important and the contribution of the matric potential Ψ_m becomes more important. The effect of

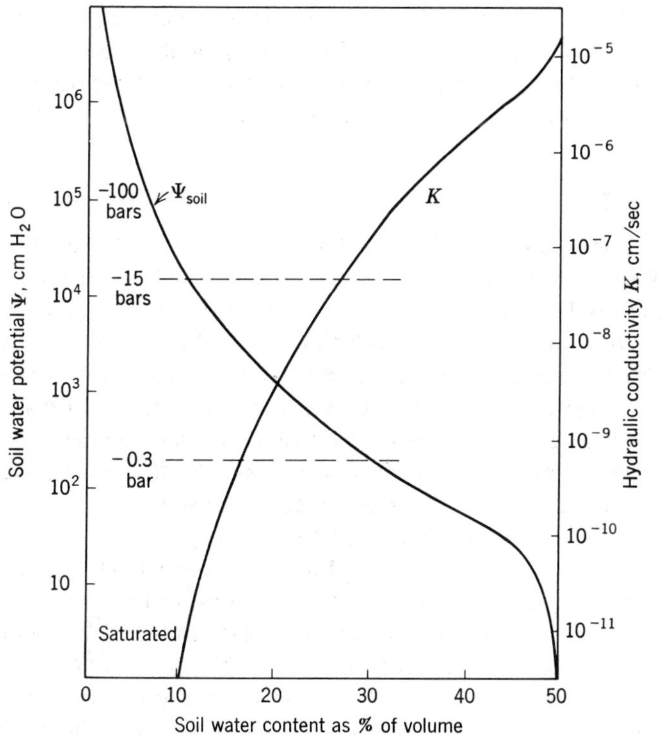

FIG. 2.12 **Decrease in hydraulic conductivity K and soil water potential Ψ_{soil} with decrease in soil water content. (*After Philip,* 1957*b*.)**

pressure is generally negligible because of the continuous air spaces, and the solute potential Ψ_s does not affect the potential gradient unless there is an unusual concentration of salt at some point in the soil. Darcy's law is applicable to unsaturated flow if K is regarded as a function of water content. As the soil water content and soil water potential decrease, the hydraulic conductivity decreases very rapidly, as shown in Fig. 2.12, so that when $\Psi_{soil} = -15$ bars, K is only about 10^{-3} of the value at saturation. According to Philip (1957*a*) the rapid decrease in conductivity occurs because the larger pores are emptied first, greatly decreasing the cross section available for liquid flow. When the continuity of the films is broken, liquid flow no longer occurs.

DIFFERENCES IN SOIL CONDUCTIVITY. The values for K vary widely, ranging from <0.0025 cm/hr in the least permeable to >25 cm/hr in the most permeable soil (Smith and Browning, 1946). Soils with a hydraulic conductivity of less than 0.25 cm/hr are poorly drained, while those with conductivities greater than 25 cm/hr do not hold enough water for good plant growth. Permeability or conductivity decreases with decreasing pore space and is sensitive to changes in cation content, which affects the degree of hydration or swelling of clay colloids. Entrapment of air greatly reduces permeability by blocking soil pores, while decrease in temperature decreases permeability by increasing the viscosity of water.

Water vapor movement

At a soil water potential of approximately -15 bars the continuity of the liquid films is broken and water moves only in the form of vapor. Although vapor movement in soil is very slow, it occurs more rapidly than would be expected from calculations of diffusion rates. Philip and de Vries (1957) attribute this to the fact that part of the path through the soil is liquid, and water condenses on one surface and evaporates from the other surface of isolated wedges of water. They also found that the temperature gradients across air-filled pores might be twice as steep as the average temperature gradient in the soil mass. These two factors result in a rate of vapor movement much more rapid than expected from theory.

The reason for negligible vapor transfer under isothermal conditions, at moderate to high water contents, is primarily due to the fact that even quite steep gradients of Ψ_w are associated with relatively small gradients of vapor pressure because of the nature of the relationship of Ψ_w to $e/e°$. For example, at Ψ_{soil} of -13 bars the vapor pressure is reduced only 0.175 mm Hg. However, steep vapor pressure gradients develop if large differences exist in pressure, solute concentration, or temperature. Although pressure is seldom of importance and salts only occasionally cause large gradients, steep temperature gradients are frequently established, particularly near the surface.

The experiments of Gurr et al. (1952) demonstrate the relative importance of liquid and vapor movement in soils of varying water content. These workers used horizontal columns of soils, each of initially uniform water and salt content, the water content in the 10 columns ranging from practically dry to practically saturated levels. A uniform temperature difference was then applied across all columns, and the net transfer of water and of salt after a period of 5 days was used as an indication of the net water movement and total liquid movement, respectively. The results showed that in general there was a net transfer of water toward the cold side but that the salt moved in the reverse direction. Moreover, the greatest net transfer of water was in the low to medium water content treatments, and the greatest transfer of salt in the medium water contents. There was little change of salt or water distribution in the very dry and very wet treatments (see Fig. 2.13).

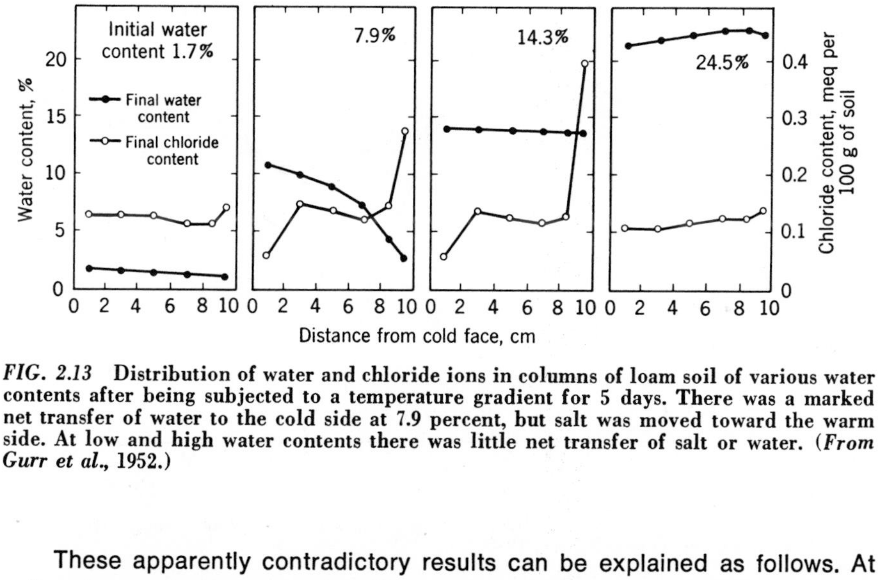

FIG. 2.13 **Distribution of water and chloride ions in columns of loam soil of various water contents after being subjected to a temperature gradient for 5 days. There was a marked net transfer of water to the cold side at 7.9 percent, but salt was moved toward the warm side. At low and high water contents there was little net transfer of salt or water. (*From Gurr et al.,* 1952.)**

These apparently contradictory results can be explained as follows. At soil water contents so low that there was no continuity in the liquid phase and transfer of water occurred solely in the form of vapor, there was no salt movement. Once liquid continuity started to develop, however, the net transfer was affected by both vapor and liquid movement, and the temperature-induced movement of vapor toward the cold face tended to be compensated for by a return flow of liquid water along water potential gradients. The net transfer of water toward the cold face could therefore be expected to decrease rapidly as water content increased, even though for a time there would still be a substantial flow of vapor in one direction and liquid in the other, leading to salt accumulation at the warm face, but no net change in water content. Once complete liquid continuity developed and the soil approached saturation, vapor flow would become negligible, and any movement of liquid water containing salt associated with temperature differences would be compensated for by a return flow associated with water potential differences. Hence, neither salt nor water content appeared to change in moist soil.

In nature there are large seasonal fluctuations in soil temperature, the surface warming up in the summer and cooling down in the winter, relative to the deeper horizons (see Fig. 2.14). As a result there can be significant

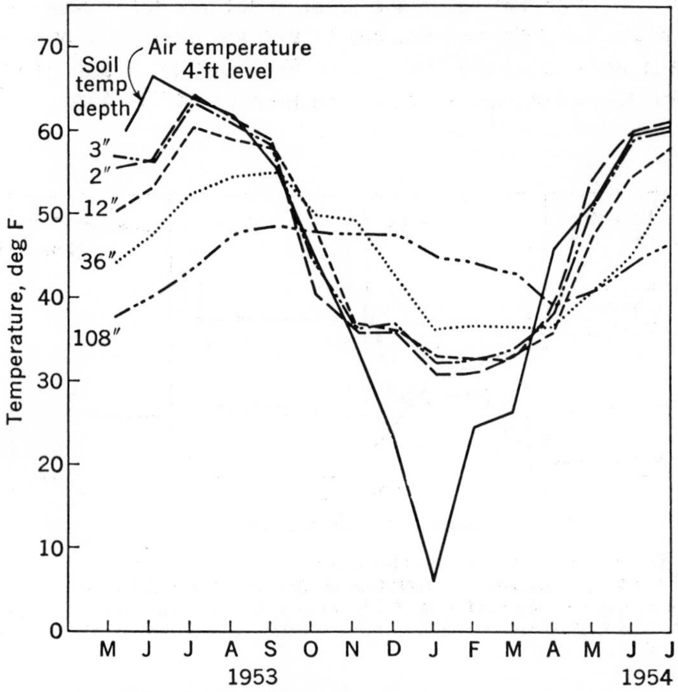

FIG. 2.14 **Seasonal trends in soil and air temperatures in a hardwood forest stand near Chalk River, Canada.** (*From Fraser,* 1957.)

upward movement of water in the winter and downward movement in the summer. Edlefsen and Bodman (1941) observed such movements in experiments at Davis, California, and Lebedeff (1928) reported this upward movement to be an important source of ground water in southern Russia, where cooling of the surface at night results in condensation in the surface layer. Such movement is usually assumed to occur as vapor, but Smith (1943) and others have pointed out that water movement along thermal gradients can occur both as vapor and as liquid. Readers are referred to Slatyer (1967, pp. 109–118) for a detailed discussion of this problem.

Upward and horizontal movement in soil

There is much interest in the amount of water which moves upward as the water content of the surface layers is depleted by evaporation and absorption by the plant cover. The question concerning the distance from which water moves through the soil toward an absorbing root is discussed in Chap. 6. The early work on upward movement of water was confusing because of failure to take into account the depth to the water table. Mathematical treatments of upward flow have been made by Gardner (1958), Gardner and Fireman (1958), and Wind (1960). The latter estimated that the upward flow in a coarse-textured soil with a water table at 60 cm is 5 mm/day, but in a fine-textured soil only 2 mm, and with the water table at 90 cm the flows were reduced to 1 mm/day in both soils. Wind (1955*a*) concluded that if the water table is no deeper than 1 m, upward flow should be adequate to subirrigate crops. Gardner and Fireman (1958) reported that lowering the water table of Pachappa fine sandy loam from 90 to 180 cm reduced evaporation to about 12 percent of the value at 90 cm because of slower upward movement to the evaporating surface. However, further lowering to 3 or 4 m had little effect on evaporation, and significant upward movement of water occurred from water tables as deep as 8 or 9 m. Patric et al. (1965) reported recharge of soil water in covered plots during the winter when the surrounding soil was moist by vertical and horizontal movement over distances of several meters. Gardner and Ehlig (1962) also observed that there was measurable upward movement of water in soil below the field capacity. Such movement of water must be taken into account in considering the amount of water available to plants rooted in the surface soil.

Summary

Soil consists of four fractions: the mineral particles and nonliving organic matter which form the matrix, and the soil solution and air which occupy the pore spaces within the matrix. It provides the anchorage which enables roots to maintain plants in an erect position, and it acts as a reservoir for water and salt. Much of the success of plants in any given habitat depends on the suitability of the soil as a medium for root growth and functioning.

Soil water is held largely by matric forces which bind water on the soil particles and, to a much lesser extent, by osmotic forces developed by the salts dispersed in the soil solution. The availability of soil water to plants depends on its potential and on the hydraulic conductivity of the soil. The water readily available for plants occurs in the range between field capacity and permanent wilting percentage. Field capacity is the water content after drainage of gravitational water has become very slow and represents a water potential of −0.3 bar or less. Permanent wilting percentage is the water content at which plants become permanently wilted and corresponds to a water potential of −10 to −20 bars. The exact value varies with the kind of plant and the conditions under which wilting occurs.

The rate of infiltration of water into soil is important in connection with recharging the soil with water by rain or irrigation. The major factors affecting infiltration are initial water content, surface permeability, internal characteristics such as pore space, degree of swelling of soil colloids, and organic matter content; also the duration of rainfall. The rate of water movement through soil, the hydraulic conductivity, decreases with decreasing pore space. Movement of water through the soil is controlled chiefly by the gravitational potential in soils above field capacity and by the matric potential in soils drier than field capacity. Hydraulic conductivity decreases very rapidly with decreasing water potential, so movement of liquid water is very slow in dry soil and practically ceases at a water potential of approximately −15 bars. In dry soils water moves only as vapor. Differences in temperature between surface soil and deeper horizons result in measurable upward movement of water vapor in the winter and downward movement in the summer.

There appears to be more movement of water in soil below field capacity than was formerly supposed. Significant movement of water occurs from depths of several meters, and if the water table is within 1 m of the surface, the upward movement should be sufficient to supply a crop.

three

Measurement & Control of Soil Water

Recognition of the importance of the water content of the soil in relation to plant growth has resulted in the development of many methods of measuring soil water. Not only is there a need for the measurement of changes in soil water storage with time, in order to estimate evapotranspiration, but there are also many situations where measurements are required at specific points, as at various depths or locations. There are also many experimental situations where careful measurement and control of soil moisture is necessary if the results are to be interpreted properly.

The following discussion is intended to give the reader an introduction to some of the techniques. Additional information may be found in the references cited in the text and in references such as Holmes et al. (1967), Rose (1966), Slatyer (1967), Slatyer and McIlroy (1961), and the bibliography on methods of determining soil water compiled by Shaw and Arble (1959).

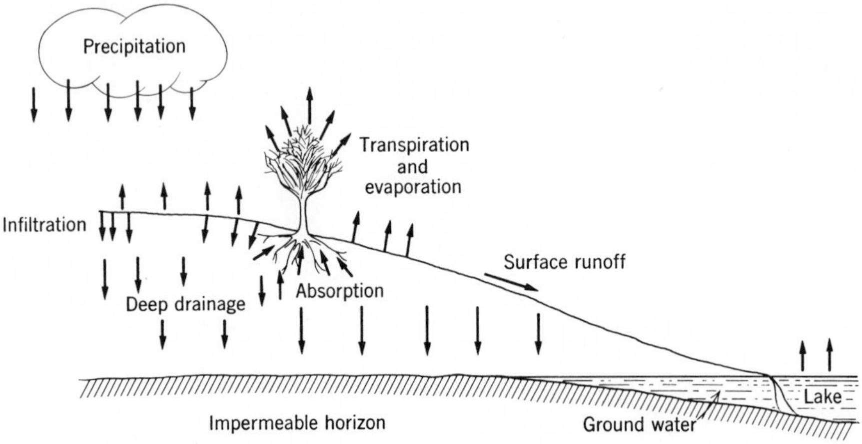

FIG. 3.1 **The hydrologic cycle. A diagram showing disposition of precipitation by surface runoff, infiltration, and deep drainage, and its removal from soil by evaporation and transpiration.**

Field measurements of soil water

Soil water balance

Measurements of change in soil water storage are commonly used to estimate evapotranspiration. The water balance equation is

$$\Delta W = P - (O + U + E) \tag{3.1}$$

where ΔW is the change in soil water storage (initial content minus final content) during the period of measurement and P, O, and U are the precipitation, runoff, and deep drainage respectively. U is defined as the amount of water passing beyond the root zone, or, for experimental purposes, as the amount passing below the lowest point of measurement. E is evaporation (including transpiration) from the plant and soil surfaces. This expression can be used on any scale, ranging from continental land masses and hydrologic catchments down to individual plants. During periods of dry weather between rains or irrigations, and neglecting (or otherwise measuring) U, $E = \Delta W$ and can consequently be measured by determining changes in soil water storage under the plant community being studied.

Measurements of ΔW are most accurately conducted by the use of weighing or floating lysimeters, provided they are properly designed and located. However, they cannot be used when the mixed nature of the species composition, the spatial distribution of the vegetation, the depth and ramification of the root system, the size of plants, or other factors make it impossible to simulate the natural environment inside the lysimeter itself. Examples of such

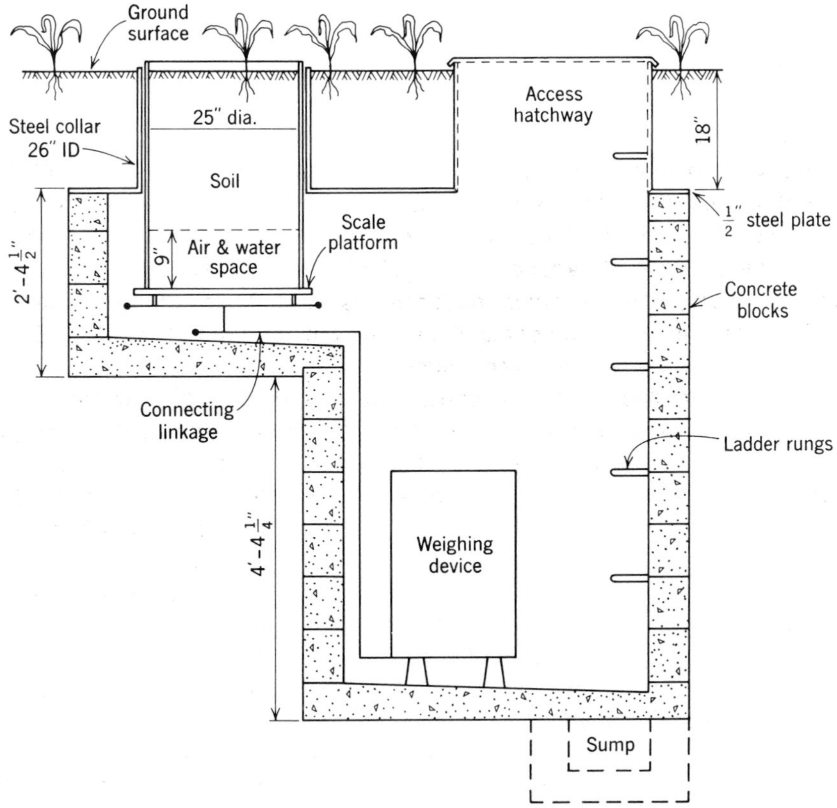

FIG. 3.2 **Diagram to show the principle of a weighing lysimeter. It consists of a container holding a mass of soil mounted flush with the soil surface and arranged so it can move up and down on a weighing device. (*Modified from England and Lesesne,* 1962.) The lysimeter pictured used a scale to follow changes in weight, but electronic weighing devices such as that used by van Bavel and Myers (1962) have advantages. It is necessary to have a lysimeter surrounded by similar vegetation if the results are to be applicable to crops or stands of vegetation. Some problems of lysimetry are discussed by Hagan et al. (1967, pp. 536–544).**

limitations are found in arid plant communities and in natural mixed forests. In such cases, determinations of soil water storage at different points in the community provide the only technique for evaluating ΔW. Even so, interpretation of the observations is difficult, and the presence of a water table near the surface can limit the application of the method unless fluctuations of water table depth themselves can be used (Holmes, 1960). Examples of water storage determination by soil water measurement are cited by Slatyer (1967). The quality of such measurements can be improved considerably if adjustments are made for the frequently significant loss of water to, or gains of water from, the soil zones below the depth of measurement (Patric et al., 1965; Rose and Stern, 1965).

Sampling problems

It is difficult to obtain reliable estimates of changes in soil water content in the entire root zone because of the great vertical and horizontal variability in soil water distribution in the field. Water content often varies drastically over short distances. This is partly because of irregularities in root distribution which cause some areas to be depleted of water sooner than others, and partly because of variations in physical characteristics of the soil, particularly its clay content, which influences its ability to retain water. Furthermore, at the same water potential, the water content of samples obtained only a few centimeters apart may vary widely.

Because of this variability, considerable replication of sampling is required to give valid estimates of the bulk soil water content. For example, Aitchison et al. (1951) found that in a typical loam soil more than 10 samples were required to show that differences of 1.0 percent in soil water content (1 g water/100 g dry soil) were significant at $p = 0.05$. More than 40 samples were required to show that differences of 0.5 percent were significant at this level. Staple and Lehane (1962) found a similar order of variability in clay loam soils which were regarded as reasonably uniform. Hewlett and Douglass (1961) also present a discussion of sampling problems.

Intensive direct sampling often disturbs the vegetation, requires care in refilling the holes from which samples are removed, is very difficult in stony soil, and requires large amounts of labor. As a consequence, field workers tend more and more to use indirect methods which permit installation of a number of sensing elements in a study area so that repeated measurements can be made at the same point. Although the equipment for such installations usually costs more initially than that for direct sampling, the saving in labor costs and the advantages of repeated measurements at the same point compensate for the greater initial expenditure.

The choice of a method to study soil water content depends partly on the objectives of the investigator. The usual objective in measuring soil water content is to determine how much water is present and available for plant growth at a particular time and place. The availability of soil water depends primarily on its potential (Ψ_{soil}), and the most useful description of the moisture characteristics of a soil is a graph showing water potential plotted against water content (Fig. 2.8).

Direct measurements of soil water content

The basic measurements of soil moisture are made on soil samples of known weight or volume; the water content is expressed as grams of water per gram of oven-dry soil or grams of water per cubic centimeter of oven-dry soil. Samples for gravimetric determinations are usually collected with a soil auger or sampling tube. Samples for volume determinations are usually collected in special containers of known volume, with a tube sampler, or from the side of an exposed soil profile (see for example, Coile, 1936, and Lutz, 1944). The bulk density of the soil can be determined separately (Vomocil, 1954). The water is usually removed from the soil by oven-drying at 105°C to constant weight. The time required can be shortened somewhat by using an oven with forced ventilation.

Several procedures have been proposed to eliminate the time required for oven-drying. One technique is to mix the soil sample with a known volume of methyl alcohol and then measure the change in specific gravity of the alcohol with a hydrometer (Bouyoucos, 1931). In another method, the water in the moist soil sample is allowed to react with a known weight of calcium carbide to form acetylene gas (Sibirsky, 1935; Magee and Kalbfleisch, 1952). The amount of water in the soil sample is calculated from the decrease in weight of sample plus calcium carbide. However, neither these nor any other suggested short-cut procedures have come into common use.

Expression of water content as a percentage of dry weight is not useful to the plant scientist unless the water potential curve or the field capacity and permanent wilting percentage are known, because a percentage water content which represents saturation in a sandy soil might be below the permanent wilting percentage of a clay soil (see Table 3.1). For some purposes it

***Table 3.1** The Water Content of Soils of Various Textures at Matric Potentials of 0.3 and 15 bars and at First Permanent Wilting.* The Values Are Water Contents as Percentages of Dry Weight.†*

Name of soil	Water content as percentage of dry weight at		
	0.3 bar	15 bars	First permanent wilting
Hanford sand	4.5	2.2	2.9
Indio loam	4.6	1.6	2.6
Yolo loam	8.4	12.6	7.1
Yolo fine sandy loam	12.6	5.5	8.3
Chino loam	19.7	8.0	10.2
Chino silty clay	40.8	21.9	23.2
Chino silty clay loam	48.9	15.0	23.3
Yolo clay	45.1	26.2	29.6

* *As determined by Furr and Reeve (1945)*
† *From Richards and Weaver (1944)*

is more useful to convert water content per unit of weight into water content per unit of volume, because the soil occupied by a root system is measured by volume rather than by weight. Also, additions and losses of water from soil are often measured in inches or centimeters, which on an area basis becomes the volume. The conversion from weight to volume units can be made by multiplying the percentage by weight times the bulk density of the soil under study. The water content can then be expressed in convenient units such as inches per foot or centimeters per meter of soil depth.

Indirect measurements of soil water content

Most of the numerous methods of measuring soil water content are indirect. This means that the property measured must be related to water content by some kind of calibration procedure.

NEUTRON SCATTERING. At the present time, the most commonly used indirect method for measuring soil water content is probably that of neutron scattering. This method is based on the fact that hydrogen atoms have a much greater ability to slow down and scatter fast neutrons than most other atoms, so that counting slow neutrons in the vicinity of a source of fast neutrons provides a means of estimating hydrogen content. Since the only significant source of hydrogen in most soils is in the soil water, the technique offers a convenient means of estimating soil water content. In soils with high root density, or high levels of organic residues, the amount of organic hydrogen may affect the estimates. However, the amount is generally small enough—compared with the hydrogen in the soil water—to be neglected.

Several descriptions of the neutron scatter method have been published (see, for example, Belcher et al., 1950; Holmes, 1956; Stone et al., 1955). Portable commercial instruments are now available and are reasonably convenient to use in the field. The instrument consists of a probe, which usually contains a source of fast neutrons, and a counter tube for detecting slow neutrons connected through an amplifier to a portable scaler. In use, the probe is lowered into a plastic-lined access tube, and counting rates are determined at the desired depths. The count rates, adjusted for background and standardized to counts made in a tank of pure water, are then calibrated against direct volumetric determinations of soil water content. Theoretical calibrations have also been developed (Holmes and Jenkinson, 1959), and Hewlett et al. (1964) discuss sampling problems.

The method has several important advantages over most other techniques, including the absence of a lag period while the soil water equilibrates with a sensing instrument and the fact that a large volume of soil (about 20 cm radius) is monitored, thus smoothing out local variability. However, it does involve the disturbing influence of an access tube, and the large volume of

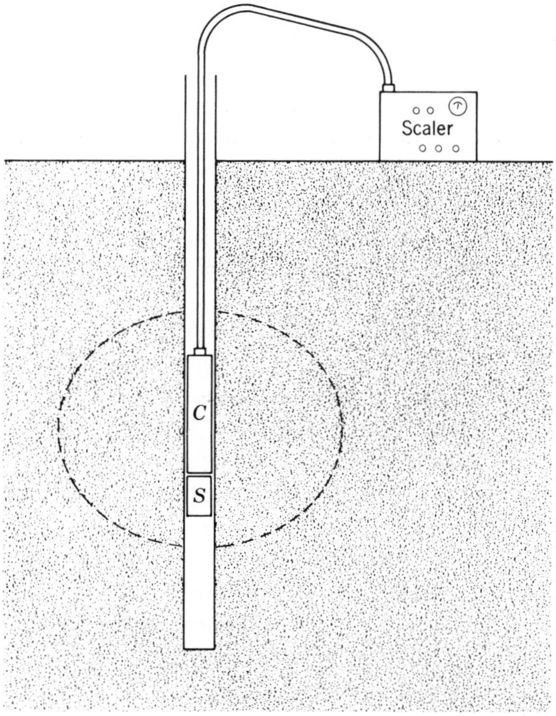

FIG. 3.3 Diagram of a neutron meter installation in a tube to measure water content at various depths in the soil. It consists of a source of fast neutrons (S) and a counter for slow neutrons (C) which can be lowered in the tube to measure soil water content at any desired depth. The slow neutrons reflected by the hydrogen in the soil, chiefly in the water, are recorded by the attached scaler or rate meter. The water content of an approximately spherical volume of soil of 15 to 30 cm in radius is measured, the radius increasing with decreasing water content of the soil. (*Modified from van Bavel et al.,* 1954.)

the sample prevents sampling at a point on or near the surface unless special precautions are taken. Also, the results are influenced by other sources of hydrogen atoms, such as in organic matter, and by other elements, notably chlorine, iron, and boron (Holmes, 1960). These factors and different swelling and shrinking patterns prevent a single universal calibration curve for all soils.

GAMMA RAY ABSORPTION. Ashton (1956) described measurement of changes in soil water content by change in amount of gamma radiation absorbed. The amount of radiation passing through soil depends on soil density, which varies chiefly with change in water content. If the soil does not shrink and swell appreciably, change in water content can be measured from change in amount of radiation passing through the soil. Ferguson and Gardner (1962) and Gurr (1962) found this method useful for measuring movement of water in soil columns. However, it is reliable only in soils where the change in bulk density is very small compared with the change in water content.

The method requires a source of gamma radiation such as cesium, a detector such as a geiger tube or scintillation probe, and a scaler. Ashton placed two plastic tubes on opposite sides of large pots and dropped the source down one tube, the detector down the other. In work with small pots or columns, the source and detector can be positioned in a jig on opposite sides of the container so the radiation can be carefully collimated.

ELECTRICAL CAPACITANCE. Measurement of soil water content by electrical capacitance is based on the fact that water has a much higher dielectric constant than dry soil (80:5); hence changes in water content should be reflected in changes in capacitance. The method is used to measure water content of grain, flour, dehydrated foods, and various industrial products. Anderson and Edlefsen (1942), Fletcher (1939), Wallihan (1946), and others have attempted to apply it to measurement of the water content of soil. However, difficulties in obtaining uniform electrical contact with the soil tend to make capacitance measurements unreliable, except perhaps as a surface probe (De Plater, 1955). Attempts were made by Anderson and Edlefsen (1942) and Wallihan (1946) to reduce the electrode contact effect by embedding the electrodes in plaster of paris blocks. The capacitance varies with temperature and requires appropriate corrections; but it is less affected by the salt concentration of the soil solution than conductivity measurements. Although the method is theoretically attractive, it is not yet practical for field measurements.

THERMAL CONDUCTIVITY. Since the conduction of heat through soil decreases with decreasing water content, it was proposed by Shaw and Baver (1939) that changes in thermal conductivity be used to measure soil water content. The method was further developed by de Vries and associates (1955, 1958). A heating element is buried in the soil and warmed by passing an electrical current through it, and the rate of heat dissipation, which varies with the soil water content, is measured. This method is independent of the salt concentration of the soil, but electrode contact effects are of considerable importance; hence, the method works best in moist soils and in sandy soils which do not shrink while drying. Johnston (1942) enclosed the units in plaster of paris blocks, but Haise and Kelley (1946) found the thermal conductivity blocks less satisfactory than electrical resistance blocks because at matric potentials lower than −4 bars the change in moisture content is too small to affect the thermal conductivity, but it does produce measurable changes in electrical conductivity.

Bloodworth and Page (1957) constructed a thermal conductivity unit by enclosing a thermistor in a porous material and using it as a combined heating element and temperature measuring unit. This unit covers the range of readily available water, but it is most sensitive near field capacity.

Measurements of matric potential

Several techniques used for estimating soil water content really measure matric potential, directly or indirectly. The two most important of these are electrical resistance units and tensiometers.

ELECTRICAL RESISTANCE BLOCKS. Near the end of the nineteenth century, attempts began to measure changes in soil moisture by means of changes in electrical resistance. The early attempts were unsuccessful because variations in electrode contact with the soil, variations in salt content of the soil, and temperature-induced changes in resistance obscured the changes produced by variations in soil moisture. The most serious of these difficulties were eliminated by Bouyoucos and Mick (1940), who embedded the electrodes in small blocks of plaster of paris. The blocks are buried in the soil and connected by well-insulated leads to a resistance bridge. The water content of the blocks changes with that of the soil; this produces measurable changes in the electrical conductivity of the solution between the electrodes. The blocks can be left in the soil for months or possibly for years, although ordinary plaster of paris or gypsum blocks tend to disintegrate rather rapidly in wet, acid soils and often last only one season. The life of commercially available blocks has been prolonged by impregnating them with resin (Bouyoucos, 1953*a*, 1953*b*, 1954). Gypsum blocks are sensitive over a range of about −0.5 to −15 bars of matric potential; hence, they are more satisfactory in dry than in very moist soil. A diagram of a gypsum block is

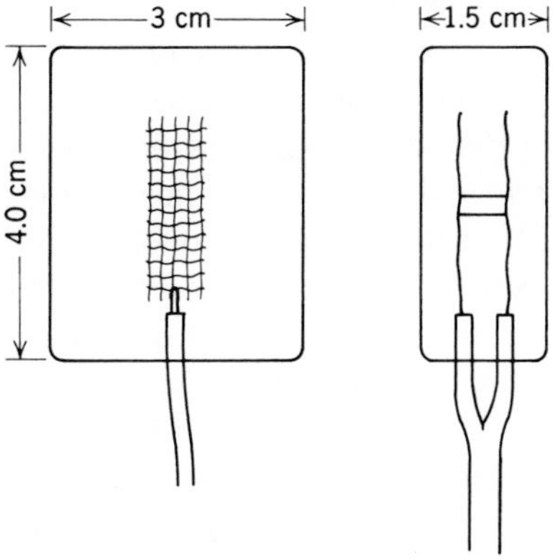

FIG. 3.4 Surface and edge views of a plaster of paris block for measurement of changes in soil moisture by change in resistance. The electrodes are pieces of stainless steel screen separated by a plastic spacer and embedded in plaster of paris. (*Bouyoucos*, 1954.)

shown in Fig. 3.4, and a graph relating resistance to available water content is seen in Fig. 3.5.

Units consisting of electrodes wrapped in nylon (Bouyoucos, 1949, 1954) or fiberglass (Bouyoucos and Mick, 1948; Colman and Hendrix, 1949) are considered to be more durable, to respond more rapidly, and to be more sensitive to soil water potentials of higher than −0.1 bar than gypsum blocks. However, they are much more sensitive to changes in salt content of the soil

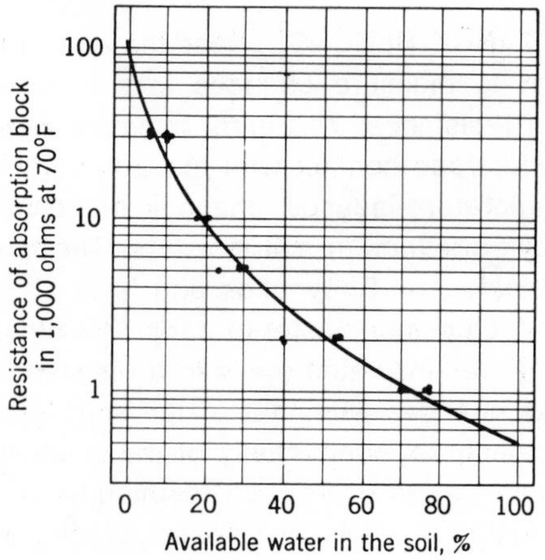

FIG. 3.5 **Resistance in ohms of a plaster of paris resistance block plotted over soil water content of a silt loam soil. (*Bouyoucos,* 1954.)**

than gypsum blocks and are less sensitive in drier soils. The gypsum blocks are less sensitive to salt because they are buffered by the dissolved calcium sulfate, and gypsum block readings are said to be unaffected by up to 2,000 lb of fertilizer per acre (Bouyoucos, 1951). The importance of the salt error can be expected to vary in different types of soil (Richards and Campbell, 1950). The nylon and fiberglass units tend to give different readings during successive wetting and drying cycles (Weaver and Jamison, 1951). England (1965) reported that field calibration was necessary every year during the first 5 or 6 years because entrance of colloidal clay and other materials caused the relation of resistance to moisture content to change. He estimated that the life of fiberglass units would probably be less than 15 years. Resistance measurements change somewhat with temperature, and thermistors are sometimes installed with the blocks to measure the temperature. All porous-block methods show hysteresis effects, and they are more reliable on drying cycles than on wetting cycles because drying is slower and gives time for a better equilibration between soil and blocks. Fortunately, the drying cycle is more important in relation to plant growth.

Resistance blocks are sometimes calibrated by placing them in a pressure membrane apparatus and measuring the resistance under various pressures (Haise and Kelley, 1946). Such calibration permits estimation of the matric potential of the soil from resistance readings of the blocks. They can also be calibrated against soil water content in the field by taking samples for gravimetric determinations from the vicinity of the blocks. Occasionally, they are calibrated in the laboratory in samples of the soil in which they are to be used (Kelley, 1944). Unfortunately, their calibration curves tend to change with time, and periodic recalibration is desirable (England, 1965; Holmes, 1956). Details of installation and calibration can be found in papers by Aitchison et al. (1951), Kelley et al. (1946), Knapp et al. (1952), and Slatyer and McIlroy (1961). In spite of certain shortcomings, resistance blocks have been used very extensively, especially in dry soil where tensiometers do not function. They are particularly useful in monitoring gross changes in soil water content between irrigations. Also, the progress of a wetting front through the soil can be followed by the sudden reduction in block resistance of a series of blocks buried at various depths in the soil.

TENSIOMETERS. Direct field measurements of the matric or capillary potential can be made only with tensiometers. These devices can also be used to

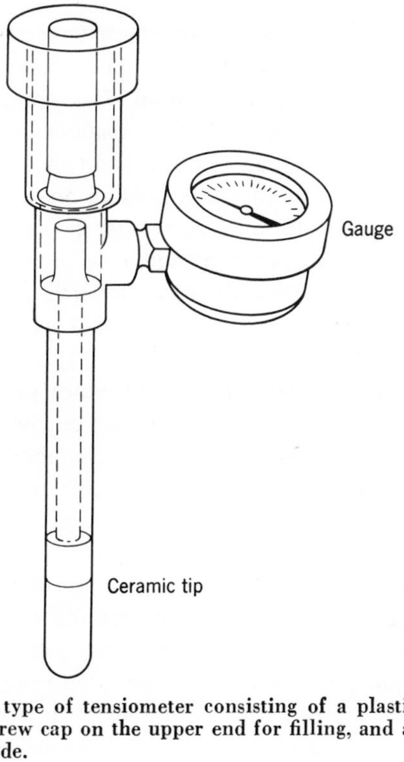

FIG. 3.6 **A commercial type of tensiometer consisting of a plastic body with a ceramic tip at the lower end, a screw cap on the upper end for filling, and a Bourdon-type vacuum gauge attached at the side.**

estimate soil water content. They consist essentially of a porous ceramic cup filled with water which is buried in the soil at any desired depth and connected by a water-filled tube to a manometer or vacuum gauge (see Fig. 3.6). The manometer indicates the pressure drop on the water in the porous cup which is in equilibrium with the matric potential of the water in the soil. The tensiometer is an excellent measuring instrument in moist soils; but when the matric potential drops to about −0.8 bar, air begins to enter the cup and it becomes useless. Although most rapid plant growth occurs within the sensitive range, lower potentials are of great interest to agriculturists and ecologists. The other limitations are relatively minor, including the necessity for recharging after entry of air, the tendency for roots to become concentrated around the porous cup, and the occasional diurnal fluctuations in reading resulting from heat conduction along the water-filled tube.

Although tensiometers read in negative pressure units, they can be calibrated against soil moisture content so that readings can be converted to percentage water content. Richards (1949, 1954) and Scofield (1945) published extensive discussions of the use and calibration of tensiometers.

Measurements of soil water potential

Field methods of measuring soil water which use porous blocks and tensiometers can be used to estimate the matric component but not the osmotic component of the total soil water potential. The pressure and gravitational potentials can usually be neglected, but the osmotic potential may be important in heavily fertilized soil (see Fig. 3.8) and in arid regions where salt accumulates. Unfortunately, there are at present no entirely suitable methods for routine measurement of the total soil water potential in the field.

It seems possible, however, that some recently described techniques may prove useful. A method described by Richards (1965) utilizes two glass bead thermistors which are built into a compact assembly for burial in the soil. One of the thermistors is exposed to soil air, and water is absorbed on its surface until equilibrium with respect to water potential exists between the water on the thermistor, in the surrounding air, and in the adjacent soil. The other thermistor is enclosed in a small chamber containing silica gel. When the thermistors are heated by an electric current, the thermistor in silica gel will heat up rapidly until air temperature is reached, at which time heat flow into the thermistor balances heat loss. The moist thermistor will, however, not approach the temperature of the dry thermistor until the water is distilled off. The length of time required for the two thermistors to reach the same temperature is a function of the water potential of the soil.

Another technique (Peck and Rabbidge, 1966) involves burying a small osmometer in the soil, the pressure inside which is monitored by a sensitive pressure transducer. If an air space is left between the membrane and the

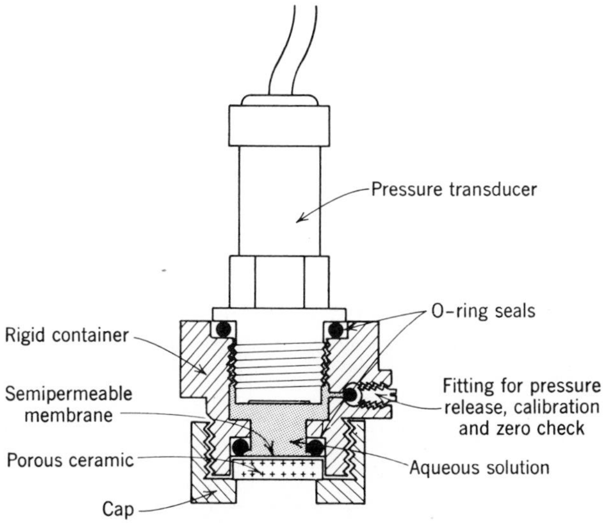

FIG. 3.7 Apparatus for direct measurement of matric soil water potential. A porous plate is covered with a differentially permeable membrane such as a dialysis membrane and fastened to a small chamber filled with a solution of polyethylene glycol with an osmotic potential lower than that of the soil. A pressure transducer is attached to the top. The difference between the equilibrium pressure and the osmotic potential of the solution is a measure of the matric potential of the soil. If an air space is left between the porous support and the soil to prevent movement of liquid, the total soil water potential can be measured. (*After Peck and Rabbidge,* 1966.)

soil, the total water potential is measured. If no air space is left and the membrane is permeable to ions, the matric component is measured. As yet, neither of these techniques has received extensive testing, but both appear to have definite possibilities.

An even more promising approach was made by Rawlins and Dalton (1967), who pointed out that if both heat and water can flow freely between the thermocouple chamber and the soil, the effect of temperature change will be negligible. By enclosing a thermocouple in a porous porcelain bulb buried in the soil they were able to measure changes in soil water potential in spite of diurnal temperature fluctuations of several degrees.

Laboratory measurements of soil water

Although the techniques already described can be used in laboratory or greenhouse work, they are intended primarily for field measurements. Considerable attention has been placed on techniques for evaluating the soil water characteristics of isolated samples and soil cores under controlled laboratory conditions. These techniques are commonly used to develop calibrations between water content and the components of soil water potential.

Measurements of soil water potential

In soils low in salt, the water potential is essentially equal to the matric or capillary potential, and field measurements with tensiometers give useful indications of the total water potential in the range higher than −0.8 bar. Likewise, estimates of water potential can be made from measurements of water content if a graph relating matric potential to water content is available. However, if enough salt is present to produce a measurable osmotic potential, the matric potential is not an adequate measure of the total water potential. This is often true of heavily fertilized soils (see Fig. 3.8) as well as saline soils.

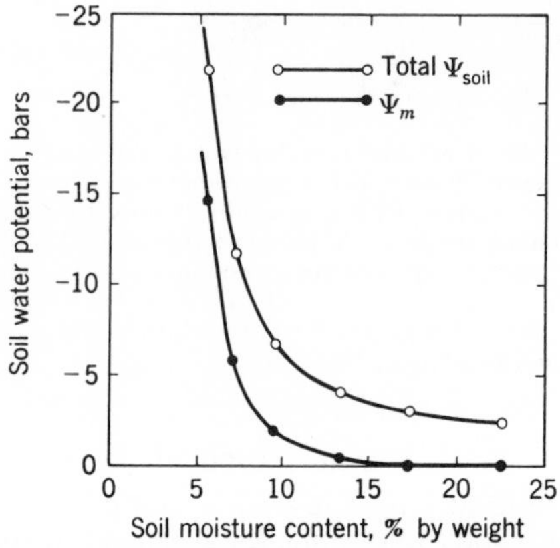

FIG. 3.8 **Total soil water potential and matric potential plotted over soil water content for a greenhouse soil consisting of a mixture of sand, loam, and peat. The relatively large difference between matric and total potential is the osmotic component resulting chiefly from addition of fertilizer. (*From Newman,* 1966.)**

VAPOR EQUILIBRATION. One of the earliest attempts to measure soil water potential was that by Shull (1916) who measured the uptake of water by dry cocklebur (*Xanthium pennsylvanicum*) seeds from soils at various moisture contents. Since he had determined how much water these seeds would absorb from solutions of various osmotic pressures, he could calculate the approximate force with which water was held by a given soil. Gradmann (1928), Hansen (1926), and other European workers measured the water potential of soil with strips of paper saturated with solutions of various concentrations. The strips were weighed before and after exposure to the soil, and the soil water potential (Saugkraft) was regarded as equal to the osmotic pressure of the strip which neither gained nor lost weight. This method can be used in the field or the laboratory, but it requires careful temperature control. The vapor equilibration method described by Slatyer (1958) can also be used for soil. Most laboratory measurements of water potential are based on measurements of relative vapor pressure, although some have been based on depression of the freezing point.

CRYOSCOPIC METHOD. This method is based on determining the freezing point depression of the water in a sample of soil (Campbell et al., 1949) and calculating the solute potential from the relationship between chemical potential and freezing point depression. Unfortunately, this method is subject to errors resulting from the low concentration of water in the soil and the nature of the adsorption forces which bind water on the solid matrix. In dry soil it often appears difficult to concentrate sufficient water at a nucleation point to get crystallization. Furthermore, the smaller the amount of water in the soil, the larger this effect tends to be; and as some water freezes out of the soil, the potential of the remaining water becomes lower. Therefore, it is difficult to get a sharp freezing point in dry soils, and the calculated potential finally obtained may be in error.

PSYCHROMETRIC METHODS. The most promising technique for measuring soil water potential involves measurement of the relative vapor pressure by a thermocouple psychrometer. This depends on the relationship between the chemical potential of water and the depression of the vapor pressure according to Eq. (1.7) developed in Chap. 1.

The two types of psychrometers in most common use are the one developed by Spanner (1951) and modified by Monteith and Owen (1958), and that developed by Richards and Ogata (1958). The operation of thermocouple psychrometers is discussed in Chap. 10. It will suffice here to point out that in addition to laboratory measurements it is now possible to make field measurements by the psychrometer method (Rawlins and Dalton, 1967).

Measurements of matric potential

Laboratory measurements of capillary or matric potential are almost always made with the pressure plate or pressure membrane equipment developed primarily by Richards (1949, 1954). In this procedure a previously wetted sample of soil is placed on a membrane permeable to water and solutes, enclosed in a container, and subjected to a pressure difference across the membrane. The pressure difference is produced either by suction beneath the membrane or by gas pressure applied above, usually from a cylinder of compressed air. When water outflow ceases, indicating that equilibrium has been reached between the capillary potential and the imposed pressure, the sample is removed and the water content is determined gravimetrically. Sometimes repeated determinations are made on the sample by measuring the water outflow associated with each increase in pressure.

A typical pressure membrane apparatus is shown in Fig. 3.9. Ceramic plates are used for suction or low pressure. For higher pressures cellulose membranes supported on metal screens have been used, but ceramic plates are now available for pressures up to 15 bars.

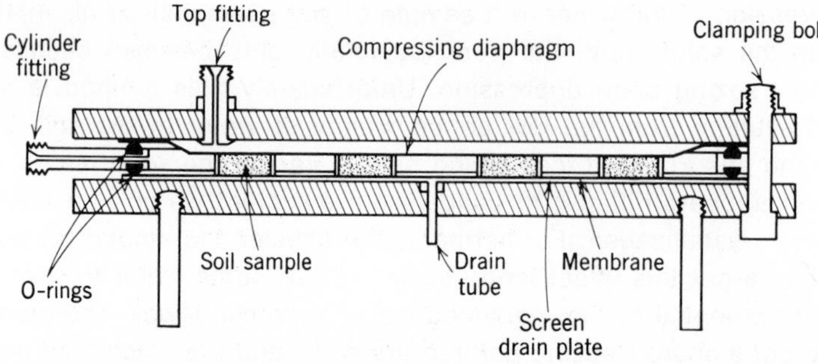

FIG. 3.9 **A pressure membrane apparatus for measuring water content of soil at various matric potentials. The soil samples are contained in metal rings about 5 cm in diameter and 1.2 cm deep laid on the membrane. Compressed air is supplied through the cylinder fitting, and air at a slightly higher pressure is supplied through the top fitting, causing the compressing diaphragm to press the soil samples firmly against the membrane. In some apparatus a porous ceramic plate is used instead of a cellulose acetate membrane.**

Measurements are usually made on small soil samples contained in rings about 5 cm in diameter and 1.5 cm in height. Approximately 20 samples can be measured at one time. The samples may be naturally occurring soil or may have been pulverized and screened to remove rock fragments. Elrick and Tanner (1955) found that the sieving procedure can result in overestimating water retention up to 30 percent, below 1 bar of matric potential, and underestimating it up to 10 percent, above 1 bar. Young and Dixon (1966) also report overestimates of field capacity from sieved samples. The possible effect of sample pretreatments should always be kept in mind in work with soil samples. Difficulty in maintaining good contact between samples and membrane is sometimes experienced with soils which shrink while drying. Richards and Richards (1962) introduced a radial cell in which the sample is the shape of a doughnut around a cylindrical membrane. This arrangement results in maintenance of good contact as the sample dries.

Measurement of osmotic or solute potential

Measurement of the osmotic pressure, or the osmotic or solute potential, can be made only on soil solution after it has been removed from the soil. Extraction is often done by adding water to the soil to a specified supersaturated level and filtering off a "saturation extract" (Richards, 1954). The osmotic potential can then be determined cryoscopically by the same method used for plant sap or in the vapor pressure psychrometers described later for measurement of total water potential. The electrical conductivity of the soil solution is often measured because there usually is a close agreement between electrical conductivity and osmotic potential of the saturation extract (Campbell et al., 1949; Richards, 1954). This relationship is shown in Fig. 3.10.

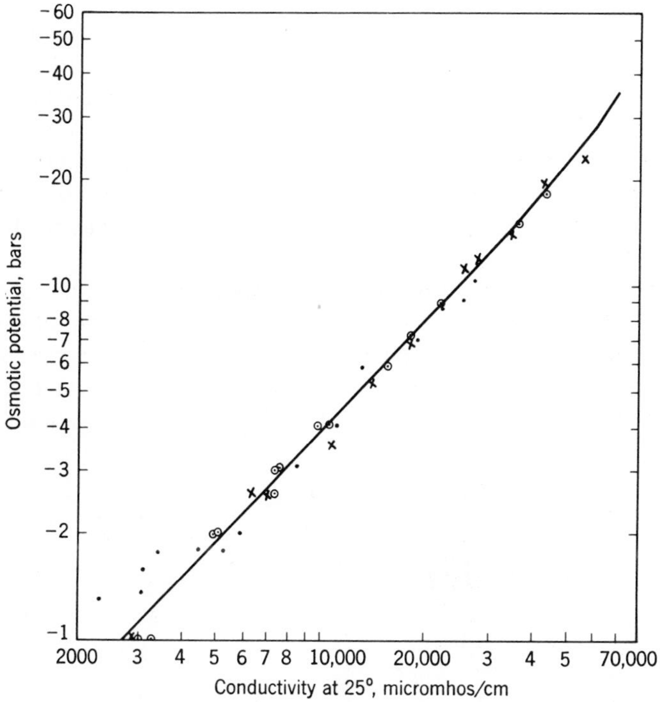

FIG. 3.10 **Relationship between osmotic potential of soil solution and its electrical conductivity expressed in micromhos per centimeter. The data include measurements made on soil extracts and nutrient solutions.** (*Adapted from Richards,* 1954, *table 6.*)

The osmotic pressure of the saturation extract is adjusted to that of the originally more concentrated field soil solution by a simple proportional correction. While this procedure is satisfactory for most purposes, it should be recognized that it gives only an approximation of the real values because the degree of dissociation and the osmotic coefficients of various salts vary with their concentration and the concentration of other salts in the solution, as well as with temperature.

Experimental control of soil moisture

There are two kinds of problems in connection with the control of water supply to plants. One involves maintenance of a uniform supply of water to plants being grown for experimental purposes. The other problem is that of subjecting plants to various known and controlled levels of soil water stress.

Maintenance of a uniform water supply

One of the most conspicuous deficiencies in control of plant environment is inability to maintain the water supply at a uniform level. In greenhouses, phytotrons, and plant growth chambers relatively large plants are often grown in small pots filled with vermiculite, gravel, or other media containing limited water reserves. As a result the water supply frequently fluctuates severely. Large fluctuations in soil moisture often occur in greenhouses, where irregular watering of benches results in marked variations in the supply of water to individual pots in an experiment, and the regular schedule of watering is seldom adjusted to compensate for differences in the rate of evapotranspiration.

The existence of these problems was recognized long ago, and Livingston (1908, 1918) attempted to remedy the situation by burying hollow porous porcelain cones, called autoirrigator cones, in the center of pots and connecting them to reservoirs which automatically supplied water as it was removed. An example is shown in Fig. 3.11. Unfortunately, the soil a few centi-

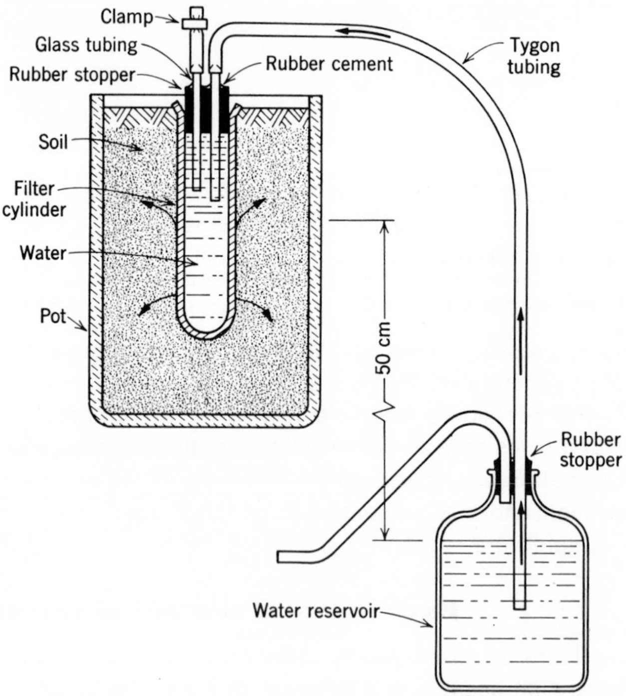

FIG. 3.11 An autoirrigation system for pots. By lowering the level of the water reservoir below the soil the matric potential can be lowered slightly. (*From Read, Fleck, and Pelton,* 1962.)

meters from the irrigators tends to dry out and roots become massed around the surfaces of the irrigators. An improvement in water distribution was provided by using double-walled pots with space for water between the glazed outer wall and the porous inner wall (Wilson, 1929; Richards and Blood, 1934). Not even this improvement can always supply enough water to maintain uniform soil moisture for large, rapidly transpiring plants (Richards and Loomis, 1942). However, Read et al. (1962) recently reported successful use of autoirrigators to maintain a constant matric potential of 50 cm of water.

Water control in greenhouses and nurseries

Greenhouse experts have given considerable attention to development of methods for watering potted plants which reduce labor and increase uniformity. Many of the earlier methods were described by Post and Seeley (1943). One simple method is to pull one end of a piece of glass rope through the hole in the bottom of the pot and spread it, thus making good contact with the soil. The other end dips in a reservoir of water (see Fig. 3.12). This

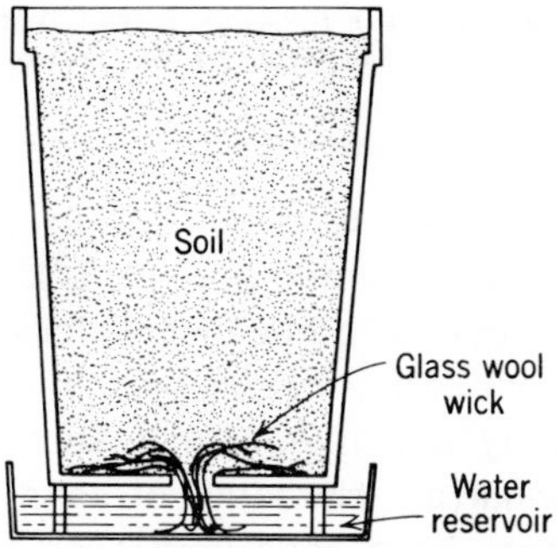

FIG. 3.12 **Section through an autoirrigated pot for small house plants. The pot and water reservoir are of plastic. A wick of glass wool conducts water from the reservoir into the soil.**

arrangement maintains the soil at approximately the field capacity and is particularly useful for house plants, which are notoriously either under- or overwatered.

Post and Seeley (1943) described several arrangements for subirrigation of greenhouse benches. The most common one uses waterproof benches with a V-shaped bottom about 2 in. deep with half tiles or pieces of eave trough inverted over the bottom to provide a channel for water movement. The V is filled with sand or fine gravel, and the pots standing on it are supplied with water by capillarity. More recently it has been found that ordinary flat-bottom benches can be lined with plastic and used just as satisfactorily as the special V-bottom benches. Water can be distributed the length of the bench through a piece of plastic tubing pierced with holes at 6-in. intervals. A

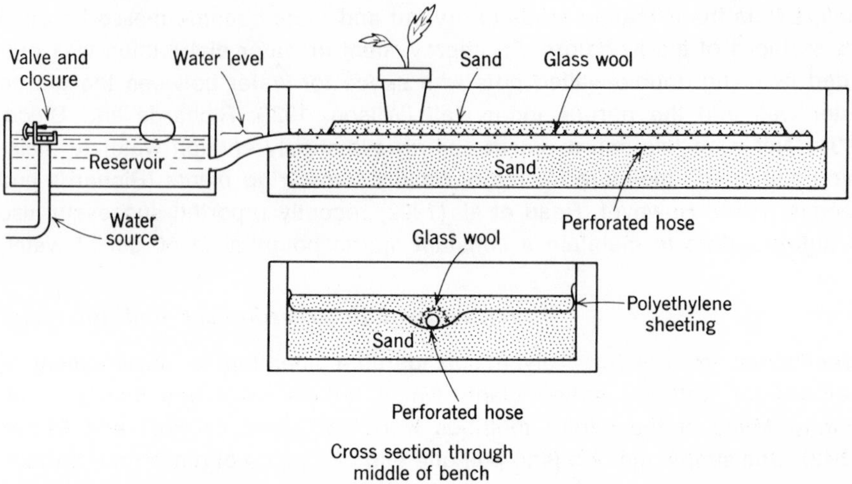

FIG. 3.13 **An autoirrigated greenhouse bench. Pots stand on sand kept wet by water from a constant-level water reservoir supplied through a piece of perforated hose extending the length of the bench. The water supply varies somewhat with the thickness and texture of the layer of sand on top of the plastic sheeting. Various forms of subirrigation were described by Post and Sealey (1943) and by the National Institute of Agricultural Engineering in England.**

diagram of such an arrangement is shown in Fig. 3.13. In some instances water is supplied manually or is turned on by a solenoid valve controlled by a tensiometer or a time clock. In other installations a constant water level is maintained in the bottom of the bench by the use of a float valve in a tank connected to the supply tube in the bench. The amount of water supplied can be controlled somewhat by the depth of sand above the water table. Modifications of this method have been developed by the National Institute of Agricultural Engineering in England, and it can be adapted for use in plant growth chambers.

Another method involves surface watering from supply tubes run lengthwise across the benches. In some cases small plastic tubes are led off to each pot so that it receives an amount of water controlled by the length of time water is supplied to the system (see Fig. 3.14). Stice and Booher (1965) described such a system, operated by a time clock, for watering container-grown nursery stock. There is an increasing trend toward using tensiometers or resistance blocks as sensing units to open solenoid valves and allow water to flow when the soil dries to a predetermined tension. Frequently a time clock is used to control the length of time that water flows and hence the amount supplied.

Aljibury et al. (1965) recently described a system used in California in which the sprinklers are activated by tensiometers placed in the containers in the zone of most active root growth. At least two tensiometers were

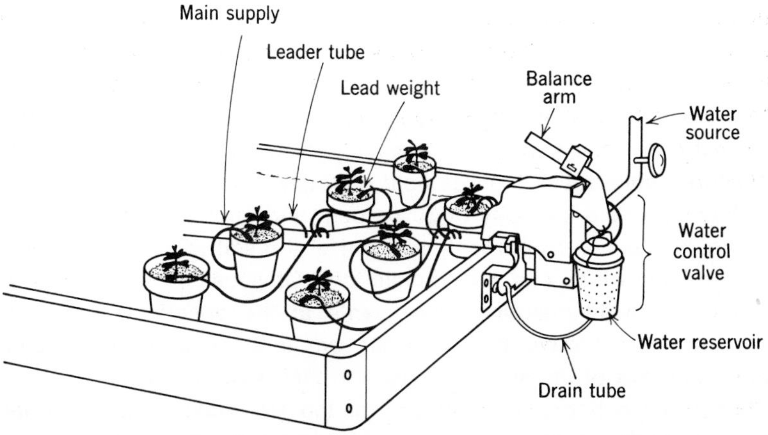

FIG. 3.14 **Apparatus for surface-watering pots on a greenhouse bench. A large plastic tube leads off from a water control valve which can be set to supply various amounts of water. The small plastic tubes leading from the central supply to individual pots have lead weights on their ends to maintain them in position. When the water reservoir fills to a weight equal to the setting on the balance arm, it trips a valve and shuts off the water supply to the pots.**

installed in each group of plants of the same variety and the same stage of development. A time clock was interposed to operate the sprinklers at 3 A.M. because the wind velocity was low at that time. This system is said to increase uniformity of water supply and consequently of growth. It also decreases labor costs and conserves water. Probably greenhouse and nursery stock watering will increasingly be controlled by soil water sensing units, which will materially increase uniformity of water supply and decrease water use compared with manual operations.

Maintenance of definite levels of soil water stress

One of the most troublesome problems in plant water relations research is that of maintaining plants growing in soil at uniform levels of water potential lower than field capacity. In the older literature, experiments were frequently described in which plants were said to have been grown in soil maintained at arbitrary water contents such as 10, 20, or 30 percent of the dry weight of the soil, or at some percentage of capillary or field capacity.

The impossibility of doing this should have been realized by all who have observed the distribution of moisture in the soil after a rain or who have considered the physical forces acting on soil moisture. Both Shantz (1925) and Veihmeyer (1927) called attention to the fact that if a small quantity of water is applied to a mass of dry soil, the upper layer is wetted to the field capacity and the remainder of the soil mass remains unwetted. Addition of more water results in wetting the soil to a greater depth, but there will always be a definite line of demarcation between the wetted and the unwetted soil. This situation has been observed by everyone who has dug in soil after a summer shower and observed the well-defined boundary between the wet soil and the dry soil beneath it. Since the field capacity is the amount of moisture held against gravity by a soil, it is obviously impossible to wet any soil mass to a moisture content less than its field capacity. If a container is filled with dry soil having a field capacity of 30 percent and enough water is added to wet the whole mass to 15 percent, one-half of the soil will be wetted to field capacity and the other half will remain dry. Obviously, the earlier investigators did not really maintain their plants at the specified soil moisture contents but merely gave them various amounts of water distributed in various proportions of the soil mass used in their experiments.

Attempts have been made to control the soil water tension or matric potential by placing autoirrigated pots at various heights above the water supply. Read et al. (1962) were able to maintain good control at a tension of 50 cm, but at 200 cm the variation in soil water tension was $\pm$ 50 percent. Livingston (1918) attempted to control the tension by inserting mercury columns at various heights between the irrigator cones and the pots, but this resulted in uneven wetting of soil in the pots. Moinat (1943) and others placed pots on top of sand columns of various heights standing in pans of water. The tension on the water in the pots increases with their height above the water table. However, it is difficult to produce tensions of more than 100 or 200 cm of water by this method.

Another method is to divide the soil mass into several layers separated by layers of paraffin, asphalt, or some other material impermeable to water but capable of being penetrated by roots (Emmert and Ball, 1933; Hunter and Kelley, 1946; Vaclavik, 1966). Water is then added to each layer separately through tubes extending above the surface. Difficulties are encountered with soil aeration, leakage of seals, and failure to obtain uniform root distribution in each layer of soil. A somewhat better method is to inject small amounts of water into various areas of the soil mass instead of applying it all to the surface. Vaclavik (1966) used a long needle attached to a large syringe to inject small amounts of water at various levels and at various points around the soil mass. By replacing the water lost each day in this manner he obtained relatively uniform distribution of water and of roots in the soil. However, this procedure does not bring the entire soil mass to a uniform water potential, but only brings masses near the point of injection approximately to field capacity.

Because of the difficulties inherent in other methods, most investigators limit the water supply to plants in containers by varying the intervals between irrigations. The plants are allowed to reduce the soil water content from field capacity to some predetermined water content or water potential, and enough water is then added to bring the entire soil mass back to field capacity. This method is used extensively. Wadleigh (1946) discussed methods of calculating what he termed the "integrated soil moisture stress" over cycles of drying and rewetting. This is essentially what occurs in nature when rain or irrigation periodically wets the soil to field capacity after it has been dried to various levels of water content by evapotranspiration. However, this method requires frequent weighing and daily replacement of water in those containers kept near field capacity, and there are wide variations in soil water potential between replacements of water if the intervals are longer than a day or two (see Fig. 3.15).

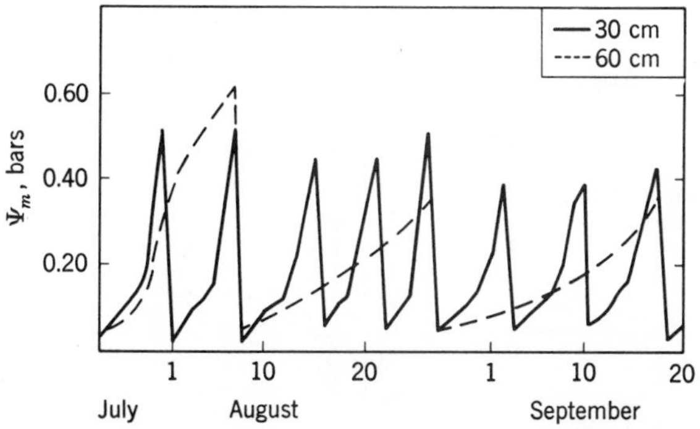

FIG. 3.15 **Variations in matric potential of an irrigated soil in an avocado orchard, as measured with tensiometers at 30 and 60 cm. The peaks indicate irrigations. Enough water was added to wet the soil to the depth of 60 cm at every third irrigation. (*Modified from Richards and Marsh,* 1961.)**

There is increasing interest in maintaining the soil at a constant water potential lower than field capacity. The most promising possibility is to grow plants in thin layers of soil in containers with side walls made of a differentially permeable membrane such as cellulose acetate. A diagram of such a device is shown in Fig. 3.16. This type of apparatus was described by Gardner (1964) and his coworkers, by Painter (1966), and by Zur (1967). The latter reported that he could maintain a relatively uniform water potential in a small chamber containing a sunflower seedling. The chief problems with this method are the limitation on size of the soil mass and rapid deterioration of the membrane. The soil layer must be relatively thin if water is to move to the center as rapidly as it is absorbed. More work is needed to establish the critical thickness, which certainly will depend on the water potential used, the capillary conductivity of the soil, and the rate of transpiration of the plants. Cellulose acetate membranes have been used in most experiments, but they are often destroyed by microorganisms in one to two weeks. The next step in improving this apparatus is to find a suitable differentially permeable membrane which resists attacks by microorganisms.

In some kinds of experiments it is possible to subject plants to water

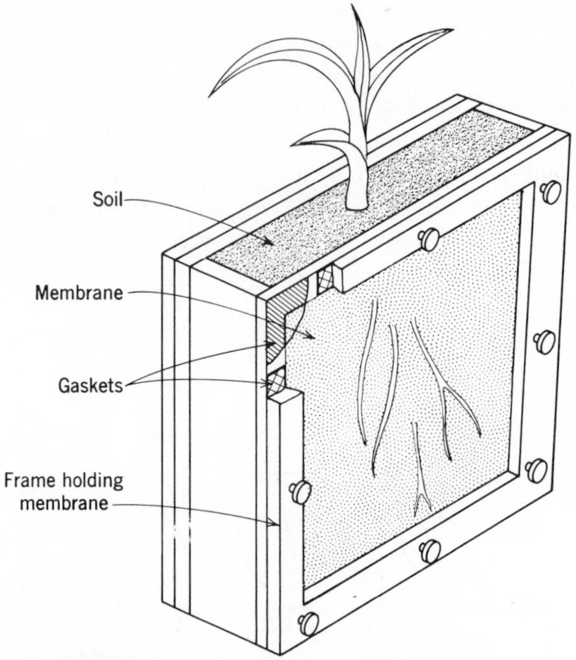

FIG. 3.16 A plastic chamber in which plants can be grown at various soil matric potentials. The flat sides are covered with cellulose membrane sometimes supported by plastic screen to provide mechanical support. The chamber is immersed in a polyethylene glycol solution to provide the desired decrease in potential. The soil mass can be only a few centimeters thick because of slow conduction of water at low potentials. Also, the membranes are often destroyed by microorganisms after a week or two. (*Designed by Dr. David Lawlor.*)

stress by growing them in nutrient solution plus additional solute to further lower the water potential. Solutes such as sodium chloride, potassium nitrate, and even sucrose are absorbed by the roots, producing a decrease in osmotic potential of the plant somewhat in proportion to the decrease in osmotic potential of the root medium (Eaton, 1942; Slatyer, 1961). For example, Boyer (1965) reported that the osmotic potential of cotton leaves decreased 1.2 to 1.5 bars for each bar of reduction in the osmotic potential of the root medium brought about by addition of sodium chloride. Mannitol is not entirely satisfactory because it is absorbed to a limited extent (Groenewegen and Mills, 1960), and it also is attacked by microorganisms. Investigators are now turning to polyethylene glycol because it can be obtained in a range of molecular weights, and it is less subject to attack by microorganisms. There have been some complaints of toxic effects (Jackson, 1962), perhaps partly because of the impurities found in some lots of polyethylene glycol. These can be removed by dialysis (Lagerwerff and Eagle, 1961). Other investigators have observed no visible toxic effects after long periods of exposure. There are reports of absorption of polyethylene glycol and its appearance on leaves (Lagerwerff and Eagle, 1961), but it is difficult to see how a substance with a molecular weight of 20,000 (that used in these experiments) can enter unbroken roots.

There should be more research on the control of water potential in the root medium and the effects of various methods on plant growth. For example, it appears easier to use solutions of various osmotic potentials than to control the water potential of the soil. However, there is some evidence that roots of plants grown in aerated culture solutions differ structurally from those grown in soil, and we need to learn whether plants grown in water culture are exactly comparable to those subjected to the same degree of stress in soil.

Irrigation problems

Irrigation of crops is an example of soil water control on the broadest possible scale. Successful and efficient irrigation involves numerous problems requir-

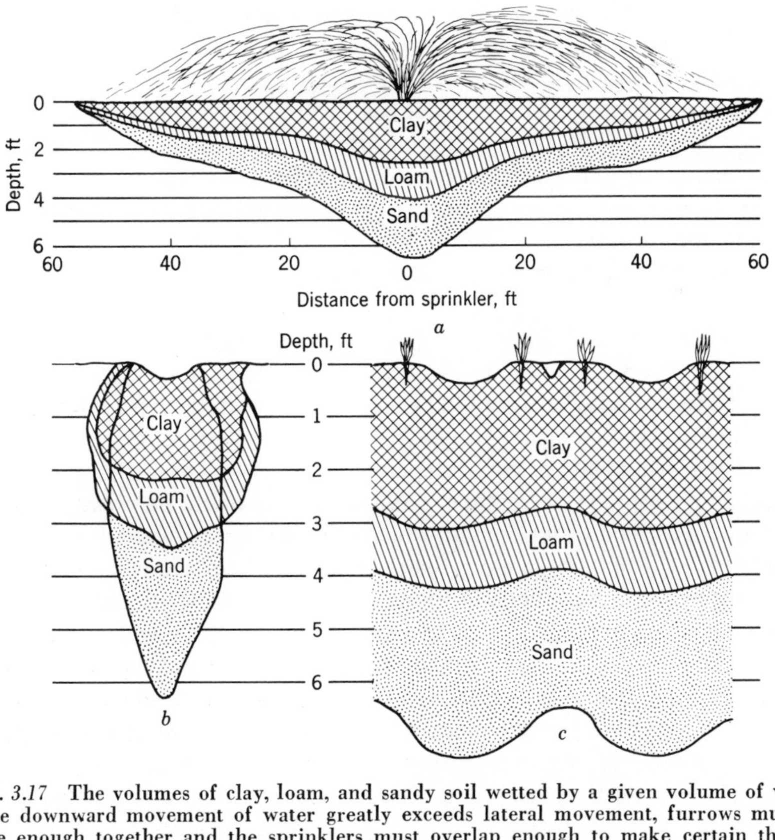

FIG. 3.17 The volumes of clay, loam, and sandy soil wetted by a given volume of water. Since downward movement of water greatly exceeds lateral movement, furrows must be close enough together and the sprinklers must overlap enough to make certain that all of the soil between them is wetted. (*From Doneen and MacGillivray*, 1946.)

ing some knowledge of engineering, soil water relations, plant physiology, and the characteristics of the particular crop being irrigated. There is a comprehensive review of these problems in the volume "Irrigation of Agricultural Lands," edited by Hagan et al. (1967). Among the important problems related to the content of this book are the timing of irrigation, aeration and drainage, and salt accumulation.

Timing of irrigation

Because of water shortages and competing demands for the limited supply of water available, there is increasing need to irrigate as efficiently as possible. This situation emphasizes the need for using scientifically sound methods for deciding when and how much to irrigate crops. Unfortunately, much irrigation still is based on rule of thumb methods instead of on our knowledge of soil-plant-atmosphere water relations. According to Taylor (1965), "Irrigation should take place while the soil water potential is still high enough that the soil can and does supply water fast enough to meet the local atmospheric demands without placing the plants under a stress that would reduce yield or quality of the harvested crop."

There are three general approaches to the determination of the proper time to irrigate. One method is to measure the soil water content, a second is to measure the plant water content or plant water stress, and the third is to measure evaporation and estimate the water needed by a crop. Each of these methods will be discussed briefly, but the reader is referred to Haise and Hagan (1967) for a more complete discussion.

SOIL WATER MEASUREMENTS. The traditional method of determining the need for irrigation is to dig into the soil and decide from its appearance and "feel" whether or not water should be applied. Determination of water content by gravimetric sampling or by use of neutron meters is not much better because the availability of water depends on its potential rather than on the content as percentage of weight or volume. A sand might be at field capacity ($\Psi_m = -0.3$ bar) at a water content which is below the permanent wilting percentage ($\Psi_m = -15$ bars) in a clay (see Table 3.1). Thus, the only reliable indication of the soil water status in terms of plant growth is its water potential. The matric potential of the soil can be measured directly with tensiometers down to about -0.8 bar, which covers the range for best growth of most vegetable crops. Use of the calibrated electrical resistance blocks mentioned earlier in this chapter permits measurements to much lower soil water potentials. Resistance blocks have been used successfully on sugarcane in Hawaii (Ewart, 1951), pastures in Oregon, sorghum and wheat in Texas, orchards in California, and in a variety of other special situations. Tensiometers also are used in agriculture and horticulture in the southwestern

United States and elsewhere, and for special purposes such as watering golf greens. The successful use of tensiometers and resistance blocks requires that they be installed in the zone of maximum root concentration, where water is removed most rapidly.

PLANT WATER STRESS. Plants themselves are the best indicators of the need for irrigation. As pointed out in Chap. 10, the growth of plants is controlled by their internal water stress, and the best indicator of the need for irrigation is the decrease in water potential of plant tissue. Since measurements of water potential and water deficit are usually impractical at the farm or orchard, other means of estimating the water status of plants must be used. Irrigation of sugarcane has been controlled successfully by measuring the water content of selected leaf sheaths (Clements and Kubota, 1942; Robinson et al., 1963). The color of beans and some other crops changes to a darker blue green at the beginning of water stress, and this change in color has been used successfully as a guide to irrigation (Robins and Domingo, 1956). Changes in leaf angle of sorghum and rolling or other movement of leaves of other species sometimes indicate the development of water stress prior to wilting (Haise and Hagan, 1967). Occurrence of definite wilting usually indicates that soil water deficiency has already developed and the plants should have been irrigated previously. The guard cells of plants are very sensitive to water stress, and premature closure of stomates has been used as an indicator of need for irrigation on a variety of plants (Alvim, 1959; Oppenheimer and Elze, 1941; Shmueli, 1953). It is well known that decrease in available soil water is accompanied by decrease in root pressure exudation (see Chap. 5), and according to Haise and Hagan (1967) this fact has been used in Russia as a guide to irrigation. However, it requires destruction of the indicator plants. Reduction in growth rate of fruits, leaves, tree trunks, and other plant parts has also been used as an indicator of the need for irrigation. Nurserymen sometimes use a very sensitive plant such as hydrangea for an indicator of the need to irrigate. The use of plant indicators has been reviewed by Hagan and Laborde (1966). Although it is impractical to measure plant water potential on the farm, its direct measurement in research on plant water relations seems essential. As pointed out by Kramer (1963*a*), most of the conflicting reports concerning the effects of soil water supply and irrigation on plant growth resulted from failure to measure the plant water potential.

USE OF EVAPORATION DATA. There is increasing interest in the timing of irrigation from meteorological data. This requires data on the available water storage capacity in the root zone of the crop under consideration and the rates of evapotranspiration and rainfall during the growing season. An example of the application of this method to tobacco was described by van Bavel (1953). In one instance it was assumed that the average rooting depth was 20 cm, the limiting soil water potential was −0.8 bar, and the available water content of the root zone between field capacity and −0.8 bar was 4.25 cm. With a calculated daily loss by evapotranspiration of about 0.5 cm, all the readily available water would be exhausted after 8 days without rain. By use of the proper values for available soil water, depth of rooting, critical soil water potential, and evapotranspiration, this procedure can be applied to other crops in other areas. Opinion varies concerning the best method of establishing the rate of evapotranspiration, but it can probably be estimated fairly reliably from the evaporation rate of properly located pans. This method is now used extensively in Hawaii, where it appears to be supplanting measurement of soil water content because it is a more convenient procedure for the growers (Chang, 1961). Haise and Hagan (1967) give other instances of its use.

Probably the weakest point in scientific control of irrigation timing is our lack of knowledge of the permissible limits of plant water stress for various crops. We need to know how much soil and plant water stress can be allowed to develop in a given crop growing on a certain type of soil before significant reduction in yield will result. In some instances it may be more economical to irrigate less frequently and obtain less than the maximum yield than to irrigate often enough to produce the maximum yield. The physiological aspects of water stress are discussed in Chap. 10 of this book, and both basic and applied aspects are discussed in the monograph on irrigation edited by Hagan et al. (1967).

Soil aeration and drainage

The importance of adequate soil aeration for good growth and functioning of roots is mentioned repeatedly in this book. Irrigation by flooding saturates the surface soil and produces at least temporary decrease in oxygen supply to roots. Basin irrigation, often used on trees, presumably is worst in this respect, but furrow irrigation also reduces the oxygen supply, as indicated in Fig. 3.18. Sprinkling at moderate rates ought to cause the least interference with gas exchange, unless continued long enough to saturate the soil. Apparently very few data on the effects of irrigation on aeration are available. Perhaps worse than the temporary saturation of the surface layer is the prolonged saturation of deeper soil horizons which sometimes occurs as a result of too frequent or too heavy irrigation combined with poor internal drainage. Letey et al. (1967) reported a case of chlorosis in a citrus orchard where sufficient water was applied to wet the entire root zone whenever tensiometer readings in the surface soil indicated need for water. As a result, the subsoil was kept too wet for adequate aeration of the roots.

Waterlogging of soil in low areas and where impermeable layers inter-

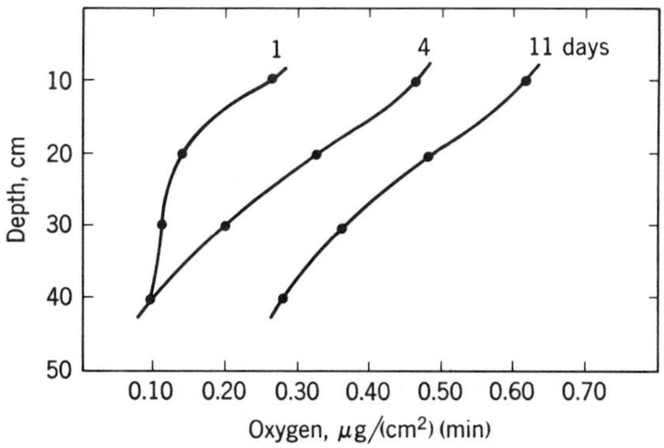

FIG. 3.18 **Oxygen diffusion rate in micrograms of oxygen per square centimeter per minute at various depths, 1, 4, and 11 days after irrigation. Oxygen diffusion rates less than 0.20 $\mu g/cm^2/min$ are regarded as inadequate for root growth; hence aeration in the root zone is probably limited for several days after irrigation. (*Redrawn from Hagan et al.*, 1967, p. 946.)**

fere with drainage often causes extensive damage to crops. In addition it increases the accumulation of salt at the soil surface. Proper drainage of irrigated lands is as important as an adequate supply of good-quality water.

Salt accumulation

One of the most serious problems in arid regions is the accumulation of salt in the soil. Large areas of land all over the world have been rendered unproductive by the accumulation of salt, either deposited from irrigation water or from natural sources. In humid areas where rainfall exceeds evaporation salt is leached out, but in arid regions there is insufficient rainfall to remove it. Irrigation water usually contains appreciable amounts of salt. For example, water used in the southwestern United States is said to contain from 0.1 to 5.0 tons of salt per acre-foot. Thus, depending on the quantity and quality of water used, 1 to 10 tons of salt might be added to the soil in a year; most of it remains when the water is removed by evapotranspiration. The only way to prevent salt accumulation is to supply enough water to leach out the salt, but this is often impossible because of either lack of water or lack of adequate drainage to remove the water. Salt accumulation is therefore a major problem in almost every irrigated area in the world (Reeve and Fireman, 1967).

There are two principal effects of salt accumulation, namely, salinity effects and the alkali effects. Mere accumulation of salt reduces the soil water potential and decreases the availability of water. Osmotic potentials equivalent to −4 bars at the permanent wilting percentage are said to reduce the growth of even salt-tolerant crops such as alfalfa, cotton, and sugar beets in soil kept near field capacity (Magistad and Reitemeier, 1943). The complex problem of how high salt concentration reduces plant growth is discussed in

Chap. 6 and by Slatyer (1967). In some areas there is an accumulation of excessive amounts of exchangeable sodium which replaces calcium and magnesium. This leads to decreased permeability, poor drainage and the development of a black surface film which causes such land to be called "black alkali." These areas support little or no plant growth. The problem of saline and alkaline soils in relation to irrigation was reviewed by Reeve and Fireman (1967); methods of diagnosing and improving such soils were discussed by Richards (1954).

Boyko (1965) discussed salinity problems in considerable detail. He reported successful growth of a wide range of plants supplied with water containing high concentrations of salt, but attempts to repeat his experiments have been unsuccessful. Perhaps Boyko was successful because his plants were grown in deep sand where little salt accumulation occurred.

Summary

The ability to measure and control the soil water supply to plants is basic to effective research on problems of soil-plant-water relations. It is also necessary for efficient irrigation of crops. The best measure of the availability of water to plants is the water potential. The principal components of the soil water potential are the matric and solute potentials. In humid regions the matric potential is the dominant component, but in soils of arid regions the solute potential or osmotic pressure of the soil solution is often an important component of the total soil water potential. The matric potential in the field can be measured directly down to about −0.8 bar by tensiometers, and lower values can be estimated from readings made on calibrated electrical resistance units. In the laboratory it can be measured to −25 bars or lower by the use of pressure plates and pressure membrane apparatus. The solute potential can be measured cryoscopically on soil extracts. Measurement of the combined matric and osmotic potentials can be made directly by use of the thermocouple psychrometer or indirectly by adding the solute and matric potentials. The most useful method of describing the water supplying characteristics of a soil is by means of a curve relating soil water potential and water content. Two useful soil water values are the field capacity and the permanent wilting percentage, because they represent the upper and lower limits of water readily available for plants. The matric water potential at field capacity is about −0.3 bar and at permanent wilting percentage about −15 bars. However, neither value is sharply defined but merely represents an approximate value on the water potential/water content curve.

Experimental maintenance of uniform soil water potential at any value lower than field capacity is impossible in large-scale experiments. In such experiments water stress is ordinarily produced by withholding water until some predetermined level of water stress is reached, then rewatering to field capacity. This produces a varying degree of soil water stress over each cycle. Small plants can be grown in thin layers of soil enclosed in differentially permeable membranes and immersed in polyethylene glycol solutions of various water potentials. In some instances a uniform level of water stress can be applied by growing plants in nutrient solution plus added solute to produce a desired level of water potential. Polyethylene glycol of high molecular weight appears to be useful for this purpose because little or none is absorbed, it is relatively harmless to plants and is minimally attacked by organisms.

Irrigation involves control of water supply on a large scale. The timing of irrigation can be improved by basing it on the available water content or water potential of the soil, the water potential of plant tissue, or estimates of available soil water obtained from measurements of evaporation or estimates of evapotranspiration. Although measurements of plant water stress are essential in some types of research, they are impractical in agriculture, and the current trend seems to be toward irrigation timing based on evaporation data. Irrigation also produces serious soil drainage and aeration problems and often results in accumulation of objectionable amounts of salt.

four

Roots & Root Growth

This chapter deals chiefly with the growth of roots and the factors which affect root growth. First, however, we shall briefly discuss the functions of roots, which are just as important as the functions of shoots.

Functions of roots

Root systems have four important functions: absorption, anchorage, storage, and the synthesis of various organic compounds. Practically all water and minerals absorbed by terrestrial plants enter through their roots. It is true that water and solutes can be absorbed through leaves and that absorption of mist and dew can be of some significance in survival (Duvdevani, 1964; Stone, 1957; Vaadia and Waisel, 1963), but foliar absorption of water is negligible compared with absorption through roots.

The role of roots in anchorage is generally taken for granted, but the success of most kinds of plants depends on their ability to remain upright. There are wide differences among plants in resistance to overthrow by wind which are related to differences in extent, depth, and mechanical strength of roots. Mechanical strength is also a factor in winter hardiness of wheat, because tough roots enable plants to withstand frost heaving with a minimum of injury. Considerable quantities of food are stored in roots, especially those of biennial and perennial species. This stored food is not only used during resumption of growth in the spring, but also is often economically important in crops such as beets, carrots, and sweet potatoes.

The role of roots as synthetic organs is usually neglected, but it is actually very important. In many species inorganic nitrogen is converted to organic nitrogen compounds in the roots before being translocated to the shoots, and organic sulfur and phosphorus are reported to occur in the xylem sap (see Chap. 7). Nicotine is synthesized in the roots and translocated to the shoots of tobacco; other alkaloids are produced in the roots of other species. It has also been claimed that roots produce hormones, sometimes called caulocalines, which are essential for shoot growth (Went, 1943). Chibnall (1939) believed that they produce a substance essential for normal nitrogen metabolism in leaves; Kende (1965), Skene and Kerridge (1967), and others reported that roots synthesize cytokinins. Phillips (1964) also suggested that roots supply some kind of growth-regulating substances to the shoots, and there are reports that root tips synthesize gibberellic acid (Skene, 1967).

It seems possible that the reduced growth of shoots which occurs when roots are subjected to unfavorable environmental conditions may result from interference with their synthetic functions as well as from interference with absorption. For example, Skene and Kerridge (1967) reported that the exudate from root systems of grapes grown at 30°C contained more of certain cytokinins than exudate from those grown at 20°C. Both roots and shoots elongated more at the higher temperature; in other experiments more fruit was set by plants with roots at 30°C. Although it was not proved that the differences in growth and yield are caused by differences in production of growth-regulating substances in the roots, the possibility seems to justify further investigation.

Although root cells possess most of the synthetic capacity of shoots, including the ability to produce functional chloroplasts, they are dependent on shoots for specific substances such as thiamin, niacin, pyridoxin, and possibly others (Street, 1966), which sometimes are spoken of collectively as rhizocalines. Roots also receive auxin from the shoots. The physiological activities of roots were reviewed recently by Burström (1965) and by Street (1966).

In addition to the synthetic activities of roots themselves, the synthetic activities of bacteria and other organisms associated with roots must be taken into account. The nitrogen-fixing activities of certain bacteria (*Rhizo-*

bium) which grow in nodules of both legumes and nonleguminous plants are well known, and the nitrogen fixed by them stimulates the growth of both host and neighboring plants. The kinds of activities of microorganisms in the rhizosphere are affected by substances escaping from roots, and presumably roots are affected by substances produced by the rhizosphere organisms (Burström, 1965; Rovira, 1965; Zak, 1964).

Primary and secondary growth

The effectiveness of roots in absorption depends on the extent of the root systems and on the efficiency of individual roots. These characteristics depend both on their hereditary potentialities and on the environment in which they develop.

Primary growth of roots

During growth and maturation roots undergo extensive anatomical changes that greatly affect their permeability to water and solutes. A detailed account of root development can be found in Esau (1965). Elongating roots are usually regarded as possessing four regions: the root cap, the meristematic region, the region of cell elongation, and the region of differentiation and maturation (see Fig. 4.1); but these regions are not always clearly delimited. Although the root cap is composed of loosely arranged cells, it is usually well defined. However, it is absent from certain roots such as the short roots of pine. Since it has no direct connection with the vascular system, it probably has no role in absorption.

The meristematic region typically consists of numerous small, compactly arranged, thin-walled cells, almost completely filled with cytoplasm. Relatively little water or salt is absorbed through this region, largely because of the high resistance to movement through the cytoplasm and the lack of a conducting system. In barley roots grown in aerated solutions at about 23°C, this region is only about 0.2 mm in length (Kramer and Wiebe, 1952). In roots of *Phleum pratense* the meristematic region may include the apical 0.4 mm (Goodwin and Stepka, 1945). Growth in the apical portion of the meristematic region is probably limited by food supply because the phloem is not differentiated to the apex and food must move to it, probably by diffusion, through a thick layer of cells.

Near the apex elongation is by cell division, but usually there is a zone of rapid cell elongation and increase in size of vacuoles a few tenths of a millimeter behind the root apex. It is difficult to indicate a definite zone of differentiation because various types of cells and tissues are differentiated at different distances behind the root apex. Sieve tubes of *Phleum* mature within 230 μ of the tip, well within the zone of cell division; but xylem elements first become differentiated about 1,000 μ behind the apex (Goodwin and

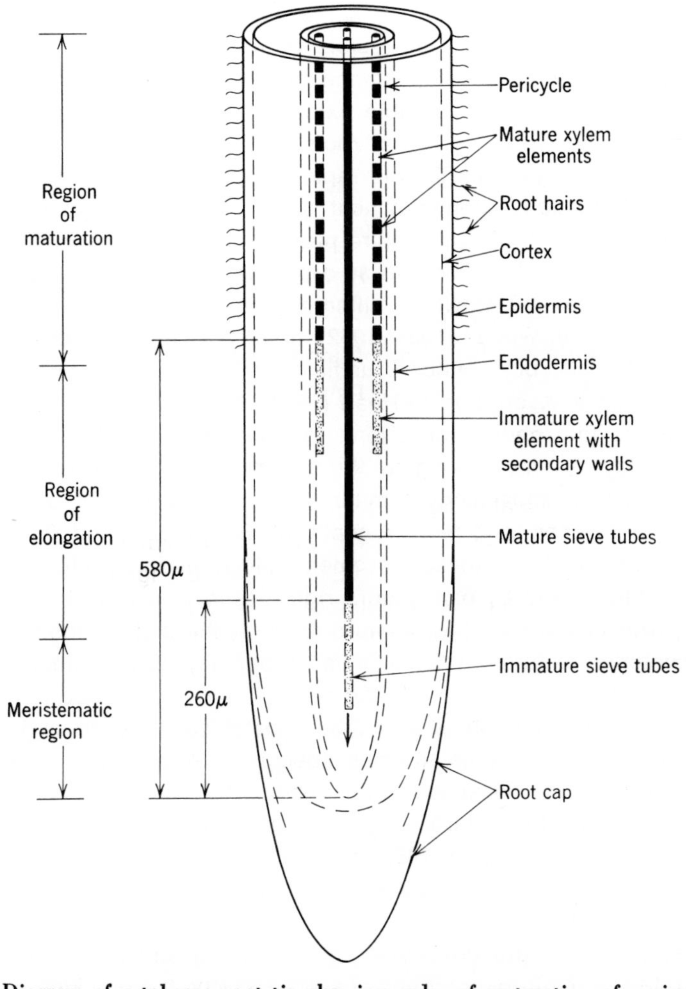

FIG. 4.1 **Diagram of a tobacco root tip showing order of maturation of various tissues. The distance from the tip at which various tissues differentiate and mature depends on the kind of root and the rate of growth. (*Modified from Esau,* 1941.)**

Stepka, 1945). Phloem also becomes differentiated closer to the tip than xylem in roots of Valencia orange (Hayward and Long, 1942) and pear (Esau, 1943); this probably is the typical situation (also see Esau, 1965, p. 503).

The control of growth in roots is obviously very complex (Burström, 1947, 1965; Esau, 1965; Street, 1966; Torrey, 1965; Wilcox, 1962), since cell division, enlargement, and differentiation occur somewhat independently. For example, tissues and cells continue to differentiate after root elongation ceases, especially when cessation occurs because of unfavorable conditions. Branch root formation also often continues down to the apex, indicating that requirements for branch root growth are somewhat different from those for the central axis. As a result it is impossible to identify the zones of elongation and differentiation in terms of distance behind the root tips because their location depends to a considerable extent on the species and the rate of root elongation. In slowly growing roots differentiation of tissues extends much closer to the root tips than in rapidly growing roots; and if elongation ceases, differentiation extends almost to the apex, leaving a very short meristematic region. In *Abies,* mature protoxylem occurs only 50 μ from the tips in dormant roots, but as much as 5 mm behind the tips in rapidly elongating roots (Wilcox, 1954), where differentiation lags behind elongation.

Friesner (1920), Wilcox (1954, 1962), and others have reported rhythmic elongation of roots, but neither Brumfield (1942) nor Goodwin and Stepka (1945) observed it. According to Head (1965), studies by time-lapse photography indicate that cherry roots tend to grow more rapidly at night than during the day. As they elongate through the soil, apple root tips show rhythmic nutational movements similar to those reported for other species by Darwin and Sachs, and more recently by Bolz (1927) and Fisher (1964). The latter suggested that these movements aid roots in penetrating the soil.

Supported behind by older, more rigid tissue, and on the sides by soil particles, root tips are pushed forward through the soil by their elongating cells, sometimes at rates of 5 cm or more per day. Usually their course is somewhat tortuous as they follow the path of least resistance between soil particles and around pebbles and other obstacles. In spite of these temporary deflections, the roots of some species of plants grow outward in a generally straight line because the tips tend to return to their original direction of growth after being turned aside by obstacles. This tendency, termed exotropy by Noll (see Wilson, 1967), was found to be very strong in roots of red maple and some other tree species which sometimes grow straight out for distances up to 25 m.

Great pressures are developed by elongating and enlarging roots, as demonstrated by the frequent lifting of sidewalks and the cracking of masonry walls. Clark (1875) described several examples of the pressures developed by growing roots and reported an experiment in which a growing squash plant supported a weight of 5,000 lb, a vivid demonstration of the power of enlarging cells. Gill and Bolt (1955) present some data on these pressures.

As the newly enlarged thin-walled cells at the base of the zone of enlargement lose their ability to elongate, they become differentiated into the epidermis, cortex, and stele, which constitute the primary structures of a root. The arrangement of the principal tissues is indicated in the diagram of a cross section through a root shown in Fig. 4.2. The conductive tissues of roots usually form a solid mass in the center, instead of being dispersed in bundles around the periphery of the pith, as in stems of most herbaceous dicotyledonous plants. The primary xylem usually consists of two to several

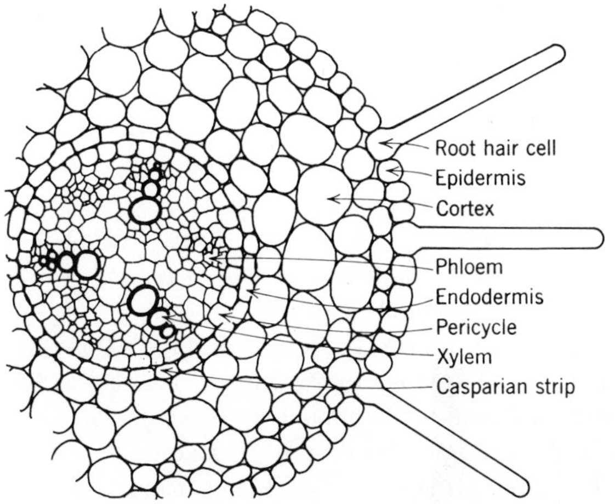

FIG. 4.2 **Diagram of transverse section through a squash root in the region where maximum water and salt absorption occur. (*After Crafts and Broyer,* 1938.)**

strands extending radially outward from the center with the primary phloem located between them.

The narrow layer of parenchyma cells separating the xylem from the phloem later becomes the cambium and produces secondary xylem and phloem. The outermost layer of the stele is the pericycle. Its cells retain their ability to divide, and they not only give rise to branch roots but also to the cork cambium found in many older roots. The endodermis usually consists of a single layer of cells. Because of its peculiar and conspicuous structure it has received much attention from anatomists and physiologists (Esau, 1965, pp. 489–493; Van Fleet, 1961). Early in development suberized thickenings, called Casparian strips, develop in the radial walls, which presumably renders them impermeable to water and solutes. As the endodermis matures, the inner tangential walls as well as the radial walls often become much thickened and lignified. The importance of these features in relation to water and salt absorption will be discussed later.

Epidermis and root hairs

Considerable attention has been given to the epidermis and root hairs because of their importance as absorbing surfaces. The epidermis is usually composed of relatively thin-walled, elongated cells which form a compact layer covering the exterior of young roots. Sometimes a second compact layer, the hypodermis, lies beneath the epidermis. The most distinctive feature of epidermal cells is the production of root hairs. These usually arise as protrusions from the external tangential walls.

In a few kinds of plants, including citrus and pine, root hairs occasionally arise from cortical cells one or two layers beneath the epidermis. In

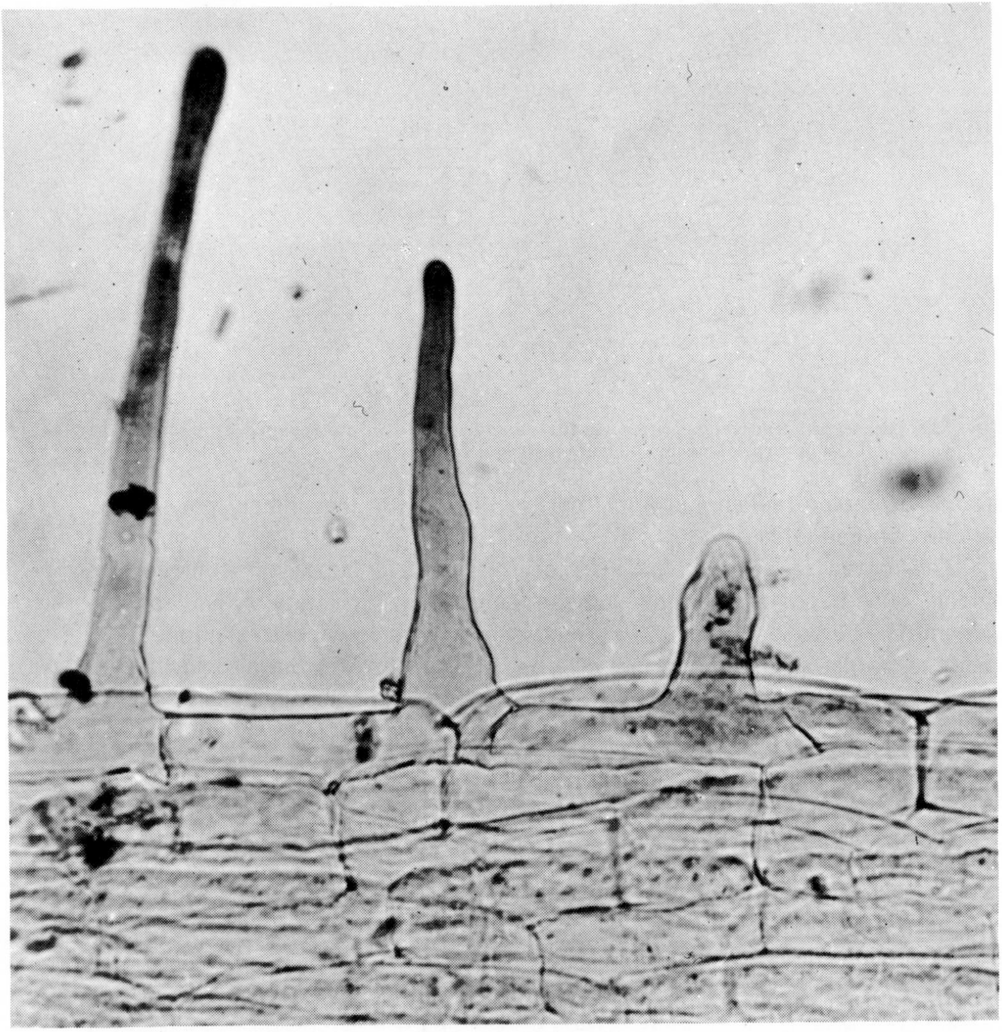

FIG. 4.3 **Epidermal cells and root hairs.** (*From Weisz and Fuller, "The Science of Botany," McGraw-Hill Book Company, Inc., New York,* 1962.)

some species, any or all of the epidermal cells can produce root hairs; but in others, the rudimentary epidermal cells divide and give rise to long cells and to short ones, known as trichoblasts, which usually produce root hairs (Sinnott, 1939; Sinnott and Bloch, 1939; Cormack, 1944, 1945). Root hairs are normally produced at the apical end of epidermal cells, probably because the walls mature more slowly and they remain extensible longer at that end.

The development of root hairs is affected by pH, the kinds and concentration of ions, temperature, aeration, and even light; but there are such wide differences among species that generalizations are difficult. Usually, however, most land plants produce the maximum number of root hairs in well-aerated soil with a water content near field capacity, and their development is inhibited in both wet and dry soil and by an excess of salt (Hayward and Blair, 1942). The literature on formation of root hairs was reviewed by Cormack (1962).

Wide variations occur in the numbers of root hairs on roots of various species. In general, they are more abundant on roots of herbaceous plants than on roots of woody plants. Dittmer (1937) reported 2,500 root hairs/cm^2 on the root surface of winter rye, but there were only 520 hairs/cm^2 on roots of black locust seedlings, and 217 on loblolly pine roots (Kozlowski and Scholtes, 1948). Root hairs are said to be rare on peanut (Reed, 1924) and absent on pecan (Woodroof and Woodroof, 1934) and certain conifers (Busgen and Münch, 1926). Nutman reported that root hairs increased the absorbing surface of *Coffea arabica* L. roots about 8.5 times, but Dittmer (1937) estimated that they increased the surface of winter rye roots only 1.6 times. However, root hairs greatly increase the volume of soil in close contact with roots because they occupy a cylinder surrounding the root which is approximately equal in radius to their length. For example, the roots of rye plants studied by Dittmer (1937) were 120 to 250 μ in diameter and bore root hairs 700 to 800 μ in length. The roots and root hairs therefore came in contact with the soil particles in a cylinder of soil about 2 mm in diameter. There is no clear evidence that the presence of root hairs results in increased absorption of water, but they may facilitate ion absorption, especially absorption of less mobile ions such as phosphorus.

Root hairs are usually assumed to be short-lived. They either die or are destroyed by changes associated with secondary growth. This is not always true, however, since even the oldest roots of the 4-month-old rye plants studied by Dittmer bore healthy root hairs. Weaver (1925) found that the root hairs on wheat roots did not begin to die until the roots were 7 weeks old, and some were still alive after 10 weeks. On some species root hairs become suberized or lignified and persist for months or years, but it is doubtful that they are of much importance in absorption. On other species the cortex collapses in a few weeks, and the root hairs are destroyed.

According to Scott (1963, 1965), the outer surfaces of the epidermal cell walls of young roots and the walls of root hairs are covered with a layer of cutin and a film of mucilaginous material. Furthermore, staining reactions indicate that the cell walls bordering intercellular spaces in roots also possess layers of cutin or some lipidlike material. Scott regards the formation of this internal lipid layer as a reaction to wounding when the intercellular spaces are formed by cell enlargement. In addition, pits and plasmodesmata occur in the cell walls of roots, including the outer walls of epidermal cells and root hairs (Scott, 1949, 1964; Scott et al., 1963). The significance of the plasmodesmata and cutin or lipid layers on water and salt absorption has not been evaluated. However, the layers are apparently so thin and contain so many pores that they probably do not constitute a serious barrier to the movement of water and solutes. Furthermore, the gelatinous surface film probably improves contact between roots and soil particles (Jenny and Grossenbacher, 1963).

Secondary growth

Secondary growth usually destroys the epidermis and root hairs. Sometimes hypodermal cells become suberized, and a cork cambium or phellogen develops in the outer part of the cortex. In roots of Valencia orange and some

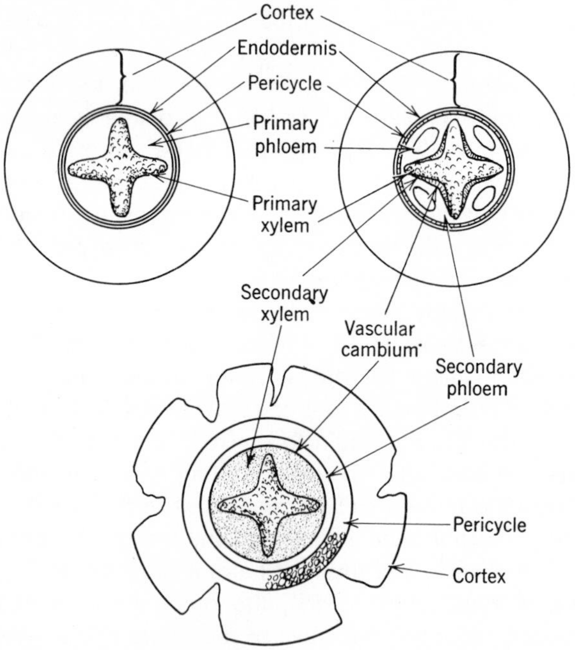

FIG. 4.4 **Secondary growth of a woody root, showing development of vascular cambium and production of secondary xylem and phloem. Enlargement by addition of secondary tissue crushes the primary phloem and endodermis and splits off the cortex. (*Adapted from Esau,* 1965.)**

other woody species the hypodermal cells produce secondary root hairs. Lenticels and areas of thin-walled, radially elongated cells also develop, and through these water is absorbed (Hayward and Long, 1942). Usually, as cambial activity increases the diameter of the stele, the cortex and epidermis collapse and disappear. Frequently a cork cambium arises from the pericycle, and the roots become covered with a layer of corky tissue containing lenticels. In woody roots successive cork cambia develop until the cortex has disappeared, and the arrangement of tissues in older woody roots is similar to that in woody stems. However, the bark of roots appears to be considerably more permeable to water than the bark of stems.

The absorbing zone

Water tends to enter roots most rapidly through those regions which offer the lowest resistance to its movement. The location of the region of most rapid absorption varies with the species, the age, and the rate of growth; also, with the magnitude of the tension developed in the water-conducting system.

Young roots

A number of studies of the absorbing zone have been made by attaching potometers at various distances behind the root tips. As mentioned earlier, little water is absorbed through the meristematic region because of the resistance offered by the dense protoplasm and lack of conducting elements to carry it away. Maximum absorption of water and salt seems to occur in the region where the xylem is well differentiated but suberization and lignification has not progressed far enough to seriously reduce permeability (Hayward and Spurr, 1943). This also appears to be the region of maximum absorption of minerals (Kramer and Wiebe, 1952; Canning and Kramer, 1958; also see Fig. 7.8).

Potometer studies indicate that water absorption through growing corn roots more than 10 cm long increases to a maximum about 10 cm behind the tip, then decreases toward the base (Hayward, Blair, and Skaling, 1942). In onion roots maximum absorption occurred 4 to 6 cm behind the root tips in roots over 7 cm long and decreased toward both tip and base. In roots less than 5 cm in length there was greater absorption toward the base than toward the tip (Rosene, 1937).

Sierp and Brewig (1935) found that maximum intake of water through roots of *Vicia faba* L. over 10 cm long occurred 1.5 to 8 cm behind the tip. When the rate of transpiration was increased, the absorbing zone was extended, and the region where most rapid absorption occurred shifted toward

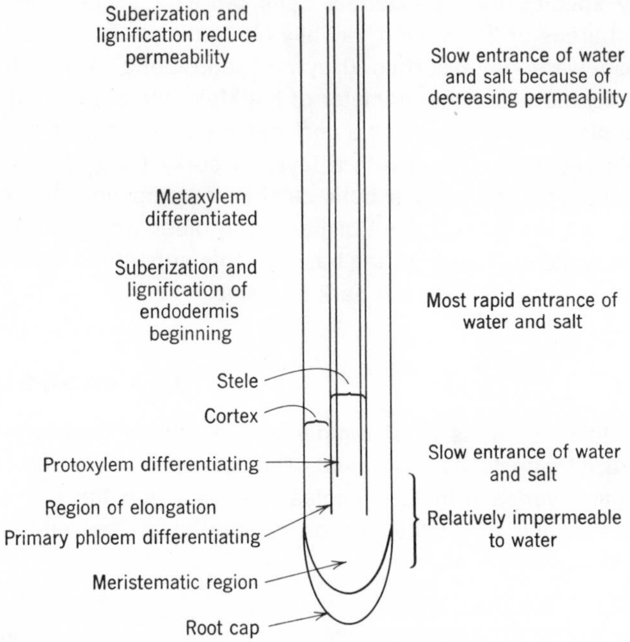

FIG. 4.5 **Diagram of a young root, showing relation between anatomy and absorbing regions for water and salt. The stages in differentiation and the location of the regions of maximum water absorption are based on a study of corn roots by Hayward and Spurr (1943). The region of maximum salt absorption is based on the work of Wiebe and Kramer (1954).**

the base of the root. Brouwer (1953) and Soran and Cosma (1962) also observed a shift in the region of most rapid absorption toward the base of the root as stress was increased. This situation is discussed in Chap. 6 in the section on permeability of roots (see Fig. 6.2*a*).

Role of older roots

Most discussions of absorption deal with young roots and leave the impression that the older suberized roots do not function as absorbing surfaces. However, it is probable that the larger part of the absorption of water and solutes by many perennial plants occurs through roots which have undergone secondary growth and are covered with layers of suberized tissue.

It has been noted by several observers that when the soil is cold or dry, or the soil solution becomes too concentrated, few or no growing unsuberized roots can be found. McQuilkin (1935) and Reed (1939) found few growing tips on various species of pine in dry or cold weather, and few or no white unsuberized root tips were found on citrus trees during the winter (Chapman and Parker, 1942; Reed and MacDougal, 1937). Pines and citrus lose large amounts of water by transpiration on sunny winter days and obviously over a

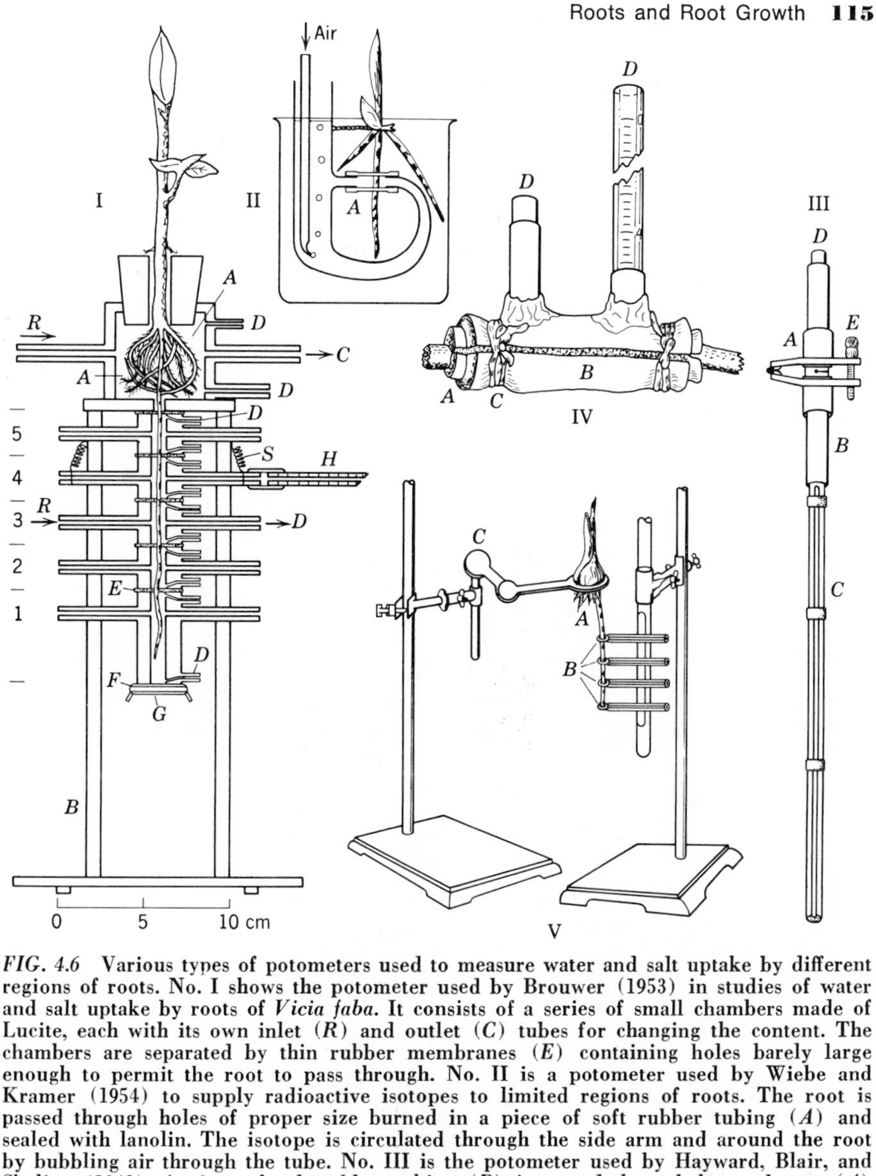

FIG. 4.6 Various types of potometers used to measure water and salt uptake by different regions of roots. No. I shows the potometer used by Brouwer (1953) in studies of water and salt uptake by roots of *Vicia faba*. It consists of a series of small chambers made of Lucite, each with its own inlet (*R*) and outlet (*C*) tubes for changing the content. The chambers are separated by thin rubber membranes (*E*) containing holes barely large enough to permit the root to pass through. No. II is a potometer used by Wiebe and Kramer (1954) to supply radioactive isotopes to limited regions of roots. The root is passed through holes of proper size burned in a piece of soft rubber tubing (*A*) and sealed with lanolin. The isotope is circulated through the side arm and around the root by bubbling air through the tube. No. III is the potometer used by Hayward, Blair, and Skaling (1942). A piece of soft rubber tubing (*B*) is passed through brass sleeves (*A*) attached to a hinge, and a hole is burned in it to accommodate the root, which is placed in the hole through a slit in the tubing. The screw (*E*) is used to compress the tubing around the root. No. IV was used by Kramer (1946) on large, attached, woody roots. Split one-hole rubber stoppers (*A*) are placed on the root and a piece of split rubber tubing is placed over the stoppers, cemented with rubber cement, and wrapped with rubber bands. An inlet tube (*D*) and graduated tube (*E*) for measuring water absorption are sealed into holes in the rubber tubing with grafting wax or a low-melting-point sealing compound. No. V shows the arrangement used by Rosene (1937, 1941) to measure uptake of water by various regions of a root. The root is passed through holes in the flattened ends of capillary tubes (*B*), and water absorption is measured by observing the meniscus in each tube. The entire apparatus is enclosed in a glass or plastic chamber to maintain high humidity around the root.

period of time must be absorbing equivalent quantities of water through their suberized roots. Crider (1933) and Nightingale (1935) also reported instances where woody plants possessing no actively growing roots were able to absorb both water and minerals through their suberized roots. Head (1967) found a reduction in production of new roots on apple and plum in midsummer and concluded that a considerable proportion of the water and salt must be absorbed through older roots.

Roberts (1948) reported that only a small percentage of the root surface under a pine stand was unsuberized, and Wilcox (1954) reported rapid shoot growth in *Abies* at times when no unsuberized roots were present. Kramer

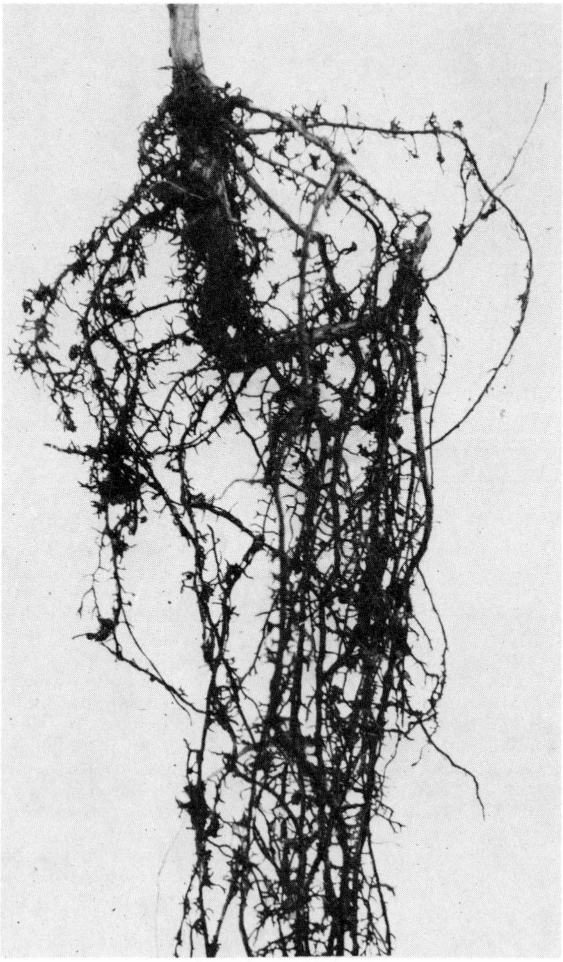

FIG. 4.7 **Photograph of a root system of a pine seedling showing the large proportion of root surface which is suberized.**

and Bullock (1966) studied the roots under a pine and a hardwood stand in North Carolina. They found that during the summer an average of less than 1 percent of the root surface under stands of *Pinus taeda* L. and *Liriodendron tulipifera* L. consisted of white, unsuberized growing roots. Even though it is admitted that the unsuberized roots are much more permeable than the suberized roots, it would be impossible to account for all the water and salt absorption through such a small part of the root system. The unsuberized root surface is too limited in extent and occupies too small a volume of soil to supply the water and salt required by a tree. It seems quite certain that absorption is not restricted to the unsuberized regions of roots, but that it occurs to some extent over the entire surface of root systems. In fact, the major part of water and salt absorption by perennial plants probably occurs through suberized roots.

Pathway of radial water movement

Having discussed the location of the absorbing zone, it seems appropriate to discuss the path followed by water in moving from the surface of the root into the xylem elements. The pathway in primary roots will be discussed first, then the pathway in roots which have undergone secondary growth.

A somewhat diagrammatic cross section of a young root of an herbaceous plant is shown in Fig. 4.2. The epidermis and the hypodermis, if present, are important because they are composed of very compact layers of cells containing no intercellular spaces. The cortical parenchyma usually consists of loosely arranged cells containing numerous intercellular spaces and even lacunae. Water movement in this region might occur through the protoplasts, and in the walls. The intercellular spaces are usually filled with gas, even in water culture, and infiltration by liquid results in abnormal growth (Burström, 1959, 1965).

The endodermis interposes a serious barrier to water and salt movement because the radial walls of the endodermal cells are rendered impermeable to water and solute movement by the Casparian strips. Thus, where the endodermis is intact, all water and solutes must pass through the protoplasts before they enter the stele; Arnold (1952) and others consider that the endodermis controls absorption. However, in many roots the endodermis is pierced by numerous root initials, which probably open up pathways for movement of materials.

In the past it has been assumed that water movement across the cortex occurs from vacuole to vacuole; the water passes through the outer wall, a layer of cytoplasm, the vacuole, another layer of cytoplasm, and the inner wall of each cell in the pathway. However, it should be emphasized that water will follow the path of least resistance and that there appears to be considerably less resistance to movement through the walls than through the cytoplasm. Thus considerable water movement might occur through walls, bypassing the vacuoles. This view is supported by Strugger's (1943, 1949) experiments with fluorescent dyes. Russell and Woolley (1961) calculated from data on the cross-sectional areas and relative permeabilities of cytoplasm and walls that the larger part of the water might move to the xylem along the walls. The situation is complicated somewhat by the fact that the protoplasts of the root cells are joined by strands of cytoplasm, the plasmodesmata, which extend through the walls and interconnect them into a common system, the symplast (Arisz, 1956; Münch, 1930; Scott, 1949). As a result, even passive movement of water through cell walls is affected by aeration, respiration inhibitors, and other factors which affect the condition of the protoplasm.

Another reason for believing that much of the water bypasses the vacuoles on its way from root surface to stele is that considerable time is required to bring about equilibration of the vacuolar water in roots with tritium supplied to the system in which the roots are immersed (Biddulph et al., 1961; Cline, 1953; Raney and Vaadia, 1965). Evidence indicating that

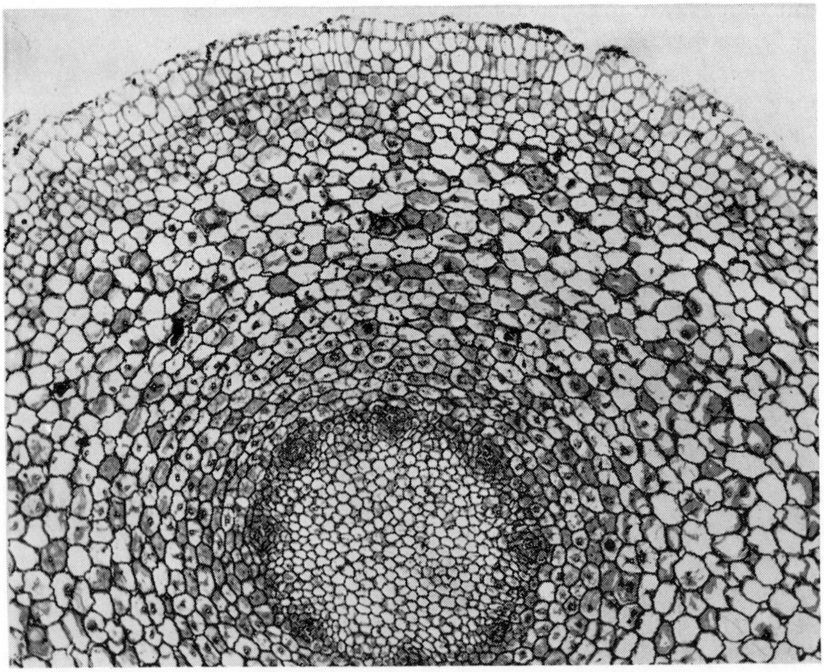

FIG. 4.8 Cross section of a young *Liriodendron* root (approx. ×80) about 0.6 mm behind the apex. Note the thick layer of tissue surrounding the stele. (*From R. A. Popham, "Developmental Plant Anatomy," 1952, Long's College Book Co., Columbus, Ohio. By permission of the author.*)

mineral elements can move across roots and enter the stele without passing through the vacuoles is also presented in Chap. 7 and possible pathways are shown in Fig. 7.9.

It has been shown in various ways that there is considerable resistance to movement of water through roots. For example, if the roots of a rapidly transpiring plant are cut off, there is an immediate increase in water absorption; and the lag of absorption behind transpiration which normally occurs almost disappears (Kramer, 1938). Most of this resistance disappears if the roots are killed (Renner, 1929; Kramer, 1932; Ordin and Kramer, 1956). On the other hand, respiration inhibitors, low oxygen, and high carbon dioxide all increase the resistance to water movement through the roots (Brouwer, 1954*a*; Kramer, 1949; Lopushinsky, 1964*a*; Mees and Weatherley, 1957).

Resistance to water movement through suberized roots must be located in the layer of suberized tissue around the outside of the root, in the cork and vascular cambia, and in the phloem (see Figs. 4.8 and 4.9). The relative importance of these tissues as sources of resistance might be established by removal of one layer at a time while water is moving through the roots under pressure, but this has never been done.

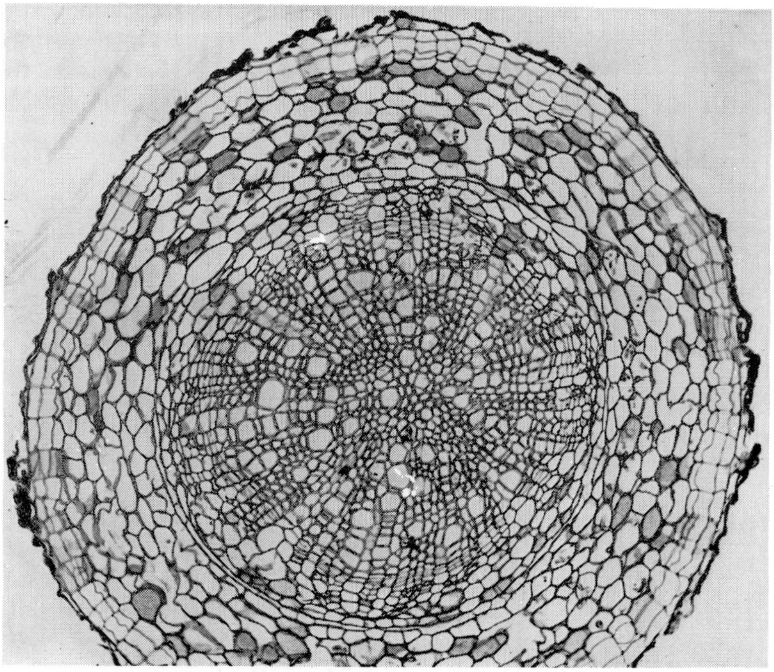

FIG. 4.9 Cross section of an older *Liriodendron* root (approx. ×60) after much of the outer parenchyma tissue has sloughed off and a layer of cork has begun to develop. (*Same source as Fig. 4.8.*)

Mycorrhizae

The root systems and absorbing surfaces of most trees and shrubs, as well as those of some herbaceous plants, are considerably modified by the presence of mycorrhizae. These structures are formed as a result of the invasion of young roots by the hyphae of various species of fungi commonly found in forest soils. Two types of mycorrhizal roots are common. In endotrophic mycorrhizae the mycelium develops within the root tissue but affects the external appearance very little. In ectotrophic mycorrhizae the mycelium penetrates the cortical tissue, causing marked hypertrophy and extensive branching. *Liriodendron tulipifera* L. roots are an example of the endotrophic type, and the roots of many conifers exemplify the ectotrophic type. Combinations of the two types also occur. A feltlike covering or sheath of mycelium is often produced over the root surface, and rhizomorphs and individual hyphae extend out into the soil from this sheath. Mycorrhizal roots are shown in Fig. 4.10.

According to Hatch and Doak (1933) and Hatch (1937), the type of root system as well as the fungus determines the type of mycorrhizae. Trees such as sweetgums, yellow-poplar, maple, and members of the Ericaceae which have roots that can be differentiated only as fine or coarse produce endotrophic mycorrhizae when invaded by fungi. Trees of the Abietineae, Salicaceae, Betulaceae, Fagaceae, and certain others produce roots of two types, long and short. Long roots grow relatively rapidly, produce root hairs which

FIG. 4.10 Photograph of mycorrhizae on *Pinus virginiana*. Note the mycelial strands and particles of earth adhering to the surfaces. (*Courtesy of Edward Hacskaylo, U.S. Department of Agriculture.*)

usually originate from a hypodermal layer, branch racemosely, are relatively long-lived, undergo secondary thickening, and seldom become mycorrhizal, possibly because they grow too rapidly to be affected by invading fungi. The short branch roots elongate slowly, bear few root hairs, do not have secondary growth, usually live one season or less, and are often invaded by fungi and become mycorrhizal. Mycorrhizal roots of this type are blunt and thickened. They often branch dichotomously and sometimes form coralloid clusters, as shown in Fig. 4.10. Slankis (1958) claims the branching is caused by auxin produced by the fungus, but the development of mycorrhizal roots

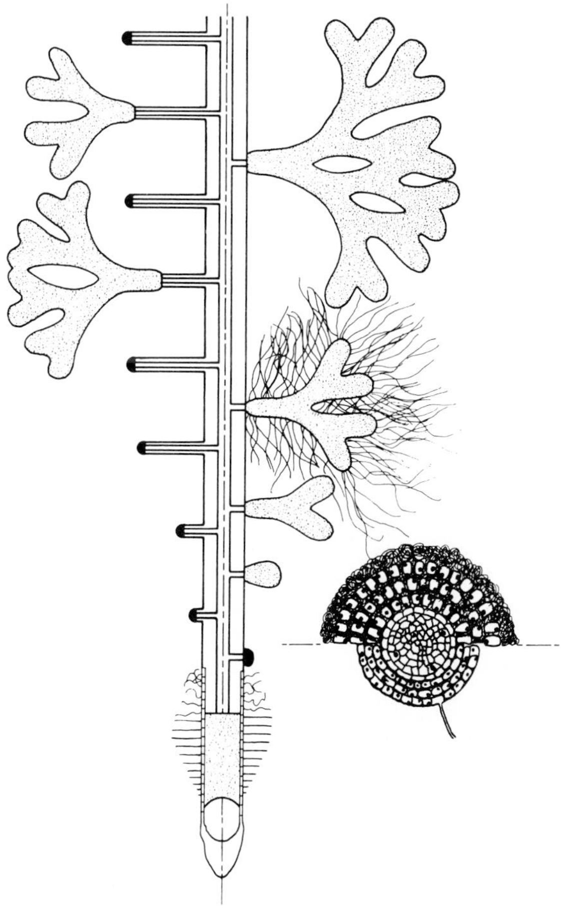

FIG. 4.11 **Diagram of development of mycorrhizae on a pine root. The main axis is a long root bearing a root cap and root hairs on the unsuberized surface behind the meristematic region. Further back, mycorrhizal branches, such as those shown in Fig. 4.10, are developing from short roots. The upper part of the cross section represents a mycorrhizal root; the lower part, an uninfected root. (*After Hatch,* 1937.)**

also seems to depend on an adequate supply of carbohydrates from the shoot.

The importance of mycorrhizal roots has been warmly debated but now seems well established. Attempts to grow conifers in grasslands, on infertile soils, and in regions where they had not grown previously have often failed, but in many instances treatments which resulted in the development of mycorrhizae, such as the addition of organic matter containing fungi to the soil, have resulted in successful growth. Some of the literature on this subject is cited by Bowen (1965), Kramer and Kozlowski (1960), and Zak (1964). A number of papers on mycorrhizae also appear in the Proceedings of the Fourteenth IUFRO Congress held in 1967.

It is generally assumed that the beneficial effects of mycorrhizal roots result from the increased absorbing surface provided by the hyphae and rhizomorphs extending into the soil from the roots (Hatch, 1937). It has been reported that mycorrhizal roots accumulate more minerals per unit of dry weight than nonmycorrhizal roots (Harley, 1956; Hodgson, 1954; Kramer and Wilbur, 1949). Bowen and Theodorou (1967) found that the rapidly elongating apical regions of pine roots accumulate phosphate as rapidly as mycorrhizal roots. However, the uninfected roots are functional for only a few days, while the mycorrhizal roots are functional for many weeks. Where extensive mycelial growth into the soil occurs, the absorbing surface is materially increased; but there is uncertainty concerning the amount of hyphal extension into the soil.

It has been demonstrated by Melin and his coworkers that various substances supplied to fungal hyphae attached to conifer roots were absorbed and translocated to the shoots (Melin and Nilsson, 1950, 1955; Melin et al., 1958). MacDougal and Dufrenoy (1944, 1946) and McComb and Griffith (1946) suggested that the synthetic activities of mycorrhizal fungi are important because they supply vitamins, growth substances, and organic nutrients to the host plants on which they grow. This seems unlikely in view of observations in Melin's laboratory which indicate that, in contrast to most of the other fungi found in forest litter, mycorrhizae-forming fungi require simple carbohydrates and accessory substances. Zak (1964) suggested that part of the benefit results from protection of young roots against parasitic organisms by antibiotic products of the mycorrhizal fungi.

Mycorrhizal roots do not always develop even when the proper fungi are present. Rayner and Neilson-Jones (1944) claimed that failure of conifers on certain English heath soils occurred because mycorrhizal development is inhibited by low oxygen and toxic substances produced by the heath vegetation. Björkman (1942) suggested that failure to develop mycorrhizal roots usually results from lack of sufficient carbohydrates in the roots. He found that girdling and shading seedlings greatly decreased development of mycorrhizae. Wenger (1955) also found that loblolly pine seedlings growing in the sun produced many more mycorrhizal roots than those in dense shade.

It appears that if an abundance of mineral nutrients, especially nitrogen and phosphorus, are present, carbohydrates are used in growth and no surplus occurs in the roots to stimulate fungal development. If moderate deficiencies of nitrogen or phosphorus exist, vegetative growth is retarded and sufficient carbohydrates accumulate in the roots to stimulate abundant formation of mycorrhizal roots. If there is a severe deficiency of nutrients or heavy shading and reduced photosynthesis, no surplus carbohydrates accumulate in the roots and mycorrhizal formation is suppressed. Harley (1965) regards this explanation as oversimplified and believes specific root secretions may be involved.

Although mycorrhizal roots seem necessary to good growth in certain soils, they are not essential when seedlings are given an adequate supply of nutrients. Vigorous seedlings can be grown in sand cultures (Addoms, 1937; Hatch, 1937) and nursery beds (Mitchell et al., 1937; Hacskaylo and Palmer, 1957) in the absence of mycorrhizal roots, if adequate nutrients are supplied. The importance of mycorrhizae is seen most clearly in soils deficient in certain nutrient elements, particularly phosphorus.

Readers interested in more detailed discussion of mycorrhizal roots are referred to publications by Björkman (1949), Bowen (1968), Harley (1956, 1959, 1965), Hatch (1937), Kelley (1950), Melin (1953), and Mikola (1965); also several papers in the Proceedings of the Fourteenth IUFRO Congress.

Root systems

The amount of water and mineral nutrients available to a plant is determined by the volume of soil with which its roots are in contact. The volume of soil depends on the amount of branching and the distances to which roots extend horizontally and vertically. Water movement toward roots is relatively slow, and the only water immediately available is that occurring within a few millimeters or, at most, a few centimeters from roots (see Chap. 2). Thus, the horizontal and vertical extent of root systems and the degree of branching are important to the success of plants.

Depth and spread of roots

The best-known studies of the development of root systems are those made by Weaver and his colleagues (Weaver, 1919; Weaver, Jean, and Crist, 1922; Weaver and Bruner, 1927; Weaver and Clements, 1938). Most of these studies were made in deep, well-aerated prairie soils, where roots penetrate to great depths. Corn and sorghum roots regularly penetrate to a depth of 2 m, alfalfa roots have been found at a depth of 10 m, and Wiggans (1936) reported that roots of 18-year-old apple trees had penetrated to a depth of at least 10 m

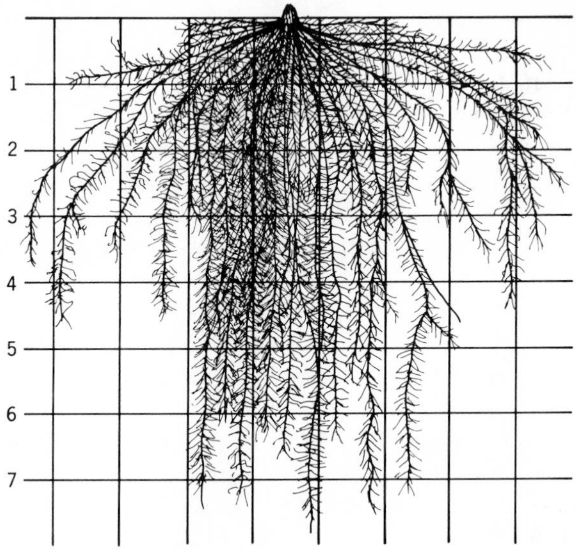

FIG. 4.12 The root system of a mature corn plant growing in a deep, well-aerated soil. Each square is 1 ft or 30 cm across. Where the subsoil is too compact or poorly aerated, roots do not penetrate so deeply. (*From Weaver, Jean, and Crist,* 1922.)

and had fully occupied the space between the rows, which were about 10 m apart. Proebsting (1943) found that roots of various kinds of fruit trees growing on a deep loam soil in California penetrated at least 5 m, and the greatest number of roots occurred at a depth of 0.6 to 1.5 m. Hough et al. (1965) studied root extension by placing ^{131}I in the soil in a forest stand and measuring radioactivity in the surrounding trees. It was absorbed in detectable amounts from as far away as 17 m in longleaf pine and 16.5 m in turkey oak. Hall et al. (1953) used uptake of radioactive phosphorus to measure root extension of various species of crop plants (see Fig. 7.4). De Roo (1957) discussed methods of excavating and measuring the extent of root systems and showed how root penetration is prevented by compact layers of soil.

The situation is very different with plants growing on heavy soils. Pears growing on a heavy adobe soil in Oregon had about 90 percent of their roots in the upper meter of soil (Aldrich et al., 1935). Coile (1937) found that over 90 percent of the roots less than 2.5 mm in diameter occurred in the top 12.5 cm under pine and oak stands in the heavy soils of the North Carolina Piedmont. Even in sandy soils trees often form mats of roots near the surface because the surface soil contains more nutrients and is wetted by summer showers (Woods, 1957). An example of restriction of root penetration by a hardpan layer is shown in Fig. 4.13*a*.

The branching and rebranching of root systems often produce phenomenal numbers of roots. Grasses develop especially large numbers, woody species fewer. Pavlychenko (1937) estimated that a 2-year-old plant of

FIG. 4.13 Root system of a *Pinus radiata* tree growing in deep pumice soil in Kaingaroa Forest, New Zealand. Note the 6-ft (180-cm) measuring stick. The water-holding characteristics of this soil are shown in Fig. 2.6*a*. (*From Will*, 1966.)

crested wheat grass possessed over 500,000 m of roots occupying about 2.5 m^3 of soil. Nutman (1934) estimated that 3-year-old coffee trees growing in the open possessed about 28,000 m of roots, 80 percent of which occurred in a cylinder 1.5 m deep and 2.1 m in diameter. Kalela (1954) reported that a 100-year-old pine produced about 50,000 m of roots with about 5,000,000 root tips. A 6-month-old dogwood seedling grown in greenhouse soil bore over 5 m of roots; a loblolly pine seedling, only 0.38 m of roots (Kozlowski and Scholtes, 1948). Much information on the extent of root systems was summarized by Miller (1938, pp. 137–148) and by Weaver (1926).

The depth and extent of branching of roots is important in choosing plants for soil stabilization and watershed cover. Deep-rooted species are preferable for stabilizing soil, but they also remove water from greater depths, so shallow-rooted species are preferable where maximum water yield is important (Hellmers et al., 1955). Further discussion of depth of rooting occurs later in this chapter in the section on hereditary characteristics of roots.

FIG. 4.13a Effect of a compacted layer of soil on root penetration by 11-week-old oat plants. Left, undisturbed soil with dense mass of roots above the compacted layer, but few below it. Right, uniform penetration of roots into soil loosened by tillage to depth of 20 in. (50 cm). Restriction of root penetration was caused by mechanical resistance as aeration was not limiting below the compacted layer. (*Courtesy of H. C. De Roo, Connecticut Agricultural Experiment Station,* 1957.)

Root grafts

The extent of the root system of an individual tree is sometimes increased by natural grafting to the roots of adjacent trees. Bormann and Graham (1959) found so many root grafts in stands of white pine that they regarded the entire stand as a physiological unit. The grafts provide effective pathways for translocation of water, solutes, and even fungus spores from one root system to another (Kuntz and Riker, 1955). Bormann and Graham (1960) found that 43 percent of the untreated trees in a 30-year-old white pine plantation were killed by "backflash" through root grafts when the plantation was thinned with ammonium sulfamate. Root grafts are also common in plantations of Monterey pine (Will, 1966). Root systems and stumps of cut trees sometimes survive on carbohydrates from intact trees to which they are grafted, and apparently such root systems function to at least a limited degree as absorbing surfaces for the attached intact trees.

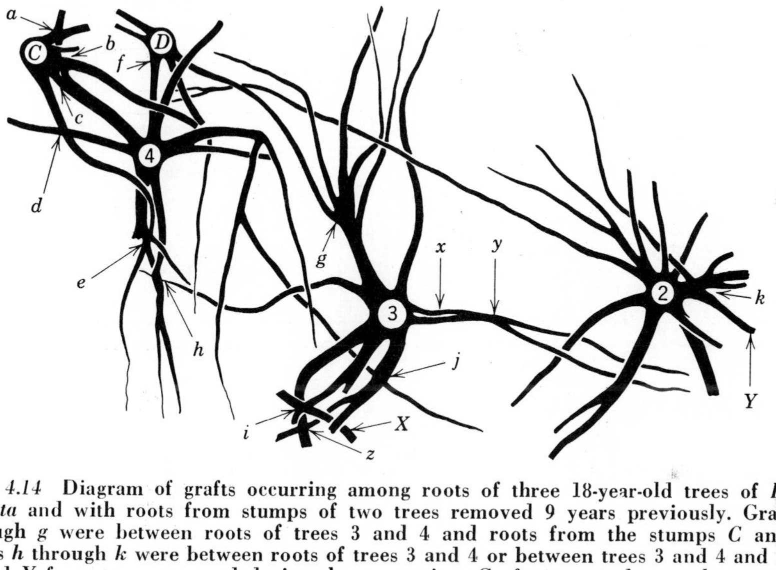

FIG. 4.14 Diagram of grafts occurring among roots of three 18-year-old trees of *Pinus radiata* and with roots from stumps of two trees removed 9 years previously. Grafts *a* through *g* were between roots of trees 3 and 4 and roots from the stumps *C* and *D*, grafts *h* through *k* were between roots of trees 3 and 4 or between trees 3 and 4 and roots *X* and *Y* from trees removed during the excavation. Grafts *x*, *y*, and *z* are between two roots of the same tree. (*From Will*, 1966.)

Rate and periodicity of root growth

The rate of extension of root systems varies widely among species and with varying soil conditions (see Figs. 4.17 to 4.19). The principal vertical roots of corn have been observed to grow downward at a rate of 5 to 6.25 cm/day for 3 or 4 weeks (Weaver, 1925). The shoots were probably growing twice as rapidly at this time. A rate of 1.25 cm/day is common in roots of grasses, but Reed (1939) found that the rate for pine roots was usually less than 2.5 mm/day. Barney (1951) found that the maximum rate of elongation of roots of loblolly pine seedlings was 3.4 to 5.2 mm/day at the optimum temperature, which was 20° to 25°C, and Rogers (1939) reported a rate of about 3 mm/day for apple roots. According to Wilcox (1962), roots of incense cedar grow about 1 to 2 mm/day in the autumn and 3 to 5 mm/day in the spring. The most rapid rate observed was 7.0 mm/day.

There has been some uncertainty concerning the existence of autonomic rhythms in root growth because it is not clear whether the cycles often observed are caused by internal or external factors. Romberger (1963) and Lyr and Hoffman (1967) reviewed the early literature in this field. Turner (1936) observed root growth of loblolly and shortleaf pine in Arkansas every month in the year. Least growth occurred in the winter, most in spring and autumn,

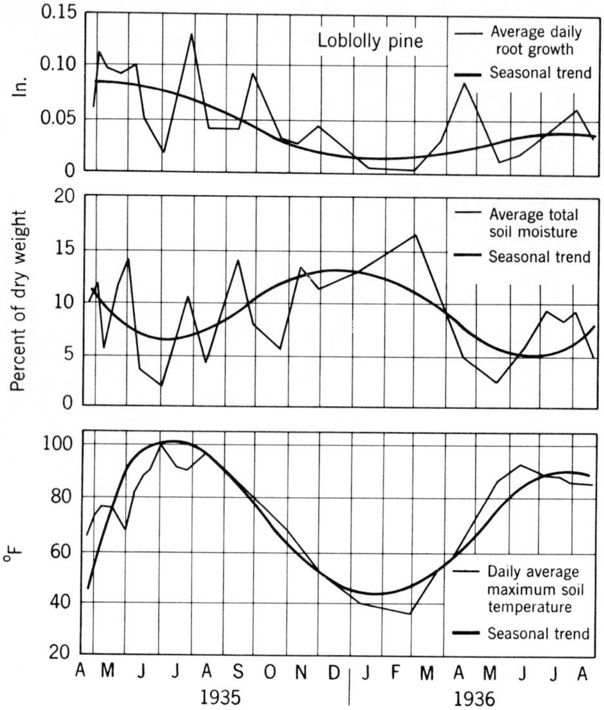

FIG. 4.15 **Effects of soil water content and soil temperature on growth of loblolly pine roots at Durham, N.C. (*From Reed,* 1939.)**

and less during dry periods in the summer. These species also grew every month in the year at Durham, North Carolina, but they made greatest growth in April and May and least in January and February (Reed, 1939). The periods of slowest root growth in winter coincided with the lowest temperatures; the periods of slowest growth in summer coincided with periods of lowest soil moisture. In colder climates there is complete cessation of root growth during the winter, although root growth sometimes continues on seedlings brought in the greenhouse (Stevens, 1931). On the other hand, seedlings of incense cedar kept under favorable temperature and photoperiod made little growth from December to April (Wilcox, 1962). This suggests the operation of an internal control.

Much of the observed periodicity can be explained by variations in soil temperature, soil moisture, and food supply from the shoots. In temperate climates there are usually two periods of rapid root growth, one in the spring when the soil becomes warm, the other in the autumn when the soil is wetted by rain. Growth during summer droughts and winter cold periods is severely reduced (Wilcox, 1962). In Israel most root growth of Aleppo pine (*Pinus halepensis* Mill.) occurs during the cool, rainy winters and ceases during the dry summers. However, it starts again in late summer before the rains begin. Summer root growth is not correlated with shoot growth but must be initiated by internal conditions. Reed and MacDougal (1937) also emphasized the importance of internal conditions in control of root growth. However, neither Leshem (1965) nor Wilcox (1962) found any correlation between root and shoot growth in the conifers they studied. In contrast, Head (1967) found that vigorous shoot growth reduced the production of new roots on apple and plum (see Fig. 4.16).

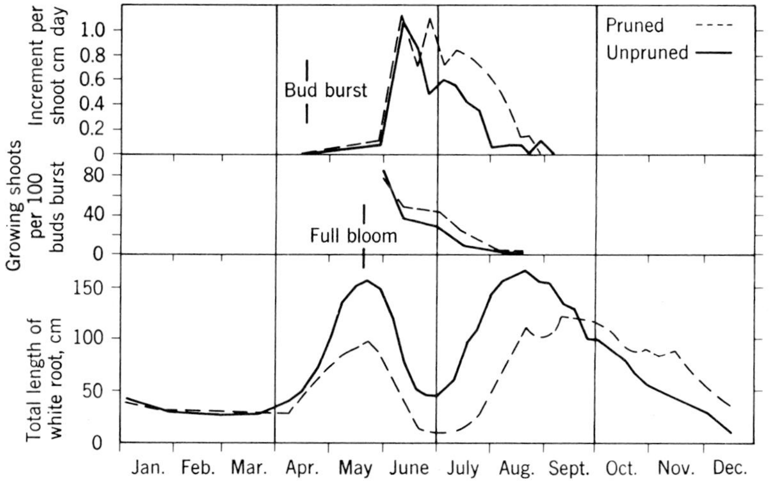

FIG. 4.16 **Relationship between root and shoot growth of apple. As shoot growth increased (upper curves), the production of new roots decreased (lower curves). Pruning stimulated shoot growth and reduced midsummer root growth.** (*From Head,* **1967.)**

The occurrence of strong seasonal periodicity in the initiation of new roots and in root extension is important in connection with transplanting trees and shrubs, because their survival usually depends on rapid growth of a new root system. Thus, transplanting should be most successful at a season when root growth is most rapid. Stone and Schubert (1959) reported definite seasonal periodicity in initiation of new roots and root elongation in ponderosa pine. There was little or no root elongation on seedlings transplanted during the summer and autumn, and new roots were initiated only on seedlings transplanted from December to June. The authors concluded that transplanting of this species would be most successful in the spring because the greatest initiation of new roots and most rapid elongation occurs at that season. However, loblolly and slash pine are transplanted successfully during the late spring and early summer in Florida, indicating that adequate summer root growth occurs in those species. The literature on rates and periodicity of root growth from Theophrastus to the present was reviewed by Lyr and Hoffman (1967).

Longevity of roots

Roots of perennial plants are usually assumed to be perennial, but this is by no means true of all such roots. Of course, the larger roots are approximately as old as the tree or shrub, but even casual examination reveals heavy mortality among the small roots. The short mycorrhizal roots of pine usually live only one season, and many of the smaller roots of fruit trees die during the winter. The small branch roots of apple and other fruit trees often live only a week or two, even when environmental conditions are favorable (Childers and White, 1942; Kinman, 1932; Rogers, 1929). Head (1966) found that the cortex of apple roots turns brown and disintegrates in one to four weeks in the summer but remains white for as long as three months in the winter.

There is also considerable variation in longevity among roots of herbaceous plants. It is often stated that the primary roots of cereals and grasses live only a few weeks and are succeeded by secondary roots, which arise adventitiously. This is by no means always true. Under certain conditions the primary roots of barley, rye, wheat, and various wild grasses are the only roots present, and they maintain the plants for an entire season. Weaver and Zink (1946) found that most of the roots of several species of perennial prairie grasses survived two seasons, and some were alive after three seasons. Stuckey (1941) observed that some species of grasses produced a new root system each year, but the root systems of other species were perennial, with few new roots being added the second year. In the latter instance the old roots must have constituted an important part of the absorbing system.

Internal factors affecting root growth

The development of root systems is controlled by their hereditary potentialities as well as by their environment. The role of heredity and the contribution of shoots are discussed in this section, the role of environmental factors in the next section.

Hereditary characteristics

The importance of hereditary factors in controlling root development is seen where a number of species grow side by side in the same soil (Fig. 4.17). Some species always develop tap root systems, others always develop fibrous root systems. Some species are always deep-rooted, others always have shallow root systems, but there are other species (Fig. 4.18) which develop different types of root systems in different kinds of soils (Toumey, 1929; Weaver and Clements, 1938).

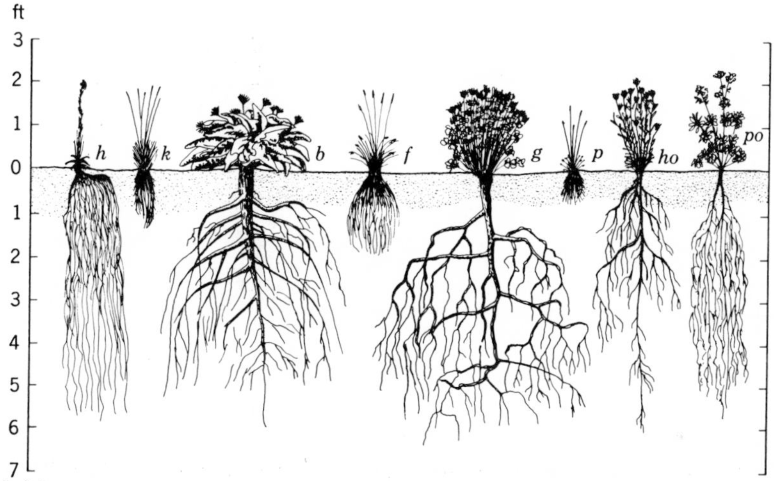

FIG. 4.17 **Differences in spread and depth of root systems of various species of prairie plants: *h, Hieracium scouleri; k, Koeleria cristata; b, Balsamina sagittata; f, Festuca ovina ingrata; g, Geranium viscosissimum; p, Poa sandbergii; ho, Hoorebekia racemosa; po, Potentilla blaschkeana.* (*From Weaver,* 1919.)**

Differences in type of seedling root systems can significantly affect establishment and survival of seedlings. Baldcypress [*Taxodium distichum* (L.) Rich.] and yellow birch (*Betula alleghaniensis* Britton) can become established only in moist soil because their shallow root systems do not enable them to survive droughts. In contrast, upland species such as oaks and hickories typically develop tap roots which penetrate deeply into the soil and provide water even after the surface soil has become dry during summer droughts (Toumey, 1929). Roots of bur oak seedlings (*Quercus macrocarpa* Michx.) which grow on dry ridges penetrated 1.7 m in the first season, but roots of linden (*Tilia americana* L.) penetrated only about 0.3 m in the same soil, although they spread out laterally (Holch, 1931). As a result of this difference, most linden seedlings in open prairie soil died during summer droughts, while the deeper-rooted bur oak seedlings survived. Albertsen

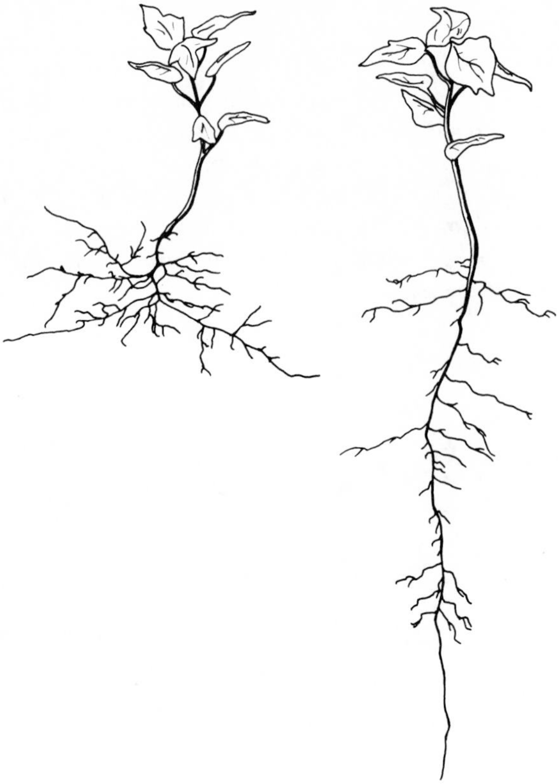

FIG. 4.18 Effect of environment on development of root systems of red maple (*Acer rubrum*) seedlings. On the left is a typical seedling from a swamp; on the right, a seedling from a dry, upland site. (*After Toumey*, 1929.)

and Weaver (1945) concluded that survival of trees in the prairie region of the United States during the severe drought of the 1930s was closely related to depth of rooting.

Drought resistance of crop plants also is often related to the possession of deep, profusely branched root systems, which absorb water from a large mass of soil. According to Burton et al. (1954), the greater drought resistance of coastal Bermuda as compared to common Bermudagrass (*Cynodon dactylon* Pers.) results from the deeper rooting of the former. Smith (1934) attributed the greater efficiency of certain inbred lines of corn in absorption of phosphorus to a larger absorbing surface resulting from more extensive branching of their roots. Slatyer (1955) showed that the greater volume of soil occupied by sorghum roots as compared with peanut roots markedly delayed onset of water stress during a drought.

It would be interesting to know the physiological basis for differences in depth of rooting of various species and varieties growing in similar soil. Per-

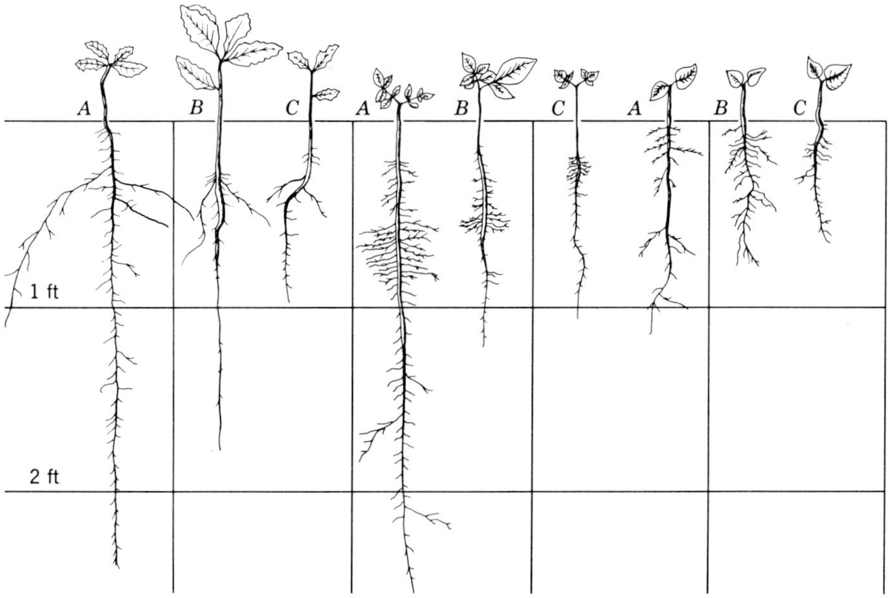

FIG. 4.19 Interaction of heredity and environment on amount of root growth produced by seedlings of three tree species grown in three different environments. Seedlings of *Quercus rubra* on left, *Hicoria ovata* in center, and *Tilia americana* on right. Seedlings A were grown in an open prairie habitat; B, in an oak forest; and C, in the deep shade of a moist linden forest. The linden seedlings in the prairie habitat were watered to prevent death from desiccation. Oak developed the deepest and largest root system in all three habitats, linden the shallowest. The shade of the linden forest greatly reduced the size of the oak root systems. (*From Holch,* 1931.)

haps there are differences in tolerance of low soil oxygen between shallow- and deep-rooted species. Another possibility is that some sort of limitation on translocation from shoots to roots exists. This interesting problem deserves investigation.

Root-shoot relationships

There is only limited information available concerning the amount of dry matter incorporated in roots as compared with shoots, largely because of the difficulty of obtaining entire root systems. Some data from Bray (1963) are shown in Table 4.1. He reported that an average of 40 percent of the dry matter of 28 species of herbaceous plants occurs in the roots, the percentage being highest in root and tuber crops such as beets and potatoes. In four species of trees only about 18 percent of the dry matter occurred in the roots. Will (1966) reported that the root system of 18-year-old Monterey pine amounts to only about 10 percent of the total weight of the trees. These data suggest that studies of net productivity may be misleading unless they include roots as well as shoots.

***Table 4.1** Amounts of Dry Matter in Metric Tons per Hectare Incorporated Annually into Roots and Shoots of Various Plant Species* *

Species	Roots	Shoots	Root/shoot ratio
Zizania aquatica	0.6	4.0	0.15
Hordeum	3.0	12.0	0.25
Andropogon scoparius (1st yr)	3.5	14.2	0.25
Triticum (average)	2.0	6.8	0.29
Medicago sativa (average)	3.2	7.4	0.43
Zea mays (average)	4.5	8.7	0.52
Solanum tuberosum (average)	4.0	2.6	1.54
Beta (average)	9.5	3.1	3.06
Pinus sylvestris (average)	1.6	8.9	0.17
Picea abies (average)	2.1	11.9	0.18
Fagus sylvatica (average)	1.6	8.2	0.19
Ghana rainforest	2.6	21.7	0.12

* *From Bray (1962)*

Roots and shoots are dependent on each other in various ways, and if the growth of one is much modified, the other is likely to be also. Since root growth depends on a supply of carbohydrates from the shoots, factors such as shading and reduction in leaf area (which reduce photosynthesis) also reduce root growth. Overgrazing of pastures and frequent cutting of hay crops can reduce the dry weight of roots to as low as 10 percent of control plants (Biswell and Weaver, 1933; Weaver and Darland, 1947).

Shading usually reduces both the absolute size of root systems and the ratio of roots/shoots. This has been reported for a variety of tree seedlings (Barney, 1951; Coile, 1940; Gast, 1937; Kozlowski, 1949; Richardson, 1953). Not all species react similarly to shading, however. Kozlowski (1949) found that the root system of *Quercus lyrata* Walt. was reduced only slightly by a decrease in light intensity which reduced the weight of roots of *Pinus taeda* L. to 25 percent of the controls and the root/shoot ratio by about 30 percent.

Development of fruits and seeds sometimes reduces root growth. Nutman (1933) reported that a heavy crop of coffee often reduces carbohydrate reserves in the roots so much that many die, thus causing serious injury to the trees. Eaton (1931) found that both total dry weight and ratio of roots to shoots of cotton plants was nearly tripled by preventing branch and boll formation. Growth of corn roots nearly ceases during ear formation, but if the ears are removed, heavy root growth continues until frost (Loomis, 1935). Head (1967) reported that vigorous shoot growth reduced the production of new roots on apple and plum (see Fig. 4.16), presumably because it leaves little carbohydrate to be translocated to the roots. Cooper (1958), Hudson (1960), and Leonard and Head (1958) found that root growth of tomato was greatly reduced during the period of fruit formation.

In addition to supplying carbohydrates, the shoots supply the roots with hormones (Street, 1966; Wilcox, 1962). These have sometimes been referred to as rhizocalines (Went, 1938, 1943). Auxin moves down from shoots to roots, and research with root cultures indicates that roots of some species require thiamin, niacin, or pyridoxin from the shoots. Richardson (1953) concluded that root growth of *Acer saccarinum* L. seedlings is dependent on chemical stimuli supplied by the leaves or buds in addition to carbohydrates. It was mentioned in the section on synthetic activities of roots that they supply organic compounds and probably hormones necessary for shoot growth. More intensive study of the interdependence between roots and shoots is needed.

Root stocks show differences in resistance to disease and cold and in tolerance of salts and flooding; trees grown on various root stocks differ in size, yield, longevity, and many other characters. The mineral content and chemical composition of leaves and fruits of certain citrus varieties differ when grown on different root stocks (Cooper et al., 1952; Haas, 1948; Sinclair and Bartholomew, 1944). Horticulturists have long been aware that the kind of root system on which a scion is grafted greatly affects the size and vigor of the tree. The dwarfing effect of Malling IX root stocks on apple is well known, and pears are commonly grafted on quince roots to dwarf them. Vyvyan (1955) claimed that the influence of the root stock on the relative growth rate of apple is greater than that of the scion, but Maggs (1964) concluded that neither root nor shoot continuously limits growth. Both usually operate considerably below their maximum efficiency. Allen (1967) found that when loblolly pine was grafted on the slower-growing shortleaf pine, shoot growth was somewhat less than when it was grafted on loblolly stock, but there was no evidence of biochemical incompatibility. Rogers and Beakbane (1957) reviewed the literature on stock-scion relationships and concluded that our understanding of the problem is very incomplete. This field deserves more research.

Environmental factors affecting root growth

The successful growth of roots and their functioning as absorbing surfaces depends on many factors in the soil environment, especially those affecting mechanical resistance to root extension, water supply, aeration, and the chemical composition of the soil solution. Impervious layers of soil or a high water table often restrict root penetration to a shallow layer of surface soil. In fine-textured soils root growth is frequently reduced by deficient aeration, and mineral deficiencies in the lower horizons sometimes inhibit root penetration. Substances produced by roots or by microorganisms associated with roots may also inhibit growth of roots of other plants. The relationship of various soil factors to root growth was discussed in detail in "Soil Physical Conditions and Plant Growth," edited by Shaw (1952). Danielson (1967) also has a good review. Some of the most common limiting factors will be discussed in this chapter.

Soil moisture

Either an excess or a deficiency of soil water limits root growth and functioning. Water is not directly injurious to roots, as shown by their vigorous growth in well-aerated nutrient solutions. However, an excess of water in the soil displaces air from the noncapillary pore space and produces an oxygen deficiency, causing the death of many roots. This will be discussed in the section on soil aeration. A deficiency of water brings about cessation of root growth, and there is probably little or no root growth in soils with a water content near the permanent wilting percentage.

In one of the few quantitative studies of the effect of soil water stress on root growth, Newman (1966) found a marked reduction in growth of flax roots at a soil water potential of −7 bars. At −15 bars root growth was 20 percent or less of control rates, but there was some growth occurring in soil drier than −20 bars. It also appeared that root growth in each layer of soil was independent of the water potential in other soil layers or in the shoot. At a stage where shoot growth and root growth in the upper soil layers was much reduced, root growth in the lowest and moistest soil layer was not at all reduced. Kaufmann (1967) found that growth of loblolly pine roots in soil with a water potential which was allowed to decrease to −6 or −7 bars was only one-fourth the rate in soil kept near field capacity. When roots were subjected to several drying cycles, less root growth occurred during the second or third cycle than during the first. Apparently, severe water stress causes roots to become more or less dormant and slow to resume growth when the soil is rewetted. In general, during the season when temperatures are high enough to permit root growth, it is often limited by deficient soil water (Reed, 1939; Turner, 1936).

In dry regions root penetration is usually limited by the depth to which the soil is wetted by rain (Cannon, 1911; Weaver, 1920; Weaver and Crist, 1922). Not only is root elongation stopped by lack of water, but roots tend to become suberized up to their tips; and suberization reduces their absorbing capacity. As a result, plants subjected to severe droughts do not recover their full ability to absorb water for several days after the soil is rewetted (Brix, 1962; Kramer, 1950; Leshem, 1965; Slatyer, 1962).

Soil aeration

Aeration of the root medium is often a limiting factor for root growth and functioning. Respiration of roots and soil organisms tends to reduce the

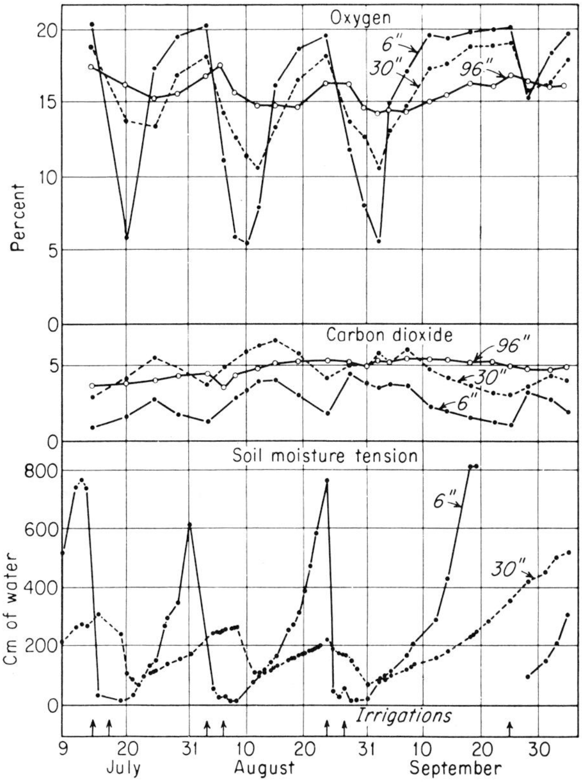

FIG. 4.20 **Variations in oxygen, carbon dioxide, and soil moisture tension of soil at various depths. An increase in soil moisture tension indicated a decrease in water content and was accompanied by an increase in oxygen content. (*From Furr and Aldrich,* 1943.)**

oxygen and increase the carbon dioxide concentration. Since this activity increases with temperature, these changes are more marked in the summer than in the winter (see Fig. 4.20). They are also more marked in soil high in organic matter, because of the greater microbial activity, than in soil containing little organic matter. There is considerable exchange of gases by diffusion between the soil and the air. Woolley (1966) estimated that diffusion alone might supply the oxygen requirements to a depth of 1 m if as much as 4 percent of the soil volume consists of interconnected gas-filled pores. This estimate seems rather optimistic. However, diffusion is aided by temperature changes, changes in atmospheric pressure caused by local air movement, and by downward percolation of water (Schroeder et al., 1965).

The effectiveness of gas exchange depends largely on soil texture and structure. Aeration is seldom limiting in sandy soil, but it is often inadequate in fine-textured soils, especially those in which less than 10 or 12 percent of their volume consists of noncapillary pore space (Robinson, 1964; Vomocil and Flocker, 1961). Compaction of soil by trampling, construction of pavements, and filling in greatly reduces gas exchange and results in low oxygen and high carbon dioxide contents in the soil atmosphere (Yelenosky, 1964). It is difficult to separate the mechanical effects of compaction on root extension from the effects on oxygen supply, and probably both are often involved (Rickman et al., 1966; Rosenberg, 1964; Tackett and Pearson, 1964).

SYMPTOMS OF DEFICIENT AERATION. Everyone is familiar with the wilting, yellowing of leaves, reduction in growth, and eventual death of most plants if the soil in which they are growing is saturated. Although these symptoms are usually attributed to reduced absorption of water caused by injury and death of roots, this does not explain the occurrence of epinasty or the development of adventitious roots, neither of which is associated with loss of turgor. The epinasty resembles that produced by ethylene (Williamson, 1950). In fact, the symptoms of flooding injury often resemble those of wilt diseases, where ethylene produced by injured tissue is said to play a part (Sadasivan, 1961). The yellowing of leaves suggests premature senescence, possibly hastened by cessation of a supply of hormones from the injured roots. As soon as adventitious roots appear at the water line on flooded plants, the shoots resume growth. This suggests that the new roots supply substances essential for shoot growth in addition to water and mineral nutrients. The development of adventitious roots on stems of flooded plants may be stimulated by blockage of downward translocation of carbohydrates and growth-regulating substances.

DIFFERENCES IN TOLERANCE OF FLOODING. There are wide variations in tolerance of flooding and poor aeration depending on duration, stage of development, and species. Dormant trees can survive many weeks of flooding in winter with little or no permanent injury, but a single day of flooding during the growing season will seriously injure some species. Grass species also vary widely in tolerance of flooding and are more tolerant when dormant than when growing (Rhoades, 1964). According to Erickson (1965), tomato yields are greatly reduced by one day of oxygen deficiency early in life, and oxygen deficiency for a single day at the early blooming stage causes a marked reduction in yield of peas. Flooding on a sunny day is more injurious than flooding on a cloudy day, and tobacco wilts so rapidly when the sun shines after the soil has been saturated by a heavy rain that farmers describe it as "flopping." Such a response must be attributed to reduced absorption of water, rather than to effects on root growth.

Although most readers are aware that there are wide differences in the tolerance of poor aeration by various species of plants, the reasons for these differences are not well understood. Cypress, willow, rice, cattails, and many other aquatic plants grow well in saturated soils where pine, dogwood, corn, and tobacco cease to grow and die. It is not clear how much of this difference results from differences in capacity of roots to tolerate anaerobic conditions and how much is due to differences in the downward movement of oxygen from shoots to roots. Possibly the roots of some aquatic species such as *Nuphar* (Weaver and Himmel, 1930; Laing, 1940*a*, 1940*b*) and rice are able to carry on respiration under anaerobic conditions, whereas corn is very intolerant of oxygen deficiency (Woolley, 1966).

The presence of alcohol was reported in the rhizomes of *Nuphar* and other aquatic plants by Laing (1940*a*). More recently Fulton and Erickson (1964) reported that as the oxygen diffusion rate of soil decreased, the amount of ethanol in xylem exudate of tomato increased. Kenefick (1962) found evidence of ethanol in sugar beets subjected to anaerobic conditions. Accumulation of alcohol and other incompletely oxidized products of anaerobic respiration may be one cause of injury to inadequately aerated roots. Apparently the roots of *Nuphar* are very tolerant of alcohol, but the roots of many other species probably are not.

Several investigators have suggested that oxygen produced in the shoots during photosynthesis moves down to the roots (Arikado, 1955; Brown, 1947; Cannon, 1932; Conway, 1940; Coult, 1964; Glasstone, 1942; Laing, 1940*b*; Leyton and Rousseau, 1958; Scholander et al., 1955; and others). Laing (1940*b*) reported a gradient of oxygen concentration from leaves to rhizomes of *Nuphar* on sunny days, and a gradient of carbon dioxide concentration from rhizomes to leaves. When the leaves were darkened, the oxygen concentration in distant parts of rhizomes fell to less than 1 percent. Experiments by van Raalte (1940, 1944), Vlamis and Davis (1944), and Barber et al. (1962) suggest that considerable oxygen moves from shoots to roots in rice. Such movement has been demonstrated in corn and barley, although to a lesser extent than in rice (Jensen et al., 1964, 1967). Woolley (1965) calculated that sufficient oxygen to supply respiratory needs could be supplied to roots only to a depth of 10 cm by diffusing through the internal air spaces of the roots. He assumed an average of 8 percent of intercellular space, probably a minimum value. Alberda (1953) stated that downward movement of oxygen in rice is restricted after the stems elongate, and oxygen is supplied to the deeper roots from a mat of fine roots formed at the surface of the water. One of the most remarkable examples of movement of oxygen from shoots to roots is in pineapple, where gas containing up to 80 percent oxygen has been observed to escape from roots in a culture solution (Ekern, 1965). This apparently originates from photosynthesis and is observed only when the leaves are illuminated. It probably occurs because the stomata of pineapples are tightly closed most of the day and the oxygen formed in photosynthesis cannot escape by the usual pathway. Incidentally, upward movement of carbon dioxide from roots to shoots may be a significant factor in the photosynthesis of certain plants (Billings and Godfrey, 1967).

LIMITING CONCENTRATION OF OXYGEN. As Woolley (1966) pointed out, there are many contradictory reports concerning the levels of oxygen and carbon dioxide which are limiting for root growth. This is not surprising in view of the fact that the experiments have been conducted on different species of plants in various stages of development, growing in a variety of media from a fine-textured soil to water cultures. The methods used to aerate root systems vary widely and produce quite different results. For example, the actual oxygen concentrations at the root surfaces are quite different in water culture, in soil aerated by forced circulation of a gas mixture, and in soil aerated by diffusion. Furthermore, measurements of samples of gas removed from the soil do not necessarily indicate the concentrations available at the root surfaces; these are usually covered with films of water through which oxygen diffuses much more slowly than through air. In fact, the diffusion coefficient of oxygen in water is approximately 2.6×10^{-5} cm^2/sec, compared with 1.9×10^{-1} cm^2/sec in air. Therefore, diffusion through water films covering roots is likely to be the limiting factor in their oxygen supply.

Attempts have been made to study the oxygen supply to roots by measuring the oxygen diffusion rate to a platinum electrode inserted in the soil to simulate a root (Lemon and Erickson, 1952). This method is described in detail by Letey and Stolzy (1964). Several other methods of measuring soil aeration are described by Letey (1966). In general, it appears that oxygen diffusion rates of less than 0.20 $\mu g/(cm^2)(min)$ are inadequate, and values greater than 0.40 $\mu g/(cm^2)(min)$ are high enough for optimum growth. However, Williamson (1964) reported good yields of several crops at oxygen diffusion rates of 0.15 $\mu g/(cm^2)(min)$. These plants were grown in much larger containers than those used by most workers; Williamson suggests that crowded root systems require a higher rate of oxygen diffusion than those in large volumes of soil. It is difficult to correlate oxygen diffusion rates with volume percentages of oxygen in the soil because of wide variations in the rate of diffusion of oxygen in different kinds of soil. However, it appears probable that oxygen concentrations above 10 percent are adequate for growth (Kramer, 1949, p. 147), and it has been suggested that high concentrations maintained by energetic aeration with air containing 20 percent oxygen can even be inhibitory (Loehwing, 1934; Letey, private communication).

CARBON DIOXIDE. There is considerable uncertainty concerning the effects of high carbon dioxide on root growth. Work in New York orchards indicates that the carbon dioxide concentration rarely rises above 12 percent (Boynton and Compton, 1944) and is usually much lower. It has been claimed that as little as 1 percent carbon dioxide is injurious (Stolwijk and Thimann, 1957), although other workers have reported no injury at concentrations of 20 percent for cotton (Leonard and Pinckard, 1946), 20 percent or more for corn and soybean (Grable and Danielson, 1965), and at least 6.8 percent for tomato (Erickson, 1946). It appears that the carbon dioxide concentration usually found in the soil is not high enough to cause injury, but the concentration of oxygen is often low enough to be inhibitory.

AERATION AND ROOT STRUCTURE. Deficient aeration not only reduces root growth but also causes modification of root structure. Roots grown in poorly aerated media, particularly in water culture, are larger in diameter and contain more and larger air spaces than those growing in a well-aerated environment (Bryant, 1934; McPherson, 1939; Schramm, 1960). McPherson suggests that the large air spaces which develop in roots growing with inadequate aeration are produced by collapse of cells injured by lack of oxygen, but Schramm (1960) concluded that they are formed by pulling apart of the better-aerated outer cells of the roots. Valoras and Letey (1966) suggest that water itself may have effects on root anatomy independent of aeration. They found that when roots of rice plants grown in drained soil were flooded, the plants suffered injury until adventitious roots developed. This has been observed in several other species (Jackson, 1955; Kramer, 1951) and is probably generally true. It is in accord with field observations that high water tables maintained at a constant level are less injurious to crops than occasional periods of complete flooding.

According to Geisler (1963, 1965), decrease in oxygen and increase in carbon dioxide both decrease root elongation but do not inhibit branch root formation. As a result, more branch roots occur per unit of length on treated roots than on control roots. It is surprising to find the lateral root meristems more resistant to unfavorable concentrations of oxygen and carbon dioxide

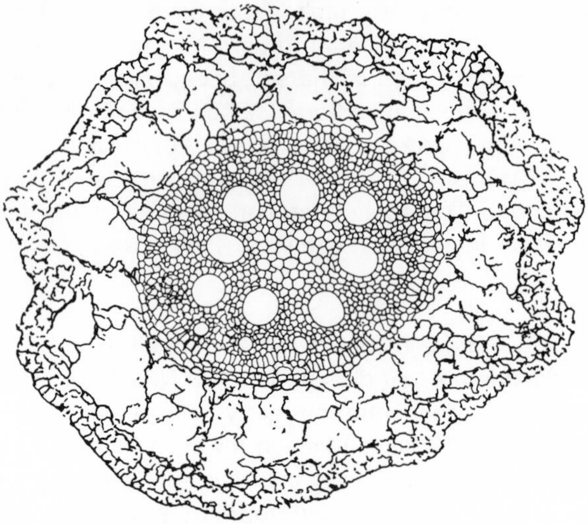

FIG. 4.21 Cross section of a corn root showing characteristic air spaces developed with inadequate aeration. (*From McPherson,* 1939.)

than the apical meristems. However, Brouwer and Hoogland (1964) state that although a temperature of 35°C stops elongation of bean roots, branch roots continue to develop almost down to the apex.

OTHER EFFECTS OF DEFICIENT AERATION. In addition to the direct effects on root growth, aeration affects water and salt absorption, water balance, photosynthesis, and susceptibility to root diseases. The effects of deficient aeration on water absorption are discussed in Chap. 6, the effects on salt absorption in Chap. 7. As mentioned earlier, a sudden decrease in aeration, such as is caused by flooding the soil, interferes with water absorption and causes wilting and leaf water deficits. As a result photosynthesis is reduced in apple (Childers and White, 1942) and tomato (Stolzy et al., 1964) and probably in most other species. Susceptibility to ozone injury is also reduced (Stolzy et al., 1964), probably because stomatal closure occurs.

Susceptibility of roots to attacks by fungi and other organisms is often increased by poor aeration of the roots. A number of pathogenic species of organisms grow well in poorly aerated soils, and this combined with reduced root growth results in injury to root systems of citrus, avocado, pine, and some other species when grown on such soils. In contrast, the Panama wilt of banana is controlled by flooding the soil to reduce growth of the causal fungus. Zentmyer (1966) recently summarized some of the literature on this subject, and various aspects of the problem are discussed in the proceedings of a conference on soil-borne plant pathogens edited by Baker and Snyder (1965).

Soil temperature

Root growth is often limited or stopped by low temperatures, and occasionally the surface soil becomes hot enough to stop root growth. The optimum temperature varies with species, stage of development, and oxygen supply, but it is probably about 20 to 25°C for most species. Root growth of apples ceases or is very slow at 4.5 to 7.0°C (Batjer et al., 1939). Pecan roots grow very slowly in this range (Woodroof and Woodroof, 1934) and are killed at −2°C. Barney (1951) found that roots of loblolly pine seedlings grew most rapidly at 20 to 25°C and that the rate of elongation at 5 and 35°C was less than 10 percent of the maximum rate (Fig. 4.22). Burström (1956) discussed the effect of temperature on root growth at the cellular level.

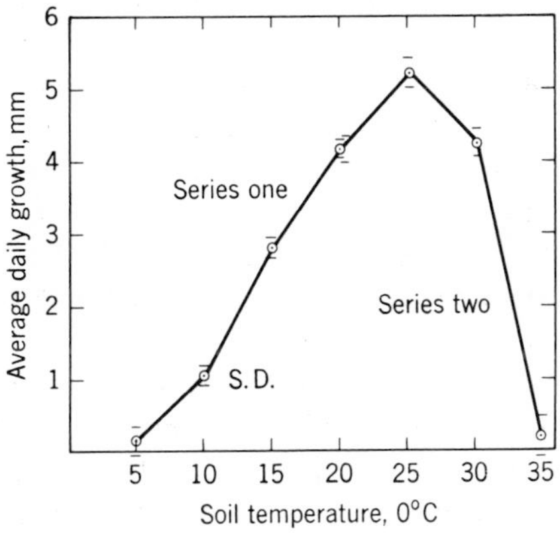

FIG. 4.22 **Relationship between rate of root growth of loblolly pine and soil temperature. (*From Barney*, 1951.)**

Roots of species native to warm climates cease growth at higher temperatures than those from cool climates. The minimum temperature for growth of roots of grapefruit, sweet orange, and sour orange in solution culture was found by Girton (1927) to be 12°C; the optimum, 26°; and the maximum, 37° Soil temperatures in the Los Angeles area are never sufficiently high for optimum growth of citrus, and for four months they are low enough to seriously limit root growth (North and Wallace, 1955). According to Arndt (1945), the minimum temperature for elongation of roots of cotton is 16 to 17°C, and the optimum decreases from 33–36 to 27°C as the roots grow downward into cooler soil. In experiments carried out by Brown (1939), roots of Bermudagrass made no growth at 4.5 and little at 10°C; and the rate increased with increasing temperature up to at least 38°. In contrast, roots of Kentucky and Canada bluegrass grew well at 4.5°C, and the optimum was only 10° for Canada bluegrass and 15 for Kentucky bluegrass. Roots of both species were severely injured by the high temperature favorable for Bermudagrass.

High temperatures can severely limit root growth, and temperatures in exposed soil can become high enough to injure or kill roots and bases of stems (Bates, 1924; Korstian and Fetherolf, 1921; Pearson, 1931; Shirley, 1936). It is claimed that the root surface of strawberries is sometimes reduced so much by high soil temperatures that the tops suffer from lack of water. The small number of roots found in the surface 30 cm of soil in many California orchards is attributed by Proebsting (1943) to high summer soil temperatures.

Injury to roots from frost heaving is an indirect effect of low temperature. Much damage is caused to alfalfa, winter cereals, and sometimes even to tree seedlings in winter because repeated freezing and thawing lifts plants and breaks off their roots below the soil surface. Mechanical strength of roots may be an important factor in winter survival (Lamb, 1936).

Brouwer and his colleagues at Wageningen intensively studied the effects of soil temperature on plant growth. It was found that the best root and shoot growth of bean (*Phaseolus vulgaris*) occurred between 20 and 30°C and that very little occurred at 5, 10, and 35°. The reduction in growth is attributed to water stress caused by reduced absorption of water at both low and high temperatures (Brouwer, 1964). On the other hand, Davis and Lingle (1961) claimed the reduction in shoot growth of tomatoes at low soil temperatures was not caused by reduced absorption of water or mineral nutrients. They suggested that low temperatures decrease translocation from shoots to roots. The result is accumulation of substances in the shoots which inhibit metabolic activity. It seems more probable that low root temperatures reduce synthetic activities in the roots in addition to causing reduced absorption of water. This possibility was discussed in the section on synthetic activity of roots.

Unfavorable temperatures also affect root differentiation and anatomy. In bean root, elongation is limited and roots become differentiated up to the apex. Branching often continues so that the branches occur almost to the root apex. The walls of endodermal cells and of the cells forming the external surfaces of roots become heavily suberized and less permeable to water and salt. Leaves of plants grown at unfavorable root temperatures tend to develop xeromorphic characters, presumably because they are subjected to high water stress (Brouwer and Hoogland, 1964). Lyr and Hoffman (1967) report a number of instances where temperature differences produced morphological differences in roots. They also comment on the fact that the optimum temperature for root growth in short-term experiments ($>$20°C) is much higher than the usual soil temperatures. According to Hellmers (1963), the optimum temperature is lower for root growth than for shoot growth of Jeffrey pine.

The extensive early literature on soil temperature and root growth was summarized by Richards et al. in Shaw (1952).

Minerals, salt concentration, and pH

It is well known that soil pH and the kind and concentration of ions have important effects on root growth. An abundance of certain essential elements, particularly phosphorus and nitrogen, stimulates root growth; but shoot growth is stimulated even more, so the ratio of roots/shoots is usually lower in fertile than in infertile soil. Not enough is known about the effects of specific ions, but it is recognized that phosphorus stimulates root growth and that deficiencies of boron and calcium in the root environment result in short stubby branches and cause many root tips to die. Failure of roots to penetrate deeply in certain soils is related more closely to lack of nutrients than to mechanical resistance or deficient aeration. Thus, it often happens that loosening the subsoil does not increase depth of rooting unless nutrients are added to it (Bushnell, 1941; De Roo, 1961; Pohlman, 1946).

Calcium and boron seem to have direct effects on root growth and must occur directly in the root environment if growth is to occur. Haynes and Robbins (1948) found that when a part of a root system was grown in a complete nutrient solution and the other part in a solution lacking calcium, the roots in the solution lacking calcium died. If boron was supplied to only a part of the root system, the other part survived but did not grow. Apparently, neither boron nor calcium can be translocated from one part of the root system to the other. In addition to specific ion effects on root growth, there is a general osmotic effect: growth is reduced due to a decrease in cell division and elongation, as osmotic potential of the substrate increases (Gonsalez-Bernaldez, et al., 1968).

In arid regions all over the world where irrigation is practiced, salt tends to accumulate in the soil until the concentration becomes too high for satisfactory plant growth (see also Chap. 3). As a result, large areas of land have become unproductive or have even reverted to desert. The direct effects of excess salt on water absorption are discussed in Chap. 6, but noticeable effects on root growth also occur. High concentrations of salt tend to slow down or stop root elongation and hasten maturation. This results in roots which are suberized to the tips and appear dormant (Hayward and Blair, 1942). Considerable differences exist among roots of different species in tolerance of salt. Wadleigh, Gauch, and Strong (1947) grew crop plants in soil containing various amounts of salt. Few bean roots penetrated soil containing 0.1 percent sodium chloride, and few corn roots penetrated soil containing 0.2 percent; but some alfalfa roots penetrated soil containing as much as 0.25 percent sodium chloride, and cotton roots were abundant in this soil. The physiological basis of these differences has not been determined.

Although much attention is given to soil pH, other factors ordinarily are more important. The pH of soil probably has little direct effect on root growth as long as it does not render essential nutrient elements unavailable or increase the solubility of elements such as aluminum or manganese until their concentration becomes toxic. Howell (1932) found that ponderosa pine seedlings survived from pH 2.7 to 11.0 and grew well from pH 3.0 to 6.0. Guest and Chapman (1944) concluded that no direct injury occurred to roots of sweet orange seedlings over a range from below pH 4.0 to above pH 9.0, although indirect effects were observed at both extremes. Arnon and Johnson (1942) also concluded that over the range from 4 to 8, pH has little direct effect on plant growth.

Light

Light is not normally a factor in the root environment because most roots develop in darkness. However, light of low intensity is necessary for the successful growth of many kinds of excised roots in culture, apparently because it is necessary for the synthesis of a growth regulator. A notable exception is pine, the roots of which grow much better in darkness (Barnes and Naylor, 1959). In general, the growth of attached roots is unaffected or slightly inhibited by light. When exposed to light, roots develop chlorophyll and even carry on photosynthesis (Fadeel, 1963).

Root competition and interaction

The size of root systems is usually much reduced when grown in competition with other systems. For example, Pavlychenko (1937) reported that root systems of barley and wheat plants were nearly 100 times as large when grown without competition as when grown in rows 15 cm apart. Apparently competition tends to reduce root growth more than shoot growth. Grass and other herbaceous species tend to reduce growth of tree root systems, probably largely because of competition for nitrogen (Richardson, 1953). It has also been claimed that grass inhibits root growth of trees by increasing carbon dioxide and depleting oxygen in the soil (Howard, 1925; McComb and Loomis, 1944). Observations by Lyford and Wilson (1964) and others suggest the existence of some mechanism which prevents roots of some species from growing very close to one another.

For centuries it has been claimed that some plants encourage increased growth by their neighbors, while others inhibit the growth of neighboring plants. For example, grapes were said to grow well with elm trees, and certain grains grow better in combination than separately, while black walnut (*Juglans nigra* L.) inhibits the growth of neighboring plants. The beneficial effects of legumes on neighboring plants can be explained by an increased supply of nitrogen, but some other effects are more difficult to explain. The extensive literature on this topic was summarized by Evenari (1961), Loehwing (1937), Miller (1938, pp. 164–174), Overland (1966), Patrick et al. (1964), and Woods (1960). A number of references to this problem can also be found in the symposium edited by Baker and Snyder (1965).

The detrimental effects of one kind of plant or crop on another which grows with it or follows it can be attributed to (1) depletion of water or mineral nutrients by the first crop, (2) the release of toxic substances from its roots or leaves, or (3) the production of toxic substances during its decomposition. In Kansas, where soil moisture is often limiting, the fact that yield of winter wheat following sorghum is lower than that following corn is attributed by Myers and Hallsted (1942) to greater depletion of soil water by sorghum. Depletion of nitrogen and other mineral nutrients might be a limiting factor in some instances, but it is doubtful if the reduction in yields observed when one kind of plant follows another can be attributed entirely to exhaustion of nutrients.

There is some support for the view that products formed during decomposition of plant parts are sometimes toxic to plants growing in the soil where this decomposition is occurring. This may be the direct effect of toxic products released during decomposition, or it may be the indirect effect of depletion of soil nitrogen by the microorganisms involved in the decomposition. Also the oxygen concentration is reduced and carbon dioxide increased by decomposition of large amounts of organic matter. It also appears that there are differences among species in respect to these effects. For example, Janes (1955) reported that lettuce and onions grown after sweet corn were badly stunted because of injury to their roots, but beets eventually made normal growth. Another example is the reduction in growth caused by quackgrass (*Agropyron repens* L.) (Welbank, 1963). Since residues of different crops stimulate different kinds of microbial populations (Patrick and Toussoun, 1965), this may also be a factor in the relative success of one crop following another.

There are numerous reports in the literature of plants producing substances toxic to neighboring plants. It was reported by Pliny that walnut trees were injurious to other vegetation, and it is claimed that the injury is caused by a compound called juglone which is produced in the roots (Gries, 1943). Bonner (1946) found that substances escaping from guayule plants (*Parthenium argentatum* L.) are toxic to guayule seedlings, and Hamilton and Buchholtz (1955) reported that living rhizomes of quackgrass inhibit growth of various species of weeds, apparently because they produce toxic substances. These and similar observations led to attempts to explain various instances of competition in terms of toxic substances (see Keever, 1950; Muller, 1966).

A related problem is the difficulty encountered in replanting orchards. For example, when peach and citrus are planted as replacements in old orchards they often grow poorly (Martin, 1950; Patrick and Toussoun, 1965). Soil fumigation to kill nematodes and other pests results in improved growth but does not entirely eliminate the difficulty, and Proebsting and Gilmore (1941) believe that toxic substances are released by decaying peach roots. This rather confused problem deserves further study because of its ecological and pathological importance.

Even healthy roots release considerable amounts of organic materials into the soil (Street, 1966). Some materials leak out, and others are released by the death and decay of root caps, root tips, and epidermal and cortical cells sloughed off during secondary thickening (Burström, 1965; Lundegårdh and Stenlid, 1946). Rovira (1965) lists at least 10 sugars, 23 amino acids, 10 vitamins, and various organic acids, flavonones, and enzymes which escape from roots. Although roots possess no glandular structures of the type found in some leaves, they at least occasionally exude drops of liquid (Head, 1965), which presumably contain solutes. The various substances escaping from the roots into the rhizosphere must have considerable effect on the microbial population, which in turn affects the root environment (Katznelson et al., 1962; Rovira, 1965). When mineral nutrients are limiting, these organisms may even compete with the roots for essential nutrients (Nichols, 1965).

Summary

The role of roots in absorption of water and minerals and in anchorage is well known, but more attention should be given to their role as synthetic organs. Inorganic nitrogen is converted to organic nitrogen in the roots of many species. Roots also produce a variety of other organic substances, including alkaloids such as nicotine, auxins, cytokinins, gibberellins, and possibly other hormones essential to the shoots. In turn, roots are dependent on shoots for carbohydrates, auxin, and certain vitamins.

Most plant scientists assume that young roots constitute the principal absorbing surfaces, but the older suberized roots must constitute an important part of the absorbing surface of woody perennials. It was formerly assumed that water and salt move inward across the vacuoles of the root cells. However, it appears that much of the water and minerals absorbed by plants moves in through the cell walls rather than through the vacuoles, at least as far as the endodermis. The root systems of many woody plants are modified by the presence of fungi to form mycorrhizae, which are very efficient in the absorption of minerals, particularly phosphorus.

The type of root system is determined by the hereditary potentialities of the species. As a result, plants of different species growing in the same soil often have widely different forms of root systems, i.e. deep taproots, widely spreading coarse roots, or shallow fibrous systems. In well-aerated soils in the temperate zone root growth is controlled chiefly by soil water supply and temperature. Most root growth occurs in the spring and early summer, when the soil moisture content is optimum, and in the early autumn, after fall rains. Growth is reduced or stopped in midsummer by soil water deficiency and in the winter by low temperature. There is also evidence of a limited degree of internal control over the growth periodicity of roots of some species regardless of their environment. There is evidence that roots of plants of one species sometimes modify the growth of roots of neighboring plants, possibly through the release of toxic substances. One crop may also affect a succeeding crop by depletion of water or minerals and by release of toxic substances during decomposition.

Numerous questions remain unanswered. For example, what controls depth of rooting? Are roots of deep-rooted species more tolerant of deficient aeration than those of shallow-rooted species? Why do roots grow out radially for long distances, and why do some of these roots send branches upward which then rebranch and form mats of small roots near the surface? More must be learned about factors controlling production of new roots following transplanting. To what extent is this related to time of year, and is it affected by mineral nutrition or other cultural practices to which the seedlings were subjected prior to transplanting? Do the roots supply the shoots with significant amounts of growth regulators such as cytokinins and gibberellins? If so, part of the effect of unfavorable soil temperatures and deficient aeration may be caused by interference with synthetic activities of roots.

five

Absorption of Water

Continuous absorption of water is essential to the growth and even the survival of most plants. Only a few xeromorphic types have such low rates of water loss that they can survive for more than a day or two without absorbing measurable quantities of water. The daily loss of water by transpiration often exceeds the water content of a plant. For example, a corn plant may lose 2 to 4 liters of water on a hot summer day, or twice the weight of water in the plant. Unless most of the water lost is replaced immediately, rapidly transpiring plants would die of desiccation in a single day.

Absorption linked to transpiration

The absorption of water is not an independent process but is related to and largely controlled by the rate of water loss in transpiration, at least when

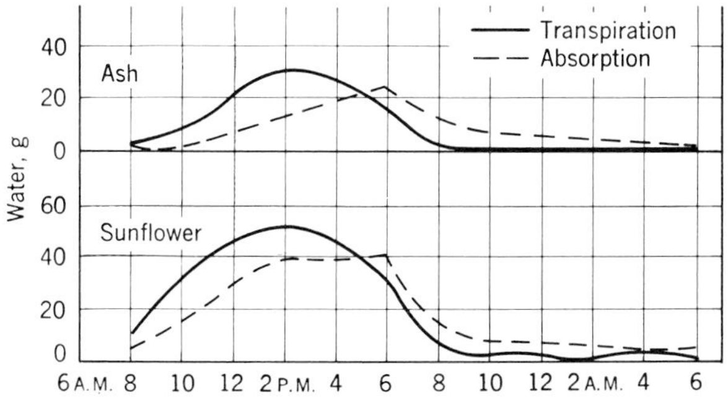

FIG. 5.1 Rates of transpiration and absorption of a woody and an herbaceous species on a bright, hot summer day. The plants were rooted in soil in autoirrigated pots. (*From Kramer*, 1937.)

water is readily available to the roots (see Fig. 5.1). Absorption and transpiration are linked by the continuous water columns in the xylem system of plants. Thus, water movement through plants from soil to air is regarded as a series of linked processes in which the overall rate is controlled by the slowest process, that is, the stage at which the greatest resistance to water movement occurs (Cowan, 1965; Gradmann, 1928; van den Honert, 1948). This viewpoint is quite different from the classical view that absorption of water is an independent process controlled by osmotic processes occurring in the roots. Actually, as we shall see later, an osmotic absorption mechanism also exists, but it becomes evident only during periods of very low transpiration or in root systems from which the tops have been removed.

Absorption by slowly and rapidly transpiring plants

Conditions in the water-conducting systems of slowly transpiring plants are quite different from those existing in rapidly transpiring plants. When the soil is moist and warm and little transpiration is occurring, water in the xylem is often under positive pressure, as indicated by the occurrence of guttation and the exudation of sap from cuts made in the xylem. When transpiration is rapid, the water in the xylem is usually under tension, and no guttation occurs. The difference can be demonstrated by immersing a portion of the stem of an herbaceous plant in a dye such as acid fuchsin and cutting into the stem beneath the surface of the dye. If the plant is transpiring even moderately rapidly, dye will rush into the cut xylem elements and almost instantly stain the stem above and below the cut to a distance of many centimeters. If the plant has been in moist soil and a humid atmosphere before making the cut, the dye will not enter the opened xylem elements, but in many species sap will begin to exude from them. Such experiments indicate that the water in the xylem of rapidly transpiring plants is at less than atmospheric pressure, but in very slowly transpiring plants it is under positive pressure, the so-called root pressure.

The existence of these differences suggests that water absorption of rapidly transpiring plants may be brought about by a mechanism different from that operating in slowly transpiring or nontranspiring plants. Long ago, Renner (1912, 1915) noted that the water absorption responsible for root pressure occurs only in healthy, well-aerated root systems and seems to depend on the activity of living cells in the roots, so he termed it "active absorption." In contrast, intake of water by transpiring shoots can occur through anesthetized or dead roots or even in the absence of roots. The role of roots in transpiring plants seems to be that of passive absorbing surfaces across which water moves by mass flow, so Renner called the process "passive absorption." Since Renner's terminology is fairly well established in the literature, we shall continue to use it. However, readers are warned that "active absorption" as used by Renner and in this book does not mean that water enters roots by active transport or nonosmotic uptake. This should be clear after reading the next paragraph.

All water absorption occurs along gradients of decreasing water potential from soil to roots, but the cause of this gradient differs in active and passive absorption. In active absorption by slowly transpiring plants the reduction in water potential of the xylem sap is caused largely or entirely by the accumulation of solutes in the xylem, and the pressure on the xylem sap is positive. During passive absorption by rapidly transpiring plants the solute concentration of the xylem sap is low, and the water potential is lowered chiefly by the decrease in pressure or by the tension developed in the xylem sap. The water potential of the xylem sap can be represented by the following equation:

$$\Psi_{\text{xylem sap}} = \Psi_s + \Psi_p \qquad (5.1)$$

During active absorption Ψ_s is relatively low, −1 or −2 bars, and Ψ_p is usually positive, amounting to about 1 bar. During passive absorption Ψ_s is relatively high (i.e. sap is very dilute), but Ψ_p is low or often negative. Hence, $\Psi_{\text{xylem sap}}$ may have a negative value of a few to many atmospheres. A wide range of intermediate conditions can exist, depending on the rate of transpiration. Examples of the values in roots of slowly and rapidly transpiring plants are shown in Table 5.1.

Passive absorption by transpiring plants

It is generally agreed that the forces bringing about the absorption of water by transpiring plants originate in the shoots rather than in the roots, and the rate of water absorption is controlled by the rate of water loss from the shoots rather than by the activity of the roots themselves.

Table 5.1 Relative Water Potentials in Soil, Root Cortex, and Xylem Sap of Slowly and Rapidly Transpiring Plants in Soil at Approximately Field Capacity
(Values are estimates in bars. The positive pressure in the xylem of the slowly transpiring plant might result in guttation.)

Soil		Slowly transpiring plant, active absorption		Rapidly transpiring plant, passive absorption	
		Cortex	Xylem sap	Cortex	Xylem sap
Ψ_s	−0.1	−5.0	−2.0	−5.0	−0.5
Ψ_m	−0.2	——	——	——	——
Ψ_p	——	4.0	0.5	1.0	−5.0
Ψ_w	−0.3	−1.0	−1.5	−4.0	−5.5

Mechanism of passive absorption

As water evaporates from the leaves, the reduction of water potential in the leaf cells causes water to move into them from the xylem of the leaf veins. Removal of water from the xylem reduces pressure on the xylem sap and reduces its water potential. This reduction is transmitted through the continuous water columns of the xylem elements and their water-saturated walls down to the root system. Reduction of water potential in the root xylem produces a gradient along which water moves from the root surface across the intervening tissues and into the xylem. At low rates of transpiration, water can be regarded as moving by diffusion across the living cells in roots and leaves. However, there is considerable resistance to water movement into the roots. Hence, as the rate of transpiration increases, absorption lags behind transpiration and diffusion becomes too slow to supply the demand for water (Levitt, 1956). The pressure on the xylem sap is reduced and finally falls below zero, so that the water is under tension. Under these conditions water can be regarded as moving through the plant by mass flow as a cohesive column, pulled by the matric or imbibitional forces developed in the evaporating surfaces of the leaf cells.

This explanation assumes that the cohesion theory of the ascent of sap is in operation. The cohesion theory is discussed in more detail in Chap. 8, and it will suffice here to state that the arguments for its operation greatly outweigh those against it. The view that the rate of water absorption by transpiring plants is usually controlled principally by the rate of water loss is supported by the relationship between the rates of the two processes shown in Fig. 5.1. When transpiration is rapid, absorption tends to lag somewhat behind transpiration but follows essentially the same course.

The lag of absorption behind transpiration is caused by resistance to water movement through the system, chiefly in the cells of the roots. Most of the resistance to water movement obviously occurs in the living cells, because the resistance to water movement through roots under a pressure gradient is greatly decreased by killing the roots (Renner, 1929; Kramer, 1932; Brouwer, 1954*b*). Ordin and Kramer (1956) demonstrated that killing roots also decreased resistance to entrance of water by diffusion. Removal of roots results in temporarily increased absorption of water and reduces the lag period of absorption when the rate of transpiration is changed, as shown in Fig. 8.17. The relative importance of the resistances in roots, stems, and leaves is discussed in Chap. 8.

Kuiper (1964*a*, 1964*b*) reported that compounds such as decenyl succinic acid greatly reduce the resistance to water movement through roots at low temperature, presumably because of their effects on the lipid layers of cytoplasmic membranes. However, it seems probable that the reported reduction in resistance is really caused by injury to the roots (Newman and Kramer, 1966). On the other hand, treatment of roots with low concentrations of respiration inhibitors (Brouwer, 1954*b*; Lopushinsky, 1964), deficient aeration, and low temperatures greatly increases the resistance to water movement through roots. This will be discussed further in Chap. 6, which deals with factors affecting the absorption of water.

Pathway of radial water movement

It seems appropriate to return to the discussion begun in Chap. 4 concerning the path followed by water and solutes in moving across the cells lying between the surfaces of roots and the xylem. The slow equilibration of the vacuolar water in roots of transpiring plants supplied with tritiated water suggests that the main stream of water bypasses the vacuoles of the parenchyma cells, instead of passing through them, as is generally supposed (Biddulph et al., 1961; Cline, 1953; Raney and Vaadia, 1965). Using quite different methods, Weatherley and his coworkers (see Weatherley, 1963) concluded that most of the water movement in roots occurs in the cell walls, except in crossing the endodermis. It appears that ions may also bypass the vacuoles of the root cells on their way to the xylem (see Chap. 7).

The idea that water movement occurs chiefly in the cell walls may seem inconsistent with observations that treatments which affect permeability of protoplasm greatly affect the movement of water into roots. This seemingly contradictory situation can be clarified by considering the following points. First, not all water movement occurs in the walls, but some obviously occurs through the protoplasts which are parallel with them. The distribution of carrying capacity between the two pathways depends on their relative cross sections and resistances. Second, cell walls are not completely nonliving but are extensively permeated with strands of cytoplasm, the plasmodesmata, whose physical condition must affect the resistance to flow in the walls. Third, at some point in most roots water must pass through a protoplasmic barrier, which in young roots may be the endodermis and in older roots may be the cambial layer. Thus, even if water moves chiefly in the walls, factors which affect the physical state of protoplasm will modify the resistance to movement.

Active absorption and root pressure theories

Few plant processes have provoked more different explanations than the causes of root and stem pressures in plants. This may have resulted in part from attempts to explain such diverse phenomena as exudation of organic solutes from nectaries of flowers and glandular leaves of insectivorous plants, bleeding from cut stems, and guttation, all by the same mechanism.

The numerous explanations for active absorption and the occurrence of root pressure can be classified in three groups. One group contains theories which assume secretory activity on the part of root cells, the second deals with electroosmotic theories, while the third explanation assumes that active absorption is a simple osmotic process in which the living root cells function as the differentially permeable membrane of an osmometer.

Secretion theories of root pressure

A number of writers have suggested that differences in permeability on the inner and outer surfaces of root cells somehow result in secretion of water into the xylem. Ursprung (1929) claimed that the suction force is greater on the inner than on the outer sides of the cells of the endodermis and vascular parenchyma. Osterhout (1947) suggested that unidirectional secretion of water can occur if a higher concentration of solutes is maintained at one end of the cell than the other, but there is no evidence that this actually occurs. Franck and Mayer (1947) described an ingenious system in which either solute or solvent might be moved against a concentration gradient by expenditure of energy. However, it is doubtful if significant differences in permeability, concentration, or chemical potential can be maintained on opposite sides of cells by expenditure of energy released by respiration. Therefore, the effectiveness of this mechanism in bringing about movement of appreciable quantities of water is questionable.

Because of the inadequacies of secretory theories of water movement, interest in them waned during the first quarter of this century. About 1940 interest was revived by new evidence that a nonosmotic factor might be involved in uptake of water by cells and roots. The extensive literature on this subject was reviewed by Kramer (1956*a*). Van Overbeek (1942) reported that the osmotic pressure of the exudate from tomato root systems was somewhat lower than the osmotic pressure of the solution required to stop exudation. He regarded this difference as a measure of nonosmotic secretion of water into the xylem. He also reported that 50 to 70 percent of the pressure could be eliminated by KCN and considered this part to be the nonosmotic fraction. Rosene (1944, 1947) claimed that root pressure is entirely nonosmotic and that the reason it is not completely inhibited by substances such as KCN is because they do not completely inhibit respiration. More recently, House and Findlay (1966) suggested that there is a small amount of non-

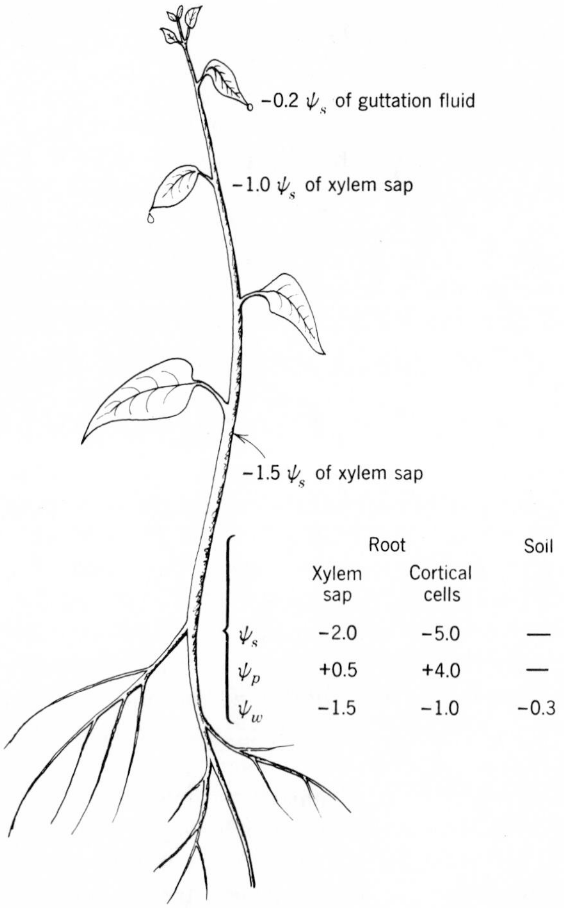

FIG. 5.2 **Decrease in osmotic potential Ψ_s of xylem sap as it moves through the plant, caused by transfer of salt to living cells bordering the xylem. The values given are hypothetical, but the decrease is based on measurements by Klepper and Kaufmann (1966) and Oertli (1966). The data for roots illustrate how a gradient in water potential Ψ_w can develop in slowly transpiring plants from soil at field capacity to root xylem across the root cortex with a much lower osmotic potential than either soil or root xylem, but an intermediate water potential.**

osmotic movement of water into roots when the external solution and the xylem sap exuding from stumps have the same osmotic potential. However, in view of the fact that considerable salt is removed from the xylem sap as it moves upward, the osmotic potential of root pressure exudate is probably higher than the potential in the region where water is being absorbed (Klepper, 1967; Klepper and Kaufmann, 1966; Oertli, 1966). The increase in osmotic potential in the xylem sap is shown diagrammatically in Fig. 5.2. Readers are reminded that osmotic potentials of solutions increase or become less negative as salt is removed. This situation may also explain the discrepancy reported by van Overbeek and others. Secretion theories of root pressure have been abandoned by most physiologists in recent years, because the high permeability of cell membranes to water movement makes appreciable nonosmotic uptake of water highly improbable (Levitt, 1947, 1948; Ordin and Kramer, 1956).

Electroosmotic theories

In past years a number of investigators attributed root pressure to the electroosmotic movement of water into the xylem (Keller, 1930; Lund, 1931; Heyl, 1933). This explanation is attractive because it is well known that water can be caused to move across a membrane under the influence of an applied electrical current. If the permeability of the membranes remains constant, the volume of flow is more or less proportional to the difference in electrical potential, and the direction of flow is toward the pole with the same polarity as the membrane. Since cellulose membranes in contact with water are negatively charged, water moves toward the negative pole, which is in the interior of roots. Lundegårdh (1940) measured a difference in potential of about 100 millivolts from surface to interior, but he regarded this as too small to produce any significant water movement. Blinks and Airth (1951) were unable to produce appreciable water movement in *Nitella* cells by applying electrical potentials of 100 to 200 millivolts.

Fensom (1957, 1958) revived interest in electroosmotic movement of water. He observed that cycles in bioelectric potentials were correlated with cycles in exudation from roots and formulated an electrokinetic theory of transport to account for these observations. Dainty (1963*a*) pointed out the limitations of electroosmosis as an effective means of moving water, and it is doubtful that it ever is an important factor.

Osmotic theories

As is pointed out in Chap. 1, water moves along gradients of water potential, also termed diffusion pressure deficit, suction force, or suction potential by various authors, which may be largely independent of the osmotic potentials of the tissues involved. As is shown in Table 5.1 and Fig. 5.2, water can move by osmosis from a dilute soil solution with a water potential of perhaps −0.3 bar to the xylem sap with a water potential of −2 bars. In moving, it crosses the turgid cortical parenchyma cells with an osmotic potential of −5 or −6 bars, but a water potential of only −1 bar. Such a plant with a hydrostatic pressure of 0.5 bar on the xylem sap might develop enough root pressure to cause guttation.

The first application of this principle to an explanation of root pressure seems to have been that of Atkins (1916) who stated (p. 203) that "the inflow of water from the ground to the elements of the wood of the roots takes place across the cortical cells of the roots. For, though the latter have a much higher osmotic pressure than have the tracheae, they function merely as a complex semipermeable membrane, as they are already fully distended." Scott and Priestley (1928) and subsequently many others have regarded the endodermis as the differentially permeable membrane of roots. The osmotic explanation treats the root as an osmometer in which water moves from the outside to the inside of roots along a gradient of decreasing water potential across a differentially permeable membrane formed by the endodermis or other tissues.

The chief problem in connection with an osmotic theory of root pressure is to explain how a sufficient concentration of solutes is maintained in the xylem vessels of the roots to produce the necessary gradient in water potential across the roots. Atkins (1916) suggested that the adjoining parenchyma cells secrete sugar into the xylem vessels. Priestley (1922) believed that an adequate supply of solutes might be provided from the contents of those cells which are differentiating into the xylem vessels, plus some leakage from surrounding cells subjected to turgor pressure. These investigators based their assumptions on analyses of sap obtained from woody species in the spring and consequently probably overestimated the importance of sugars in maintaining the osmotic pressure of xylem sap. Van Overbeek (1942) found no sugar in the xylem sap of tomato, and there appears to be little in the xylem exudate of most herbaceous species. The concentration of salt may be relatively high, however. Eaton (1943) reported osmotic potentials of −1.5 to −2.4 bars in the sap exuding from stumps of cotton plants, Stocking (1945) observed osmotic potentials of about −1.9 bars in squash, and van Overbeek (1942) recorded −1.3 bars in tomatoes in Hoagland solution, but only −0.4 bar in sap from tomato root systems in distilled water.

Hylmö (1953) suggested that newly formed xylem elements might continue to accumulate salt even after the protoplasts are ruptured, thus providing a supply of salt to bring about osmotic movement. This theory, which resembles that proposed earlier by Priestley (1922), would be attractive if enough new xylem were being formed to supply salt. Scott (1949, 1965) claims that living protoplasts occur in xylem elements in the absorption zone of some roots, and Anderson and House (1967) reported that there is a relationship between the number of living protoplasts in the xylem and the amount of salt absorbed by corn roots. However, it seems unlikely that enough salt can be absorbed in this manner. Furthermore, completely dormant roots sometimes exhibit root pressure, which suggests that active absorption occurs even when no cell division is occurring. This problem is also discussed in Chap. 7.

Hoagland (1944, pp. 84–92) emphasized the close relationship between accumulation of salt in roots and the occurrence of root pressure and guttation. These processes occur only when root systems are healthy, well aerated, and immersed in dilute salt solutions maintained at a moderate temperature. Root pressure and guttation cease if the roots are immersed in distilled water or are subjected to deficient aeration or low temperature. Eaton (1943) argued that simple osmosis is adequate to explain the occurrence of root pressure. He reported that the rate of exudation from cotton root systems is proportional to the difference in osmotic pressure between the xylem sap and the external solution. Van Andel (1953) came to a similar conclusion. Rapid reversal from exudation to absorption through the stumps and back to exudation occurs if root systems are transferred from water to a solution with an osmotic pressure of about 2 atm, then back to water. Such reversals require less than a minute, can be repeated indefinitely, and are more compatible with an osmotic mechanism than with any other mechanism.

It is sometimes argued that decrease of root pressure exudation by respiration inhibitors and deficient aeration, along with the often observed diurnal periodicity in root pressure, indicates a mechanism more directly dependent on metabolism. For example, Gračanin (1964) argued that there is a nonosmotic component in root pressure and that root pressure and guttation are not identical in origin because they are not stopped by the same concentrations of respiration inhibitors. Such arguments overlook the fact that the accumulation of salt is dependent on expenditure of metabolic energy and that the permeability of the cell membranes through which salt and water move is modified by factors which affect metabolism. The fact that uptake of salt and water is much reduced by respiration inhibitors does not by itself prove that either is moved directly by active transport. Passive water movement through roots caused by a pressure gradient is greatly reduced by respiration inhibitors (Brouwer, 1954*b*; Lopushinsky, 1964; Mees and Weatherley, 1957, and others) and by deficient aeration (Kramer, 1940*b*, 1951; Kramer and Jackson, 1954).

In the light of existing information, the osmotic explanation of root pressure seems the most probable one, although some troublesome problems remain to be explained. For example, there is no fully satisfactory explanation of how salt is accumulated in the xylem of roots. This problem is discussed in Chap. 7.

Root pressure phenomena

Root pressure and the resulting exudation of sap has attracted the attention of observers from early times to the present.

Species differences

No recent attempts have been made to list the kinds of plants in which root pressure occurs, but it has been observed in hundreds of species. Wieler (1893) collected references to "bleeding" or "weeping" in 126 species belonging to 93 genera distributed among ferns, angiosperms, and gymnosperms. He added 62 additional species from his own observations. Unfortunately, Wieler appears to have included examples of exudation from glandular hairs and from wounded stems along with examples of true root pressure. Several conifers were included in his list, but observations of root pressure from conifers are rare. White et al. (1958) reported exudation from the basal ends of roots severed from white spruce and pine, and O'Leary and Kramer (1964) reported exudation from root segments of loblolly pine and white spruce. Thut (1932) reinvestigated root pressure in aquatic plants and found considerable amounts of sap exuding from cut stems of several species. The volume was greatest in species with well-developed root systems, and no exudation occurred from species with few or no roots.

The reasons why some species seldom or never show exudations are unknown, although some suggestions are now possible in view of recent observations by O'Leary (1965). O'Leary was unable to obtain exudation from stumps of healthy loblolly pine, white spruce, or sugar maple seedlings grown under favorable conditions. However, he found that apical root segments of all three of those species and of all others studied produced measurable exudation. Failure to exhibit root pressure did not result from inability to accumulate salt in the xylem sap. However, it is possible that the high internal resistance to longitudinal water movement through some kinds of roots or root-shoot transitions, combined with a tendency for sap to leak out of older roots under pressure, prevents transmission to the shoot of the sap pressure developed in apical areas of roots. Readers are warned that failure to observe sap flow from one or two plants is not conclusive evidence that it never occurs in the species, because there are wide individual variations in amount of exudation from apparently similar root systems.

Volume of exudate

The volume of root pressure exudate naturally varies with the size and condition of the plant and with various environmental factors, such as temperature and availability of soil water. A paper birch (*Betula papyrifera* Marsh) tree 37.5 cm in diameter produced 28 liters of sap in one day and 675 liters during the spring, according to Clark (1874). Johnson (1944) reported sap yields of from 20 to over 100 liters in a season from paper birch trees 20 to 38.5 cm in diameter. Yield was not proportional to size. Some smaller trees produced more sap than larger ones.

Weller (1931) observed sugarcane stools to exude over 400 ml of sap in 24 hr and 1,000 ml during the first week after detopping. Corn plants detopped in the milk stage of the grain yielded over 500 ml of sap in 3 days and over 1,700 ml in 15 days. Crafts (1936) transferred squash plants grown in nutrient solution to tap water and cut off their tops. Four root systems yielded 550 ml of sap in 24 hr, a volume greater than the total volume of the root systems. Sperlich and Hampel (1936) reported that the volume of exudate from sunflower root systems increased with their age until flowering occurred, then it decreased. Úlehla (1963) observed a similar cycle in exudation from maize. The volume of sap flow is affected by weather, cultural treatments, season, and age and condition of plants. Sometimes inexplicable variations occur among plants which appear identical.

Composition and concentration of exudate

The liquid exuding from stumps and wounds in stems differs widely in composition and concentration. The sucrose content of maple sap ranges from 1 to 7 percent, but usually it is 2 to 3 percent (Jones et al., 1903). It also contains small amounts of glucose and very small amounts of organic acids, inorganic salts, and nitrogen compounds such as ammonia, peptides, and amino acids (Pollard and Sproston, 1954), amylases (Meeuse, 1952), and other unidentified organic constituents (Taylor, 1956). Birch sap contains only about half as much sugar as maple sap, and the sugar consists chiefly of fructose and glucose (Johnson, 1944). Priestley and Wormall (1925) reported that the xylem sap of grape contained 1.56 percent solids, which consisted of one-third ash, one-third organic acids, some reducing sugar, and small quantities of organic and inorganic nitrogen.

Bollard (1960) cited much of the extensive literature on the composition of xylem sap and discussed its role in the translocation of nitrogen and minerals. The mineral composition of xylem sap of woody species changes with the supply of salt available to the roots and with the season (Bennett et al., 1927; Carter and Larsen, 1965; Denaeyer-DeSmet, 1967; Dimbleby, 1952; Ladefoged, 1948). Several investigators have attempted to use the composition of xylem sap as a guide to the mineral nutrient needs of plants (Lowry et al., 1936; Pierre and Pohlman, 1934). However, more satisfactory samples of soil solution can be obtained by extraction with pressure plate or pressure membrane equipment (see Chap. 3).

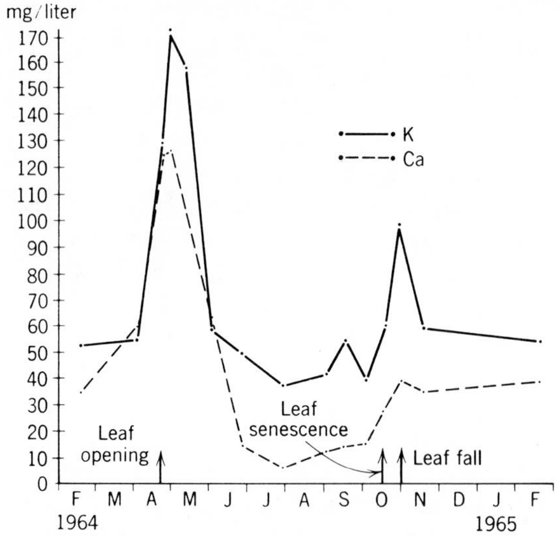

FIG. 5.3 **Seasonal changes in calcium and potassium content of xylem sap of *Corylus avellana* L., growing in Belgium. (*After Denaeyer-DeSmet,* 1967.)**

It should be remembered that the xylem sap from maple exudes as a result of stem pressure, but birch sap results from root pressure exudation (see Fig. 5.4). Furthermore, the sap studied by Bennett, Bollard, and Carter and Larsen was extracted from cut branches or twigs under pressure and is not strictly comparable with that exuding because of naturally occurring pressures.

The organic matter content of xylem sap from herbaceous species appears to be lower than that of sap from woody species. Several investigators have reported negligible amounts of sugar in the xylem exudate from herbaceous species (van Die, 1958; van Overbeek, 1942; Skoog et al., 1938). Van Die reported only 2.57 mg of dry matter per milliliter of sap in cucumber and tomato exudates. Minshall (1964) found from 2.9 to 3.6 mg/ml in tomato sap, of which 0.9 to 2.0 mg was ash. Several investigators have found small amounts of amino acids in xylem exudate (van Die, 1958; Hofstra, 1964; Minshall, 1964), and Wiegert (1964) stated that 98 percent of the organic matter in tomato root exudate was amino acids. Growth regulators have also been reported. Cytokinins have been found in the root pressure exudate from grape (Loeffler and van Overbeek, 1964; Skene and Kerridge, 1967) and sunflower (Kende, 1965). Skene (1967) found a significant amount of gibberellin in grape root exudate and it also has been reported in the xylem sap of apple and pear (Jones and Lacey, 1968).

It appears that the xylem sap of roots contains considerable amounts of organic nitrogen compounds and growth regulators and may constitute an important source of these compounds for the shoots. How these compounds

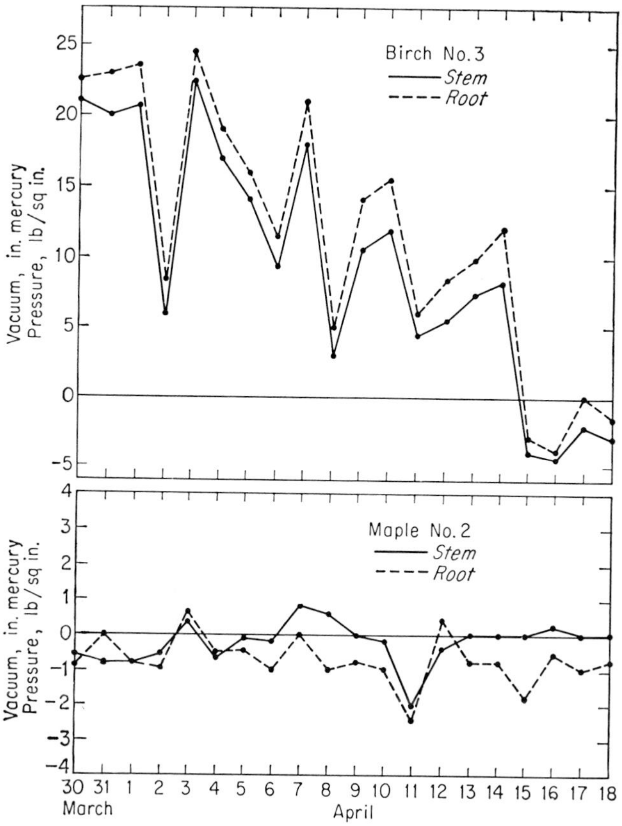

FIG. 5.4 Simultaneous measurements of root and stem pressures in river birch and red maple. Root pressures in birch exceed stem pressures, and the two are closely correlated; but root pressure in maple is usually absent, even when positive pressure occurs in the stems. (*From Kramer and Kozlowski,* 1960.)

get into the xylem sap constitutes a puzzling problem which deserves further study. Presumably they are synthesized in the root cells, but how do they get out of those cells into the xylem elements?

Magnitude of root pressure

Hales (1727), who made the first recorded measurements of root pressure, observed a pressure of about 1 bar in grape. Pressures of 2 to 3 bars were reported for birch in New England by Clark (1874) and Merwin and Lyon (1909). The maximum pressure measured for sugarcane was 1.9 bars (Weller, 1931), and Sabinin (Maximov, 1929, pp. 53–55) calculated by indirect methods pressures of 0.5 to 1.5 bars in various species. Leonard (1944) observed that cotton plants in the field developed root pressures of 0.5 to 0.8 bars. There was an immediate reduction when the root systems were injured by cultivation, but recovery usually occurred within a week. Barrs (1966) reported data from which it was inferred that root pressures in excess of 3 bars occur in sunflower and pepper. However, further study of these experiments indicates that although guttation occurred from plants with roots in solutions having an osmotic potential of −3 to −6 bars, because of salt absorption the hydrostatic pressure developed was less than 3 bars (private communication).

White (1938) reported that excised tomato roots growing in culture solution developed root pressures of over 6 bars, perhaps the highest values ever recorded. Boehm (1892) and Figdor (1898) observed pressures of over 8 bars in stems of woody plants, but Molisch (1902) and MacDougal (1926) claimed these pressures were of local origin, caused by wounding, and not true root pressures.

Periodicity

A century ago Hofmeister (1862) observed diurnal periodicity in root pressure, and White (1938) observed it in tomato roots growing in nutrient solution. Grossenbacher (1938) reinvestigated this problem and found a well-defined periodicity in the root pressure of sunflower root systems, with the maximum pressure occurring during the day and the minimum at night (see Fig. 5.5). When plants were grown with artificial light at night and kept in darkness during the day, the cycle was controlled by the artificially imposed light and dark periods. Plants grown with continuous light and constant temperature showed a cycle which was related to the time of detopping. Hagan (1949) even observed periodicity in negative exudation or uptake of water through

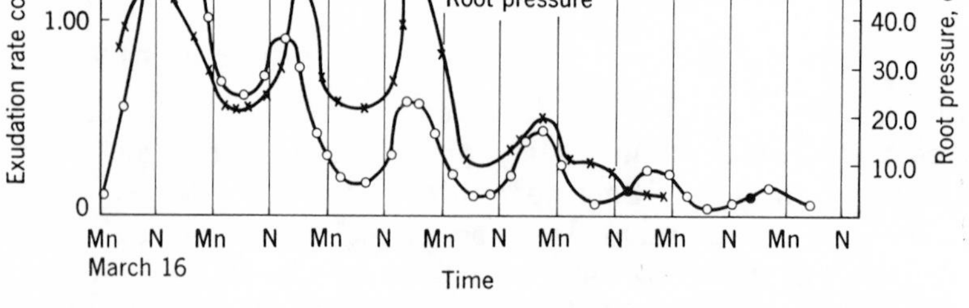

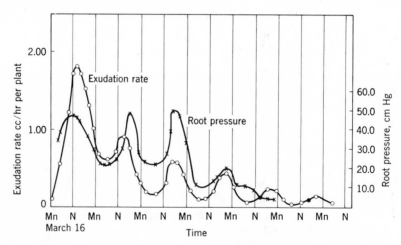

FIG. 5.5 **Diurnal fluctuations in rate of exudation and root pressure of sunflower plants kept at constant temperature. Plants had been grown in a greenhouse in Hoagland solution and were in half-strength Hoagland solution during the experiment.** (*From Vaadia,* 1960.)

stumps of detopped root systems of sunflower plants wilted before detopping. Greatest uptake occurred at midnight and least uptake at noon, just the reverse of the cycle for positive exudation.

The diurnal periodicity in root pressure exudation is usually attributed to variations in salt release or transport into the xylem. Hanson and Biddulph (1953) and Wallace et al. (1966) reported greater translocation of ions to shoots during the day than at night. Vaadia (1960) also attributed periodicity in root pressure of sunflower plants to periodicity in the rate at which salt is transported into the xylem. Another possible cause of diurnal variations in root pressure exudation is a diurnal variation in permeability of roots. According to Skidmore and Stone (1964), resistance to water flow through cotton root systems increased fivefold between noon and night. Barrs and Klepper (1968) also found evidence of the existence of a diurnal change in root resistance with a minimum near midday and a maximum near midnight. The cause of this diurnal variation is unknown, and the phenomenon deserves further study.

Guttation

The most common evidence of root pressure is guttation, the exudation of drops of liquid from the edges and tips of leaves. Burgerstein (1920) reported that guttation had been observed in plants of 333 genera, to which Frey-Wyssling (1941) added 12 more. Usually guttation occurs through hydathodes, which are stomate-like pores over intercellular spaces in the epidermis. The xylem of a small vein usually terminates among the thin-walled parenchyma cells below each hydathode; when root pressure develops, water is forced out into the intercellular spaces and flows out of the hydathodes. Guttation

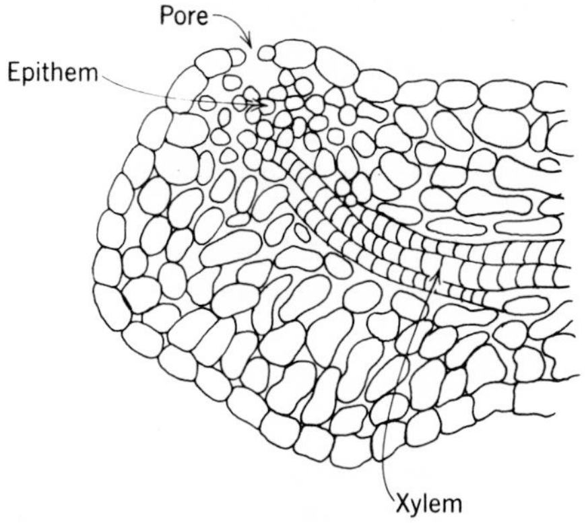

FIG. 5.6 Diagram of a hydathode, showing a pore, the underlying epithem, and termination of xylem elements. The epithem is a mass of thin-walled parenchyma with large intercellular spaces. Hydathodes usually resemble incompletely differentiated stomata with nonfunctional guard cells. Adapted from several sources.

may occur through ordinary stomates, as in some grasses; Sachs stated that water occasionally exudes from epidermal cells where no hydathodes occur. It is sometimes difficult to distinguish between guttation caused by root pressure and secretion from glandular structures (Haberlandt, 1914; Frey-Wyssling, 1941). However, Lepeschkin (1923) claimed that secretion from glandular structures can occur even when plants are wilted, while true guttation occurs only in turgid plants. Guttation is not restricted to leaves, but sometimes occurs from twigs and branches of trees (Büsgen and Münch, 1926; Friesner, 1940; Raber, 1937). Exudation of liquid from roots and root

FIG. 5.7 A pine seedling caused to guttate by forcing water into the xylem under pressure. The author has never observed natural guttation from pine seedlings, probably because of lack of strong root pressure. (*From Klepper and Kaufmann,* 1966.)

hairs has also been reported (Breazeale and McGeorge, 1953; Head, 1964; Rogers, 1939). Since this probably results from root pressure, it may be termed root guttation.

The quantity of liquid exuded from hydathodes varies from a few drops per leaf to many milliliters. The composition of the liquid varies from almost pure water to a dilute solution of organic and inorganic solutes. Curtis (1944) reported an instance where the leaves of grass became encrusted with a deposit of glutamine following heavy fertilization of a lawn. Leaves of certain saxifrages become encrusted with calcium left by evaporation of guttation water (Schmidt, 1930). Eaton (1943), Höhn (1951), and Oertli (1966) found the salt concentration of guttated liquid to be much lower than that of exudate from stumps of similar but detopped plants. They attributed the decrease in concentration to removal of salt by surrounding cells from the ascending sap stream. Klepper and Kaufmann (1966) recently demonstrated a decrease in salt concentration as dilute solutions were forced through leaves under pressure. Klepper (1967) also reported that salt is removed from the xylem in the upper part of the root.

Guttation, like root pressure, sometimes seems to show an endogenous rhythm (Engel and Friederichsen, 1952). It also shows a strong diurnal periodicity, because it always ceases during the day when transpiration increases enough to produce an internal water deficit. Its occurrence is promoted by conditions favorable to water absorption (such as moist, warm soil) and by conditions unfavorable to water loss (such as humid air). For this reason, guttation nearly always occurs at night. Mineral deficiencies are said to inhibit guttation (Raleigh, 1946), and Minshall (1964) found that heavy fertilization with nitrogen materially increases the volume of exudation from tomato root systems.

Guttation is of negligible importance to plants. Occasionally injury is caused to leaf margins and tips by deposits of salts left by evaporation of guttated water (Curtis, 1943; Ivanoff, 1944; Turrell et al., 1952). Curtis thinks that injury sometimes occurs when spray materials are drawn back into the intercellular spaces by reabsorption of guttated liquid. Johnson (1936) claimed that guttated water creates conditions favorable for the entrance and spread of pathogenic organisms, and Bald (1952) regards this as an important factor in the invasion of gladiolus leaves by fungi and bacteria. According to Lewis (1962), the germination of spores of *Claviceps purpurea* is much lower in guttation fluid from plants of species resistant to this organism than in guttation fluid from plants of susceptible species. Goatley and Lewis (1966) recently published data on the substances found in the guttation fluid from barley, rye, and wheat. Differences were found among the species, but the exudate from all contained aspartic acid and asparagine, several sugars, and various other organic compounds, plus a dozen mineral elements.

Relative importance of active and passive absorption

There are two views concerning the importance of root pressure. Many writers in the past have regarded it as an important factor in water absorption, and some still consider its importance to be underestimated (Fensom, 1957; Minshall, 1964; White, 1938).

In the opinion of the author, active absorption of water is of negligible importance in the water economy of most or perhaps all plants. Root pressure and the associated phenomena are merely the fortuitous result of the fact that salt is accumulated in the stele of many kinds of roots, producing a difference in water potential which brings about inward movement of water and development of pressure in the xylem sap. Active absorption and associated phenomena have received more space in this chapter than passive absorption, but this is because of interest in the mechanism and manifestations of the process rather than because of its intrinsic importance in the water relations of plants.

There are several reasons for regarding active absorption as unimportant: (1) The volume of exudate from stumps is rarely more than a small percentage of the volume of water lost in transpiration by similar intact plants under conditions favorable for transpiration. (2) Intact plants can absorb water from more concentrated solutions and from drier soils than can detopped root systems. (3) No root pressure can be demonstrated in roots of transpiring plants, but instead water is absorbed through their stumps if the tops of transpiring plants are removed (see Fig. 5.8). (4) There is a large group of plants, including the conifers, in which root pressure is rarely observed at the stump, although it now appears that it may be developed in individual roots (O'Leary, 1965). In view of the reluctance of many botanists to abandon active absorption as an important cause of water uptake, some of the evidence will be presented in more detail.

There is much evidence that the volume of sap exuding from the stump of detopped plants is usually much less than the water lost in transpiration

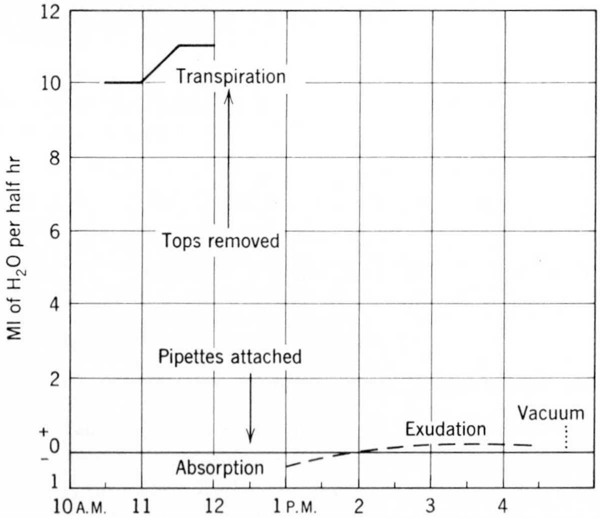

FIG. 5.8 Behavior of transpiring tomato plants following removal of tops. The root systems absorbed water through their stumps for over an hour after removal of the tops. The maximum rate of exudation was less than 5 percent of the rate of transpiration. The rate of water movement through the same root systems when attached to a vacuum pump and subjected to a pressure gradient of 64 cm Hg was 3.5 times the rate of exudation. Plants were rooted in soil at field capacity. (*From Kramer*, 1939.)

by plants of similar size. Some data are shown in Fig. 5.8 and Table 5.2. Minshall (1964) obtained much higher rates of exudation from tomato plants given large amounts of nitrogen. The data in Table 5.2 are from plants receiving normal greenhouse fertilization, and the rate of exudation would probably have been increased if they had received more nitrogen. These observations indicate that even if the active absorption mechanism were operating, it would have contributed relatively little of the water needed by the transpiring shoot.

Table 5.2 A Comparison of Exudation with the Rate of Transpiration Prior to Removal of the Tops*

(Rapidly transpiring plants usually show absorption of water through the stumps during at least the first half hour after the tops are removed, exudation beginning only after the water deficit in the root system is eliminated.)

Species	Number of plants	Transpiration, milliliters of water per plant per hour		Exudation, milliliters of water per plant per hour		Exudation as percent of transpiration †
		First hour	Second hour	First hour	Second hour	
Coleus	6	8.6	8.7	0.30	0.28	3.2
Hibiscus	5	5.8	6.7	−0.01‡	0.05	0.7
Impatiens	6	2.1	1.9	−0.22	−0.06	
Helianthus	8	4.3	5.0	0.02	0.02	0.4
Tomato (1)	6	10.0	11.0	−0.62	0.07	0.6
Tomato (2)	6	7.5	8.7	0.14	0.27	3.1
				Tops removed		

* *From Kramer (1939)*
† *Percentage relations are based on transpiration and exudation rates for the second hour.*
‡ *A minus sign indicates absorption of water by the stump instead of exudation.*

It is sometimes claimed that even though active absorption is not important quantitatively, it occurs all the time and supplements passive absorption. This is unlikely for two reasons. During periods of rapid transpiration salt is swept out of the root xylem until the concentration drops so low that no inward osmotic movement of water can be expected to occur (Lopushinsky, 1964; O'Leary, 1965). Furthermore, rapid transpiration usually reduces the water potential of the cortical cells in the roots to a level where osmotic movement of water from cortex to xylem could not be expected to occur, even if the salt were not flushed out of the xylem.

It is easy to demonstrate that intact transpiring plants can absorb water from drier soil and more concentrated solutions than root systems without tops. Renner (1929) found that although intact plants of sunflower developed absorbing forces of 4.1 to 11.7 bars, root systems alone could develop only 1.6 to 2.2 bars. Tagawa (1934) noted that bean plants could absorb water from solutions with osmotic potentials down to −14.6 bars, but root systems could absorb from solutions with osmotic potentials no lower than −1.9 bars, and this was verified by others (see Fig. 6.8). Intact plants can absorb water from much drier soil than can detopped root systems, as shown in Fig. 5.9 (Army and Kozlowski, 1951; Gračanin, 1963*a*; Jantti and Kramer, 1956; Kramer, 1941; McDermott, 1945).

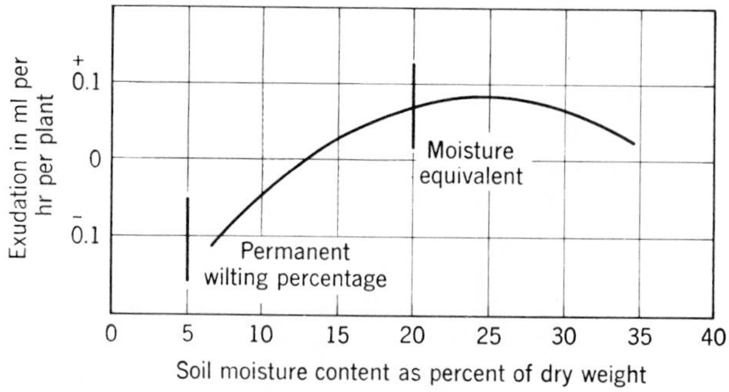

FIG. 5.9 **Relation between rate of exudation from sunflower root systems and water content of the soil. Plus values indicate exudation; minus values indicate absorption of water through stumps.** (*From McDermott*, 1945.)

The evidence available indicates that most or perhaps all of the water absorbed by transpiring plants is pulled in through the roots by passive absorption. Active absorption is measurable only in slowly transpiring plants in which active absorption tends to exceed transpiration. This fact does not decrease the importance of healthy, rapidly growing root systems. Young plants with rapidly expanding, rapidly transpiring shoots could not survive unless their root systems also expanded. Water movement in soils below field capacity is relatively slow, and a static root system would soon remove all of the available water in the soil mass it occupies (see Chaps. 2 and 6).

Only by continuous root extension into new soil can sufficient water be provided to rapidly growing plants during periods between rains or irrigations. The situation is somewhat different in well-established plants which already possess large root systems. Trees, for example, already occupy a large mass of soil, and root growth is chiefly important in replacing roots which die. There are exceptions to this, such as situations where trees are removing water faster than it is being supplied. An extreme example was the apple orchard observed by Wiggans (1936, 1937, 1938) which was removing 25 to 37 cm more water per year than was being supplied by precipitation. This was possible only as long as the roots were able to extend deeper into the soil.

In conclusion, it may be said that generally roots function as absorbing surfaces through which water is pulled passively into the plant by forces developed in the evaporating surfaces of the shoot. However, under certain circumstances—as in warm, moist soil and with low rates of transpiration—roots absorb water by an osmotic mechanism caused by accumulation of salt in the root xylem. This so-called active absorption results in the occurrence of root pressure and guttation.

Stem pressures

Considerable confusion has occurred because of failure to distinguish between exudation caused by root pressure and exudation caused by local stem pressure, wounding, or other special causes. Exudation from birch and grape is caused by root pressure, while that from maple and palm is caused by stem pressure.

Maple sap flow

The best-known example of exudation from stems in North America is the flow of sap from maple trees, chiefly from *Acer saccharum* Marsh and *A. nigrum* Michx. W. S. Clark (1874, 1875), Jones et al. (1903), Stevens and Eggert (1945), Johnson (1945), Marvin (1958), and many others have written on maple sap flow. Sap flow can occur from late autumn to early spring, any time that freezing nights are followed by warm days with temperatures above freezing; but the best flows are obtained in the spring. Over 60 percent of the flow is obtained before noon, and it often ceases in the afternoon. The yield varies widely, usually falling in the range of 35 to 70 liters per tree in a season, although twice that much is produced occasionally. The sugar content, all sucrose, is usually 2 to 3 percent, but varies from 1 to 7 percent. The distinctive flavor of maple syrup is produced by heating, which changes certain nitrogen compounds present in it (Pollard and Sproston, 1954).

The flow of maple sap is caused by stem pressure, not by root pressure. Sap flow will occur from isolated stem segments, and Stevens and Eggert (1945) obtained sap flow from isolated tree trunks placed in tubs of water and subjected to alternating temperatures. Simultaneous measurements of root and stem pressures indicated that no root pressure can be detected at the time maple sap is flowing. In contrast, root pressure is always observed at times when sap flow occurs in birch and grape, and it appears to be caused directly by root pressure. Marvin (1958) reviewed the literature on maple sap flow and added additional observations. Although it is clearly related to temperature changes, he doubts if it can be explained solely by thermal expansion in warming stems.

Sap flow in other species

In the tropics large quantities of sap are obtained from palms, usually by cutting out the inflorescence. Molisch (1902) attributed such flow to the effects of wounding. In Mexico the young inflorescence is cut out of agave plants, and a cavity is scooped out of the stem in which sap collects. Although Sachs attributed this sap flow to root pressure, MacCallum (1908) states that *Agave americana* flowers in soil so dry that no water could be absorbed, and the flower bud removes water from other parts of the plant. Apparently, water is moved to the stem tip from the leaves. The mechanism of sap flow in *Agave* deserves further investigation. MacDougal (1925, 1926) found that pressures of several bars were developed for a few days after holes were bored in stems of cacti and *Pinus radiata,* but these pressures soon disappeared. He regarded these as local in origin, and not caused by root pressure. It is evident that we need to learn more about the causes of the exudation pressures developed in wounded stems.

Gas pressure

Occasionally, flow of liquid from cracks and wounds, called slime flux, is observed. This is associated with a water-soaked condition of the heartwood, caused by bacterial activity. Carter (1945) observed gas and liquid pressures of 0.3 to 1.0 bar in trees suffering from slime flux. The gas contained as much as 46 percent methane. Abell and Hursh (1931) and others have reported instances in which gas pressures were encountered in oak trees high enough to blow the core out of an increment borer. Liquid and combustible gases were forced out of the holes. Gas pressures apparently are caused by microorganisms in the decaying heartwood.

Summary

The absorption of water occurs when there is a decreasing gradient in water potential from the soil or solution surrounding the root to the root xylem. The rate at which water absorption occurs depends on the magnitude of the gradient in water potential and the resistances to water flow in the soil and the roots. Resistance to water movement in the soil depends chiefly on its water content; resistance in roots depends chiefly on the degree of suberization and on the physical condition of the protoplasm and its resistance to water movement. The latter depends on factors such as aeration and temperature.

In rapidly transpiring plants the pressure in the xylem sap is often below atmospheric pressure, and water is pulled in through the roots, which act as passive absorbing surfaces. This is termed passive absorption. In slowly transpiring plants growing in a medium with a high water potential, roots often behave as osmometers and develop root pressure which sometimes results in guttation. This is termed active absorption, although it does not involve active transport of water.

The occurrence of root pressure has been attributed to secretion or active transport of water into the stele, electroosmosis, and osmosis. It is believed that root pressure exudation is a purely osmotic process, depending on the accumulation of sufficient salt in the xylem to lower the water potential of the xylem sap below that of the substrate. The reduction in root pressure exudation caused by respiration inhibitors, deficient aeration, and low temperatures is attributed to reduction in salt accumulation and changes in permeability rather than to direct effects on any nonosmotic water absorption mechanism. Among the unsolved problems are the cause of diurnal periodicity in exudation and the mechanism by which salt is accumulated in the xylem. The synthesis of organic compounds in roots and their transfer into the xylem also presents some interesting problems.

Root pressure is not regarded as an essential factor in the absorption of water, because the volume absorbed by this mechanism is much less than the volume required by transpiring plants and the pressure in the xylem sap of transpiring plants is usually below atmospheric pressure. Furthermore, many kinds of plants never exhibit positive pressure in the stem xylem, even though it may occur in individual roots.

It is believed that all the water absorbed by moderately and rapidly transpiring plants is absorbed passively because of the reduction in water potential caused by the tension produced in the xylem sap.

Stem pressures sometimes cause exudation from wounds in sugar maples, palms, and a few other species, quite independently of root pressure. The cause of this type of exudation is not fully understood.

six

Factors Affecting the Absorption of Water

The various factors affecting the absorption of water can be classified into two groups: (1) those affecting the driving force or gradient in water potential from soil to roots and through the root tissues into the xylem, and (2) those affecting the resistance to water movement in the soil and through the roots. Thus water absorption can be described by the following equation:

$$\text{Absorption} = \frac{\Psi_{\text{soil}} - \Psi_{\text{root surface}}}{r_{\text{root}}} = \frac{\Psi_{\text{root surface}} - \Psi_{\text{root xylem}}}{r_{\text{soil}}} \tag{6.1}$$

Factors such as soil texture and hydraulic conductivity affect water absorption because of their effects on the resistance to water movement through the soil to the root surface. Factors such as soil aeration, temperature, and degree of suberization of roots affect absorption chiefly by modifying the resistance in the roots. The driving force is the difference in water potential between the bulk soil and the root surface and between the root surface and the xylem sap.

Efficiency of roots as absorbing organs

The effectiveness of roots in water and salt absorption depends chiefly on their extent and their permeability.

Extent of root surface

There are wide differences among plants of different species with respect to the depth, spread, and amount of branching, and therefore, the extent of their root surfaces (see Fig. 4.17). The larger the volume of soil occupied by a root system, the larger the volume of water available to it and the longer the plant can survive without replenishment of soil water by rain or irrigation. Miller (1916) attributed the greater drought resistance of sorghum to the fact that it has nearly twice as many fine roots as corn. Slatyer (1955) likewise reported that sorghum has a better-developed root system than cotton or peanuts. Sorghum also maintains a higher level of turgor than cotton or peanuts when subjected to drought. This, presumably, is largely because sorghum has a more extensive absorbing surface, although better control of water loss may also be a factor.

There has been some question whether roots at a distance from the stem of a plant are as effective in absorption as those nearby. It is usually assumed that the removal of water from soil is more closely related to the concentration of roots in the soil than to distance from the plant (Aldrich, Work, and Lewis, 1935; Veihmeyer and Hendrickson, 1938*b*), and if the soil is uniformly occupied by roots, water should be absorbed uniformly throughout the root zone. This assumption is difficult to prove because of the difficulty of determining whether root distribution really is uniform. Davis (1940) reported that after the soil near the plant was reduced to permanent wilting, roots in moist soil at a distance of 1 m or more were unable to prevent the plants from wilting. Wind (1955*b*) also claimed that water could move through the soil more rapidly than through the roots of certain grasses, but this seems to be an exceptional situation. In general, it appears that distance from the plant is not as important as extent of contact of roots with soil containing available water. Veihmeyer and Hendrickson (1938*b*) found that water 6 or 7 m from fruit trees is absorbed. Research with forest trees also indicates that water is absorbed from distances of several meters as readily as from nearby if roots extend far out from trees. Reimann et al. (1946) reported that although by early August corn had exhausted all available water in the upper meter of soil and was absorbing from a depth of 1 to 2 m, it was still growing well. McWilliam and Kramer (1968) found that plants of *Phalaris tuberosa* L.

survive when the water potential in the upper meter of soil is below −15 bars because some of the roots penetrate to deeper horizons which contain readily available water. They also found that when the deep roots of plants growing in dry surface soil were cut 1 m below the soil surface, the shoots died. It was demonstrated in laboratory experiments that both water and ^{32}P were absorbed from a depth of more than 1 m and translocated to the shoot (see Fig. 6.1).

The majority of the evidence indicates that the entire root system need not be in soil above the permanent wilting percentage for survival or even for

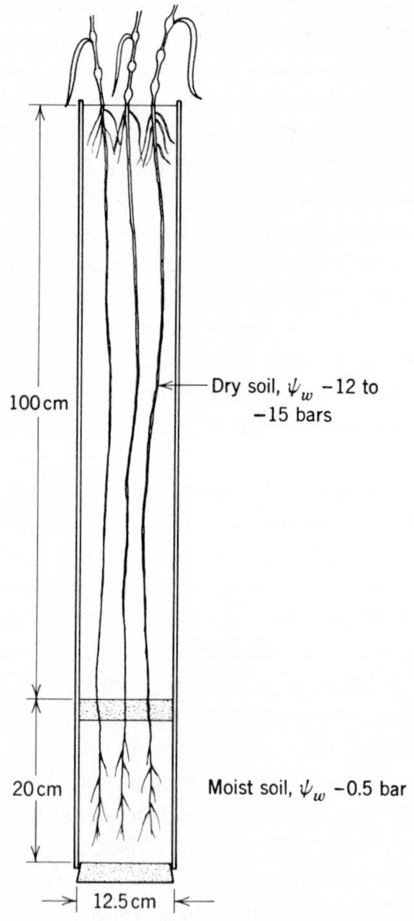

FIG. 6.1 Equipment used to study uptake of water and salt by deep roots of *Phalaris tuberosa.* The lower compartment of the tube was separated from the upper compartment by a rubber stopper containing holes through which roots grew into the lower compartment. Watering the upper compartment was discontinued, and after the soil in that part of the tube had dried to −12 or −15 bars, tritiated water or ^{32}P was supplied to the roots in the lower compartment. (*McWilliam and Kramer,* 1968.)

limited growth. Taylor et al. (1934) concluded that plants can absorb enough water to replace transpiration losses if one-fourth to one-half of their root system is in soil above the permanent wilting percentage. Furr and Taylor (1939) stated that to maintain good growth of lemon trees the soil in half of the root zone must be kept above the permanent wilting percentage. This is in accord with the view expressed later in this chapter that most plants have root systems larger than are necessary to provide them with water.

The pattern of water absorption from soil is different for rapidly growing annuals and established perennials. Annual plants tend to absorb the soil moisture near the base of the plant approximately down to the permanent wilting percentage, and the volume of drying soil enlarges as the roots extend out laterally and vertically. If no water is added, absorption occurs from increasing depths beneath the crop as the season progresses. This pattern of water absorption was found for corn by Russell et al. (1940) and is shown for tomato in Fig. 6.2. In unirrigated soil continual extension of roots into previously unoccupied soil is essential to survival because it makes water available. Trees and other perennial plants have a different pattern of water absorption because they start the season with a root system which already occupies a large volume of soil. Observations by Hendrickson and Veihmeyer (1942) and others indicate that trees in soil near field capacity may begin to extract water from depths of 2 m or more. However, because the root concentration is usually higher near the trunk and in the surface soil, soil water content is often reduced in those regions before much is absorbed from

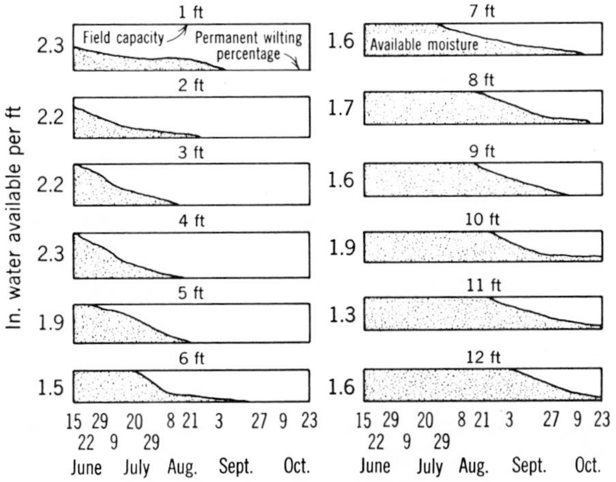

FIG. 6.2 Rate of water extraction from an unirrigated soil by tomato plants. Water was progressively extracted from deeper levels as the roots penetrated into them. Growth slowed by early August when most of the readily available water had been removed from the upper 6 ft of soil. A total of 21 in. of water was removed by October 23. This plot yielded only half as much as an irrigated plot. (*From Doneen and MacGillivray*, 1946.)

deeper soil horizons. The presence of roots in soil horizons of various water potentials makes it very difficult to give an accurate summary of the potential of the water which is being absorbed by a given plant.

Deep-rooted species often survive droughts which cause serious injury to shallow-rooted species, and soil conditions favorable to deep penetration of roots are important factors in increasing drought resistance. Trees on shallow soil suffer more serious injury during droughts than those on deep soil because their roots do not occupy a sufficiently large reservoir of soil to keep them supplied with water. Several of the "dieback" diseases of forest trees seem to result from attempts to grow them on shallow soil or on soil unfavorable for root penetration and development.

In terms of mere survival, many vigorously growing plants have an excess of root surface. Trees sometimes survive removal of half or more of their root systems, and considerable portions of the root systems of crop plants are often removed by cultivation without obvious permanent injury. Weaver and Zink (1946) were able to remove half or more of the root systems of certain species of grasses without serious injury. Such drastic surgery usually reduces growth temporarily because of the increased water stress caused by the reduced absorbing surface. Many roots are lost during transplanting, and the leaf areas of transplanted trees and shrubs are usually reduced by pruning or application of transpiration-reducing substances to compensate for the reduced absorbing surface. However, root pruning is not always disastrous. Andrews and Newman (1968) found that removal of 60 percent of the roots from wheat plants did not measurably reduce growth, either in drying soil or in soil kept at field capacity.

Nutman (1934) regarded the root surface of coffee trees as being a limiting factor in water absorption. This is true in the sense that the resistance to water movement into roots is always so high that even very extensive root systems may not prevent occurrence of transient midday wilting in rapidly transpiring plants (Fig. 10.1). It has been suggested that because of its high resistance to water flow the limiting surface for water absorption is really the endodermis or surface of the roots (Scott and Priestley, 1928; Nutman, 1934). Because of the small area of the xylem vessels in roots of coffee trees, Nutman (1934) calculated that water must enter the xylem 170 times as rapidly as it crosses the root surface. This seems excessively high. A suberized yellow-poplar root with a diameter of 2.8 mm had a stele with a diameter of 0.8 mm. In this root the velocity at the surface of the stele was 3.5 times that at the root surface. In secondary roots of *Vicia faba,* the circumference of the epidermis is about four times that of the endodermis, and water must cross the endodermis four times as rapidly as it crosses the epidermis. Nearer the tips of these roots the tangential surface of the xylem points is only about 6 percent of the root circumference, and water would presumably enter the xylem at this point with a velocity 16 times that at which it crosses the epidermis. However, the resistance to entrance into xylem vessels is quite low and should not restrict flow seriously.

Permeability of roots

A typical root system consists of roots in various stages of differentiation, from newly formed tips to fully matured secondary roots which have lost their epidermis and cortex and are enclosed in a layer of suberized tissue. Obviously roots varying so much in structure must also vary widely in permeability. Some measurements of rates of water entry into various kinds of roots are given in Table 6.1. These data were obtained by enclosing the

Table 6.1 Rates of Entrance of Water into Roots

Investigator	Material and condition	Observed rate	Rate, $mm^3/(cm^2)(hr)$
Hayward, Blair, and Skaling (1942)	Corn, young roots in water	$0.2\ mm^3/(mm^2)(hr)$	20.0
Rosene (1941)	Onion, young roots in water	Max. of $84 \times 10^{-4}\ mm^3/(mm^2)(min)$	50.4
Rosene (1943)	Radish, root hairs	Max. of $31 \times 10^{-4}\ mm^3/(mm^2)(min)$	18.6
Hayward, Blair, and Skaling (1942)	Sour orange, suberized roots in water	$0.3\ mm^3/(mm^2)(6\ hr)$	5.0
Kramer (1946)	Shortleaf pine, suberized roots in water	$3.37\ mm^3/(cm^2)(hr)$	3.37
Nutman (1934)	Coffee tree, entire root system in soil	$2.5\ ml/(m^2)(hr)$	0.25

roots in potometers, except those for coffee, which were calculated from the rate of transpiration and the estimated root surface. These are not really measurements of permeability because they were made under widely varying conditions and rates of transpiration, and in most instances the water potential gradient or driving force across the roots is unknown. Nevertheless, with one exception, the highest rates of water entry found were for root hairs and unsuberized roots, the lowest rates being those for suberized woody roots, as would be expected.

Textbooks usually emphasize the role of young, unsuberized roots in absorption and ignore the possibility of absorption through suberized roots. However, as pointed out in Chap. 5, considerable absorption of water and salt must occur through suberized roots, especially in woody perennials which possess large permanent root systems with a small percentage of unsuberized root surface. Observations made by Kramer and Bullock (1966) indicated that usually less than 1 percent of the root surface in the upper 12.5 cm of soil under pine and yellow-poplar was unsuberized. Even though the suberized roots are less permeable than unsuberized roots, they must play an important role in absorption simply because of their extensive surface.

The permeability of suberized roots varies widely because of variations in thickness and structure of bark, breaks caused by death of small branch roots, and variations in numbers of lenticels. Bullock and Kramer (1966) found the water uptake rate of suberized yellow-poplar roots under a pressure of 0.4 bar to vary from essentially zero to 30,000 $mm^3/(cm^2)(hr)$. The permeability of loblolly pine roots varied from 6.6 $mm^3/(cm^2)(hr)$ for roots 1.33 mm in diameter to 36.6 $mm^3/(cm^2)(hr)$ for roots 3 mm in diameter and 178 $mm^3/(cm^2)(hr)$ for unsuberized roots.

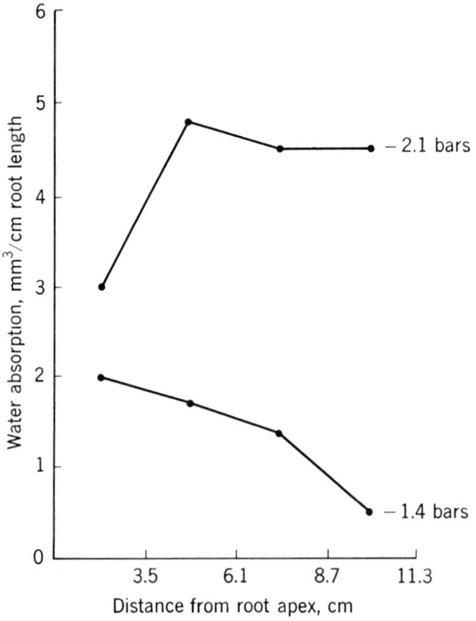

FIG. 6.2a Shift in region of root through which most rapid absorption occurred when the rate of transpiration was increased by turning on a light. In darkness the water potential in the root was estimated to be −1.4 bars; in the light, −2.1 bars. The rate of absorption is given in cubic millimeters per centimeter of root length per hour. (*After Brouwer*, 1953.)

It was noted in Chap. 4 that when the rate of transpiration increases, the region through which maximum water absorption occurs is shifted toward the base of the root (Brouwer, 1953; Sierp and Brewig, 1935; Soran and Cosma, 1962). This shift is shown in Fig. 6.2*a*. It has been suggested that increased tension in the xylem causes increased permeability of the basal regions of roots. No doubt the application of an increasing pressure gradient across the roots causes movement of water through smaller and smaller pores until it is moving through all possible pores (Hylmö, 1958). Apparently, more pathways are available for water movement under high pressure in the basal regions than near the tips of roots. It also seems likely that much of the water absorption by very slowly transpiring plants is active absorption, but an increasing fraction is by passive absorption as the rate of absorption increases. The change from active to passive absorption as the dominant mechanism may also be a factor in the shift in the region of maximum absorption.

Root systems subjected to severe water stress show decreased permeability to water for several days after rewatering (Kramer, 1950). Root systems exposed to high concentrations of salt also show reduced permeability to water (Klepper, 1967). In both instances permeability is probably reduced by inhibition of root extension and increased suberization of the existing roots. Skidmore and Stone (1964) and Barrs and Klepper (1968) reported

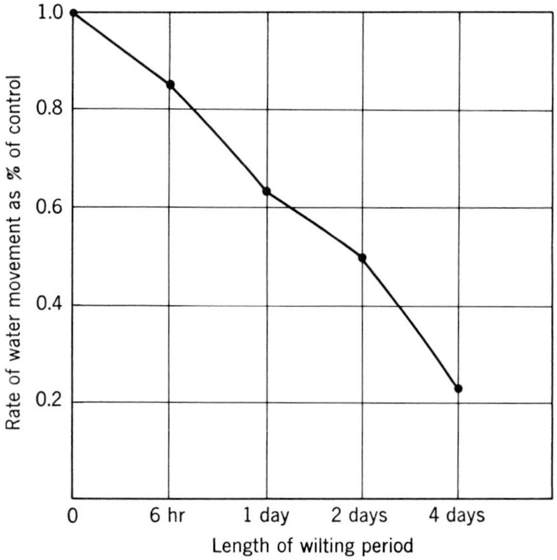

FIG. 6.3 **Effect of wilting for various periods of time on root resistance as measured by rate of water movement through tomato root systems under a vacuum of 40 cm Hg, expressed as percentage of rate through unwilted control plants. Plants were subjected to wilting for the indicated time periods, then detopped and attached to a vacuum line; water movement through them was then measured.** (*Based on data of Kramer*, 1950.)

puzzling diurnal variations in root permeability of cotton, with highest permeability near noon and lowest near midnight. No satisfactory explanation of this variation has been proposed.

Metabolic activity and water absorption

Some writers have suggested that there is a direct relationship between root respiration and water absorption. For example, Henderson (1934) claimed that there was a relationship between carbon dioxide production, oxygen uptake, and the absorption of water by roots of corn seedlings. On the other hand, Wilson and Kramer (1949) found no correlation between respiration and water absorption by roots of tomato plants and concluded that Henderson's data showed no correlation, either. Loweneck (1930) likewise found no direct relationship between root respiration and water absorption. Experiments by Skoog et al. (1938) showed no correlation between rate of root respiration and rate of exudation from stumps of detopped plants, indicating that active absorption was not directly correlated with root respiration.

It seems clear that passive absorption of water by roots of rapidly transpiring plants is not directly dependent on the expenditure of energy released by respiration. Neither is active absorption by very slowly transpiring

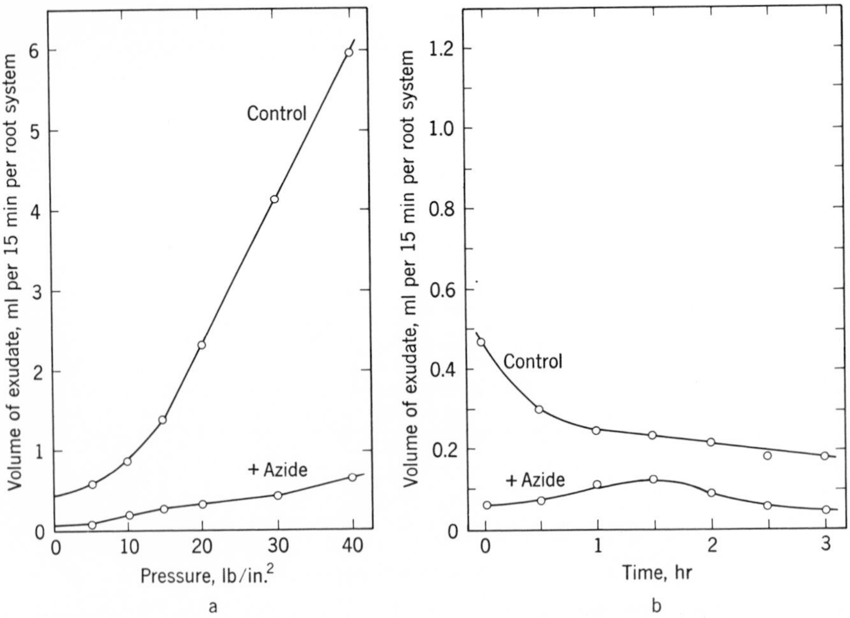

FIG. 6.4 Effect of $10^{-3}M$ sodium azide on water movement through tomato root systems. Treated roots were in azide for 1 hr before the measurements started. (*A*) Rate of water movement through root systems subjected to pressure. (*B*) Rate of root pressure exudation. (*From Lopushinsky*, 1964.)

plants. However, active absorption does depend on the accumulation of salt in the root xylem, which depends on expenditure of respiratory energy, as described in Chap. 7.

It has been shown by a number of workers that treatment of roots with respiration inhibitors such as azide, cyanide, and dinitrophenol drastically reduces both active and passive absorption of water, as shown in Fig. 6.4 (Brouwer, 1954; Lopushinsky, 1964). The reduction in passive absorption apparently results from increased resistance to water movement into roots. Active absorption is reduced by the reduction in salt accumulation in the xylem as well as by increased root resistance. An excess of carbon dioxide in the root environment also drastically reduces passive water absorption (see Fig. 6.19), chiefly because it increases the resistance to water flow into roots, as shown later in this chapter.

Protoplasmic differences

There are significant differences in the reaction of the protoplasm of roots of various species to aeration, temperature, and other factors. The protoplasm of roots of some species seems much more tolerant of deficient aeration than that of other species. Roots of cypress and tupelo gum can grow in permanently saturated soil, but roots of dogwood, yellow-poplar, and many other species are killed by relatively short periods of flooding. Deficient aeration

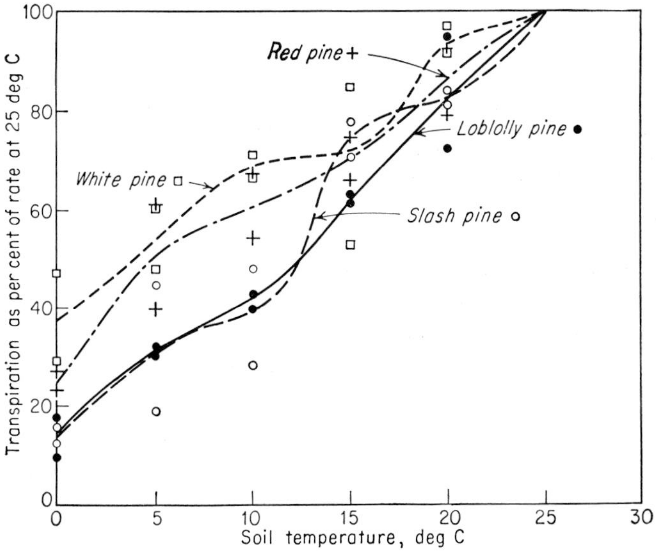

FIG. 6.5 **Effect of soil temperature on water absorption by four species of pine seedlings as determined from the rate of transpiration. Absorption was reduced more by low temperature in the southern species (slash and loblolly pine) than in the northern species. (*From Kramer and Kozlowski,* 1960.) See also Fig. 6.14 for other examples of species differences.**

reduces absorption of water much more in some species than in others, as shown in Fig. 6.17. The complex nature of these differences in tolerance of deficient aeration are discussed in Chap. 4 and later in this chapter.

The differences in effect of low temperature on water absorption through roots of various species, shown in Fig. 6.5, suggest that low temperature affects the protoplasm of cotton and watermelon roots more than that of collard roots. It has also been reported that if root systems are cooled slowly to 5°C over a period of several days, water absorption is reduced less than if they are cooled this much in a few hours, as shown in Fig. 6.6 (Böhning

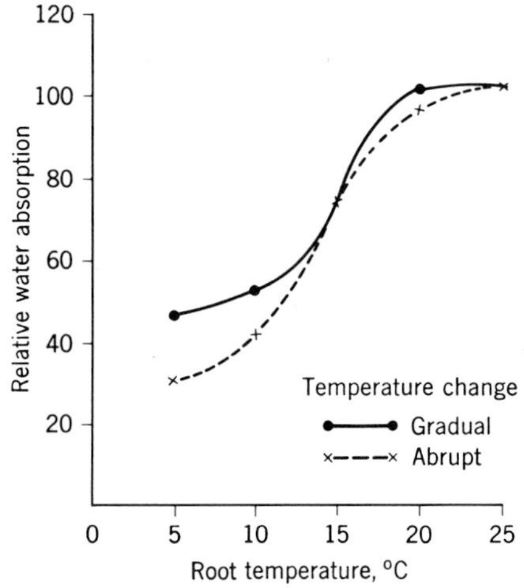

FIG. 6.6 Comparison of effects of gradual and abrupt changes in root temperature on water absorption of red kidney bean plants. Each point on the curve for an abrupt change shows absorption by different groups of four plants, whose roots were abruptly cooled from 25°C to the indicated temperatures. The curve for gradual cooling represents a single group of plants cooled gradually over a period of 13 days from 25 to 50°C. (*From Böhning and Lusanandana,* 1952.)

and Lusanandana, 1952). Kuiper (1964) reported that bean root systems grown at 17°C showed greater water uptake at low temperature than those grown at 24°C, and this increased ability to absorb water at low temperatures appeared in about 36 hr after transfer of the root systems from 24 to 17°C. This suggests that compensatory changes occur in the protoplasm of slowly cooled roots which prevent the severe reduction in permeability observed in rapidly cooled roots. These might be similar to the changes in permeability during cold hardening reported by Levitt and Scarth (1936).

Availability of soil water

Although midday plant water deficits are common even in soil near field capacity when atmospheric conditions favor rapid transpiration, long-term water deficits result from decreasing soil water potential. The availability of soil water depends chiefly on its potential and on the hydraulic conductivity of the soil, both of which are closely related to the water content of the soil, as shown in Chap. 2 (see Fig. 2.12).

Soil water

As stated earlier, the water readily available to plants is often designated as that occurring in the range between field capacity and permanent wilting percentage. The limitations of this terminology are discussed in Chap. 2, but it remains a useful concept for practical purposes. As shown in Fig. 6.7, the

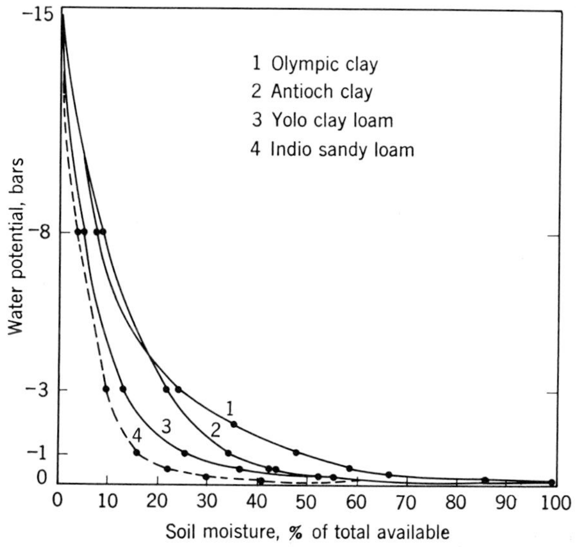

FIG. 6.7 **Percentage of available water remaining in four soils at various soil water potentials. Curves were obtained from curves for soil water potential over soil water content by assuming that available water occurs in the range from −0.15 to −15.0 bars. (*After Richards and Wadleigh,* 1952.)**

amount of available water varies widely in different soils. This is also shown in Table 6.2, where it will be noted that the readily available water storage per 30 cm of depth varies from 4.3 to 8.6 mm in fine sand to 77.0 mm in a clay. Obviously, plants growing in soils that have a low storage capacity, such as Oakley fine sand or Aiken clay loam, will exhaust the readily available water and suffer from drought much sooner than plants growing in soils with a high storage capacity, such as the Catherine loam or Wooster silt loam. The limitation in water storage capacity becomes particularly important for shallow-rooted plants or plants growing on shallow soils underlain by rock or hardpan layers impermeable to roots.

*Table 6.2 Storage Capacity of Various Soils for Readily Available Water**

Soil type	Moisture equivalent	Permanent wilting percentage	Mm of available water per 30-cm depth
Oakley fine sand	3.29	1.33	8.6
Yolo fine sandy loam	16.80	8.93	32.0
Aiken clay loam	31.12	25.70	17.7
Salinas silt clay loam	28.33	12.49	63.2
Salinas clay	34.50	16.80	70.7
Catherine loam	37.90	19.03	77.0
Wooster silt loam	23.36	6.12	72.2
Brockton clay loam	24.51	11.55	49.5
Plainfield fine sand	2.40	1.36	4.3

* *From MacGillivray and Doneen (1942)*

If the water content of soil exceeds its field capacity, air is displaced from the remaining pore space, and aeration can become a limiting factor. If the water content falls much below field capacity, water begins to be held so firmly by the matric forces that its potential decreases rapidly and it becomes less and less available to plants. However, soil water does not become completely unavailable to all plants at the same soil water potential. Some kinds of slowly transpiring plants can absorb water at potentials much lower than the −15 bars usually set as the permanent wilting percentage (Slatyer, 1957).

There has been considerable unprofitable discussion concerning the relative availability of water to plants over the range from field capacity to permanent wilting percentage. Veihmeyer and his colleagues claimed that water either is or is not available and that growth and transpiration are not reduced materially until the soil water content falls nearly to the permanent wilting percentage (Veihmeyer, 1956; Veihmeyer and Hendrickson, 1950). The more soundly based view, that water becomes decreasingly available as the soil moisture content decreases below field capacity, has been presented by Richards and Wadleigh (1952), Stanhill (1957), and others.

Hagan, Vaadia, and Russell (1959) discussed the difficulties of interpreting studies of the effect of soil water on growth and yield of plants and pointed out that water stress affects various plant processes differently. For example, total yield of dry matter is usually decreased by even moderate water stress, but the yield of seed or some other desired product may be little affected, or even increased, as is the rubber content of guayule (Fig. 10.21). Much of the discussion concerning the effects of moderate soil water stress on plant growth resulted from the mistaken belief that plant growth should be consistently reduced by some predictable level of soil water stress. As Hagan (1955) pointed out, this is very improbable. Gardner and Nieman (1964) state that no common lower limit of water availability can be defined for transpiration, cell division, or cell enlargement. Furthermore, water content at permanent wilting was not a lower limit for any of these processes. The physiological processes which control plant growth are controlled directly by the plant water stress and only indirectly by soil and atmospheric water stress (see Chap. 10). It is not safe to assume that a certain level of soil water stress is accompanied by an equivalent level of plant water stress. Plants in moist soil may be subjected temporarily to high stress during periods of rapid transpiration, and plants in relatively dry soil may be subjected to relatively low water stress in atmospheric conditions in which the rate of transpiration is low. Thus the effects of various levels of soil moisture on plant growth can be evaluated accurately only if plant water stress is measured during the experiment, as was done by Boyer (1965), Brix (1962), and Slatyer (1957). The measurement of water stress is discussed in Chap. 10.

Gradient of water potential from soil to roots

The factors affecting the steepness of the gradient of water potential from soil to roots will now be discussed in more detail.

ROOT WATER POTENTIAL. The roots of slowly transpiring plants often behave as osmometers, and their water potential depends chiefly on the

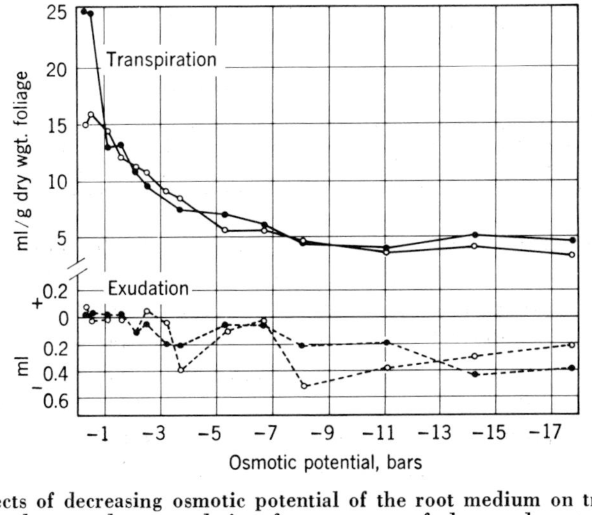

FIG. 6.8 Effects of decreasing osmotic potential of the root medium on transpiration of intact tomato plants and on exudation from stumps of detopped root systems. Root systems in an osmotic potential lower than −2 bars absorbed water through their stumps instead of exuding sap. (*After Army and Kozlowski,* 1951.)

concentration of salt in the xylem sap. The osmotic potential of the xylem sap is rarely lower than −2 bars. As a result, active water absorption seldom occurs from soils or solutions with a water potential lower than −2 or −3 bars (see Fig. 6.8).

As explained in Chap. 5, passive absorption by actively transpiring plants is brought about by a decrease in water potential developed at the evaporating surfaces of the shoots and transmitted through the cohesive water columns of the transpiration stream to the root surfaces. The magnitude of the decrease in water potential varies from 1 to 2 bars when transpiration is negligible to perhaps 15 or 20 bars at noon on sunny days, or even 50 to 100 bars in xerophytic plants subjected to severe water deficits. Some measurements of leaf and root water potentials of pine seedlings growing in drying soil are shown in Fig. 6.8*a*.

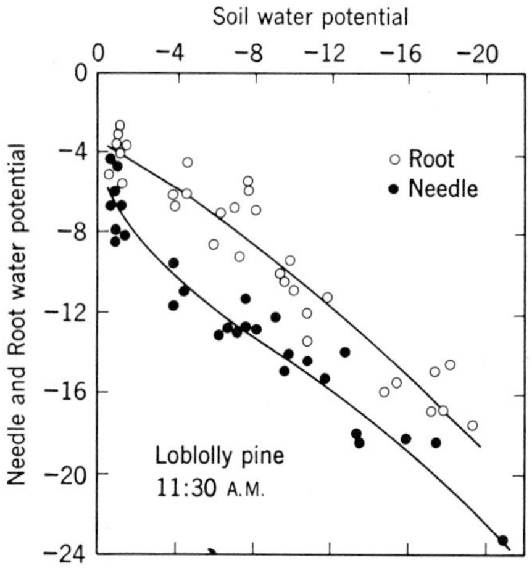

FIG. 6.8a **Relationship between soil water potential and needle and root water potentials of loblolly pine sampled at 11:30 A.M. Shoot water potentials were always lower than root water potentials, but in drying soil, root and soil water potentials were similar, indicating conditions unfavorable for water absorption.** (*From Kaufmann*, 1968.)

Measurements of root water potential have been made by Slavikova (1964, 1967), who reported that the water potential in roots of plants in moist soil usually increases significantly from the base toward the apex. However, the gradient is reversed in a root in dry soil attached to a plant with the remainder of its roots in moist soil, indicating that water is translocated from roots in moist soil to roots of the same plant in dry soil. There are many reports in the literature of the translocation of water through root systems from moist to dry soil (Breazeale and Crider, 1934; Hunter and Kelley, 1946; Müller-Stoll, 1965; and others). Bormann (1957) even demonstrated transfer of sufficient water through intertwined roots to keep sunflower plants in dry soil alive. McWilliam and Kramer (1968) reported the presence of small living roots in dry soil at the base of the crown of *Phalaris tuberosa* L. which are apparently supplied with water from deep roots. These roots are surrounded by a thin layer of moist soil, indicating that water moves out of them to the surrounding dry soil.

SOIL WATER POTENTIAL. As stated in Chap. 2, the water potential of the soil is affected by gravitational, hydrostatic, and surface forces and by the presence of solutes. The factors can be indicated by the following equation from Chap. 2:

$$\Psi_{soil} = \Psi_m + \Psi_p + \Psi_s + \Psi_g \qquad (2.1)$$

where Ψ_m is the matric potential, Ψ_p the pressure potential, Ψ_s the solute potential, and Ψ_g the gravitational component. In agricultural soils the concentration of solutes is usually low enough so that Ψ_s is a minor component of the total soil water potential, which is controlled by the matric potential. The role of solutes in connection with water absorption will be discussed later in this chapter.

The water potential at the root-soil interface is assumed to be the principal soil characteristic controlling the availability of water for plant growth (Gardner, 1960; Kramer, 1956; Slatyer, 1957). This value is influenced both by the water potential in the soil mass, Ψ_{soil}, and by the gradient from the soil mass to the soil in immediate contact with the root, which develops as a result of water absorption by the roots. Its magnitude is controlled by the relative rates of flow to and removal by the roots, and with a steady rate of water uptake it reaches a constant value inversely proportional to the hydraulic conductivity of the soil. Some measurements of soil and plant water potentials are shown in Fig. 6.8*a*.

Hydraulic conductivity of soil

Different soils exhibit different hydraulic conductivities. Consequently, the same bulk soil water potential is often associated with different rates of flow of water to roots. For example, Peters (1957) attempted to separate the effects of water potential from water content by preparing mixtures of various proportions of sand and silty clay loam which contained quite different amounts of water at the same potential. He found that root elongation of corn seedlings was reduced with decrease in water potential over the range used (−0.33 to −8.0 bars), but it was also reduced as the soil moisture content at a given potential decreased. Miller and Mazurak (1958) grew sunflowers in a series of 20 soil fractions ranging in size of particles from 4,760 to 2.3 μ in diameter, with pore diameters from 529 to 2.3 μ. The various particle sizes were all maintained at the same water potential. Best growth occurred in particles of intermediate size, in a series maintained at a poten-

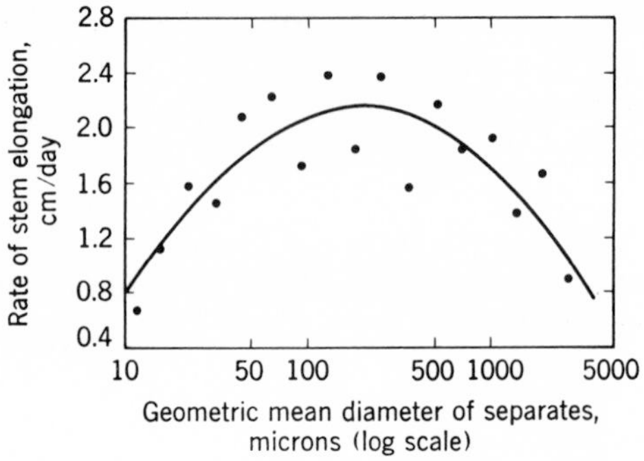

FIG. 6.9 Rate of elongation of stems of sunflower plants growing in soils of different average particle sizes, ranging from 3,360 to 9.25 μ in average diameter. All systems were maintained at a water tension of 20 cm. (*From Miller and Mazurak*, 1958.)

tial of −20 cm of water, as shown in Fig. 6.9. In very coarse particles growth was limited by inadequate supply of water to the roots resulting from low hydraulic conductivity. This is caused by lack of sufficient contacts between the roots and the relatively small volume of solution surrounding the particles and filling the smaller pores. In the fine particles where nearly all the pore spaces were filled with solution, inadequate aeration became a limiting factor.

The essential fact is that as the soil dries, the number and size of channels through which water can move toward roots decreases rapidly (Philip, 1957*a*), reducing the amount of water which can be moved per unit of time with a given gradient of potential. As transpiring plants wilt permanently when $\Psi_{\text{plant}} = \Psi_{\text{soil}}$, the degree to which the water potential gradient from soil to root surface can be increased is limited by the value of Ψ_{plant} at which wilting occurs.

Movement of water to roots

So many factors are involved in the movement of water to roots that it has been difficult to develop an equation to describe it. The root systems of actively growing plants continually occupy new regions of soil, and there is more rapid absorption in the more permeable regions of roots, hence the soil mass occupied by a root system often consists of several zones of different water contents. As a result most of the water absorption by a given root system may occur from a small fraction of the soil mass occupied by it. This situation is accentuated if rain or irrigation recharges only part of a root zone from which most of the readily available water has been removed. Under such conditions the roots are in contact with a wide range of soil and root water potentials, and it is probably impossible to determine directly the effective value for the root system as a whole. However, the root system itself tends to integrate the range of soil water potentials encountered, and the value of Ψ_{root}, measured at the base of the stem, probably gives a reasonable estimate of Ψ_{root} in the wettest area of the root zone where most of the water is being absorbed (Slatyer, 1960).

Flow rates in soil

Although the problem of soil water flow to roots is complicated, simple models such as those proposed by Philip (1957*a*) and Gardner (1960) appear to give good agreement with experimental data. The model developed by Gardner (1960) treats the root as an infinitely long cylinder of uniform radius and water-absorbing properties, and soil water is assumed to move only radially toward the root. He assumes that as soil water content decreases and Ψ_{soil} falls, Ψ_{root} must decrease still more to maintain water movement. This is because the hydraulic conductivity also declines and a larger gradient from soil to root is required to maintain the same rate of flow in dry soil than in moist soil. Finally Ψ_{root} declines to an injurious level, and absorption practically ceases when Ψ_{soil} declines to this value. By solving equations based on this model, Gardner (1960) obtained the information shown graphically in Figs. 6.10 to 6.13.

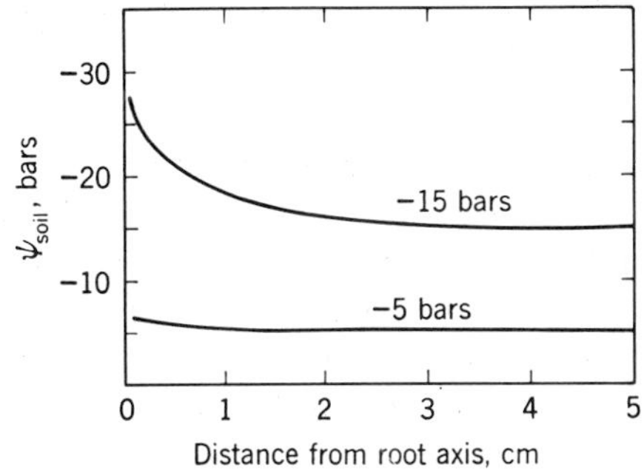

FIG. 6.10 **Estimated gradient in soil water potential Ψ_{soil} from soil mass to root surface in a Pachappa sandy loam at soil water potentials of −5 and −15 bars at an assumed rate of absorption of 0.1 ml/cm of root length per day. (*After Gardner*, 1960.)**

In Fig. 6.10 the change of Ψ_{soil} as a function of distance from the root is shown for Pachappa sandy loam. The initial values of Ψ_{soil} are taken as −5 and −15 bars, and a flow rate of 0.1 ml/cm of root length per day is assumed (Ogata, Richards, and Gardner, 1960). Gardner has shown that the distance from the root at which any given value of Ψ_{soil} occurs increases as the square root of time. Consequently, the distance from which water can be expected to move can be estimated from the total time allowed for movement. For a time period which would be effective in supplying the assumed plant demand, water can be expected to move about 4 cm to a root surface. The hydraulic conductivity of the soil and the transpiration demand jointly determine the optimum root density for a given situation. As shown in Fig. 6.10 a gradient of only about 2 bars is required to move water at the specified rate for $\Psi_{soil} = -5$ bars, but at −15 bars a gradient of 13 bars is required because of the decreased hydraulic conductivity.

The estimated difference in potential from soil to root ($\Delta\Psi$) required to produce a given rate of flow at various water potentials is shown in Fig. 6.11.

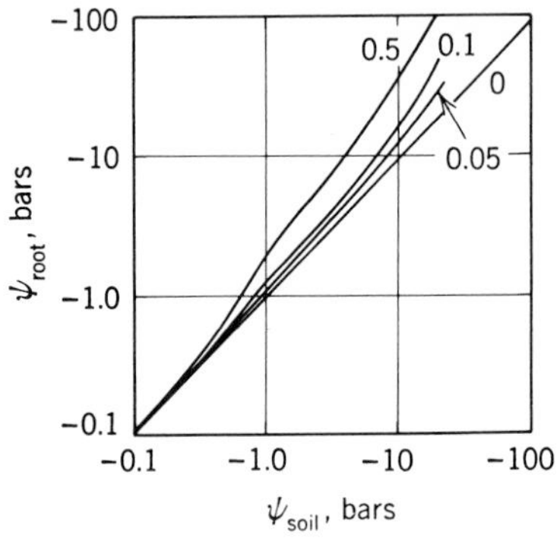

FIG. 6.11 Estimated root water potentials required to cause water movement from soil to roots at rates of absorption of 0, 0.05, 0.1, and 0.5 ml/cm of root length per day at various soil water potentials in a Pachappa sandy loam soil. (*After Gardner*, 1960.)

The value of $\Delta\Psi$ depends on the hydraulic conductivity *K* and on the actual flow rate required. Because *K* decreases with decreasing Ψ_{soil}, the value of $\Delta\Psi$ required to increase the flow above 0.1 ml/day, although initially small, increases rapidly as Ψ_{soil} drops below −5 bars. This estimate may be based on a root density much lower than that usually found beneath crop plants. With increasing root density, both the flow rate per unit of root surface and the distance water must move decrease; hence the $\Delta\Psi$ required to maintain flow is lower.

LIMITING INFLUENCE OF PLANT WATER POTENTIAL. Assuming that wilting occurs when Ψ_{root} decreases to a certain value, set here at −20 bars, the relationship of the flow rate to the onset of wilting can be demonstrated, as shown in Fig. 6.12. Zero flow rate and wilting will occur when Ψ_{soil} is −20 bars, as in the ideal determination of the permanent wilting percentage. However, if transpiration is occurring, wilting will occur at progressively higher values of Ψ_{soil} as the rate of transpiration increases and the magnitude of $\Delta\Psi$ increases. This phenomenon is shown in nature by temporary wilting at the time of most rapid transpiration. As there is a gradient in potential through the plant ($\Psi_{soil} > \Psi_{root} > \Psi_{leaf}$), wilting of leaves will not be associated with a unique value of Ψ_{root} but will be influenced by variations in magnitude of the difference in water potential $\Delta\Psi$ between root and leaf. Some idea of

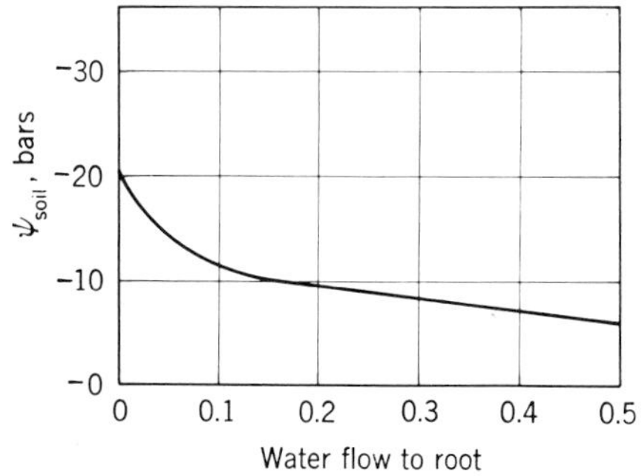

FIG. 6.12 Rate of water flow to roots in milliliters per centimeter of root length per day at various soil water potentials when the root water potential is set at −20 bars. (*After Gardner*, 1960.)

the magnitude of $\Delta\Psi$ through the plant can be obtained by observing wilting of rapidly transpiring plants rooted in water culture of a known potential.

If transpiration rate is controlled by the vapor pressure gradient from the evaporating surfaces of the leaf to the outside air and the resistance to flow in the path of the diffusing water vapor, then decreasing values of Ψ_{leaf} do not directly limit transpiration, because their effect on vapor pressure at the evaporating surfaces will be small under most conditions (see Chap. 9). Instead, they presumably operate by causing decreased turgor of guard cells and stomatal closure, which greatly increases resistance to diffusion. Since the original cause of low Ψ_{leaf} values is soil water stress, the latter is the cause of reduced transpiration even though it operates indirectly.

Flow in soil and plants

Gardner (1960) applied his equation to the experimental data of Slatyer (1957) with the results shown in Fig. 6.13. The smooth lines are calculated values demonstrating the decline of water flow if the lowest values of Ψ_{leaf} are −40 and −100 bars. In the −40 bar situation the inability to maintain the same flow rate as Ψ_{soil} drops below −1 bar is shown by the divergence of the curve from that for −100 bars, which shows little reduction until Ψ_{soil} approaches −10 bars. The experimental data shown as points on the diagram indicate that the calculated values are close to those actually obtained for privet and cotton, with lowest measured values of Ψ_{leaf} of −100 bars, and for tomato with −40 bars. Figure 8.24 from Denmead and Shaw (1962) shows a similar relationship for a field experiment with corn in which actual transpiration on days with various levels of potential transpiration and Ψ_{soil} were compared.

It appears that simplified models of the type developed by Gardner (1960) give a good description of water flow from soil to root. Under most conditions it seems that if the average Ψ_{soil} is higher than −1 bar, only small

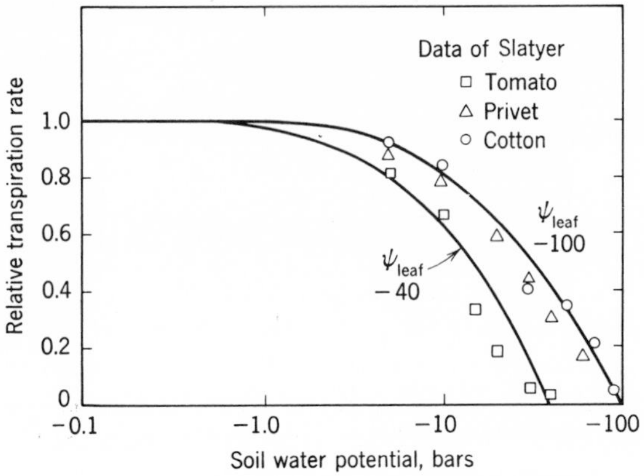

FIG. 6.13 **Relative transpiration of tomato with a minimum leaf water potential of −40 bars and cotton and privet with minima of −100 bars, plotted over soil water potential. The lower leaf water potentials are accompanied by more absorption from drying soils and longer maintenance of a high rate of transpiration. (*After Gardner*, 1960.)**

gradients of water potential will develop between soil and root, even at high rates of absorption. However, as Ψ_{soil} and hydraulic conductivity decrease, $\Delta\Psi$ increases rapidly, and the slowing water movement into roots causes a decrease in Ψ_{leaf}, which may cause loss of turgor, stomatal closure, and reduced transpiration. The value of Ψ_{soil} which will be associated with such responses depends on the rate of absorption, the hydraulic conductivity of the soil, and the extent of branching of the root system.

It may be added that Gardner and Ehlig (1962) found that at a Ψ_{soil} higher than −0.6 bar the resistance to water movement in plants is greater than the soil resistance, but at a Ψ_{soil} of less than −1 or −2 bars resistance to movement through the soil became the limiting factor in water supply. Cowan (1965) came to a similar conclusion, but Newman (private communication) questions this. He claims that with the root density usually found under crop plants soil resistance will not limit water supply until Ψ_{soil} approaches the permanent wilting percentage. Gardner, and Cowan and Monteith have made further contributions to this problem in Kozlowski (1968, chaps. 5 and 6).

In general it appears unlikely that sufficient water will move more than about 4 cm through soil near field capacity to a root surface to be of physiological significance, although where a water table is close to the surface the distance may be 1 m or more. Gardner and Ehlig (1962) observed upward movement of appreciable amounts of water from as deep as 180 cm under alfalfa. Wind (1955*a*, 1955*b*) concluded that there is less resistance to water flow through the soil than through the fine metaxylem elements of grass roots. Thus under some conditions there is simultaneous upward movement of water through roots and soil.

It has been proposed that as the water content of soil and roots decreases, shrinkage may occur, resulting in an air gap between soil and roots across which water can move only as vapor. Occurrence of such a gap was proposed by Philip (1958*a*) to explain the failure of some plants to absorb large quantities of salt from saline soils. Bernstein et al. (1959) pointed out that rates of vapor flow are too slow to account for observed rates of water absorption, except in soils much drier than field capacity. Also, it has been shown that the particular salinity effects referred to by Philip (see Chap. 9) can occur in water culture, where no possibility of a vapor gap exists (Slatyer, 1961). Soil shrinkage, however, is probably accompanied by considerable damage to root hairs and fragile roots, and the decreased contact between soil and roots will materially decrease the transfer of water. Cowan and Milthorpe recently revived the idea that a considerable amount of water might be supplied to roots as vapor if the gap between roots and soil is no greater than the radius of the roots (Kozlowski, 1968, chap. 6).

Soil temperature and water absorption

It has been known at least since the time of Hales in the early eighteenth century that low soil temperatures reduce the absorption of water by plants. Soil temperature not only has direct effects on water absorption, but also has indirect effects through its influence on root growth and possibly on synthetic activity in roots. The effects on root growth are discussed in Chap. 4, and the effects on water absorption will be discussed in this section. Temperatures are sometimes high enough in agricultural soils to be a limiting factor, and perennial plants of the Temperate Zone are subjected annually to the hazard of low temperatures. Considerable "winter killing" of plants results from desiccation caused by transpiration at times when the soil is frozen or so cold that absorption is too slow to replace the water lost by transpiration.

Effects of low temperatures

Cold soil is an important ecological factor. The decreased availability of water in cold or frozen soil is a limiting factor for vegetation in the Arctic and near the timberline on mountains (Whitfield, 1932; Clements and Martin, 1934; Michaelis, 1934). Local differences in soil temperature also affect plant growth and distribution. Heavy, poorly drained soils are slower to warm up in the spring than sandy, well-drained soils, and Firbas (1931) and Döring (1935) state that the cold soils of European high moors are a limiting factor for plant growth. Also, cold soil sometimes becomes a limiting factor for crop plants. Schroeder (1939) found that cold soil, aggravated by watering with cold water, caused serious injury to greenhouse cucumbers in the winter, and Cameron (1941) reported that orange trees in California often wilt during cold weather. There is also evidence that irrigation water from the high dams of the Sierras is too cold for good growth of rice in the Sacramento Valley of California, and procedures such as warming basins and removal of the warmer water from the tops of reservoirs are being used to bring the cold water up to physiologically satisfactory temperatures. Low soil temperatures are also said to limit growth of rice in Japan and Italy. On the other hand, low soil temperatures caused by irrigation increase the yield of some crops such as potatoes. The role of irrigation water temperatures was reviewed by Raney and Mihara (1967).

SPECIES DIFFERENCES IN TOLERANCE OF LOW TEMPERATURES. Sachs, about 1860, observed that tobacco and gourd plants wilted more severely than cabbage and turnip plants when the soil in which they were growing was cooled to 3 to 5°C. According to Brown (1939), water absorption of Bermuda-grass (*Cynodon dactylon* Pers.), which is a native of warm climates, is reduced so much by cooling the soil to 10°C that it wilts, while bluegrass (*Poa pratensis* L.), which thrives in cooler regions, is not affected.

The results of some studies on the effects of soil temperature are summarized in Table 6.3. The striking differences between collards (which are grown as a winter crop in the southern United States) and cotton and watermelon (which are warm-season crops) are shown in Fig. 6.14. Kozlowski (1943) found that water absorption by loblolly pine (*Pinus taeda* L.), a southern species, was reduced much more by low soil temperatures than absorption by white pine (*P. strobus* L.), a northern species. These differences suggest that the permeability of roots of various species is reduced to different extents by low temperature. There are many other accounts of reduced

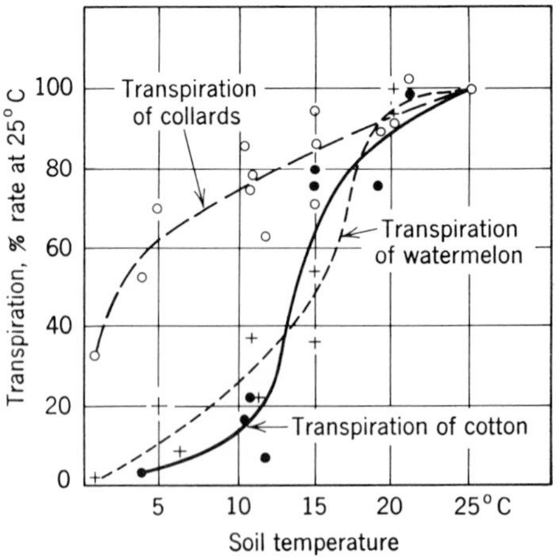

FIG. 6.14 Effects of low soil temperatures on water absorption by four species of plants, as measured by rates of transpiration. (*From Kramer*, 1942.)

*Table 6.3 Effects of Soil Temperature on Water Absorption by Plants of Various Species**

(Rates of transpiration were measured, and it was assumed that absorption was approximately equal to transpiration over 24-hr periods.)

Experiment	Species	Number of plants per experiment †	Final soil temperature	Transpiration of cooled plants as percent of controls at 25°C
1	Collards (*Brassica oleracea acephala* DC)	6	12.0	63.0
	Cotton (*Gossypium hirsutum* L.)	6	12.0	7.4
2	Collards	6	4.3	53.0
	Cotton	6	4.3	4.3
3	Collards	6	1.0	33.0
	Watermelon (*Citrullus vulgaris* Schrad.)	6	1.0	1.4
4	Loblolly pine (*Pinus taeda* L.)	4	0.5	13.7
	Slash pine (*P. elliotti* Engelm.)	4	0.5	13.9
	White pine (*P. strobus* L.)	4	0.5	37.7
	Red pine (*P. resinosa* Ait.)	4	0.5	25.0
5	Elm (*Ulmus americana* L.)	14	0.5	25.0
6	Privet (*Ligustrum japonicum* Thunb.)	12	2.5	47.0
7	Sunflower (*Helianthus annus* L.)	12	1.0	27.0

* *Modified from Kramer (1942)*

† *The plants were divided into two groups, one of which was cooled about 5° per night while the other was kept at 25°C.*

absorption by plants at low soil temperatures, including Vesque (1878), Kosaroff (1897), and Bode (1923), also Duncan and Cooke (1932) working on sugarcane, Clements and Martin (1934), Rouschal (1935), Arndt (1937) with cotton, Raleigh (1941) with muskmelons, and Ehrler (1963) working on alfalfa. In general, species which normally grow in warm soil are affected much more by cooling the soil than species which normally grow in cooler soil. Also, rapid cooling reduces absorption more than gradual cooling (Böhning and Lusanandana, 1952), as shown in Fig. 6.6.

CAUSES OF DECREASED ABSORPTION AT LOW TEMPERATURES. There are several reasons why decrease in soil temperature reduces water absorption. The most important ones are:

(1) Decrease in root growth. This is most important in soils so dry that water movement toward roots becomes a limiting factor. The effects of soil temperature on root growth are discussed in Chap. 4.
(2) Increased viscosity of water. The viscosity of water is about twice as great near 0°C as at 25°C.
(3) Increased resistance to movement of water into roots, caused by decreased permeability of cell membranes and the effects of increased viscosity.
(4) Decreased metabolic activity of root cells. This has some effect on active water uptake because it results in decreased salt accumulation (see Chap. 7). It affects passive absorption only so far as it affects permeability of the root cell membranes.

Decreased water absorption and wilting of plants in cold soil cannot be attributed to decrease or cessation of active absorption because, as stated in Chap. 5, this process supplies only a small percentage of the total amount of water required by the shoot. However, passive movement is greatly reduced at low temperatures. As shown in Fig. 6.15, water movement through root systems attached to a vacuum pump and subjected to a pressure difference of about 0.85 bar is about 20 percent as great at 0° as at 25° C. When root systems killed by immersion in hot water were cooled, water intake under the same pressure gradient was about half as rapid near 0° as at 25°C. Killing the roots greatly decreased the resistance to water movement by mass flow, presumably because of collapse of the protoplasts and cell membranes. Ordin and Kramer (1956) and Woolley (1965) also showed that killing roots increased their permeability to movement of water by diffusion. Curves for movement of water through dead roots and collodion and porcelain membranes at various temperatures are very similar, suggesting that the rate of water movement through nonliving membranes is controlled primarily by its viscosity (Kramer, 1940*a*).

Kuiper (1964*a*) reexamined the effects of low temperature on water absorption. In very short exposures to low temperature, water intake was controlled by the viscosity of water, as in nonliving membranes, but after an exposure of 30 min the added effect of the change in permeability of protoplasm appeared. Kuiper also restudied data of Kramer and others and corrected them for changes in the viscosity of water. By this procedure he showed that above some critical temperature water absorption appears to

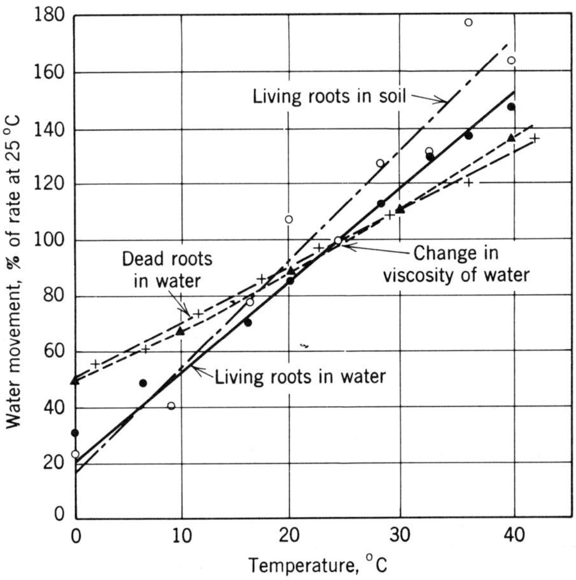

FIG. 6.15 **Effects of temperature on rate of water movement through living and dead roots attached to a vacuum pump and subjected to a pressure differential of 64 cm Hg. Living root systems were under vacuum for 1 hr, and a new set of roots was used for each temperature. Rates are plotted as percentages of the rate at 25°C to facilitate comparison between living and dead roots. Viscosity is expressed as percentage of the reciprocal of the specific viscosity at 25°C. Note that change in temperature affected water movement through living roots more than through dead roots, probably because temperature affects the resistance to flow through living protoplasm. (*From Kramer,* 1940.)**

be limited only by change in viscosity of water, but below it absorption is also limited by decrease in permeability of root cell membranes. The results are summarized in Fig. 6.16. The effect on permeability was shown below 5°C for cabbage and white pine, 10°C for citrus, 20°C for loblolly pine, and about 22°C for cotton and watermelon. Kuiper suggested that above this critical temperature there are permanent water-filled pores in the root cell membranes, but below it these do not exist and there is a much higher resistance to water movement because changes in metabolic activity at lower temperatures bring about changes in structure of the cytoplasmic membranes.

Kuiper reported that the critical temperature could be shifted downward in beans by growing the root systems at lower temperatures (17 instead of 24°C) or decreasing aeration during growth. The gradual adjustment of root systems to low temperature was mentioned earlier in this chapter. Kuiper (1964*b*) also claimed to have almost eliminated the increased root resistance of young bean root systems at low temperatures by adding 5×10^{-4} M solution of decenyl succinic acid to the root medium. He believed that this substance was incorporated into the lipid layers of the cytoplasmic membranes and increased their permeability to water. However, further experiments indi-

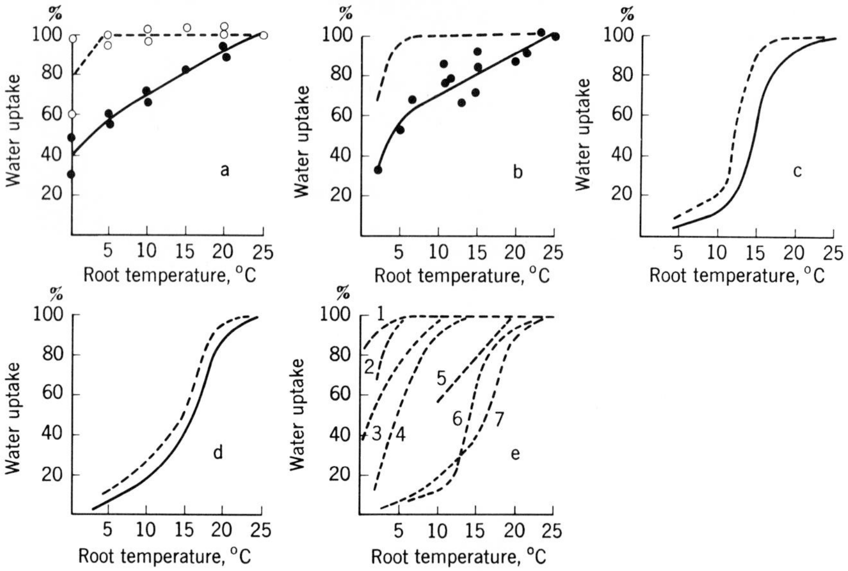

FIG. 6.16 The effect of temperature on water uptake through roots of white pine (*a*), cabbage (*b*), cotton (*c*), and watermelon (*d*). The solid curves are after Kramer (1942), and the dashed lines are corrected for differences in the viscosity of water at various temperatures. Graph *e* combines corrected curves from various sources, white pine (1), cabbage (2), citrus (3, from Bialoglowski, 1936), sunflower (4, Clements and Martin, 1934), loblolly pine (5, Kramer, 1942), cotton (6), watermelon (7). The rate at 25°C was set as 100 percent. The critical temperature for reduction in root permeability is much lower in white pine (1) and cabbage (2) than in cotton (6) and watermelon (7). (*From Kuiper*, 1964.)

cate that the increased permeability in the presence of decenyl succinic acid actually results from injury to the roots (Newman and Kramer, 1966).

It seems clear that the principal cause for decreased absorption of water at low temperatures is the physical effect on root resistance of increased viscosity of water and decreased permeability of the root cell membranes. Other factors such as decreased root extension, hydraulic conductivity, and root metabolism are of distinctly secondary importance.

Effects of high temperatures

Relatively few data are available concerning water absorption at high temperatures. Bialoglowski (1936) found that root temperatures above 30°C reduced water absorption by lemons, and Haas (1936) reported a similar condition in lemons, grapefruit, and Valencia oranges. Exudation from detopped tomato root systems attained its maximum at about 24°C and decreased at higher temperatures (Kramer, 1940). However, only those roots in surface soil, in greenhouse benches, and in containers exposed to full sun are usually subjected to temperatures high enough to interfere with water absorption.

Soil aeration and water absorption

It has been known since early in the nineteenth century that deficient aeration decreases the absorption of water. Much of the early literature was summarized by Clements (1921) and by Kramer (1949). Russell (1952) wrote a review on soil aeration in relation to plant growth, and there are many recent references in the *Proceedings of the Drainage for Efficient Crop Production Conference* (van Schilfgaarde, 1965). There is a discussion of aeration in relation to root growth in Chap. 4, and some information given in that chapter is repeated in this section in connection with the overall effects of aeration on plant growth.

Many factors affect the severity of injury caused by inadequate aeration. Among them are the species, the condition of the plants, the temperatures, duration of flooding, type of soil, and kinds of microorganisms present.

Species differences

Considerable differences in tolerance of deficient aeration occur among species. Livingston and Free (1917) reported that plants of *Coleus blumei* Benth. and *Heliotropium peruvianum* L. soon wilted when the soil air was replaced by nitrogen, but *Salix nigra* Marsh was unaffected by the absence of oxygen. Absorption of water and ions by corn and wheat roots is inhibited severely if the nutrient solution is saturated with carbon dioxide, but rice is little affected (Chang and Loomis, 1945; Vlamis and Davis, 1944).

As mentioned in Chap. 4, field observations indicate that although cypress, tupelo gum, and cattails may thrive in swamps, other species such as dogwood, yellow-poplar, and some species of pine are severely injured or killed by relatively short periods of flooding. In laboratory experiments the absorption of water by cypress was reduced very little by flooding, but that of several species of oak, *Pinus taeda* L., and *Juniperus virginiana* L. was reduced to less than 50 percent of controls (also see Fig. 6.17). Loblolly pine is more resistant to flooding than some other species of pine, and much more resistant than yellow-poplar or dogwood (Hunt, 1951; Kramer, 1951; Parker, 1950). Yelenosky (1964) compared resistance to flooding of five species of shade trees and found that American elm in flooded soil retained its leaves for over eight weeks, and then resumed growth when drained, but yellow-poplar lost its leaves within two weeks and died.

Large differences in tolerance of flooding and inadequate aeration are also found among herbaceous species. Everyone is aware that rice thrives in flooded soil, but if the soil in which tobacco is growing is flooded for more than one or two days, it is permanently injured. In one series of experiments tobacco was injured most severely by flooding the soil, sunflower least, and

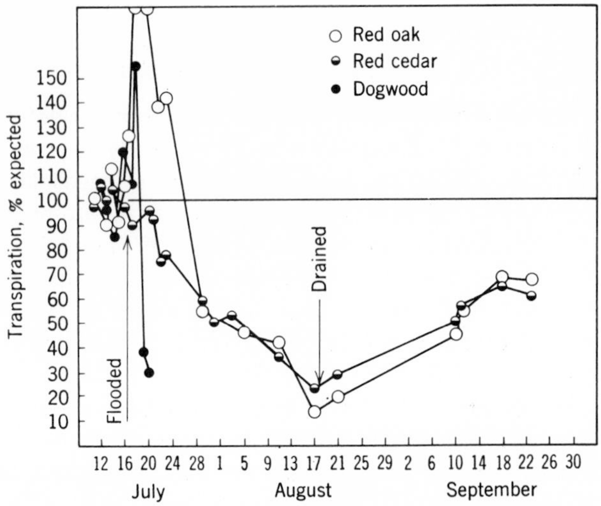

FIG. 6.17 **Transpiration of red oak, red cedar, and dogwood in flooded soil. Dogwood was killed, but the other species survived and began to recover when drained.** (*From Parker*, 1950.)

tomato was intermediate (Kramer, 1951). In experiments by Williamson (1964) poor aeration decreased yields of several crops in the following order from least to greatest reduction: grain sorghum, soybeans, cabbage, sweet corn, dwarf field corn.

Causes of species differences

As stated in Chap. 4, the reasons for the differences among species in tolerance of flooding and insufficient aeration are not completely understood. They are related either to differences in structure, resulting in a better oxygen supply to roots of some species, or to physiological differences, resulting in greater tolerance of products of anaerobic respiration by cells of some species, or a combination of the two.

As pointed out in Chap. 4, roots produced in a poorly aerated environment are usually larger in diameter and contain more and larger air spaces than roots developed in a well-aerated environment, and there is considerable evidence that oxygen diffuses down from the shoots to the roots through these air spaces. Attention has been given to the role of pneumatophores or aerial roots in supplying oxygen to submerged roots of mangroves and cypress. Scholander et al. (1955) reported that air is sucked in through lenticels in the pneumatophores of *Avicennia* when the tide falls and is forced out when it rises, providing a ventilation system for the submerged roots. Many claims have been made that cypress "knees" serve as aerating organs for submerged roots, but Kramer et al. (1952) found that oxygen uptake by cypress knees was increased, rather than decreased, when they were detached from the root system. For this and other reasons it is doubted if they play an essential role as aerating organs, though it must be admitted that gas can move fairly readily through the porous tissue of cypress roots and knees.

Another explanation for differences in tolerance of flooding is that root cells of some species are tolerant of anaerobic respiration. Crawford (1967) suggested that plants of some species are excluded from wet soil by accumulation of ethanol in their roots under anaerobic conditions. According to Laing (1940*a*) the roots of some aquatic plants are able to carry on anaerobic respiration, as is rice, but corn roots are quickly injured under anaerobic conditions (Woolley, 1966). Possibly this tolerance of anaerobic conditions is more common than generally supposed. Roots which extend several meters into heavy soils certainly grow with lower concentrations of oxygen than those near the surface.

Another factor in tolerance of flooding seems to be the development of adventitious roots near the water surface. Several studies indicate that plants which promptly produce adventitious roots suffer less injury from flooding than those which produce few or none (Kramer, 1951). Jackson (1955) reported that adventitious roots did not entirely prevent injury to shoots of flooded tomato and sunflower plants, but growth was resumed if adventitious roots were allowed to develop and was not resumed if they were removed.

Factors affecting severity of injury

It has often been demonstrated that flooding of dormant plants causes much less injury than flooding of growing plants. This presumably is related to the greater oxygen needs of growing roots and the higher water use of growing shoots. High temperatures also increase the oxygen consumption of roots, and flooding of the soil is much more serious in warm weather than in cool weather (Letey et al., 1962). As shown in Fig. 6.18, flooding is more serious

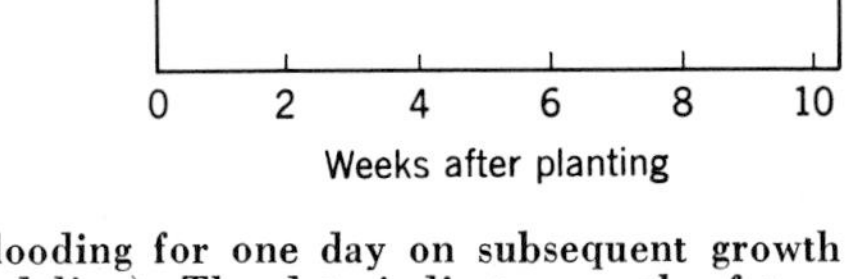

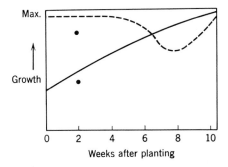

FIG. 6.18 Effect of flooding for one day on subsequent growth of tomato (solid line) and pea plants (dashed line). The dots indicate growth of tomato plants flooded on a cloudy day (upper) and a sunny day (lower). (*From Erickson,* 1965.)

on sunny days than on cloudy days, because the higher rate of transpiration on sunny days produces more severe water deficits when absorption is reduced (Erickson, 1965). Another example is the "flopping" or sudden wilting of tobacco which sometimes occurs when the sun shines immediately after the soil has been saturated by a rain.

In some experiments plants growing in pots of loam soil were injured more severely by flooding than those growing in pots of sand. Removal of soil from around the roots also decreased injury by flooding (Kramer and Jackson, 1954). It seems possible that the number and kind of soil organisms present may also affect the severity of injury.

Causes of flooding injury

Injury from flooding is usually attributed to desiccation caused by interference with the absorption of water, resulting from increased root resistance [r_{root} of

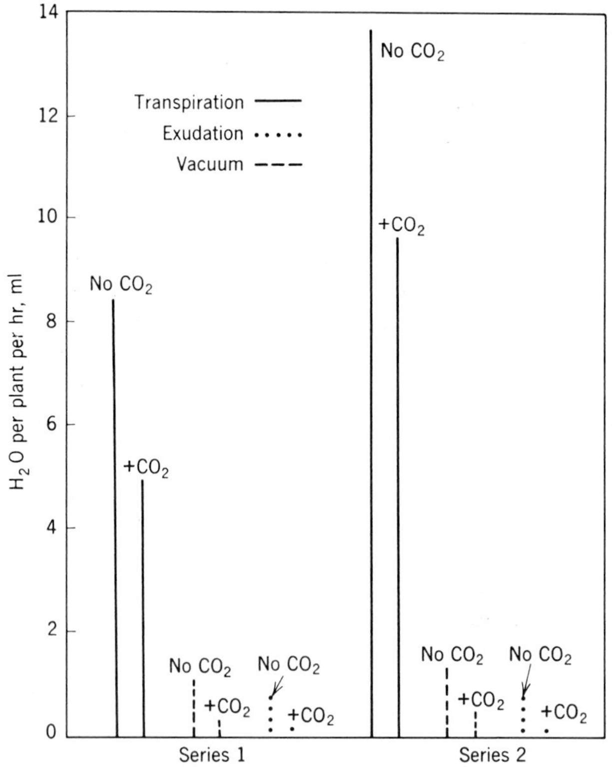

FIG. 6.19 Effects of carbon dioxide on water movement through tomato root systems. While the percentage reduction in exudation is much greater than the percentage reduction in transpiration, complete cessation of exudation could account for less than 15 percent of the reduction in transpiration. (*From Kramer,* 1940.)

Eq. (6.1)] and injury to roots. The effects of aeration and flooding on root resistance are shown in Figs. 6.19 and 6.20. However, some of the symptoms of flooding cannot be attributed to simple dehydration (Kramer, 1951). It is true that if the soil is flooded or the air displaced by carbon dioxide or nitrogen when conditions are favorable for transpiration, plants usually wilt. However, equally characteristic is the yellowing of lower leaves which usually develops in less than a week. This is often followed by their death and abscission. Leaves of some woody species develop characteristic discolorations. Less obvious is the epinasty of leaves which is often seen in herbaceous species and has also been reported for apple and yellow-poplar. It is probably quite general, but is often confused with wilting. Formation of adventitious roots often begins within a week, even on some woody species (Yelenosky, 1964). Other species which do not form adventitious roots develop callus tissue and hypertrophy of the submerged portion of the stem (Kramer, 1951; Jackson, 1956). It seems probable that the formation of hypertrophies and adventitious roots is caused by accumulation of auxin and food in the stem near the water line where oxygen deficiency stops downward translocation.

The occurrence of epinasty and the development of adventitious roots is more characteristic of turgid than of flaccid tissue. Thus one suspects that several factors are involved. The epinasty resembles that seen in plants exposed to ethylene gas (Furkova, 1944; Williamson, 1950), and the yellowing and peculiar discoloration of leaves of some species might be caused by toxic substances escaping from dying roots or produced in saturated soil and carried to the shoot in the transpiration stream. However, attempts to isolate these factors have been unsuccessful, and Jackson (1956) concluded that reduced water absorption caused by injury to the roots from oxygen deficiency is the major cause of flooding injury. Probably there is also interference with synthetic activities in the roots (see Chap. 4), which might deprive the shoots of minerals or of essential growth substances (Went, 1943). It has been reported recently that cytokinins and gibberellins are produced in roots, and interference with their synthesis might also result in injury to the shoots (Jones and Lacey, 1968; Kende, 1965; Skene, 1967).

The decrease in transpiration and the wilting which occur immediately after soil is flooded (see Fig. 6.17), or when the soil air is displaced by carbon dioxide or nitrogen, result from increased resistance to water movement through the roots. A deficiency of oxygen or an excess of carbon dioxide greatly reduces the permeability of roots to water, as shown in Figs. 6.19 and 6.20. Brouwer (1954), Hagan (1950), Hoagland and Broyer (1942), Kramer (1940), and Kramer and Jackson (1954) all reported that carbon dioxide reduces permeability of roots to water, usually more rapidly than a deficiency of oxygen. However, it appears probable that under field conditions oxygen deficiency may be limiting more often than an excess of carbon dioxide.

Generally, the first effect of flooding or inadequate aeration is a decrease in permeability of root tissues to water (see Fig. 6.20), which decreases absorption and results in water deficits and even wilting of shoots. Yelenosky (1964) found that flooding the soil in which yellow-poplar seedlings were growing reduced transpiration to 68 percent of the control rate and produced a leaf water deficit of 47 percent in three days compared with 11 percent in the controls. It seems adequately established that poor aeration reduces water absorption by decreasing the permeability of roots, as suggested by Livingston and Free (1917), Hunter and Rich (1925), and Kramer (1940). It also appears that excess carbon dioxide produces this effect even more rapidly than oxygen deficiency, perhaps because carbon dioxide has specific effects on protoplasm (Chang and Loomis, 1945; Fox, 1933; Seifriz, 1942).

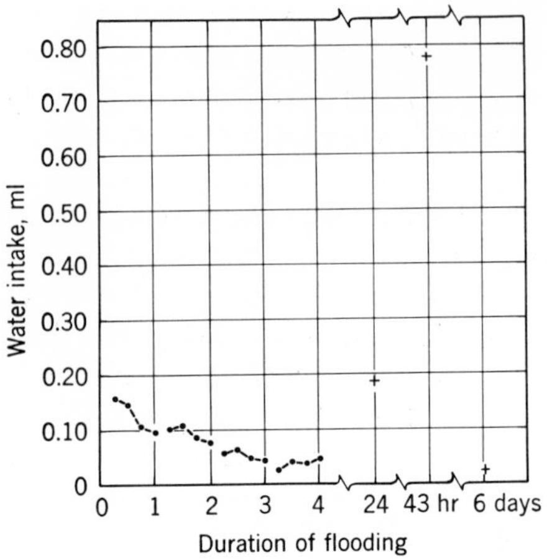

FIG. 6.20 **Effects of flooding for various periods of time on water intake through tobacco root systems attached to a vacuum line and subjected to a pressure of 0.2 bar. Each point is the average of six root systems. Permeability increased after 24 to 48 hr because of injury to the roots, but after 6 days the root systems were almost destroyed. (*From Kramer and Jackson,* 1954.)**

The long-term effect of inadequate aeration is reduction in size of root systems or even their death. It also encourages the development of undesirable fungi such as *Phytopthora cinnamomi,* which is associated with "little leaf disease" of pine and various root rots and "decline" diseases of citrus and avocado, as mentioned in Chap. 4.

This section has dealt chiefly with the effects of severe deficiencies in aeration such as occur in flooded soil. However, it should be remembered that many heavy clay soils with limited pore space are chronically deficient in aeration because of low rates of oxygen diffusion. There is not space to cite the literature in this field, but such conditions reduce the productivity of many agricultural soils. In other instances soil is compacted by traffic, cattle, or cultivation when too wet. Construction of paved roads and sidewalks and filling in of land also reduce the oxygen supply to roots. The effects of such treatments on gas composition are shown in Fig. 6.21. These activities often result in unthrifty conditions or death of shade trees. Readers are referred to Chap. 4 for further discussion of this topic.

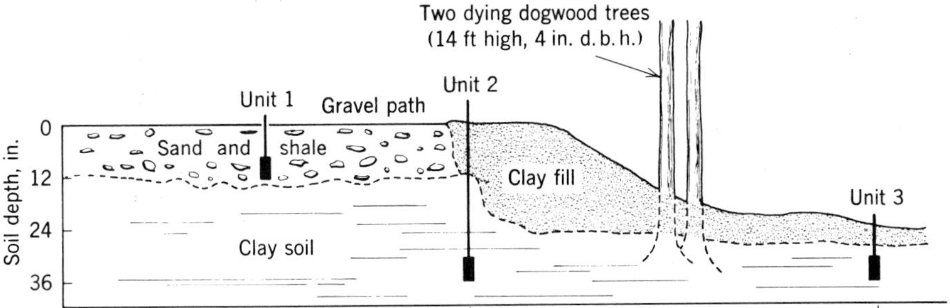

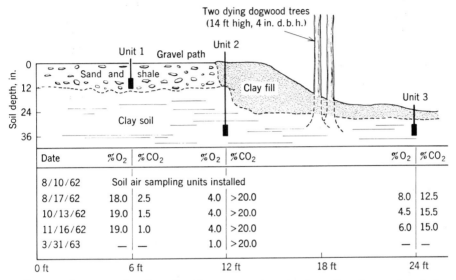

Date	%O_2	%CO_2	%O_2	%CO_2	%O_2	%CO_2
8/10/62	Soil air sampling units installed					
8/17/62	18.0	2.5	4.0	>20.0	8.0	12.5
10/13/62	19.0	1.5	4.0	>20.0	4.5	15.5
11/16/62	19.0	1.0	4.0	>20.0	6.0	15.0
3/31/63	—	—	1.0	>20.0	—	—

0 ft 6 ft 12 ft 18 ft 24 ft

FIG. 6.21 **Oxygen and carbon dioxide concentrations in the soil under sand and clay fills.** (*From Yelenosky*, 1964.)

Effects of concentration and composition of the soil solution

As pointed out in Chap. 3, in arid and irrigated soils the osmotic potential of the soil solution often falls so low that growth of crop plants is inhibited or prevented and in extreme situations all vegetation is eliminated except for a few species of halophytes. Osmotic potentials of −2 to −4 bars at the permanent wilting percentage inhibit the growth of most crops, and soils with osmotic potentials approaching −40 bars are usually barren of vegetation. Seed germination is particularly sensitive to high salt concentrations (Uhvits, 1946), and seedlings are more susceptible to injury than well-established plants. Addition of fertilizer occasionally results in osmotic potentials too low for good plant growth. For example, application of 1,200 lb/acre of 3-9-3 fertilizer to Norfolk sandy loam temporarily decreased the osmotic potential of the soil to −14 bars, but a similar application to Cecil clay loam only decreased the osmotic potential to −3 bars (White and Ross, 1939). Excessively high concentrations of salt sometimes develop in heavily fertilized greenhouse soils, and this is said to be a common cause of reduced growth.

The osmotic potential of a heavily fertilized greenhouse soil is shown in Fig. 3.8.

Causes of reduced growth

There has been much discussion of the causes of reduced plant growth in the presence of even moderate concentrations of salt. Early investigations of the effects of high salt concentration on plant growth gave much attention to the relative effects of various ions and proportions of ions in terms of antagonism, physiological balance, toxicity, and effects on permeability. The results were frequently contradictory and confusing, and generally unsatisfactory. During the late 1930s and early 1940s attention shifted to the overall osmotic effects of the total concentration of ions and away from specific effects of particular ions. Eaton (1942) pointed out that neither general observations nor specific experiments showed any evidence of critical concentration above which injury occurred. Instead, above an initial low concentration each increase in salt concentration produces a decrease in growth. Magistad et al. (1943) found a practically linear reduction in growth of a dozen species of crop plants with decreasing osmotic potential of the soil solution over the range from −0.4 to −4.5 bars. Reduction in growth with decreasing osmotic potential of the soil or culture solution has been reported by a number of investigators, including Ayers et al. (1943), Hayward and Long (1943), Wadleigh and Ayers (1945), and Wadleigh et al. (1946). The work of these investigators indicated that similar reductions in growth occurred in plants subjected to similar water stress, whether the stress was produced by allowing the soil to dry (decreasing matric potential), by adding salt to the soil or nutrient solution (decreasing osmotic potential), or by a combination of the two treatments. An example of this situation is presented in Fig. 6.22. A similar situation with respect to the vegetative growth of guayule was found by Wadleigh et al. (1946) and is shown in Fig. 10.21.

REDUCED WATER ABSORPTION. The reduction in growth observed with increasing salt content of the substrate has usually been attributed to reduced water absorption caused by the osmotic effects of reduced water potential in the root environment. If the water potential is reduced in the soil or solution in which the plants are growing, the potential difference—which is the driving force for absorption—is assumed to be reduced, resulting in decreased water absorption. This explanation assumes that the roots behave as perfect osmometers, but it has proved to be an oversimplification of the actual situation. When plants are transferred from dilute to concentrated solutions, they usually wilt at first, but then recover after a short time because they absorb salt, which causes a decrease in osmotic potential of the plant tissue. Slatyer (1961) found recovery of turgor after 28 hr for tomato plants in

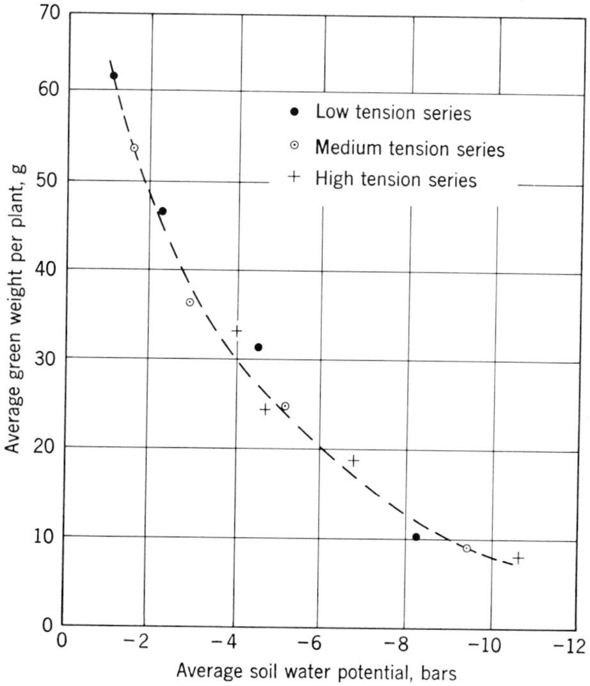

FIG. 6.22 Effect of decreasing soil water potential on growth of bean plants. Plants in the low-tension series were watered when 40 to 50 percent of the available water was removed. Medium-tension plants were watered when 60 to 65 percent of the available water was removed, and high-tension plants were not watered until 90 to 100 percent was removed. Each moisture series was subdivided into four groups which received no salt, 0.1 percent, 0.2 percent, or 0.3 percent NaCl. Reduction in yield was proportional to decrease in average water potential, whether the decrease was caused by low soil water content, by high salt, or a combination of the two. For example, the highest yield was in low-tension soil containing no salt, and the lowest yield was in high-tension soil containing 0.3 percent salt. (*From Wadleigh and Ayers,* 1945.)

nutrient solution to which was added enough potassium nitrate, sodium chloride, or sucrose to decrease the osmotic potential to −10 bars. The uptake of solutes was so great that the osmotic potential of the plants in the concentrated salt solutions was lower than that of their substrate by the same amount that the potential of the control plants exceeded that of the nutrient solution. The initial wilting observed after an abrupt change in concentration of the substrate can be avoided by increasing the substrate concentration in small daily increments.

In long-term experiments, Eaton (1942) grew six species of plants in sand cultures with osmotic potentials adjusted over a range from −0.7 to −6.0 bars. The difference between the osmotic potential of the substrate and that of the plant sap averaged over 11 bars in all substrate concentrations. Thus, the plants in the most concentrated solution maintained the same difference in water potential from substrate to roots as the plants in the least concentrated solution (see Table 6.4). Nevertheless, growth and transpiration were materially reduced in the more concentrated solutions. In view of these observations, the reduced growth of plants in media having an osmotic potential lower than −2 or −3 bars cannot be attributed solely to reduced water uptake caused by a reduced driving force from substrate to root xylem.

Actually there is some reduction in water uptake through roots in concentrated solutions. In short-term experiments Hayward and Spurr (1943, 1944) found water uptake by corn roots from a solution with Ψ_s of −4.8 bars was only 12 percent of uptake from a solution with Ψ_s of −0.8 bar. Solutions of sucrose, mannitol, sodium sulfate, and calcium chloride of the same Ψ_s reduced water absorption to the same extent. However, roots preconditioned

Table 6.4 Effect of Osmotic Pressure of Culture Solution on Osmotic Pressure of Plant Sap*

Osmotic pressure of culture solution in atmosphere	0.72	2.52	6.0
Species	Difference between osmotic pressure of solution and plant sap		
Milo	10.3	10.8	11.1
Alfalfa	13.0	12.6	10.4
Cotton	13.1	11.8	9.7
Tomato	8.8	8.3	8.2
Barley	9.2	12.4	14.7
Sugar beet	12.8	13.8	15.0
Average difference	11.2	11.6	11.5

** From Eaton (1942)*

for several days in concentrated solutions absorbed considerably more water than those transferred from dilute to concentrated solutions. Experiments with split root systems show reduced absorption from more concentrated solutions. Eaton (1941) grew corn and tomato plants with roots divided between nutrient solutions with Ψ_s of −0.3 and −1.8 bars and reversed the roots daily to eliminate differences in root growth. Over a period of days, 1.8 times as much water was absorbed from the dilute as from the concentrated nutrient solution. Long (1943) approach-grafted tomato to obtain two root systems which were immersed in solutions of different concentration. He found that most of the salt and water were absorbed through the roots in the more dilute solution.

It appears from these and other experiments that the permeability of roots to water must be decreased in salt solutions. There are probably two effects of high substrate concentration, one short-term, the other long-term. Immersion of roots in solutions with an osmotic potential of −2 or −3 bars probably tends to slightly dehydrate the walls and cell membranes, increasing the resistance to water flow. Klepper (1967) found evidence that a small decrease in osmotic potential of the substrate causes a measurable decrease in root permeability. Over longer periods of time root elongation is reduced, and suberization is increased (Hayward and Blair, 1942), resulting in smaller root systems with higher resistance to water movement than are found in roots produced in dilute solutions. Nevertheless, reduced absorption of water is not the principal cause of reduced growth in saline substrates. Plants grown in such substrates are often more succulent than controls (Boyer, 1965; Kreeb, 1965; Meyer, 1931), although both fresh and dry weight are reduced (see Fig. 6.23). This indicates that they are not suffering from dehydration in the same manner as are plants growing in dry soil. Furthermore, plants subjected to high salt concentrations do not recover promptly when restored to normal conditions, although plants subjected to water stress in drying soil may grow faster than normal when rewatered (Greenway, 1962).

It appears that the similarity in effect on growth of the same levels of matric and osmotic potential shown in Fig. 6.22 may be misleading. Low matric potential results in reduced succulence, but low osmotic potential is often accompanied by increased succulence. Slatyer (1967) suggested that perhaps the average matric potential in these experiments was higher than the osmotic potential and that a given degree of growth inhibition is produced at a higher value of matric potential than osmotic potential. This possibility ought to be investigated as soon as we have a good method for maintaining uniform levels of matric potential for considerable periods of time (see Chap. 3).

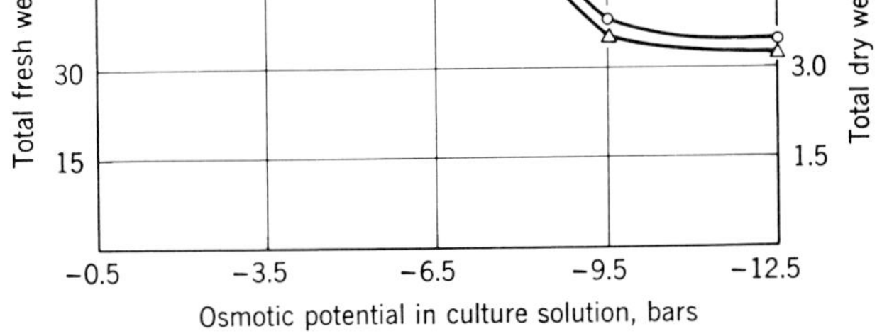

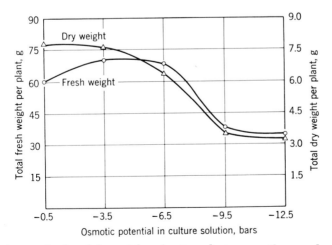

FIG. 6.23 Average fresh and dry weights of cotton plants grown the same length of time in Hoagland solutions plus additional NaCl to reduce the osmotic potential to various levels. Note that at very low osmotic potentials the dry weight is reduced more than the fresh weight, i.e., succulence increases. (*Boyer*, 1965.)

EFFECTS OF INTERNAL SALT CONCENTRATION. Apparently the reduction in growth of plants caused by saline substrates is related more closely to the uptake of abnormal amounts of salt than to the reduced absorption of water. The high concentration of salt reduces the osmotic potential of the cell sap, and the cell water potential is kept low because it can rise no higher than the osmotic potential of the root medium. Possibly osmotic effects develop which result in decreased hydration of proteins including enzymes (Honda et al., 1958; Chen et al., 1964; Laties, 1954) and thereby produce changes in rates of metabolic processes. In addition there are effects of specific ions such as Cl^- versus SO_4^- and K^+ versus Na^+ which are involved in changes in hydration and in succulence of tissue. The specific ion effects may be more important than the general osmotic effects (Hackett, 1961).

The effects of high salt content on metabolic processes are not well understood, but Kessler et al. (1964) reported suppression of RNA and DNA accumulation, Nieman (1962) reported increased respiration in several species, and Boyer (1965) reported decreased photosynthesis and respiration in cotton. In fact the decrease in growth is attributed by Boyer (1965) to decreased photosynthesis or to decreased use of photosynthesis in growth and by Nieman (1962) to both effects. It seems possible that almost every aspect of cell metabolism might be affected by an excess of salt, and it will therefore be difficult to attribute the inhibition of growth to interference with any single process. Obviously much more research should be done on this problem.

In general, tolerance of saline substrates depends on how much salt can be tolerated by the protoplasm (Repp et al., 1959). Those plants with a low tolerance fail to survive in saline soils because of protoplasmic injury from salt accumulation rather than from desiccation.

Specific ion effects

The discussion thus far has emphasized the osmotic and general effects of high salt concentration. However, there are also specific effects of ions which need to be taken into account. For example, growth of some species is reduced more by sulfate than by chloride (Haas, 1945; Hayward et al., 1946), and it is believed by many farmers that chlorides increase succulence and sulfates decrease it. As a result they prefer potassium chloride for succulent leaf crops such as lettuce or cabbage, and potassium sulfate for firmer tomato fruits. This view is supported by van Eijk (1939), who attributed the succulence of halophytes to an excess of chloride. Boyce (1954) also reported that chloride increases the succulence of leaves. It is said that plants subjected to salt spray from the ocean are more succulent than plants of the same species which are not exposed to salt spray. Strogonov (1964) also emphasizes the differences in effects of various kinds of ions, especially chlorides and sulfates, with respect to salt injury.

Determination of salt concentration in soil

As total salt concentration and osmotic potential are usually more important to plant growth than the kinds of ions present, it is possible to evaluate the salt status of soil from measurement of the total salt concentration of the soil solution. The electrical conductivity of soil solutions is well correlated with the osmotic potential and is easier to measure than the osmotic potential. The relationship is shown in Richards (1954, p. 15) and in Fig. 3.10.

Measurements can be made on samples of saturated soil or on extracts made by adding 1 volume or 5 volumes of water per volume of soil (1:1 or 1:5 extracts), filtering off the water, then measuring the conductivity in a standard conductivity cell. A special cell is required for measurements of the conductivity of saturated soil. Detailed descriptions of the methods can be found in Richards (1954).

Summary

The absorption of water is controlled by two groups of factors, those which affect the difference in water potential from soil to roots and those which affect the resistance to water movement through the soil and in the roots. Factors such as soil texture and hydraulic conductivity control water movement to root surfaces. Factors such as soil aeration, temperature, and degree of suberization of roots operate chiefly by modifying resistance to water movement into the roots. The driving force is the difference in water potential between the bulk soil and the root surface and between the root surface and the xylem.

Often various parts of a root system are in soil at different levels of water potential, and apparently many kinds of plants survive and even grow with only part of their root system in soil above the permanent wilting percentage. The minimum soil water potential at which absorption can continue is limited by the minimum water potential which can be developed in the roots. Soil water will become limiting to rapidly transpiring plants at a higher soil water potential than that at which it becomes limiting to slowly transpiring plants.

Large differences exist among species with respect to their tolerance of low temperature and deficient aeration. Decrease in osmotic concentration of the substrate also increases root resistance. However, the reduction in growth of plants subjected to high concentrations of salt is probably more closely related to their high salt content than to reduced water absorption. The physiological dryness is in the plant tissues rather than in the environment.

seven

The Absorption of Solutes

The absorption of mineral nutrients is as important for the successful growth of plants as the absorption of water. However, it is not as well understood, perhaps largely because more of the research on salt uptake has been concerned with accumulation of ions by slices of beet or carrot tissue, algal cells, and excised roots than with absorption through the root systems of intact plants. The absorption of salt and its transport to the shoots of entire plants involves considerably more than its accumulation in the root cells. In fact, ion accumulation in the vacuoles of root cells is probably an incidental aspect of cell metabolism which is independent of movement into the xylem (Bowling and Weatherley, 1965; Epstein, 1960).

The absorption of salt by intact plants includes several steps or processes. The principal ones are (1) the movement of ions from soil to root surfaces, (2) ion accumulation in root cells, (3) the radial movement of ions from root surfaces into the xylem, and (4) translocation of ions from the roots to the shoots. These processes will be discussed separately.

The most troublesome problem in salt absorption by intact plants concerns the mechanism and pathway by which ions move radially from the root surface into the dead xylem elements. Do ions move from vacuole to vacuole, through the symplast, through the free space of the walls, or by some combination of these pathways? What is the relative importance of active transport, diffusion, and mass flow in ion movement? Does water movement through root systems increase ion uptake directly, or only indirectly? What tissue constitutes the ion barrier of roots within which ions are accumulated?

It is impossible to review the voluminous literature on salt absorption. Readers are referred to books by Briggs, Hope, and Robertson (1961), Jennings (1963), and Sutcliffe (1962), also reviews in the *Annual Review of Plant Physiology* and *Fortschritte der Botanik.* There are several chapters in vol. 4 of the "Encyclopedia of Plant Physiology," also chapters by Steward and Sutcliffe (1959) and by Epstein (1965) in other books which supply many details concerning various aspects of this complex field.

Terminology

Before discussing the problem of salt absorption we shall define a few terms, because they are not always used with exactly the same meaning by all writers. Absorption and uptake are general terms applied to entrance of substances into cells or tissues by any mechanism. Accumulation refers to the accumulation of a particular substance within a cell or tissue against a gradient of concentration or electrochemical potential by the expenditure of metabolic energy. Movement of materials against such gradients by a process requiring expenditure of metabolic energy is called active transport. In contrast, passive movement refers to movement by diffusion along gradients of decreasing concentration or electrochemical potential, or by mass flow caused by pressure gradients such as those produced across roots and in the xylem by transpiration.

Accumulation of ions can be detected only behind relatively impermeable membranes, because substances leak out through permeable membranes by diffusion as rapidly as they are moved in by active transport. A membrane, according to Ussing (1954), is a boundary that is less permeable, i.e., presents a higher resistance to movement of materials than the phases separated by it. Membranes which permit some substances to pass more readily than others are termed differentially permeable or, less accurately, semipermeable membranes. Collander (1957, 1959) reviewed the subject of permeability in plant cells, and Dainty (1963*a*, 1965) summarized some recent advances in the treatment of movement through cell membranes.

Plant membranes in the broadest sense are very diverse in structure, ranging from monomolecular layers at interfaces to the membranes of organelles and individual cells, multicellular membranes such as the epidermis and endodermis, and the many layers of suberized cells forming the surfaces of woody roots and stems. The entire cortex of roots has been treated by some workers as a multicellular membrane with respect to water and salt movement. As pointed out in Chap. 1, there are well-defined membranes in cells at the outer and inner surfaces of the cytoplasm, the plasmalemma, and the vacuolar membrane or tonoplast. Differentially permeable membranes also surround organelles such as nuclei, plastids, and mitochondria. Studies with the electron microscope are yielding information concerning their structure (Frey-Wyssling and Mühlethaler, 1965; Thompson, 1965). The differential permeability of these membranes is maintained by expenditure of metabolic energy and is greatly modified by respiration inhibitors, oxygen deficiency, excess carbon dioxide, and other factors which modify metabolism.

An interesting question exists concerning the relative permeability of the plasmalemma and the vacuolar membrane. Experiments with *Nitella* and *Chara* (MacRobbie, 1962) suggest that the plasmalemma constitutes the chief resistance to ion movement, but in *Nitellopsis* it appears to be the vacuolar membrane (MacRobbie and Dainty, 1958). On the other hand, Dainty and Ginzburg (1964*b*) found the plasmalemma of *Nitella* to be more permeable to urea than the vacuolar membrane. Laties and his associates (Torii and Laties, 1966) recently proposed that at high external concentrations of salt the plasmalemma is more permeable than the vacuolar membrane. They propose that ions move into the cytoplasm by diffusion, then are moved into the vacuole by active transport. In contrast, MacRobbie (1964) presented data on the concentration of potassium and chloride ions in the cytoplasm of *Nitella* indicating that there must be active transport into it. She proposes two steps in ion uptake, active transport into the cytoplasm, followed by transport into the vacuoles. Arisz (1964) also supports the view that there is accumulation behind the plasmalemma in the symplast, followed by accumulation in the vacuoles. It has been proposed by several writers that cells be regarded as consisting of three compartments—the walls, which are largely free space; the cytoplasm, into and out of which ions pass less freely; and the vacuoles, from which the escape of accumulated ions is very slow. This is convenient, but probably an oversimplification, because there are ion-binding sites in the walls and free ions in the cytoplasm. Brouwer (1965, pp. 243–249) recently reviewed the literature on the location of ions in plant cells.

An important concept with respect to the movement of ions into roots is that of apparent free space or outer space (Briggs and Robertson, 1957; Briggs et al., 1961; Butler, 1953; Epstein, 1955, 1956; Hope and Stevens, 1952; Kramer, 1957). Apparent free space or outer space is that part of a cell or tissue into and out of which ions can move freely by diffusion, contrasting with the nonfree, inner, or osmotic space which is inaccessible to ions by diffusion, but in which they are accumulated more or less irreversibly by active transport. Since early measurements of apparent free space in roots gave values of 20 to 35 percent of the root volume, a volume far greater than the volume of the walls, it was suggested that the cytoplasm is a part of free space. However, correction of errors in the early estimates reduced the values to approximately 10 percent and restricted free space to the cell walls (Levitt, 1957; Ingelsten and Hylmö, 1961). The recent proposal of Laties (Torii and Laties, 1966; Luttge and Laties, 1966) that at high external concentrations ions enter the cytoplasm by diffusion across the plasmalemma would result in the cytoplasm again being included in free space. This view seems to create more problems than it explains. Attempts to distinguish between water free space and Donnan free space, where nonmobile ions occur, discussed by Briggs and Robertson (1957) and reviewed by Laties (1959) and Brouwer (1965), are outside the scope of this chapter. Crafts (1961) distinguishes between the apoplast (equivalent to free space) and the symplast, which includes the protoplasts of many adjacent cells, interconnected by plasmodesmata. Arisz (1956, 1964) has put much emphasis on movement of salt through the symplast of both roots and leaves. Salt movement in the symplast of roots is also a basic feature of the Crafts-Broyer theory of root pressure and its modification by Laties and his coworkers (Crafts and Broyer, 1938; Luttge and Laties, 1966).

Accumulation of salt in cells

It is generally assumed that the first step in salt absorption through roots is accumulation of ions in the root cells. Although this may not be universally true, salt accumulation is such a basic characteristic of all kinds of plant cells that it deserves considerable attention. The accumulation of ions by individual cells is a highly selective process, some ions (such as potassium) being accumulated energetically while other ions (such as sodium) are often excluded or possibly even removed by some kind of secretory process (MacRobbie and Dainty, 1958; Kylin, 1966; Pitman and Saddler, 1966).

It is generally assumed that two kinds of processes are involved in salt accumulation: nonmetabolic adsorption exchange and metabolic or active uptake. These two processes can be described as follows:

Adsorption Exchange	*Active Transport*
Equilibrium attained rapidly	Equilibrium attained slowly
Not dependent on metabolism	Dependent on metabolism
Little affected by temperature, aeration, and respiration inhibitors	Much affected by temperature, aeration, and respiration inhibitors
Reversible	Relatively irreversible
Nonselective, but mostly concerned with cations	Highly selective; involves both anions and cations

The amounts and kinds of ions accumulated vary with the kinds of cells and environmental conditions. Aeration, amount of carbohydrate in the roots, concentration of ions in the cells and in the substrate, pH, light, temperature, and the species of plant are among the factors which affect the rate and amount of accumulation. Some of these factors will be discussed later; the reader is also referred to Sutcliffe (1962, chap. 4).

Most plots of the time course of ion uptake are curvilinear, showing rapid uptake for a short time followed by a long period of slower uptake at a steady rate (see Fig. 7.1). The rapid uptake during the early part of the time course is attributed to very rapid nonmetabolic uptake by exchange. However, the importance of this phase with respect to ion uptake by roots over long periods of time has probably been overemphasized. It is often assumed that the adsorption phase is an essential preliminary step for metabolic uptake, but Epstein and Leggett (1954) pointed out that the rate of metabolic absorption is largely independent of the amount of ion adsorbed on root exchange surfaces. Laties (1959) also presented strong arguments that accumulation does not necessarily follow adsorption exchange, but occurs simultaneously with it, as suggested by Fig. 7.1.

Epstein et al. (1962) suggest that the curvilinear plots can be regarded as artifacts resulting from the immersion of roots in single salt solutions containing no calcium. Their experiments with roots in solutions containing rubidium plus 0.5 m*M* calcium chloride show uptake of rubidium to be a strictly linear function of time from the very beginning of the experimental period (see Fig. 7.2). Furthermore, there is evidence of temperature sensitiv-

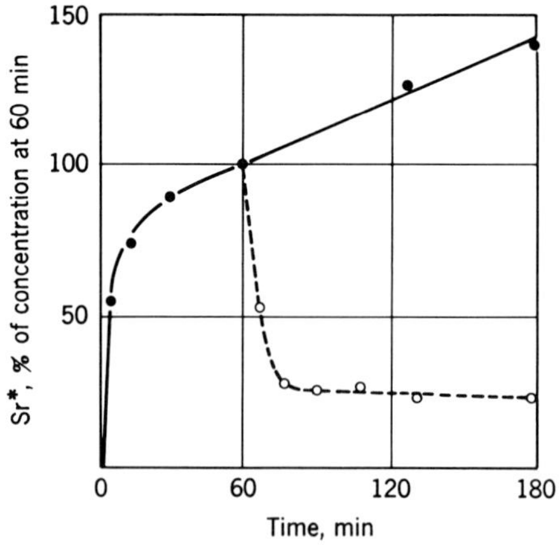

FIG. 7.1 Time course of radioactive strontium uptake by excised barley roots is shown by the solid line. The broken line shows loss of strontium when the roots were transferred to a solution of nonradioactive strontium. Uptake during the first 30 min was largely nonmetabolic, and this fraction was quickly lost by exchange with nonradioactive strontium. Uptake after the first 30 min was metabolic, and this fraction of the strontium was largely nonexchangeable. (*From Epstein and Leggett,* 1954.)

ity even during the first few minutes. Thus, the time course of absorption of the cation rubidium is exactly like that of an anion such as chloride. In the presence of calcium the initial diffusion into free space is so rapid that it is not apparent as a rate-limiting factor even during the first few minutes of exposure to the rubidium solution. Apparently the calcium saturates the cation exchange capacity of the roots, leaving no sites for nonmetabolic adsorption of rubidium. It seems reasonable to suppose that in nature the exchange sites on growing roots are occupied as rapidly as they are produced and that metabolic uptake alone is responsible for the steady, long-term movement of ions into root cells.

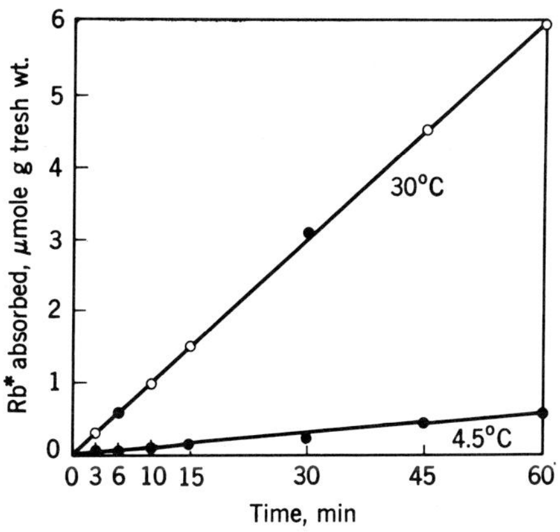

FIG. 7.2 **Absorption of rubidium by excised barley roots at 4.5 and 30°C from a solution containing calcium. In the presence of calcium, uptake was linear instead of curvilinear, as shown in Fig. 7.1. Low temperature practically stopped metabolic uptake. (*From Epstein et al.*, 1962.)**

A variety of theories have been developed to explain the transport of salt into the vacuoles of cells. These have been discussed in the articles and books cited at the beginning of this chapter, and we shall discuss only the carrier hypothesis, which is the most widely accepted theory today (Brouwer, 1965*b*; Epstein, 1960, 1965; Laties, 1959). This hypothesis assumes that ions form combinations with special molecules called carriers at the outer surfaces of the membranes, much as enzymes form temporary combinations with substrate molecules. The resulting compound can move through the membrane which the ions alone could not penetrate. The ions separate from the carrier and move into the vacuole or some other component of inner space, and the carrier becomes available to move more ions.

Various possible carrier mechanisms are described by Epstein (1965), Laties (1959), and Sutcliffe (1962, chaps. 3 and 5). Selectivity in ion accumulation is presumably controlled by differences in ability of carriers to form

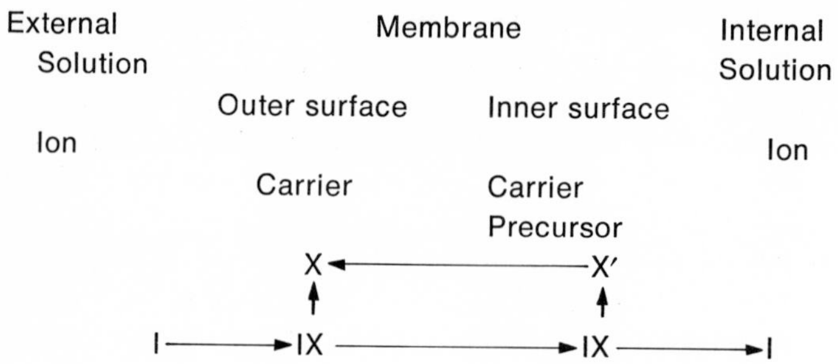

combinations with various ions. Potassium absorption is inhibited competitively by rubidium, indicating that the two ions use the same carrier, and at low concentrations sodium does not compete with either. However, at high concentrations sodium does compete with potassium, indicating that a second carrier or site is used by both at high concentrations. It has recently been proposed that there are two or more levels of carrier system, one being saturated at a much lower concentration than the other (Epstein, 1966). This will be discussed in a later section dealing with the effects of concentration on ion uptake. Some of the difficulties of the carrier hypothesis are discussed by Brouwer (1965).

Since carriers are products of cell metabolism, their formation is affected by a variety of internal and environmental conditions. Freshly cut slices of tissue often show little ability to accumulate salt, but after a period of incubation under suitable conditions they begin to accumulate. Presumably, carriers are synthesized during this period (Laties, 1964). It has been suggested that they might be enzymes (permeases or translocases of some writers), high-energy phosphorylated compounds, or even cytochrome (Lundegårdh, 1954), but their exact nature remains unknown.

Although it is agreed that salt accumulation is related to metabolism, there has been some difference of opinion whether the respiratory system functions directly in ion accumulation or indirectly by supplying energy for synthesis of carriers and binding sites. Lundegårdh (1940, 1955) suggested that anion accumulation is linked directly to respiration, with the cytochrome system acting as the carrier. According to this scheme, anions are picked up by cytochrome oxidase and move inward along the cytochrome system, while electrons move outward. At the same time, cations move in passively along the gradient created by active transport of anions. This view is advanced in modified form by Briggs et al. (1961). Epstein (1965) points out several objections to this theory, including the fact that cytochrome occurs in the mitochondria and is very inconveniently located to function as an ion carrier. Bowling et al. (1966) also suggest that anions are moved by active transport, while cations move passively.

It seems more likely that phosphorylated compounds function as carriers.

Perhaps carrier molecules are enzymatically phosphorylated at the outer surface of the membrane, and the ion-carrier-phosphate complex crosses the membrane and dissociates at the inner surface, releasing the ion and phosphate. The carrier is then rephosphorylated as it migrates back to the outer surface of the membrane, where it picks up another ion. In green tissues exposed to light, photophosphorylation might directly supply the energy for uptake (Jeschke, 1967; MacRobbie, 1965), but this could not occur in roots which grow and function in darkness.

Energy from respiration must be used to synthesize carriers and binding sites and to maintain membrane structure. Since maintenance of normal membrane structure depends on a supply of metabolic energy, both permeability and active transport are modified by treatments which reduce respiration. For example, low concentration of respiration inhibitors such as azide or cyanide reduce the permeability of roots to both water and salt (see Chap. 6). Therefore, reduction in uptake of water or salt uptake by a respiration inhibitor is not proof that it occurs by active transport.

Although the carrier theory is the most generally accepted explanation of salt accumulation, readers are reminded that it is simply a convenient theory which explains many of the observations on salt accumulation. It is already undergoing modifications, and questions are being raised concerning its general validity. For example, Oertli (1967) suggests that an equally effective mechanism might consist of an ion pump actively transporting salt inward, combined with a passive leak which would allow salt to move in either direction according to the internal and external concentrations. This scheme would explain observations not explained by the classical carrier theory. There is also an intimate relationship between ion uptake and organic acid metabolism, inequality in uptake of cations and anions being balanced by changes in organic acid metabolism (Hiatt, 1967).

Absorption of salt through roots of plants

Some writers appear to assume that absorption of salt by intact plants involves only the accumulation of salt in the root cells. This is not correct. Two processes involving ions occur simultaneously in roots, (1) the accumulation of ions in root cells and (2) the radial movement of ions inward across the living cells to the xylem. The former is independent of water movement, but the latter is clearly affected by the rate of water uptake, as will be shown later in this chapter. In addition, in at least some instances, inorganic ions such as nitrate are converted into organic compounds which are then released into the xylem.

In 1951 Biddulph wrote that the radial movement of minerals through roots into the xylem is one of the least understood processes in the realm of mineral absorption and translocation, and a decade later Russell and Barber

(1960) expressed the same opinion. Although much has been published on this question, it remains impossible to give a fully satisfactory explanation of the process. However, the various theories will be presented as clearly as possible.

Movement of ions from soil to roots

Before discussing ion uptake through roots, the movement of ions from soil to root surfaces must be considered. The concentration of free ions in the soil solution is generally low, and many of the cations are adsorbed with varying degrees of firmness on negatively charged clay particles and organic matter in the soil. Nutrient anions such as nitrate and sulfate mostly occur in the soil solution and are relatively mobile, except phosphate, which is firmly bound to the soil matrix. This led Bray (1954, 1963), Voigt et al. (1964), and Wiersum (1961, 1962) to differentiate between the total soil volume occupied by the root system and the root sorption zone, which is the thin layer of soil in immediate contact with the roots (see Fig. 7.3). Mobile ions

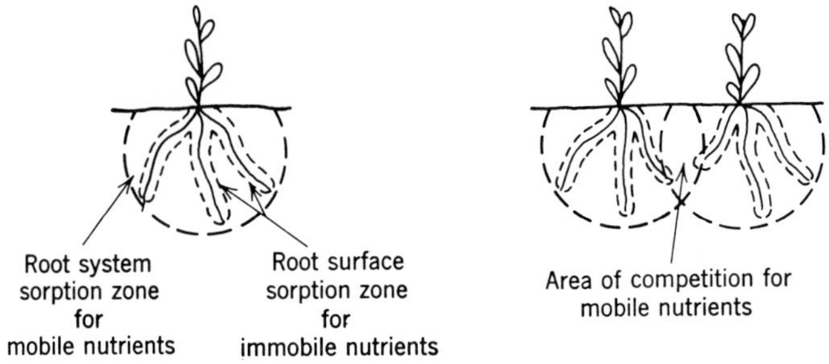

FIG. 7.3 The soil volume occupied by a root system is much larger than the root sorption zone (rhizosphere) or volume of soil actually in contact with the roots. Immobile ions such as phosphate can be absorbed only from the rhizosphere, but mobile ions such as nitrate can be absorbed from the entire mass of soil occupied by the root system. (*After Bray*, 1954.)

such as nitrate and potassium can be utilized from the entire root zone, but immobile ions such as phosphate are available only from soil in the immediate vicinity of the roots. Voigt et al. (1964) estimated that less than 10 percent of the total soil volume occupied by roots supplied calcium and potassium to young pine trees, if the soil within 1 cm of the root surfaces was regarded as the sorption zone. Lewis and Quirk (1967) concluded from successive autoradiographs that the zone of depletion of ^{32}P extends out only about 1 mm from the roots. However, mobile ions in the soil solution are presumably available to roots from greater distances since water moves to roots from distances of several centimeters (see Chaps. 3 and 6).

Another approach is to compare the nutrient requirements of plants with the amount supplied in the soil solution absorbed to replace water lost by transpiration. Fried and Shapiro (1961) estimated that corn plants transpiring 25 cm of water in a season can obtain all the calcium, magnesium, and potassium they require from a typical soil solution. This was questioned by Olsen et al. (1962), who assigned a major role to diffusion. It seems probable that the movement of ions by diffusion is aided by mass flow of soil solution toward roots of transpiring plants. Data of Barber (1962) and Lagerwerff and Eagle (1962) indicate that water enters the roots more rapidly than ions, resulting in accumulation of ions in the immediate vicinity of roots. Such an accumulation ought to result in a situation favorable for rapid movement of ions into roots by diffusion and active transport.

Root extension is also an important factor in the absorption of ions, especially those which are not very mobile. Wiersum (1962) described an experiment in which the degree of root branching was varied in several species by growing them in substrates composed of particles of various sizes. The uptake of nitrate was not affected by the amount of branching, but phosphate uptake was greatly reduced when root branching and density were reduced. The rate of root extension of maize and its role in phosphate absorption are shown in Fig. 7.4. The importance of root surface and root growth for mineral absorption by perennial plants is shown in the "little leaf" disease of shortleaf pine. This disease develops in pine trees growing where a combination of soil conditions unfavorable for root growth and damage to roots by *Phytopthora cinnamomi* reduce the absorbing surface so much that symptoms of severe nitrogen deficiency develop (Campbell and Copeland, 1954). The importance of the increased surface for mineral absorption provided by mycorrhizal roots was discussed in Chap. 4.

Jenny and Overstreet (1939) proposed that plant roots absorb ions directly from the soil surfaces with which they come into contact, without the ions passing into the soil solution. This theory is based on observations that roots absorb ions more rapidly from a suspension than from a solution of similar ion concentration. Such a mechanism would be effective only over very short distances. Thus, the contact exchange theory seems to be of limited usefulness and applicable chiefly to immobile ions.

Readers are referred to the book by Fried and Broeshart (1967) for a more detailed discussion of the complex relationship between soil and roots with respect to ion absorption.

Absorbing zone of roots

Considerable change in viewpoint has occurred with respect to the location of the region where ions are absorbed in largest quantities. Early research on barley roots by Steward and others (Prevot and Steward, 1936; Steward, Prevot, and Harrison, 1942) suggested that the greatest accumulation of salt typically occurs near the apex, as shown in Fig. 7.5. However, recent studies

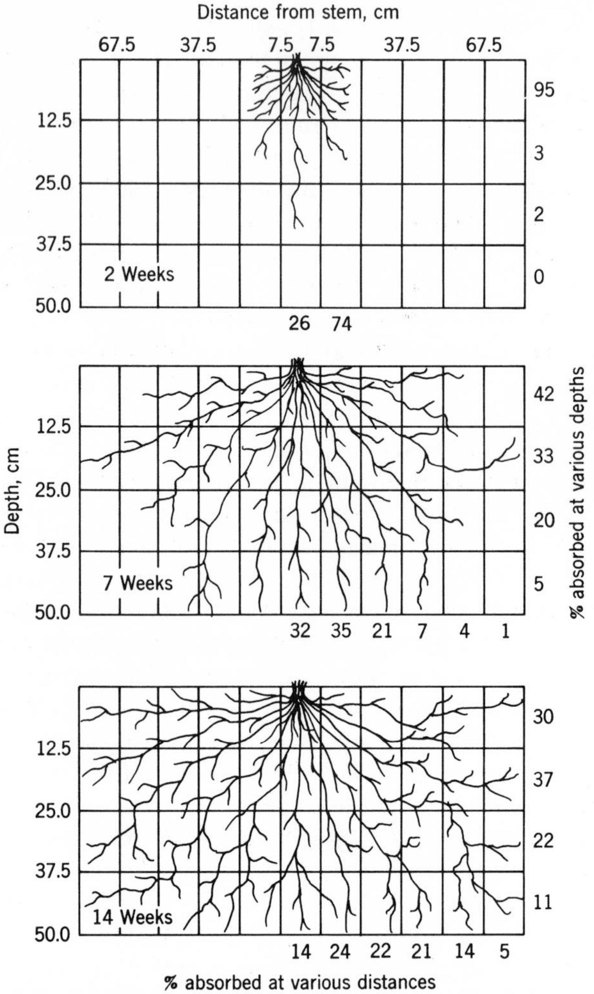

FIG. 7.4 Diagrams showing the effect of extension of corn root systems growing in Cecil clay loam on amounts of ^{32}P absorbed at various depths and distances from the plant by root systems of various ages. The percentages at the right of each figure are the percentages of ^{32}P absorbed at each depth, and those across the bottom are the percentages absorbed at various distances from the corn plants. (*From Hall et al.*, 1953.)

of the accumulation of radioactive isotopes show that this is not always true. Work by Canning and Kramer (1958), Kramer and Wilbur (1949), Wiebe and Kramer (1954), and others shows that roots often accumulate surprisingly large amounts of various ions several centimeters behind the root tips, as shown in Fig. 7.6. (See also Fig. 7.6*a*.) Typically, there is heavy accumulation just behind the apex, less accumulation in the region of elongation, and heavy accumulation 1 cm or more behind the root tip. However, many excep-

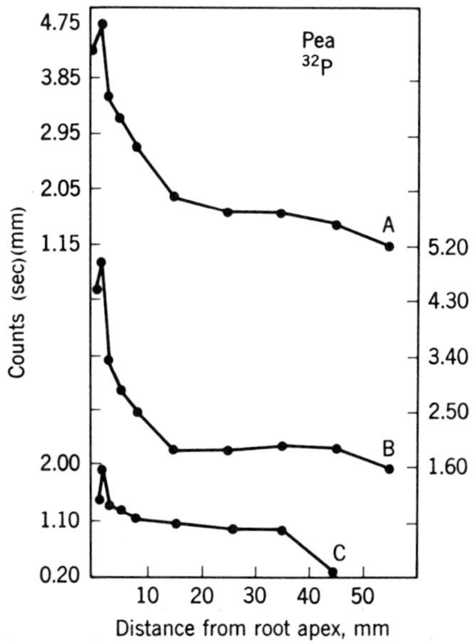

FIG. 7.5 **The amounts of ^{32}P accumulated in various regions of pea roots, expressed as counts per second per millimeter of root length. Lines A, B, and C are averages for different groups of roots. (*From Canning and Kramer*, 1958.)**

tions to this pattern occur which cannot be related to any observable differences in structure or physiological conditions.

More important than the region of accumulation is the region of roots through which most salt enters and passes to the shoots. Wiebe and Kramer (1954) studied this by supplying radioactive isotopes to a 3-mm segment at various distances behind the tip of barley roots, while the remainder of the root system was immersed in nonradioactive nutrient solution (see Fig. 7.7). Later, Canning and Kramer (1958) performed similar experiments on corn, cotton, and pea roots. These experiments showed that although ions are accumulated freely in the apical regions of attached roots, less translocation occurred to other parts of the seedlings from this region than when the isotope was supplied 20 to 50 mm behind the root tips. This view is supported by the observations of Russell and Sanderson (1967). The xylem is usually well differentiated at this distance behind the root tips, and it is also the region through which water enters most readily. A generalized diagram of root development in relation to salt absorption and accumulation is shown in Fig. 7.8.

Absorption through suberized roots

Practically all published material on salt absorption deals with young, unsuberized roots which have undergone no secondary growth. The fact that

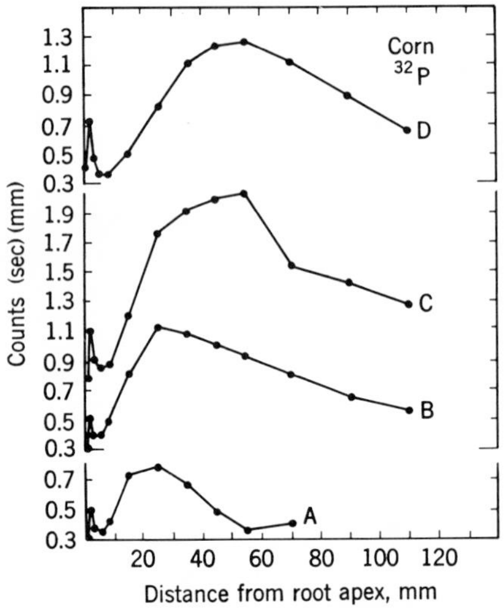

FIG. 7.6 The amounts of ^{32}P accumulated in various regions of corn roots of different ages and lengths, expressed as counts per second per millimeter of length. *A* is the average of 8 roots with an average length of 80 mm, *B* is the average of 8 roots with an average length of 131 mm, *C* is the average of 12 roots with an average length of 113 mm, and D is the average of 9 roots with an average length of 151 mm. (*From Canning and Kramer*, 1958.)

by far the larger part of the root systems of most trees and shrubs has undergone secondary growth and is completely suberized has often been ignored. It was pointed out in Chap. 4 that often as much as 95 to 99 percent of the root surface of forest trees is suberized and that much of the water absorption must occur through suberized roots (Kramer and Bullock, 1966). It seems equally probable that considerable amounts of salt must be absorbed through suberized roots of trees. Crider (1933) and Nightingale (1935) reported the absorption of minerals through suberized roots of woody plants, and recent work in the author's laboratory indicates measurable uptake of salt through suberized roots of several species. The importance of mycorrhizal roots in salt absorption was mentioned in Chap. 4.

Radial movement of salt into the stele

There are three possible pathways for ion movement from the root surface to the xylem: (1) movement from vacuole to vacuole, (2) movement through the symplast, bypassing the vacuoles, and (3) movement through the free space of the walls, or some combination of these pathways. It is questionable whether much movement occurs through the vacuoles because ions accumulated in vacuoles move out very slowly. Furthermore, experiments by Broyer (1950) and others indicate that shoots tend to absorb ions directly from the root medium rather than from the vacuoles of the root cells. Thus, the vacu-

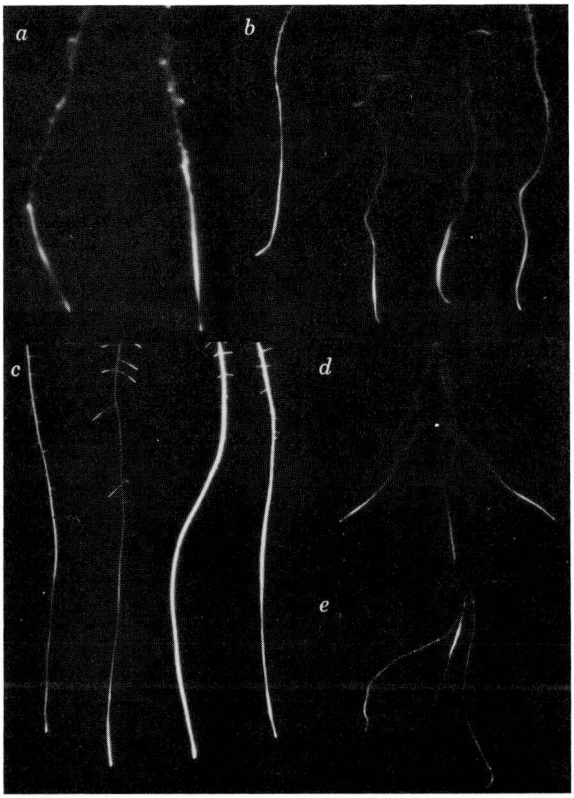

FIG. 7.6a Autoradiographs of roots of four species showing various patterns of accumulation of ^{32}P. (*a*) Two large roots of red pine. (*b*) Four small roots of barley. (*c*) Four adventitious roots of tomato. (*d*) and (*e*) Roots of barley seedlings. Most roots show heavy accumulation in or just behind the meristematic region, lower accumulation in the region of elongation, and relatively heavy accumulation in older regions. *a* and *d* show typical patterns; *b*, *c*, and *e* show variations from the typical pattern. (*From Kramer and Wiebe*, 1952.)

oles probably function as diversionary sinks for salt, at least in low-salt plants, rather than as a major pathway for movement (Broyer, 1950; Epstein, 1960). A comparison of bromide uptake of barley seedlings and excised barley roots led Böszörmenyi (1965) to conclude that uptake in both is limited by accumulation in the symplast.

The work of a number of investigators suggests that ions are accumulated in the cytoplasm and move from cell to cell through the plasmodesmata, which connect the cytoplasm of adjacent cells to form a continuous system, the symplast. Movement is assumed to occur by diffusion, assisted by cytoplasmic streaming. Various versions of this transport system have been proposed by Arisz (1956, 1964), Crafts and Broyer (1938), Lundegårdh (1955),

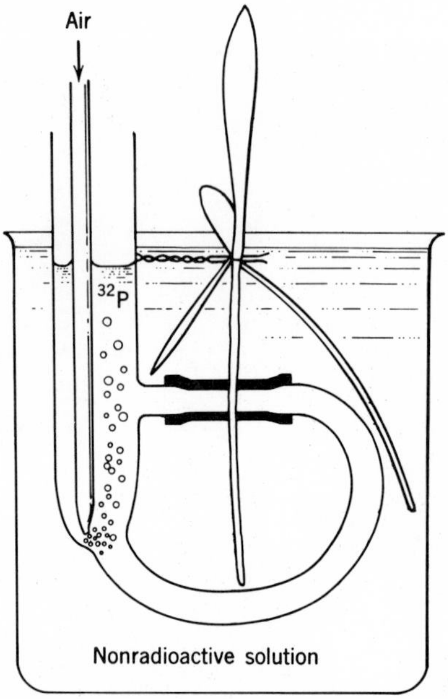

FIG. 7.7 A potometer to supply radioactive tracers to a specific location on roots. The root is passed through holes of proper size burned in a piece of soft rubber tubing and sealed with lanolin. The tracer is circulated through the side arm and around the root by bubbling air into the large tube. A 3-mm segment of root was exposed in this apparatus. (*From Wiebe and Kramer,* 1954.)

Luttge and Laties (1966), and others. Readers might refer to figures in Chap. 4 to refresh their memory concerning root structure.

It seems probable that ions also move in through the free space of the cell walls by diffusion and mass flow in the transpiration stream, at least until they reach an ion barrier. This idea was proposed long ago by Scott and Priestley (1928), who stated that "there is no reason to assume any obstacle to the free diffusion inwards of this solution [the soil solution] along the cellulose walls of the cortex up to, and including, the outer tangential walls of the endodermis," where the Casparian strips force it to move through the protoplasts. This view tends to transfer the absorbing surfaces of roots from the epidermis to the outer surface of the ion barrier, in contrast to the views of certain Swedish workers who seemed to regard the epidermis as an important ion barrier in roots (Lundegårdh, 1955; Sandstrom, 1950). Incidentally, Leggett and Gilbert (1967) reported that 90 percent of the calcium in soybean roots occurs in the epidermis, while potassium is uniformly distributed throughout the roots.

The chief problem in connection with salt absorption is to explain how ions are transferred from the living cells in the stele to the sap in the dead conducting elements of the xylem. The extensive literature on this problem

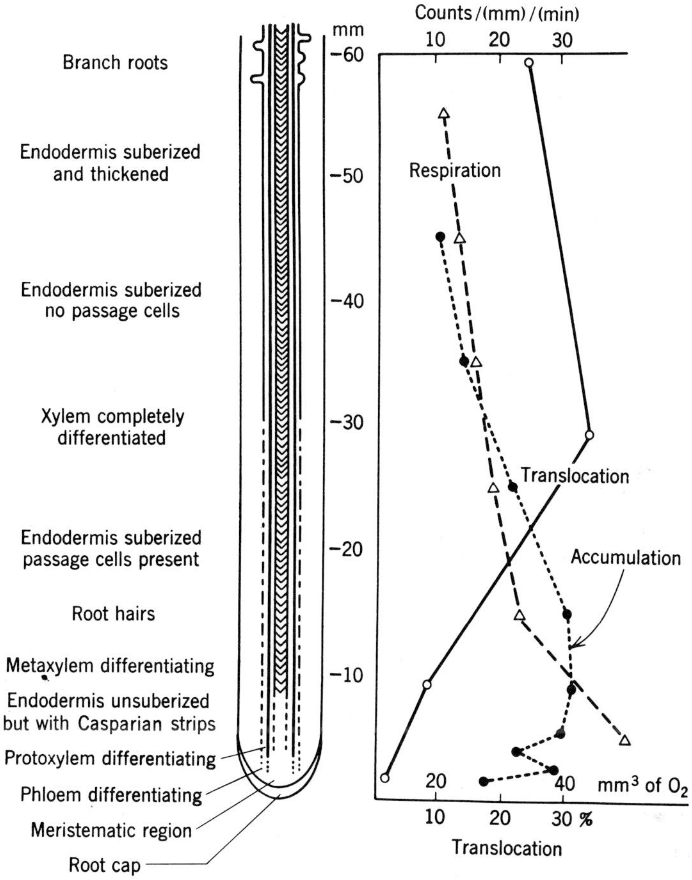

FIG. 7.8 Diagram of apical region of a barley root showing the relationship between root structure and the regions through which salt is absorbed. The curve for respiration is based on data by Machlis (1944). The curve for accumulation is based on data of Wiebe (1956). The curve for translocation represents the percentage of ^{32}P absorbed which was translocated away from the region where it was absorbed (*Wiebe and Kramer,* 1954). The relative position of the various regions is probably similar in all growing roots, but the actual distances behind the root apex vary greatly among species and among roots growing at different rates.

was reviewed recently by Brouwer (1965). It is difficult to summarize the various theories, except for the general agreement that some kind of metabolic activity or secretion is involved on the part of the endodermis or other living cells in the pathway which act as an ion barrier (Arnold, 1952; Arisz, 1956; Luttge and Weigl, 1962; Lundegårdh, 1955). Hylmö (1953) suggested that the salt in the xylem sap was first accumulated by protoplasts which die during the formation of xylem elements and release the accumulated salt, a view advanced earlier by Priestley (1922). This would restrict salt accumulation to those regions of roots where conducting elements are differentiating. Also, the amount of salt present in root exudate seems much greater than

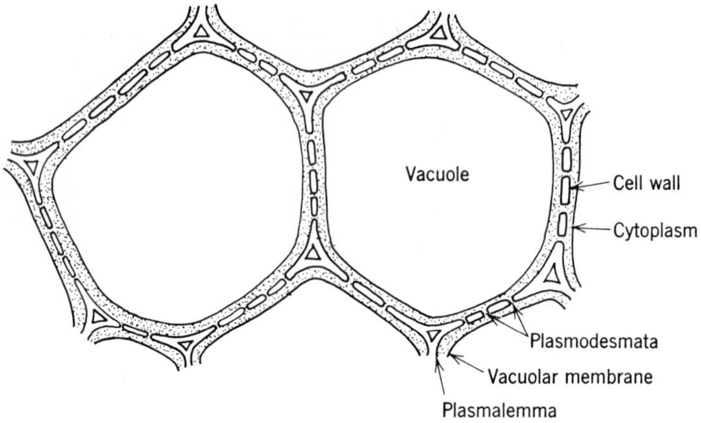

FIG. 7.9 **Diagram of cells in the cortical parenchyma indicating possibilities for salt movement through the walls and the symplast formed by plasmodesmata connecting the protoplasts of adjacent cells. The size of the plasmodesmata is greatly exaggerated; see Fig. 1.14. Resistance to salt and water movement through the cell walls and symplast is probably lower than for movement across the vacuoles.**

could be supplied by differentiating xylem elements. However, according to Scott (1963, 1965), the protoplasts of xylem elements in the absorbing zone of branch roots remain alive, and solutes entering such roots move into living protoplasts. Thus, a limited amount of salt might be accumulated directly in the xylem protoplasts and released when they disintegrate. This possibility deserves further investigation, in view of the report by Anderson and House (1967) that there is a relationship between the number of living protoplasts in the xylem and the amount of salt absorbed by corn roots.

One of the most frequently cited theories of salt accumulation is that of Crafts and Broyer (1938), who suggested that ions are accumulated actively in the well-aerated cells of the epidermis and outer cortex. The ions are carried inward across the endodermis and into the stele through the symplast by diffusion and cytoplasmic streaming. The poorly aerated cells of the stele are supposed to have less capacity to accumulate ions, which therefore tend to leak out and accumulate in the xylem vessels. They are unable to escape from the stele because the suberized Casparian strips of the endodermis prevent back diffusion along the walls.

Laties and Budd (1964) suggested that the inability of the cells of the stele to retain ions is caused by presence of an inhibitor rather than by inadequate aeration. They separated the stele from the cortex of corn roots and found that initially the stele had little capacity to accumulate ions. However, after it was incubated for 24 hr, it accumulated ions as rapidly as the cortex. This work supports the hypothesis of Crafts and Broyer, except that the increase in capacity to accumulate ions is attributed by Laties (1962) and Laties and Budd (1964) to loss of a volatile inhibitor rather than to improvement of aeration. This view also avoids the difficulty of assuming that the endodermis secretes ions into the stele, but it does not detract from its importance as a passive ion barrier. Removal of the cortex from pea roots interfered with ion uptake and translocation to the shoot, suggesting that transport to the shoot is dependent on activity of root cells (Branton and Jacobson, 1962). On the other hand, Yu and Kramer (1967) found that freshly separated stele of corn roots accumulated ^{32}P as actively as the cortex from which it was separated. The isolated stele had a higher oxygen uptake than the cortex from which it was separated. These observations raise doubts concerning theories of salt transport based on the assumption that the cells of the stele are unable to accumulate or retain salt. Perhaps different mechanisms dominate salt uptake in different kinds of roots and for different kinds of ions.

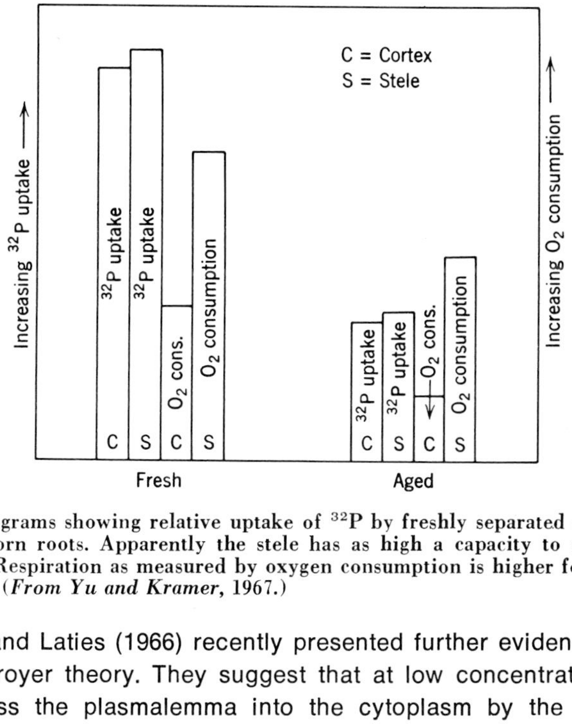

FIG. 7.9a Diagrams showing relative uptake of ^{32}P by freshly separated and aged cortex and stele of corn roots. Apparently the stele has as high a capacity to accumulate ions as the cortex. Respiration as measured by oxygen consumption is higher for the stele than for the cortex. (*From Yu and Kramer,* 1967.)

Luttge and Laties (1966) recently presented further evidence supporting the Crafts-Broyer theory. They suggest that at low concentrations ions are moved across the plasmalemma into the cytoplasm by the accumulation mechanism operating at low salt concentrations, mentioned earlier in this chapter. However, they propose that at higher external concentrations, salt enters the cytoplasm by diffusion and is carried across the vacuolar membrane into the vacuoles by the salt absorption mechanism operating at high concentrations. They suggest that the operation of two salt absorption mechanisms, one active, the other partly passive, resolves the conflict between those who argue that salt transport is entirely active and those who claim it is partly passive. However, if the plasmalemma is permeable to ions at high concentrations, it seems difficult to explain how a high concentration can be accumulated and maintained in the stele. Furthermore, this opens the cytoplasm as a pathway for mass flow of ions in the transpiration stream. The results of these experiments caused Luttge to abandon the idea that the endodermis secretes salt into the stele (Weigl and Luttge, 1962). On the other hand, Welch and Epstein (1968) claim that the two ion transport mechanisms operate in parallel across the plasmalemma.

All hypotheses concerning salt transport were developed from research on young primary roots. However, root pressure has been observed in suberized root systems of woody species which no longer possess an endodermis (O'Leary, 1965). There is no direct evidence concerning the location of the ion barrier in roots which have undergone secondary growth, but it is possible that the cambium performs this function. Probably, the mechanism which brings about salt transport across the phloem and cambium of suberized secondary roots is similar to that operating in younger primary roots, but the problem of radial salt transport in suberized roots requires more study.

None of the theories of salt transport explains the diurnal variations in root pressure reported by various investigators (Grossenbacher, 1938; MacDowall, 1964; Vaadia, 1960; Wallace et al., 1967). Usually the maximum rate of exudation occurs about noon and the minimum rate during the night, as shown in Fig. 5.5. If root pressure depends on the osmotic pressure of the sap in the root xylem, there must be diurnal variations in movement of salt into the root xylem. Hanson and Biddulph (1953) and Wallace et al. (1966) reported greater translocation of ions from roots to shoots during the day than at night in beans. Vaadia (1960) concluded that there is greater transport of salt into the xylem during the day than at night. However, no one has offered an acceptable explanation of this phenomenon. The autonomic variations are not closely related to variations in respiration (MacDowall, 1964; Skoog, Broyer, and Grossenbacher, 1938; Vaadia, 1960), although they must be related to metabolism in some way. Wallace et al. (1967) claim that the periodicity in salt transport is regulated by the tonoplast. Apparently they believe that salt is accumulated in the vacuoles and is later released, an explanation quite different from the Crafts-Broyer theory, which assumes that accumulation and transport occur in the symplast. The more rapid release of salt during the day causes greater root pressure exudation during the day (Wallace et al., 1966). It seems possible that another factor in periodicity is the diurnal change in root permeability mentioned in Chap. 5.

Current theories of salt transport do not fully take into account the fact that inorganic ions are sometimes incorporated into organic compounds in the roots. For example, all the nitrogen in the xylem sap of some species occurs as organic compounds, and organic compounds containing phosphorus and sulfur have also been reported (Bollard, 1960). Somewhere in its passage through the roots of many plants, nitrate is apparently converted into amides, ureides, and amino acids, but the site of these enzymatically mediated reactions has not been localized. Neither is it known how the organic molecules get into the xylem sap. More will be said about this problem in the section on absorption of organic compounds.

Factors affecting salt absorption

Salt accumulation in root cells and transport to shoots are affected by a variety of internal factors, such as species, condition of tissue, rate of respiration, and sugar and salt content, also by environmental factors, such as aeration, temperature, composition, concentration, and pH of the root medium.

SPECIES DIFFERENCES. There are wide differences among plants of different species in their ability to absorb various ions from the same soil or culture solution. Miller (1938, pp. 283–291) summarized considerable data on this subject, and other data are given in a review by Bollard and Butler (1966). A classic example is the study made by Collander (1941) of cation absorption by 21 species of plants from various families and habitats when grown in the same nutrient medium. Some of the data are presented in Fig. 7.10 to show the wide range of differences in uptake of sodium by various species. The genus *Astragalus* is well known as an accumulator of selenium, but

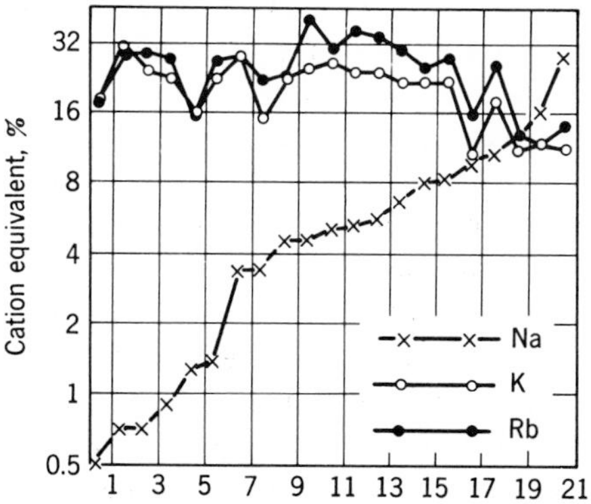

FIG. 7.10 Differences in amounts of various ions absorbed by different species of plants. The species are arranged according to increasing sodium content. Although there were wide differences in amount of sodium absorbed, there was not much difference in the amounts of potassium and rubidium absorbed. (*From Collander,* 1941.)
Arranged according to increasing sodium content, the plants are:

1. Fagopyrum
2. Zea
3. Helianthus
4. Chenopodium
5. Salsola
6. Pisum
7. Nicotiana
8. Solanum
9. Spinacia
10. Avena
11. Aster
12. Papaver
13. Lactuca
14. *Plantago lanceolata*
15. Melilotus
16. Vicia
17. *Atriplex litorale*
18. Sinapis
19. Salicornia
20. *Plantago maritima*
21. *Atriplex hortense*

A. missouriensis accumulated only 3.1 ppm from a soil containing 2.1 ppm, while *A. bisulcatus* accumulated 1,250 ppm. In another study, the ash of *Andropogon* contained 65.4 percent of silica, while the ash of *Prunus pumila* growing on the same soil contained only 1.5 percent; but *Prunus* accumulated three or four times as much calcium as *Andropogon.*

Some species of mangroves which grow in sea water exclude the salt almost entirely, but others absorb 10 to 20 percent of the sodium chloride present and excrete it from salt glands in the leaves (Scholander et al., 1966). A number of halophytic species growing in saline habitats are notable for the excretion of salt. These include *Tamarix* and some species of *Atriplex.* Even cotton secretes small quantities of salt from its leaves. More puzzling is the fact that in some species of plants the concentration of salt in the roots is much greater than in the shoots (Black, 1956; Collander, 1941; Huffaker and Wallace, 1959; and others). There is no obvious mechanism by which salt entering the roots can be excluded from the shoots. The only plausible explanation offered thus far is that sodium is accumulated so strongly in the tissues of the roots and lower part of the stem that it never reaches the shoot (Jacoby, 1964, 1965). However, such a mechanism should eventually become saturated and allow sodium to pass on to the leaves.

In some instances, the differences in uptake of specific ions by different plants growing in the same environment are known to be controlled genetically. The control may be either on the uptake mechanism directly or on the roots and vascular system. Differences in uptake of phosphorus by corn have been related to genetic differences in extent of branching of roots or differentiation of the vascular system (Rabideau et al., 1950; Smith, 1934). In some instances, the differences appear to involve the absorption mechanism; in others, translocation. Brown (1963) concluded that iron absorption and translocation could be separated in soybeans. Iron deficiency of one variety studied resulted from the failure of iron to be translocated to the shoots. Oranges grown on trifoliate orange rootstocks are said to be unusually susceptible to zinc and iron deficiencies because of restricted translocation from roots to shoots. Differences in tendency for magnesium deficiency to develop in inbred lines of corn are attributed by Foy and Barber (1958) to differences in amounts of magnesium immobilized in the stems.

Other examples of genetically controlled differences in ion uptake are cited by Bollard and Butler (1966), Epstein and Jefferies (1964), and Wallace (1963). They describe several instances of differences among species in tolerance of heavy metals and total salt concentration which must have a genetic basis. For example, a strain of *Agrostis tenuis* which has apparently grown on an abandoned lead mine in Great Britain for approximately 1,000 years has extremely low calcium and phosphate requirements and does not grow well in garden soil. The physiological basis for differences in tolerance of heavy metals has not been explained, but it might be related to chelation by organic compounds.

Epstein and Jefferies (1964) suggest that the existence of genetically controlled differences in salt uptake provides the basis for a productive plant-breeding program. Bernstein (1963) is less optimistic, because of the small differences found among varieties of carrot and lettuce. However, it seems that further study of breeding possibilities should be made in view of the differences in ion uptake and tolerance of salt which have been observed.

EXTENT OF ROOT SYSTEMS. The importance for water absorption of an extensive root system occupying a large volume of soil has been mentioned repeatedly. A wide-spreading, many-branched root system is equally important for salt absorption. For example, differences in the ability of various lines of corn to absorb phosphorus are correlated with differences in amount of root surface (Smith, 1934), and the greater success of one variety of barley over others in obtaining nutrients can be attributed to its early production of a dense mass of roots (Lee, 1960). The importance of an extensive root surface for absorption of ions of limited mobility was discussed earlier in this chapter in the section on movement of ions from soil to roots.

METABOLIC ACTIVITY OF TISSUE. Accumulation of salt is usually closely related to growth and metabolism. Enlarging cells have a high accumulation rate, but mature cells which have lost their ability to enlarge also lose their ability to accumulate ions. Some mature cells, such as those of slices of carrot or potato, regain their ability to accumulate as they regain ability to grow. During cell enlargement, there is an increase in absorbing surface and vacuole volume, but absorption of water keeps the salt concentration low in the vacuole. There also is synthesis of new binding sites and carrier molecules and incorporation of inorganic ions into organic compounds. Cessation of growth brings an end to these activities; the salt content of the vacuole increases, and salt accumulation slows or ceases. As cells become senescent, they lose their ability to retain salt.

As a result of these cellular relations, there is considerable redistribution of salt within a growing plant. The period of rapid vegetative growth is accompanied by rapid uptake of salt; but as leaves grow older, they often lose salt to newly enlarging leaves, stem tips, young fruits, and other rapidly growing structures. Thus, as pointed out by Steward and Sutcliffe (1959), there are gradients of concentration in plants related to the level of metabolism. In *Cucurbita pepo* it was found that the average concentration of bromide was leaf > stem > root. Furthermore, concentration was higher in young than in old leaves. In a shoot of *Populus nigra,* bromide was accumulated in the highest concentration in the leaves, to a much lower concentration in the phloem, and to a still lower concentration in the xylem. Minerals also tend to

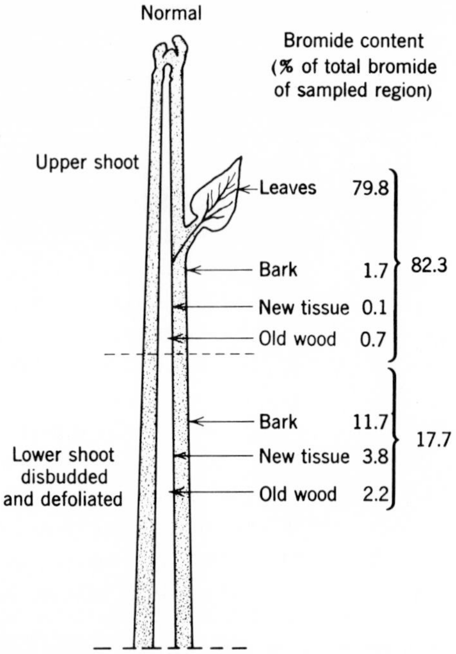

FIG. 7.11 **Relative amounts of salt accumulated in various regions of a growing shoot of *Populus nigra* which was disbudded and defoliated below the dashed line. Where leaves were allowed to grow, most of the salt was accumulated in them at the expense of other tissues. (*After Steward and Sutcliffe in Steward,* 1959.)**

move from vegetative structures to developing fruits and seeds. The internal redistribution of minerals in plants was reviewed by Biddulph (1959) and by Williams (1955).

RESPIRATION. As would be expected, conditions which lower the rate of respiration, such as inhibitors, oxygen deficiency, and low supply of sugar, usually reduce the accumulation of ions. The effect of oxygen concentration on salt accumulation is shown in Fig. 7.12 and that of KCN in Fig. 7.13. One puzzling feature of salt uptake is the fact that some kinds of tissue show a marked increase in respiration and accumulation of ions only if they are washed for a considerable time in water or a dilute solution before transfer to salt solution. Sutcliffe and others have suggested that this is related to production of carriers and binding sites, while Skelding and Rees (1952) and Laties (1959) attribute it to removal of an inhibitor. Perhaps both phenomena are involved.

A number of investigators have reported that respiration inhibitors at low concentrations decrease or inhibit both the transport of salt into the root xylem and the accumulation of salt in the root cells (see Brouwer, 1965). However, it is difficult to evaluate the effect of respiration inhibitors on salt

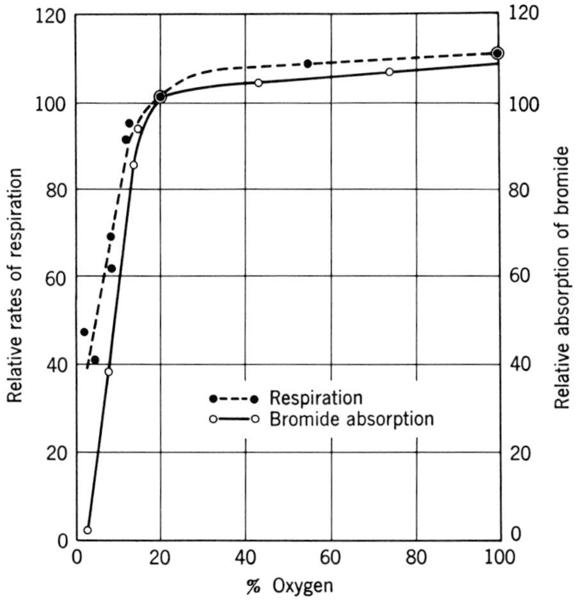

FIG. 7.12 Effect of oxygen concentration on respiration and bromide uptake by slices of potato tissue. (*From Steward*, 1933.)

transport because, as mentioned earlier, they also decrease the permeability of roots to water (Brouwer, 1954*b*, 1965; Mees and Weatherley, 1957; Lopushinsky, 1964) and probably to salt. This is shown graphically in Fig. 7.13.

INTERNAL CONCENTRATION OF SALT. As the concentration of salt in cells increases, their capacity to absorb additional salt decreases, and the curve for absorption, plotted over time, tends to flatten (see Fig. 7.14). The time required for absorption to level off may be only a few hours in a concentrated solution, but it may be several days in a dilute solution. It has been suggested that net uptake of salt represents the difference between influx and efflux, and that efflux increases with increased internal concentration until it balances influx. This is unlikely, because tests of efflux made with radioactive tracers indicate that leakage out of vacuoles is usually very slow. It seems more likely that an increasing proportion of the carrier molecules are rendered unavailable because they are combined with ions, and this blocks the transport system. The decrease in rate of absorption with time is more characteristic of mature cells than of those which are enlarging. In the latter, increase in volume of the vacuole keeps the salt concentration low, and continuing synthesis of new carrier molecules keeps a supply available.

It appears that salt usually moves more rapidly to the xylem through root cells high in salt than through those low in salt, because less is intercepted by the accumulatory activity of the cells in roots high in salt. For example,

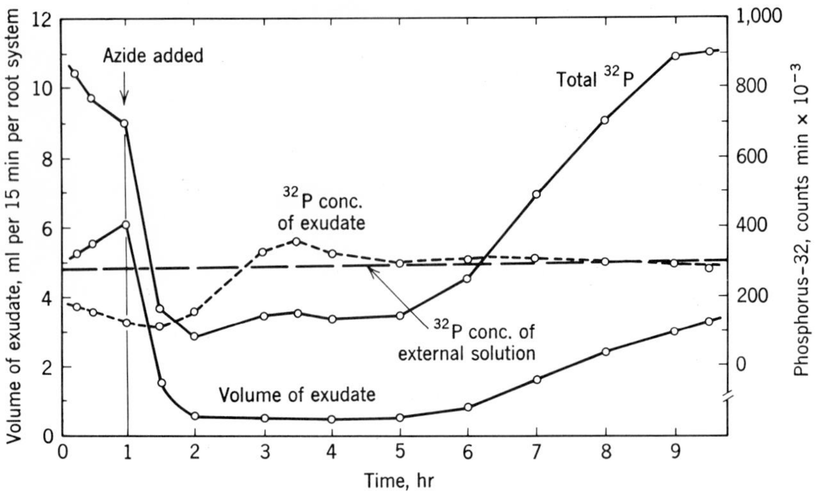

FIG. 7.13 Effect of $10^{-3}M$ sodium azide on movement of water and ^{32}P into the xylem of tomato roots. Pressure was maintained at 30 lb/in^2. Concentration of ^{32}P (dashed line) is in counts per minute per milliliter of exudate. The pH was adjusted to 4.9. After 3 hr salt began to leak out of the root cells, causing a temporary increase in concentration of the exudate. After 6 hr there was a large increase in permeability. (*From Lopushinsky*, 1964.)

Branton and Jacobson (1962) found less iron moved to shoots of iron-deficient pea plants than to shoots of plants adequately supplied with iron. Some exceptions occur, however. Hodges and Vaadia (1964) reported that roots already high in chloride not only accumulated less tracer chloride but also

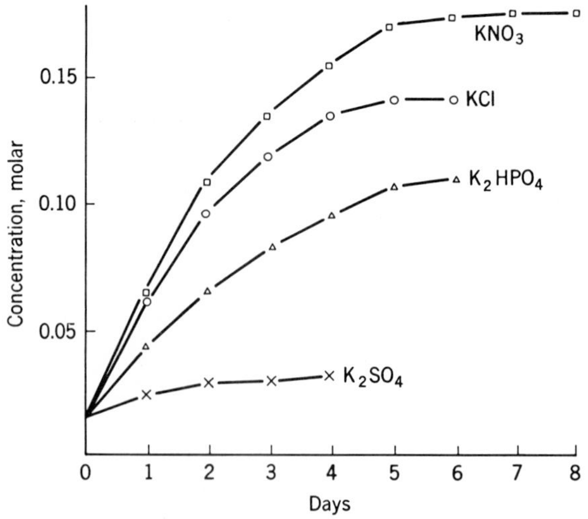

FIG. 7.14 Decrease in rate of uptake by beet root tissue of various ions with time and increasing internal concentration. The effect of various anions on potassium uptake is also shown. (*From Sutcliffe*, 1952.)

transported less to the xylem. They suggested that this was because chloride already present occupied the transport sites or carriers and reduced active transport of tracer chloride.

CONCENTRATION OF SUGAR. In view of the relationship between metabolism and salt uptake, it is not surprising to find that an adequate supply of sugar is necessary for salt uptake. It has been pointed out by Steward and others that failure to consider the past history of tissue, especially its sugar and salt status, can explain some of the contradictory conclusions in the literature. In 1940, Phillis and Mason reported that preventing translocation of carbohydrate to roots by girdling the stems reduced salt uptake by cotton roots. Eaton and Joham (1944) suggested that the decrease in mineral uptake which accompanies heavy fruiting of cotton can be attributed to reduced root metabolism caused by reduced movement of carbohydrates to the roots. However, it is difficult to determine how much of the reduced salt uptake is caused by reduced metabolism and how much results from decreased root extension.

AERATION. The inhibitory effects of inadequate aeration on root growth and water absorption were discussed in earlier chapters. The close relationship of salt accumulation to metabolism makes it even more susceptible than water absorption to inhibition by deficient aeration. Numerous investigators have shown (see Fig. 7.12) that salt accumulation of a variety of plant tissues is reduced when the oxygen supply is decreased (Hoagland, 1944; Hoagland and Broyer, 1936; Steward, Berry, and Broyer, 1936; Vlamis and Davis, 1944; and others). There are also numerous reports of decreased uptake of nutrients under field conditions attributed to deficient soil aeration, but it is often difficult to separate the effects of salt absorption from other effects. Arnon and Hoagland (1940) concluded that inadequate aeration is frequently a limiting factor for salt absorption by crop plants; Lawton (1946) reported similar reductions of mineral uptake by corn, when aeration was reduced by flooding and by compacting the soil.

It was pointed out in Chap. 4 that tolerance of inadequate aeration varies widely among different species. For example, rice absorbs sufficient water and nutrients from flooded soil where barley and tomato suffer severe injury (Vlamis and Davis, 1944). According to Chang and Loomis (1945), absorption of minerals by corn and wheat is reduced more by an excess of carbon dioxide than by a deficiency of oxygen.

TEMPERATURE. Low soil temperature can affect growth in so many ways that it is difficult to identify specific effects on salt absorption. For example, Knoll et al. (1964) thought that low uptake of phosphorus by corn at low soil temperatures was caused largely by decreased root growth. In addition to reducing root growth, water absorption, and metabolic activity, low temperature reduces nitrification and other types of microbial activity in the soil. It also reduces measurably the conversion of inorganic to organic nitrogen and other synthetic activities in roots. Soil temperature is often limiting for mineral absorption by rice, and Ehrler and Bernstein (1958) showed that a root temperature of 17.5°C decreases salt uptake, growth, and yield. Increase in soil temperature from 12 or 13°C to 17 or 18° increased the absorption of potassium, calcium, and magnesium, but not of sodium (Lingle and Davis, 1959). Some effects of temperature on salt absorption by entire plants are shown graphically in Fig. 7.15. Laboratory experiments show that although metabolic accumulation is reduced or prevented near freezing, adsorption exchange occurs unhindered (see Fig. 7.16).

As mentioned in Chap. 6, low soil temperatures may be of some ecological significance in Arctic and Alpine habitats where soils are normally cold and often frozen for long periods. However, they may be even more important when tropical plants are grown in regions where soil temperatures are relatively low. Soil temperatures on the Hawaiian island of Oahu are said to be barely high enough for growth of pineapple, and they are often limiting for mineral absorption by rice and citrus in California and other regions.

SOIL MOISTURE CONTENT. Like abnormal temperatures, abnormally high or low soil water content has so many effects on roots and their processes

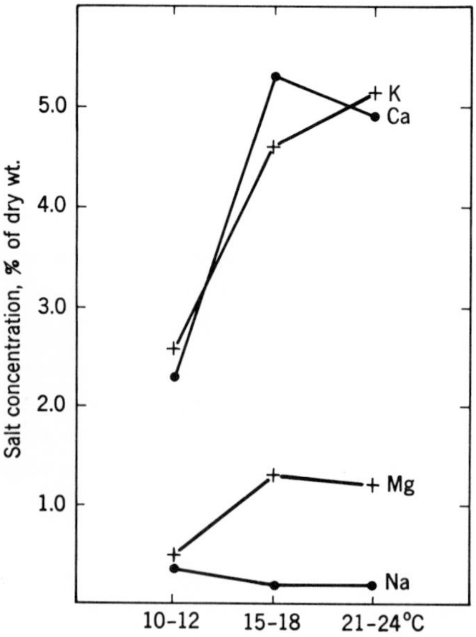

FIG. 7.15 Effect of soil temperature on amount of salt accumulated in tops of tomato seedlings after 6 weeks in sand culture. (*From Lingle and Davis,* 1959.)

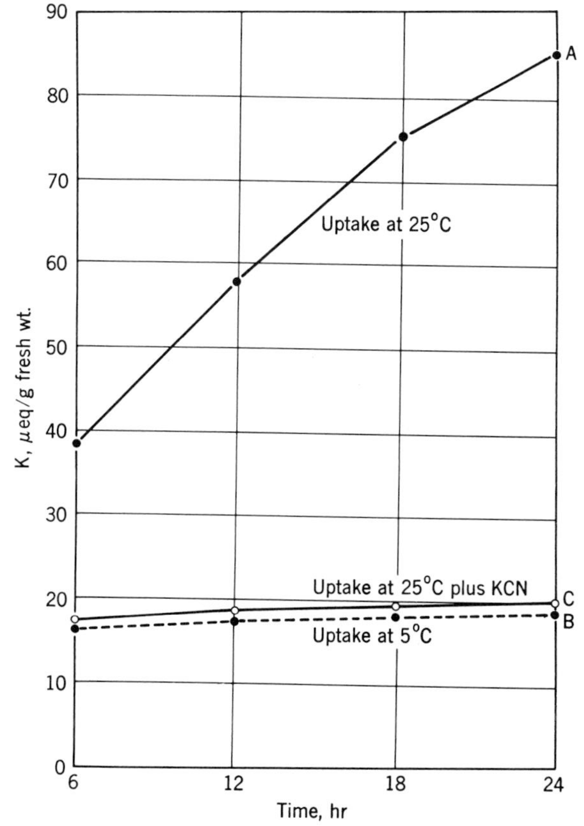

FIG. 7.16 **Effects of low temperature and a respiration inhibitor on uptake of K from KBr by slices of washed beet root. Uptake at 5°C and in the presence of KCN can be regarded as nonmetabolic. (*From Sutcliffe,* 1954.)**

that it is difficult to distinguish direct effects on mineral absorption from more general effects on root growth and differentiation. The complexity of these relationships was discussed by Richards and Wadleigh (1952). They pointed out that the deficient aeration resulting from an excess of soil water not only interferes with root growth and functioning (see Chap. 4) but also affects microbiological activity. For example, denitrification occurs rapidly in waterlogged soils, and much nitrogen is lost as ammonia. Dry soil also reduces microbiological activity. The reduced root growth of plants subjected to water stress reduces the volume of soil from which mineral nutrients can be absorbed and increases suberization of root surfaces. Richards and Wadleigh (1952) cite research indicating that drying of soil results in fixation of both phosphorus and potassium so that they become unavailable. As a result, plants in dry soil tend to be low in both phosphorus and potassium. For example, Mason (1958) and Mederski and Wilson (1960) reported lower percentages of phosphorus and potassium in plants subjected to water stress than in well-watered plants. Water stress also affects the physiology and biochemistry of plants, reducing and perhaps changing their ion requirements. Thus, changes in mineral composition of plants growing in dry soil may result from decreased availability in the soil, decreased absorbing surface, decreased use in the plant, or a combination of all three.

CONCENTRATION AND COMPOSITION OF THE EXTERNAL SOLUTION. It is well known that the uptake of a given ion is affected not only by the concentration of that ion, but also by the presence and concentration of other ions in the root medium. The more kinds of ions present, as in soil or complete nutrient solution, the more complicated the interactions. Not only are there the usual interactions between ions depending on electrical charge and degree of hydration, but there are also effects on the protoplasmic membranes of the root cells and the carrier systems. For example, calcium is believed to be essential for the normal operation of the selective absorption mechanism (Epstein, 1961; Viets, 1944). In addition, there are the well-known competitive effects between ions for carrier or binding sites, as between potassium and rubidium, or chloride and bromide ions.

There is increasing evidence that different carrier systems are involved at low and high concentrations of an ion. According to Epstein (1965, 1966), at low concentrations up to about 0.20 m*M* of potassium, uptake is related to external concentration according to Michaelis-Menten kinetics. However, at higher concentrations, above 1.0 m*M*, absorption of potassium is much greater than would be expected from the relationship between concentration and uptake found at low concentrations. The mechanism for absorption of

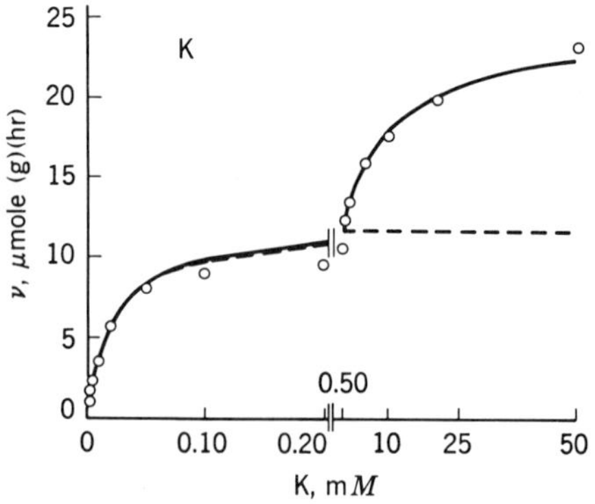

FIG. 7.17 Evidence for the existence of different mechanisms for salt absorption at low and high concentrations. The left curve is for absorption of potassium from concentrations varying from 0.002 to 0.2 m*M*, the right curve for absorption from solutions with concentrations of 0.5 to 50 m*M*. The dashed line indicates the absorption expected if the mechanism operating at low concentrations also operated at high concentrations. (*From Epstein,* 1966.)

potassium at low concentrations has a high affinity for potassium, is not affected by rubidium or sodium or by the kind of anion present, and is stimulated by the presence of calcium. The mechanism operating at high external concentrations of potassium is competitively inhibited by sodium and is inhibited more by sulfate than by chloride as the associated anion, and absorption is decreased by increasing the concentration of calcium. Böszörmenyi and Cseh (1964) reported that there are two absorption systems for halide ions, and Epstein (1966) summarized data indicating the existence of a dual mechanism for many elements in many different kinds of plants.

There is a troublesome question concerning the location of the two mechanisms. Do they operate in parallel across the same membrane or in series—one in the plasmalemma, the other in the vacuolar membrane? Locating one mechanism in the plasmalemma, the other in the vacuolar membrane as proposed by Luttge and Laties (1966) places the cytoplasm in the free space at high ion concentrations and creates difficulties. Welch and Epstein (1968) presented evidence that the two mechanisms operate in parallel across the plasmalemma. However, they consider the possibility that the two mechanisms might deliver ions into two different compartments in cells. Some ions might be delivered to the cytoplasm and some to the vacuole by way of the endoplasmic reticulum. Evidently more research must be done on this problem.

When the osmotic potential of the root medium is decreased by adding salt to it, plants absorb additional salt and show decreases in osmotic potential almost in proportion to the decrease in osmotic potential of the medium (Bernstein, 1961; Eaton, 1942; Slatyer, 1961). In fact, Slatyer reported that the absorption of ^{36}Cl from labeled sodium chloride increased exactly in proportion to increase in its concentration in the root substrate. This problem

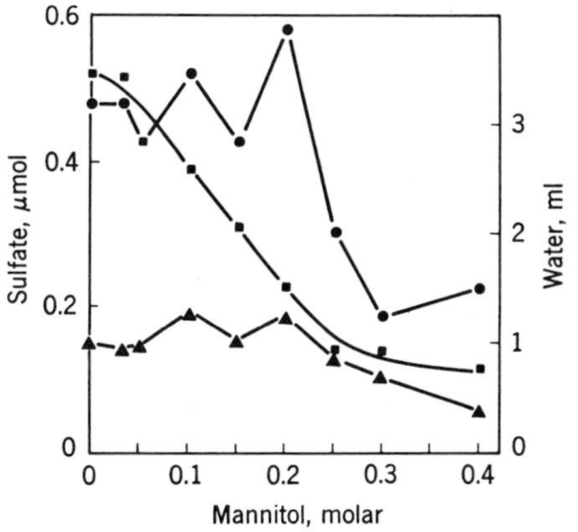

FIG. 7.18 **Effect on absorption of water and sulfate by wheat plants of decreasing the osmotic potential of the substrate by addition of mannitol. Sulfate transported to shoot, ●; sulfate retained in root, △; water uptake, ■.** *(From Ingelsten, 1966.)*

is discussed in Chap. 6 in more detail. Decrease in osmotic potential of the substrate by use of organic solutes is reported to decrease water uptake much more than it decreases salt uptake (Brouwer, 1954*b*; Long, 1943). Figure 7.18 from work by Ingelsten (1966) indicates that this is correct at low concentrations. However, sulfate transport to the shoots decreased abruptly when the concentration of the substrate was increased so much that the root cells were plasmolyzed.

Water absorption and salt absorption

The views of plant scientists concerning the relationship between water absorption and salt absorption have undergone interesting changes over the years. Near the beginning of the nineteenth century, de Saussure (1804) observed that plants absorb proportionally more water than salt from a dilute solution and that they absorb more of some elements than of others. Writers later in the century apparently neglected de Saussure's experiments, and it was assumed that there was a close relationship between the uptake of water and the uptake of ions. During the first quarter of the present century a number of experiments were performed, most of which indicated that salt and water were not absorbed in the same ratio. Miller (1938) reviewed most of the early work, so that it need not be discussed here. However, many of these experiments were poorly designed and produced inconclusive results because the treatments intended to reduce water absorption also affected plant growth in other ways.

Hoagland (1944) and his associates gave considerable attention to this problem. They found that various ions can be absorbed from a culture solution either more or less rapidly than water is absorbed, depending on the kind of ion, the salt and sugar content of the plant, and the metabolic activity of the roots. They concluded that transpiration has no direct effect on salt absorption and that any increase in salt uptake accompanying increased transpiration is an indirect effect in which more rapid removal of salt from the xylem in the roots causes more rapid active transport into the xylem. This view was generally accepted until 1953, when Hylmö reopened the question by publishing the results of experiments which he interpreted as indicating that transpiration directly affects salt uptake.

It is impossible to review the voluminous literature which has appeared since 1953, and the reader is referred to reviews by Russell and Barber (1960) and Brouwer (1965*b*) for details. There is no doubt that in many instances increased water uptake is accompanied by increased absorption and translocation of salt to the shoots. However, it remains uncertain whether transpiration causes the increase directly by increasing mass flow of ions into roots in the transpiration stream, or indirectly by stimulating active transport into the stele, or both.

Under conditions of low transpiration, ions are accumulated in the xylem sap to a much higher concentration than in the root medium, indicating the operation of an active transport system. This has been demonstrated by many investigators, including Broyer (1950), Eaton (1943), Russell and Shorrocks (1959), and Smith (1960). Russell and Barber (1960) summarized data which showed concentrations of phosphate and rubidium in the xylem sap of slowly transpiring plants varying from 0.7 to 156 times that of the root medium. In rapidly transpiring plants, the ratios ranged from less than 1 to over 30, the lowest ratios occurring in sap of plants absorbing from relatively concentrated solutions. Eaton (1943) reported that the osmotic pressure of the exudate from stumps of detopped tomato plants was 2.4 bars, while that of the solution in which the roots were growing was only 0.49 bar.

When rapidly transpiring plants are allowed to absorb from fairly concentrated solutions, the concentration in the xylem sap often falls below that of the root medium. The same situation exists when dilute salt solutions are forced through the roots under pressure (see Fig. 7.19). This indicates that there is a region in the root, sometimes termed an ion barrier, which is less permeable to ions than to water, and that ions are not crossing this barrier as rapidly as they are being carried to the shoot by the transpiration stream

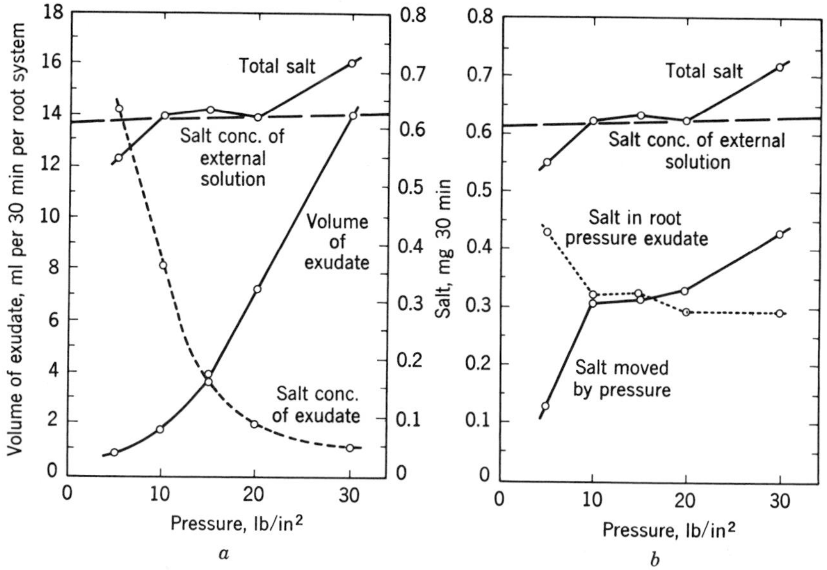

FIG. 7.19 (a) Effects of pressure on rates of water and salt movement into the xylem of tomato root systems. Salt concentrations are in milligrams per milliliter. Although salt concentration of xylem sap decreased as rate of water increased, total salt moved increased. (b) Amount of salt in root pressure exudate and amount moved under pressure. The value for salt in root pressure exudate is an average of measurements made before and after application of pressure. Salt moved by pressure is the difference between that in root pressure exudate and the total moved under pressure, and it is the fraction associated with water movement under pressure. (*From Lopushinsky,* 1964.)

(Bernstein and Nieman, 1960; Emmert, 1961; Kylin and Hylmö, 1957). The presence of this barrier makes the accumulation of ions in the xylem possible. However, this ion barrier is probably not completely impermeable to movement of ions by diffusion and mass flow. As Laties and Budd (1964) state, the ion barrier need not be an absolute barrier to back diffusion of salt from the vessels, but only a barrier which slows up diffusion considerably. In fact, Wiersum (1948) observed that measurable amounts of ions introduced into the xylem of root segments leaked out through the cortex, and this was verified by Brouwer (1954). It would be interesting if the reflection coefficient as defined by Dainty (1963*a*) for membranes could be determined for roots. This would provide a measure of the permeability of roots to leakage of ions by diffusion.

Experiments such as those of Lopushinsky (1964) and Jackson and Weatherley (1962), in which water and salt were moved through the roots under pressure to simulate the effects of transpiration, indicate that considerable amounts of salt are probably carried through the ion barrier by mass flow in rapidly transpiring plants. Lopushinsky (1964) found that the amount of salt moved across tomato roots under a pressure of 30 lb/in.^{2} was 2.0 to 2.4 times the amount moved by active transport in the absence of pressure (Fig. 7.19). Treatment of root systems with $10^{-3}M$ sodium azide initially reduced the amount of water and salt movement to about 10 percent of the control rates (see Fig. 7.13). However, after several hours' exposure to sodium azide, the roots lost their differential permeability, and the rate of water and salt movement began to increase. Jackson and Weatherley (1962) reported similar effects of pressure and metabolic inhibitors on radial movement of potassium across roots. They suggested that water and salts move across the cortex into the xylem independently at one stage and together at another stage. Jensen (1962) concluded that three-fourths of the nitrate uptake by transpiring tomato plants is passive.

In contrast, Crafts (1961, p. 146) states that it is absurd to speak of movement of ions from soil to xylem in free space because it ignores what is known about root structure and function. Apparently Crafts is willing to ignore both the results of many studies of the past decade and the fact that much absorption occurs through suberized roots. Brouwer (1965*b*, pp. 256–261) reviewed the rather confusing information on this topic and concluded that there is some kind of passive movement of salt, but that it probably plays no essential role in the mineral supply of plants. Ingelsten (1966) and Petterson (1966) agree that increased uptake of water is accompanied by increased uptake of salt, but they (like many others) attribute this to stimulation of active transport into the xylem. However, they do not explain how this stimulation might occur.

The possibility that passive movement of solutes into the root xylem occurs can be disregarded only if it can be assumed that the stele is surrounded by a perfect differentially permeable membrane which prevents passage of any solutes except those moved by active transport. This is highly improbable, as the endodermis and cortex are usually pierced by numerous branch roots and eventually slough off during secondary growth. The corky layer developed on the outer surface of older roots often contains numerous lenticels, and in some species openings are left by many short-lived branch roots which die and collapse.

It seems probable to the writer that in young plants with rapidly growing root systems and in slowly transpiring plants most or perhaps all of the salt movement into the xylem is by active transport. However, in older plants and particularly in rapidly transpiring plants a considerable fraction of the salt intake may occur by mass flow through the leaky ion barrier of the roots. This is likely to be particularly important in perennial plants, where more than 95 percent of the root surface is often suberized, but many permeable spots occur in the form of lenticels and of openings caused by branch roots, both dead and living, through which ions can enter by mass flow.

If we grant that some ions enter plants by mass flow through leaks in the ion barriers, this provides an explanation for the fact that every element found in the environment has been found in plants, including many not known to be accumulated by plant cells. It may be said that plants are less selective than their cells, because any solute which can move through cell walls might become distributed in the free space of plants, but only certain ions are accumulated in the cell vacuoles. It would be interesting to know how many of the unusual elements found in plants occur in the free space rather than in the vacuoles. Considerable amounts of even the essential elements must occur in free space, as indicated by the fact that appreciable amounts of salt are leached out of leaves by rain or sprinkling (Stenlid, 1958; Tukey et al., 1958).

In conclusion, it appears that active transport, cytoplasmic streaming, diffusion, and mass flow are all involved in the radial movement of ions across the tissue lying between the root surface and the xylem. Although active transport is probably the principal mechanism by which ions enter the xylem, it seems likely that considerable movement by mass flow occurs in roots of rapidly transpiring plants, especially those of perennials.

Movement of ions from roots to shoots

It is generally assumed that once ions or other solutes have reached the xylem, they are carried upward in the sap stream. A number of investigators have supplied radioactive tracers to roots and measured the rate of translocation to the shoot. Rates of up to 60 m/hr have been reported, the higher rates occurring in rapidly transpiring trees (Kramer and Kozlowski, 1960, pp. 198–201). Movement in some instances follows the same spiral patterns reported for movement of dyes (Fraser and Mawson, 1953). It has also been observed that tracers supplied to the roots of herbaceous plants often reach the shoot in less than an hour.

All these observations support the view that ions move upward chiefly by mass flow in the xylem. This has been questioned for some ions by Biddulph and his coworkers, who obtained data on the movement of calcium which they regarded as incompatible with mass flow in the transpiration stream. Biddulph et al. (1961) suggested that the xylem cylinder of a bean stem operates as a chromatographic column in which calcium moves by exchange along the walls of the conducting cells. Bell and Biddulph (1963) also obtained data on calcium movement which they regarded as indicating movement by exchange rather than mass flow. On the other hand, O'Leary (1965) observed the translocation of radioactive calcium paired with phosphorus or rubidium and concluded that they moved by mass flow. Perhaps when only a small amount of calcium is supplied, all or most of it is adsorbed, but when larger amounts are supplied, all adsorption sites are saturated, and part of the tracer is free to move upward by mass flow (see Thomas, 1967).

Although some ions tend to be carried directly to the young leaves and stem tips, a considerable amount of salt moves laterally out to the adjacent tissues, particularly to the cambium and phloem. There appears to be considerable exchange of materials between the xylem and phloem (Stout and Hoagland, 1939; Biddulph and Markle, 1944). Sometimes so much salt is removed from the xylem that its concentration is significantly lowered by the time it reaches the leaves. For example, Eaton (1943) found the salt concentration of guttated liquid from cotton leaves to be only about one-third that of the exudate from detopped stumps of similar plants. Klepper and Kaufmann (1966) and Oertli (1966) also reported that the concentration of the guttation fluid decreased with increasing distance above the roots. The composition of the xylem sap is affected by differential removal of ions by the leaves along a stem and by retranslocation out of the leaves. Experiments by Klepper and Kaufmann (1966) in which artificial guttation was produced by forcing Hoagland's solution through leaves indicate that petioles and leaf blades, as well as stems, remove salt from the xylem sap.

The composition and concentration of the xylem sap varies with the species, the season, and even the time of day. For example, Anderssen (1929) found in apricot and pear that the salt concentration was much higher in May than in November, and Bollard (1958) reported a tenfold increase in most elements in the xylem sap of apple from the dormant season to blossoming. Seasonal variations are shown in Fig. 7.20. There is also a tendency toward decreased concentration during the day, when maximum movement of water is occurring and the rate at which water enters roots tends to exceed the rate at which ions enter (Petritschek, 1953).

Understanding of the upward movement of ions is complicated by the

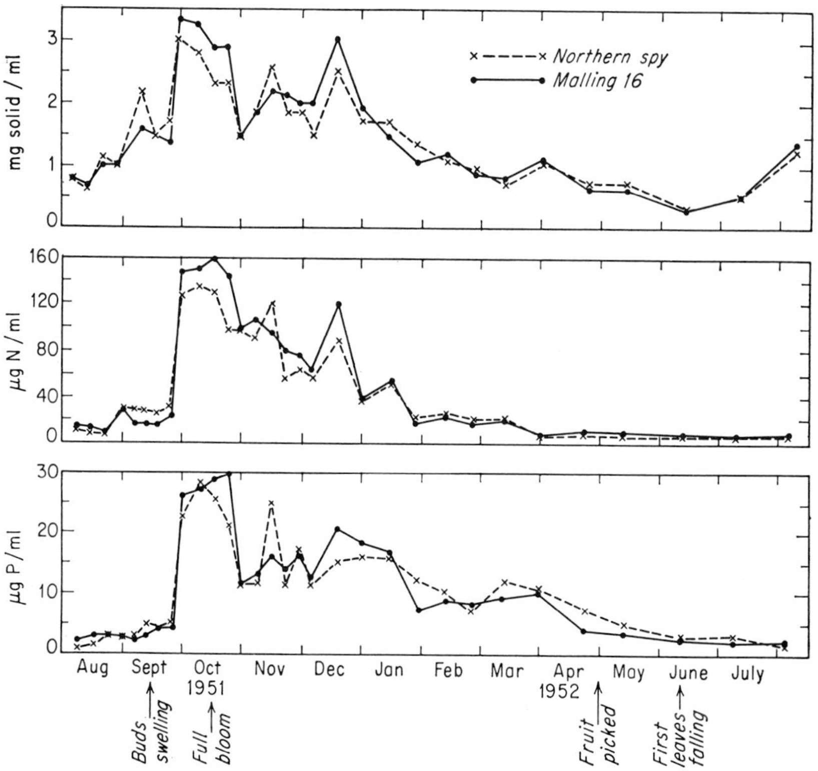

FIG. 7.20 **Seasonal changes in concentration of phosphorus, nitrogen, and total solids in the xylem sap of two varieties of apple trees growing in New Zealand. (*From Kramer and Kozlowski*, 1960, *after Bollard*, 1953.)**

fact that not all the transport of salt occurs as inorganic ions. In many species, most or all of the nitrogen is probably transported as organic compounds such as amides, amino acids, and ureides. There is also limited evidence for the transport of at least small amounts of sulfur and phosphorus as organic compounds in the xylem. Pate (1965) reported small amounts of organic sulfur compounds in xylem sap. The presence of alkaloids such as nicotine and related compounds in the xylem sap of tobacco is well known, and cytokinins and gibberellins have also been reported. Bollard (1960) reviewed most of the literature on this subject. The occurrence of organic nitrogen, phosphorus, and sulfur compounds in xylem sap means that ions are involved in enzymatically catalyzed reactions as they cross the roots. The site of these reactions and the mechanism by which the resulting compounds enter the xylem are unknown. However, the existence of extracellular enzymes in the free space of roots may provide an explanation. A variety of enzymes occur in or on the walls of root cells (Chang and Bandurski, 1964; Hellebust and Forward, 1962; Street and Lowe, 1950). The roles of these extracellular enzymes in conversion of inorganic compounds to organic compounds and in the absorption of large organic molecules constitute interesting problems for future research.

The movement of certain ions from roots to shoots is also complicated by their tendency to be precipitated in the xylem. It appears that under some conditions so much iron is precipitated in the vascular system, chiefly as ferric phosphate, that deficiencies occur in the leaves (Brown et al., 1959; Rediske and Biddulph, 1953). Hewett and Gardner (1956) reported that considerable amounts of zinc are absorbed on the walls of the xylem vessels in grape. Thus, deficiencies of certain elements can occur in the leaves because of their failure to be translocated, even though adequate quantities are being absorbed.

A number of investigators have observed that certain species absorb more salt than others when grown in a saline substrate. For example, the leaves of *Atriplex hastata* contain much less salt than the leaves of *A. vesicaria* or tomato grown under comparable conditions. The chloride content of the roots of *A. hastata* was found to be higher than that of the leaves over a wide range of chloride concentrations in the root medium (Black, 1956). Certain varieties of grape (Ehlig, 1960) and barley (Greenway, 1962) translocate less salt to their leaves than other varieties when grown in high concentrations of sodium chloride. Lagerwerff and Eagle (1962) reported that bean roots accumulated much sodium and little calcium, the leaves little sodium and much calcium. Cooil et al. (1965) think that downward movement of sodium from the leaves partially explains its failure to accumulate in the shoots of some plants. Jacoby (1964, 1965) concluded that the roots and lower stems of bean plants remove so much sodium from the xylem sap that only small amounts are left to reach the leaves. However, these tissues should eventually become saturated and allow salt to pass to the leaves. It has frequently been reported that the rootstock affects the amount of various mineral elements found in the shoots. Generally, rootstocks which decrease salt uptake increase the salt tolerance of plants (Bernstein et al., 1956).

Redistribution of ions

The continual movement of salt into leaves in the transpiration stream results in an increase in salt content during the growing season. Plants growing in saline substrates sometimes accumulate so much salt in their leaves that injury occurs. Normally, however, not all the ions reaching the leaves remain there. It is well known that considerable amounts are translocated out of senescent leaves before abscission occurs. Even in young, vigorously growing plants there is considerable redistribution of minerals, chiefly a movement of the more mobile ions out of older leaves to young leaves and meristematic regions. This movement evidently occurs in the phloem, as it is prevented by girdling or killing the phloem. However, there is considerable lateral transfer from phloem to xylem, which sometimes results in simultaneous movement upward in the xylem and downward in the phloem.

Redistribution of salt seems to be controlled largely by the relative metabolic activities of various parts of the plant. According to Williams (1955), minerals tend to move from older leaves to younger leaves, reproductive structures, and other metabolically active regions. Smith and Wareing (1966) reported that in birch seedlings showing weak apical dominance, ^{32}P injected at the base was rather evenly distributed among the lateral and leader stem tips. However, when a lateral branch was reoriented so it became strongly dominant, it accumulated a large amount of ^{32}P. The oriented translocation of minerals to metabolically active regions is probably related to the effects of increased hormone production on translocation. The manner in which internal redistribution is controlled deserves further study. Several investigators have reported a sharp decrease in salt and water absorption at the time of flower differentiation (Hall, 1949). There is no obvious explanation for this.

Some ions are redistributed more readily than others. According to Bukovac and Wittwer (1957), Rb, Na, and K are most mobile, and Ca, Sr, and Mg are least mobile. As a result, the latter accumulate in the leaves during the growing season, but the former usually do not. Deficiencies of mobile elements usually appear first in the older leaves of a plant, while deficiencies of nonmobile elements such as calcium usually appear first in the youngest leaves.

Exact evaluation of the extent of recirculation of minerals in plants is complicated by the fact that considerable amounts of salt are lost from leaves by leaching and from roots into their substrate. Leaching from leaves by rain and sprinkling was discussed by Long et al. (1956), Tukey and Amling (1958), Tukey et al. (1958), Mecklenberg et al. (1966), Stenlid (1958), and others. Alberda (1948) and Helder (1952) suggested that at least a part of the ions not used in metabolic processes in the shoots return to the roots and leak out into the root medium. Loehwing (1937) summarized the earlier literature on loss of minerals from roots. There are several reports of experiments in which tracers supplied to one side of a split root system soon appeared in the medium containing the other half of the root system. Biddulph (1959) states that redistribution ranges from almost continual recirculation of phosphorus to almost complete immobility of calcium. Stenlid (1958) summarized the literature on this topic. Some of the steps in mineral circulation are shown diagrammatically in Fig. 7.21.

Absorption of organic compounds

Some absorption of organic compounds occurs through roots. For example, roots in culture absorb sugar from their substrate, and uptake of a wide range of growth regulators, antibiotics, systemic insecticides, herbicides, and mis-

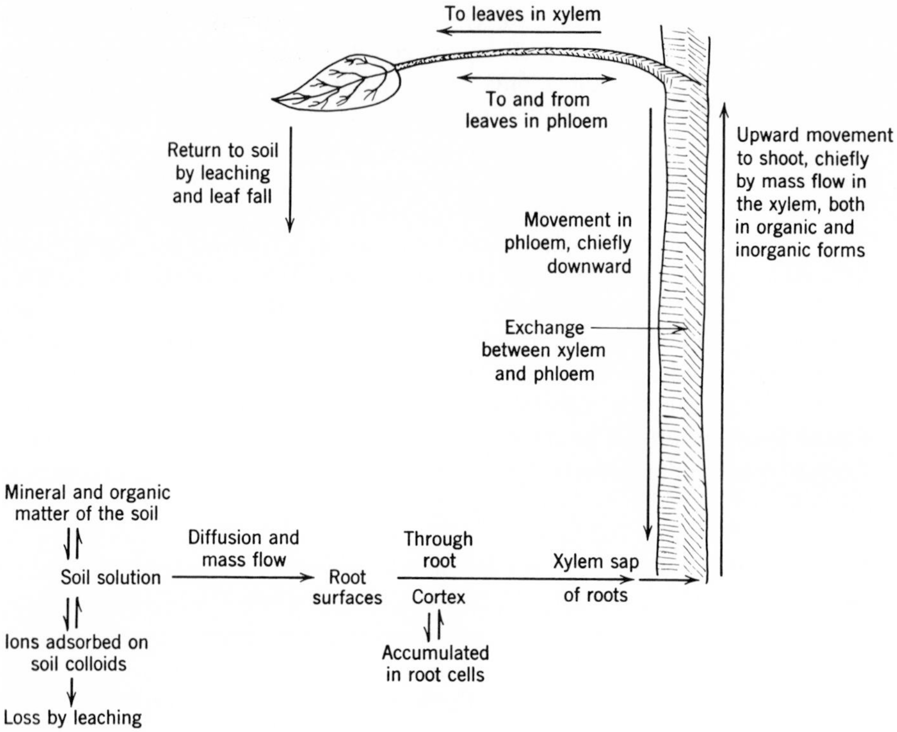

FIG. 7.21 **Diagram showing the circulation of minerals within a plant and their return to the soil by leaching and leaf fall.**

cellaneous other substances has been reported (Bollard, 1960). In most instances, detection of the compounds in the shoot by the presence of radioactivity is taken as evidence of absorption and translocation. However, it has been shown that some compounds such as sulfanilamide and aminotriazole undergo changes in the roots before moving to the shoots. Sucrose is converted to monosaccharide before it is absorbed by roots (Street and Lowe, 1950), and extracellular enzymes probably play a role in the absorption of other organic molecules. If the uptake and distribution of a compound is affected by changes in rate of transpiration, it seems reasonable to suppose that the compound is moving in the transpiration stream. According to Bollard (1960), final distribution of organic compounds is uneven and unpredictable. As with ions, some organic compounds are absorbed but not translocated to the shoot. Cucumber seedlings absorb both streptomycin and chloramphenicol from solutions (Pramer, 1954), but the concentration of streptomycin in leaves increases to a level higher than that in the nutrient solution, while chloramphenicol remains at a lower concentration. Tomato and broad bean absorb and translocate chloramphenicol throughout the plant, tomato absorbs streptomycin but does not translocate it, and broad bean will not even absorb it (Crowdy, 1959). Crafts and his colleagues report many anomalies in the absorption and translocation of various herbicides by different species of plants (Crafts, 1961; Crafts and Yamaguchi, 1960).

Mannitol and polyethylene glycol are used extensively to decrease the osmotic potential of the root substrate and produce water stress because they are supposed not to be absorbed, as are ions. However, some absorption of mannitol occurs, and it is translocated through plants and even escapes in the guttation water (Groenewegen and Mills, 1960; Kozinka and Klenovska, 1965). Carbowax or polyethylene glycol is also said to move through plants and accumulate on the surface of leaves (Lagerwerff et al., 1961). The possibility of polyethylene glycol (mol. wt. 20,000) entering uninjured roots seems small, and it is probable either that it entered through broken roots, that the solution contained a small fraction of a much lower molecular weight which entered the roots, or that it moved from roots to shoots through free space.

There have been reports in the literature of the uptake of viruses (Murphy and Syrerton, 1958), proteins (McLaren et al., 1960), and other large molecules through the roots of plants, but there is always the possibility that they enter through broken roots or breaks in the root surfaces where branch roots have died. It seems probable that they usually occur in the free space of the cell walls rather than within the protoplasts (Street, 1966, p. 338).

Absorption through leaves and stems

The entrance of solutes is not restricted to roots, but also occurs through leaves and even through stems. Ordinarily, the amounts which enter by these pathways are small, being limited to the minerals derived from dust, fumes, aerosols, rain, and sprinkler irrigation water. In recent years, there has been considerable interest in so-called foliar fertilization, in which nitrogen, phosphorus, and other nutrients are sprayed on the leaves.

According to Tamm (1958), the amount of salt carried down in rainwater is surprisingly large, especially where the air is contaminated by industrial fumes and smoke. An average of 60 kg/(hectare)(year) of ash was found in the rainwater in a rural district in England and larger amounts in industrial areas. Often there is so much salt in the dusty air of a greenhouse that it is difficult to demonstrate deficiencies of certain elements. It is said that vegetation growing near enough to the ocean to be exposed to salt spray is affected by sodium and chloride (Boyce, 1954; Oosting, 1954). The absorption by plants of radioactive strontium, ruthenium, and cesium in "fallout" from nuclear explosions is well known, though what fraction enters through the leaves is uncertain.

Certain advantages of foliar application of mineral nutrients to crop plants has led to extensive investigation of foliar fertilization (Wittwer, 1964;

Wittwer and Teubner, 1959). Although its advantages have probably been overpublicized, foliar fertilization has been used successfully for the application of urea nitrogen, phosphorus, magnesium, and various micronutrients to pineapple, sugarcane, citrus, various deciduous fruits, forest trees, and some herbaceous species. Foliar applications are particularly effective for elements such as iron and zinc which are often immobilized in the soil or in the conducting system. However, in general, foliar fertilization should be regarded as an adjunct to rather than a substitute for fertilization through the soil.

Some elements such as nitrogen, potassium, magnesium, and zinc are said to be absorbed through leaves more rapidly than others such as phosphorus, sulfur, and iron. However, all water-soluble substances can presumably be absorbed through leaves. The first step usually is penetration of the cuticle, because the high surface tension of aqueous solutions ordinarily prevents their entrance through stomata, unless a surfactant is added (van Overbeek, 1956; Dybing and Currier, 1961). The problems connected with penetration of leaves were reviewed by van Overbeek (1956) and Franke (1967). Apparently, the cuticle is heterogeneous in composition and structure and possesses hydrophilic areas which swell and increase its permeability to substances in aqueous solutions. Having reached the cell walls, substances may either move by diffusion through the free space or enter the protoplasm and move in the symplast. According to Smith and Epstein (1964), the cells of leaves accumulate ions in the same manner as the cells of roots.

The role of ectodesmata in foliar absorption remains uncertain. Ectodesmata are fine structures in the outer walls of epidermal cells which were at first regarded as homologous to plasmodesmata but are now regarded as quite different, because they contain no protoplasm. Franke (1967) thinks foliar absorption is localized chiefly in areas where ectodesmata occur. He also thinks guard cells are special points of entry for aqueous solutions into leaves.

Most of the absorption occurs within a few hours after application, and more is absorbed through young than through old leaves, either because the latter are more heavily cutinized, or possibly because they have a lower level of metabolism and less capacity to accumulate ions. Ions tend to be lost from older leaves by leaching; hence, the moistened cuticle can scarcely be a serious barrier to movement either in or out. There is considerable difference in the extent to which the absorbed elements are translocated out of leaves to other parts of the plant. Bukovac and Wittwer (1957) classify such elements as potassium, sulfur, and phosphorus as freely mobile; copper, iron, manganese, and zinc as moderately mobile, and calcium, strontium, and magnesium as immobile. Movement out of leaves occurs in the phloem and is reduced by low temperature and other factors which affect metabolism. Urea is quickly converted to amides and amino acids and organic phosphorus compounds are formed, but apparently most of the phosphorus and other elements are translocated out of leaves in the inorganic form. It should be recalled, however, that there is considerable exchange between phloem and xylem; hence, substances moved out of the leaves in the phloem might be transferred to the xylem and then moved upward again in the transpiration stream.

Absorption of solutes can also occur through stems, at least to a limited extent. In earlier times, peasants in some parts of Europe are said to have plastered dung on the stems of fruit trees to produce more vigorous growth. Wittwer and Teubner (1959) mention a number of reports of absorption of minerals through twigs and branches. Much of the entrance is believed to occur through lenticels, leaf scars, pruning wounds, and other breaks in the bark.

Salt secretion

Some plants secrete considerable amounts of salt from their leaves. The presence of solutes in guttation fluid was mentioned in Chap. 5, but in some species salt is secreted from the epidermal cells or through specialized groups of cells or "salt glands" in the leaves. This method of disposing of surplus salt is particularly common in halophytes, and the leaves of *Tamarix* and some species of *Atriplex* often become covered with salt. Even cotton leaves secrete enough salt to interfere with determination of their leaf water potential. Sections through salt glands are shown in Figs. 7.22 and 7.23.

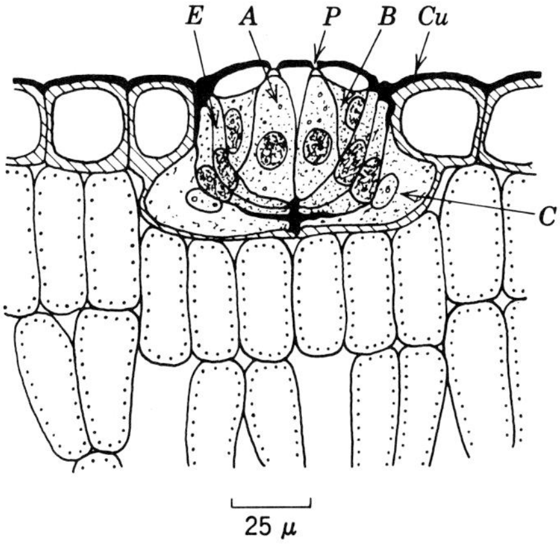

FIG. 7.22 A salt-secreting gland on a leaf of *Statice Gmelini.* Salt solution is secreted through pores (*P*) in the outer wall and cuticle from the secretory cells (*A*) which are surrounded by smaller cells (*B* and *E*) and lie above so-called collecting cells (*C*). The latter seem to be the subbasal cells (*SBC*) of Fig. 7.23. Also note that the cuticle (Cu) extends downward and encloses the glandular tissue. This also is shown in Fig. 7.23. (*After Ruhland*, 1915.)

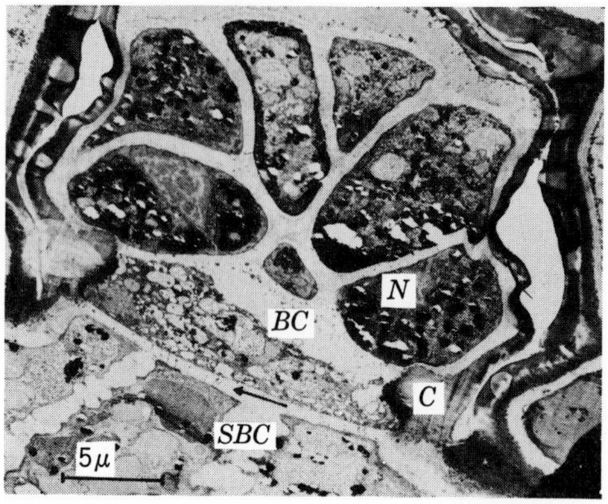

FIG. 7.23 Electron micrograph of an oblique section through a salt-secreting gland on a leaf of *Aegiliatis annulata.* The seven cells of the gland lie over a highly vacuolate basal cell (BC) under which are a group of subbasal cells (SBC). Note plasmodesmata in wall between basal and subbasal cells, indicated by arrow, also the cuticle (C), and nucleus (N). These glands are somewhat similar to those of *Statice,* shown in Fig. 7.22. (*From Atkinson et al., by permission of the authors and the Australian Journal of Biological Sciences.*)

The salt glands vary from simple two-celled structures in *Spartina* to the complex many-celled structures of *Limonium.* Ruhland (1912) estimated that in *Limonium* there are about 700 glands/mm^2 of leaf surface, which excrete up to 1 ml/hr of liquid containing 0.05 mg of sodium chloride. The concentration of salt in the liquid was no higher than in the leaf tissue; hence it appeared that no complex secretory process was involved, but salt solution was simply squeezed out under pressure. Scholander et al. (1966) reported 2 to 7 percent of sodium chloride in secretion from mangrove leaves. Arisz (1955) studied salt excretion from salt glands of another species of *Limonium* and found the concentration of salt to be higher than the average concentration of the leaf. Arisz proposes that the excretory cells accumulate salt and develop a high turgor pressure which forces salt solution out. Sutcliffe (1962, p. 162) suggests that the salt is accumulated in small vesicles in the cytoplasm which are then discharged by a kind of reverse pinocytosis. Franke (1967) suggests that the ectodesmata may provide a pathway for outward movement of substances escaping from protoplasts. The general problem of salt loss and salt secretion was reviewed by Helder (1956).

Summary

The absorption of salt by intact plants involves the movement of ions from soil to root surfaces, ion accumulation in root cells, radial movement of ions from root surfaces into the xylem, and finally their translocation from roots to shoots. The ability of cells to accumulate ions to a concentration higher than that in their environment is an outstanding characteristic of living organisms. One theory of accumulation proposes that anion accumulation is by active transport linked directly to metabolism and that cations move passively along the gradient of anions. A more widely held theory assumes that ions form temporary combinations with special organic molecules called carriers, and that the resulting complex can move through cell membranes which are impermeable to ions alone. The ions then separate from the carriers at the membrane surface and move into the vacuoles, while the carrier molecules become available to move more ions. Metabolic energy is required both for the synthesis of carrier molecules and also to maintain membrane structure. Organic acid metabolism is also involved in maintaining ionic balance.

The kinds and amounts of ions accumulated by root cells vary widely among different species of plants, with the metabolic activity of roots, the kinds and concentrations of ions in the soil, soil temperature, and aeration. Mobile ions such as nitrate and potassium move toward roots in the soil solution, but immobile ions such as phosphate are available only from soil in the immediate vicinity of roots. Root extension is therefore an important factor in ion absorption, and the increase in surface provided by mycorrhizae is especially beneficial in soils low in phosphorus.

The most troublesome problem in connection with salt absorption is to explain how ions move radially from the root surface into the stele and enter the dead xylem elements. The classical explanation is that ions are accumulated by the cells at the root surface and move inward through the symplast to the stele by diffusion and cytoplasmic streaming. However, it is not clear how they are transferred from the living cells of the stele into the sap of the dead conducting elements. The writer believes that in transpiring plants there is measurable mass movement of ions into the stele in the transpiration stream. However, this view is not generally accepted. More research is needed on the radial movement of ions in roots.

Ions which reach the xylem sap of the roots are usually carried to the shoots in the transpiration stream. However, much salt is removed from the xylem sap by adjacent living cells, and the salt concentration of the xylem decreases from roots to shoots. Also iron, zinc, and other ions are sometimes precipitated in the xylem, resulting in deficiencies in the leaves. The failure of sodium to move to the shoots in some species and varieties also is puzzling. Explanation of the movement of ions to the shoots is complicated by the fact that some inorganic ions, notably nitrate, are converted to organic compounds in the roots. Furthermore, there is considerable redistribution of ions in plants. Some mobile ions are moved from old to young leaves, and some ions carried to the shoot in the transpiration stream are moved downward again in the phloem. Thus there is a continual recirculation and redistribution of ions in plants.

eight

Movement of Water Through Plants

Introduction

The continuous movement of water from the absorbing roots to the transpiring leaves is essential to the survival of plants. This chapter deals with the structure of the water-conducting system, the continuity of water in plants, the driving forces, and the resistances encountered in various parts of the conducting system.

The existence of tall land plants became possible only after a vascular system evolved which permitted rapid conduction of water from roots to shoots. In fact, it is very difficult for land plants lacking a vascular system to attain a height of more than 20 or 30 cm because movement of water by diffusion from cell to cell is too slow to keep the tops supplied. Furthermore, the continuity of water in the conducting system provides a communication system between roots and shoots which keeps the rates of absorption and transpiration in balance. Thus, when transpiration increases, the demand for an increased water supply to the leaves is transmitted to the roots by a decrease in water potential in the xylem sap, which causes an increase in absorption. Conversely, when water absorption is reduced, the information quickly reaches the leaves as a decrease in water potential in the xylem sap, which causes loss of guard cell turgor and closure of stomata. This results in a compensating decrease in water loss by transpiration. Actually, there is some lag in response, which will be discussed in Chap. 10. On the whole, however, the continuous water system in plants functions fairly effectively as a control system (Rawlins, 1963).

The conducting system of roots

The absorbing zone of roots and the radial movement of water and solutes from the root surface to the xylem have been discussed in preceding chapters. It will suffice here to remind the reader that in young roots water must cross the compact layer of cells formed by the epidermis, and sometimes a second compact layer, the hypodermis. Within this is a cylinder of cortical parenchyma, usually many layers of cells in thickness, bounded at the inner surface by another very compact layer, the endodermis. The endodermis has received much attention because the thickness of its cell walls and the cutinized or suberized layers forming the Casparian strips suggest that it is an important barrier to the movement of water and solutes. However, thin-walled passage cells often occur opposite the xylem strands, and in many roots the endodermis is pierced by numerous branch roots. Furthermore, when secondary growth occurs, the epidermis, cortical parenchyma, and endodermis all disappear. Roots which have undergone these changes are covered by an outer layer of suberized tissue over a layer of secondary phloem. In some kinds of roots this layer is thinner than the cortex which it replaced, so that older roots sometimes have a lower resistance to water movement than younger roots (Kramer and Bullock, 1965; Queen, 1967).

After water reaches the xylem, it encounters relatively little resistance to longitudinal movement. Xylem elements are ordinarily differentiated within a few millimeters of the root tip, although protoplasts have been reported in some xylem elements several centimeters behind the tip (Anderson and House, 1967; Scott, 1963). Longitudinal movement may occur outside the xylem, but it is limited by the high resistance to movement through parenchyma tissue (see the section on conduction outside the vascular system later in this chapter).

The conducting system of stems

The xylem has been recognized as the principal pathway for the upward movement of water at least since the time of Hales in the early eighteenth century. Superficially, the xylem of the main stem of a tree might be compared with a rope in which the strands at the upper and lower ends are separated (the xylem of large root and stem branches) and then unraveled into threads (small branches), finally into individual fibers corresponding to the xylem in the smaller leaf veins. However, the xylem is not a continuous conductor, as is a pipe, but a collection of overlapping vessels and/or tracheids in which water must often pass through hundreds or thousands of crosswalls on its way to the leaves. In spite of the numerous crosswalls, xylem offers so much lower resistance to water movement than the other tissues of roots and stems that practically all longitudinal movement of water occurs in it. The presence of crosswalls is important because if the system were continuous, it would often be completely blocked and rendered useless for conduction by gas bubbles. However, because of the crosswalls, gas bubbles are confined within the individual elements instead of spreading.

Structure of the conducting system

ROOT–SHOOT TRANSITION. There is a marked change in arrangement of the vascular tissues in the transition region from root to shoot, especially in herbaceous plants (see Fig. 8.1). The xylem occurs in the center of roots, but it splits into a number of vascular bundles which occur in a ring outside the pith in herbaceous dicot stems. Complicated structural patterns often develop in the vascular system at the transition region. In some instances there appear to be physiological differences above and below the transition zone. For example, in some species the salt content is much higher in the roots than in the shoots (see Chap. 7).

WOODY STEMS. Kozlowski (1961) reviewed the extensive literature on water movement in woody stems. In conifers water moves mostly through tracheids, which are spindle-shaped cells seldom more than 5 mm long and 30 μ in diameter. Their protoplasts die and disintegrate as they mature; the walls are lignified and contain pits. Most of the water movement in angiosperms occurs through vessels formed by the destruction of the end walls and disappearance of the protoplasts of long rows of cells. The resulting tube-like structures have diameters ranging from 20 to 800 μ and lengths varying from a few centimeters to many meters. Vessels are relatively short in diffuse-porous species and quite long in ring-porous species, especially in lianas (Greenidge, 1952; Kramer and Kozlowski, 1960). It is believed that single, continuous vessels are often differentiated from top to bottom in ring-porous trees (Priestley, 1935), but in many species their effective length is soon reduced by tyloses (Liming, 1934), air bubbles, masses of gum, and other blockages.

Usually the xylem of ring-porous species ceases to function effectively

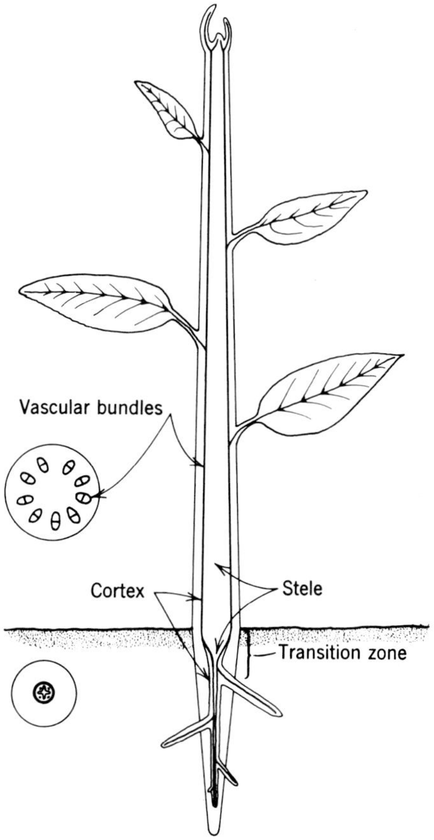

FIG. 8.1 Diagram of a young dicotyledonous plant showing the continuity of the conducting system. Cross sections show the arrangement of vascular tissues in shoots and roots. Adapted from various sources.

in conduction after a year or two, while that of diffuse-porous species often continues to function for many years. As a result, most of the ascent of sap occurs in the outermost annual ring of ring-porous species, but several to many annual rings are functional in diffuse-porous species and in conifers (Kramer and Kozlowski, 1960, pp. 333–336; Kozlowski and Winget, 1963). The outermost annual ring might be expected to function as the chief path for sap flow because it is most directly connected to the new leaves and there has been less time for it to become blocked. However, this is not always true. In dormant conifer seedlings up to four years of age, water movement occurred in all annual rings. However, experiments with the dye basic fuchsin indicated that less movement occurred through the outermost annual ring than through the second ring. More water movement also occurred through the large-diameter tracheids of the early wood than through the smaller tracheids of the late wood (Kozlowski et al., 1966). Swanson (1966) reported

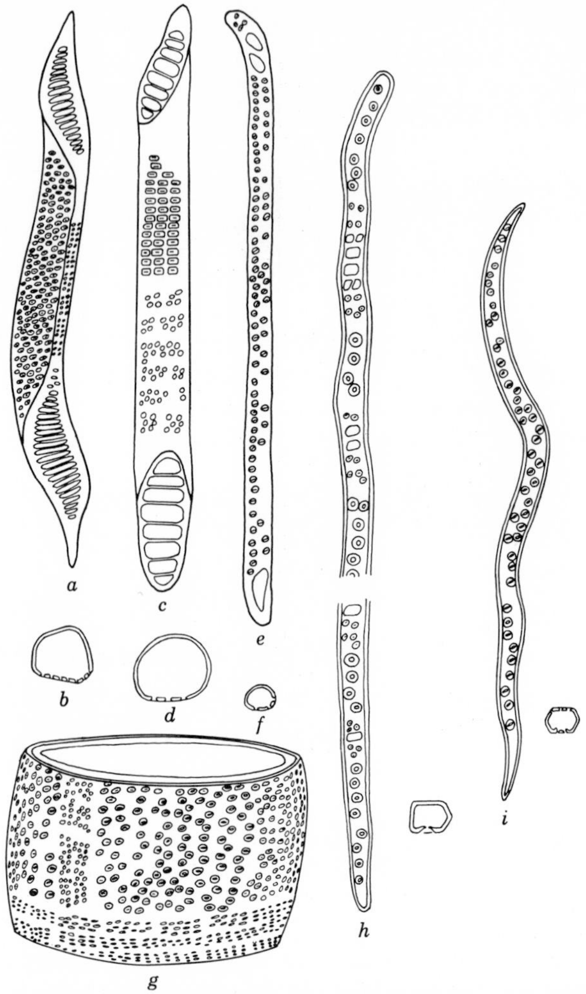

FIG. 8.2 Vessel elements and tracheids. Many vessel elements are joined end-to-end through their perforated end walls to form vessels which vary in length from a few centimeters to many meters, with diameters of 20 to 800 μ. Vessels *a, b,* from *Betula alba*; *c, d,* from *Liriodendron tulipifera*; *e, f,* from *Lobelia cardinalis*; *g,* from *Quercus alba*. Tracheid *h,* from *Pinus strobus* (one-third of length shown); *i,* from *Quercus alba*. Tracheids seldom exceed 5 mm in length and 30 μ in diameter. Vessel elements and tracheids are drawn to the same scale. (*After Eames and MacDaniels,* 2d ed., 1947.)

that in lodgepole pine and Engelmann spruce most rapid conduction occurs 15 to 20 mm within the surface, and there is relatively little upward movement of water in the outer 5 mm of the wood.

The localized conducting system of ring-porous species makes it more susceptible to blockage by air bubbles and by mechanical injury such as girdling than the xylem of diffuse-porous species. Huber (1935) states that the evergreen oaks of the Mediterranean region are diffuse-porous and questions whether a ring-porous evergreen tree could survive because of early blockage of the large xylem vessels. In monocots such as palms, the conducting system is very complicated (Tomlinson, 1963).

Persons unfamiliar with plant anatomy should consult a text such as Esau (1965) or the sections on plant structure in Meyer and Anderson (1952) to obtain more detailed information concerning the structure of the conducting system.

SPIRAL GROWTH OF XYLEM. An interesting feature of the xylem of trees of many species is its tendency to grow in a spiral so that water often moves upward in a spiral path rather than in a straight vertical path. Kozlowski and Winget (1963) reviewed the literature on this topic and presented figures showing the path of dyes in trees of several species. They state that species

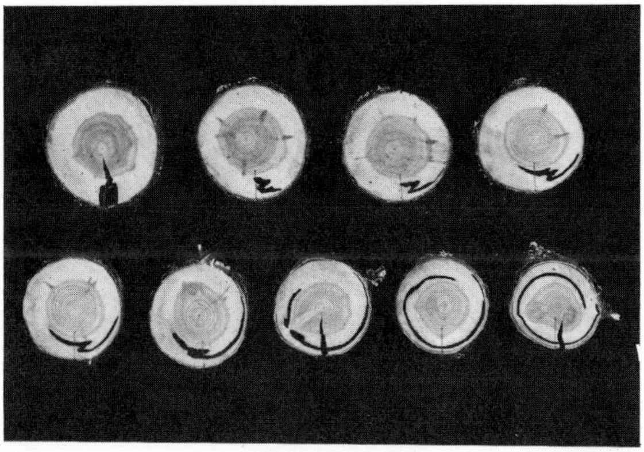

FIG. 8.3 **Spiral path of ascent of sap in larch. Acid fuchsin was injected into a root and rose in spiral pathway. As the dye ascended, it moved inward and sometimes completely encircled the stem. The vertical line is above the point of injection. The sections were cut at intervals of 60 cm, the lowest section being at the upper left.** *(Courtesy of T. T. Kozlowski.)*

such as pin oak, in which the transpiration stream spreads out in the top, suffer more severely from the oak-wilt fungus than do the white oaks, in which water moves more nearly straight upward and the fungus remains localized in a limited part of the tree top. This seems to support the claim of Rudinsky and Vite (1959) that a spiral pattern provides more effective distribution of water to the crown than straight vertical ascent. Thomas (1967) reported that dye moves spirally around as much as 90° of the circumference per meter of ascent in dogwood.

NODES. There are often very complicated patterns in the vascular systems at nodes where connections are made to leaves and the vascular bundles are interconnected. A diagram of a node in potato is shown in Fig. 8.6.

INJURY TO THE CONDUCTING SYSTEM. The conducting system of most plants provides a much larger capacity than is required for survival, and a considerable fraction of the xylem can usually be removed without killing plants. In fact, Jemison (1944) observed that trees on which over 50 percent of the circumference near the base had been killed by fire made as much growth during the following 10 years as nearby uninjured trees. New xylem laid down after an injury is often oriented so that it provides a very effective pathway for water flow around wounds.

It has been demonstrated that two horizontal cuts made half way through the stem one above the other from opposite sides of the stem do not prevent the ascent of sap (Elazari-Volcani, 1936; Greenidge, 1955; Preston, 1952). In fact, Postlethwait and Rogers (1958) demonstrated movement of radioactive phosphorus past as many as four cuts made only 6 in. apart in trunks of trees

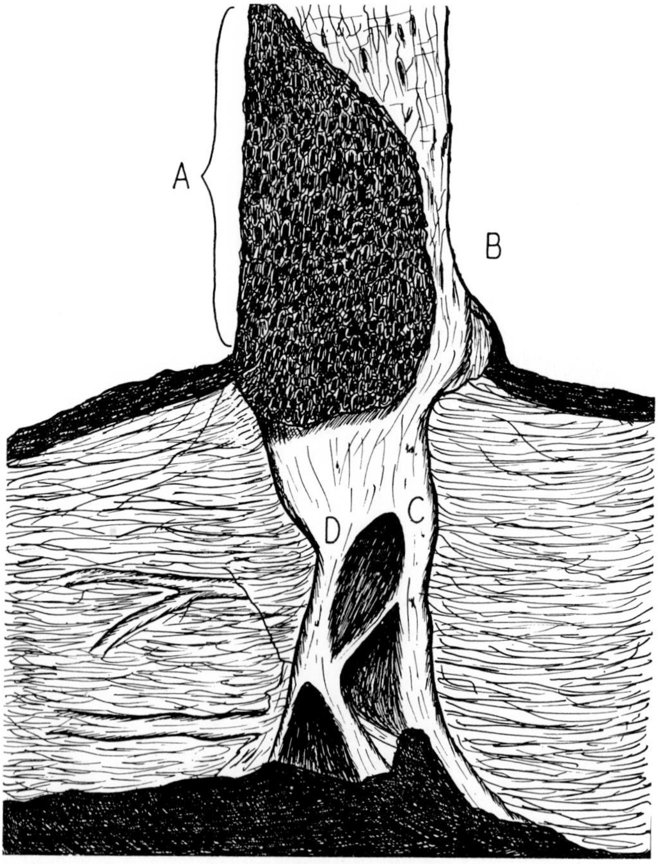

FIG. 8.4 New xylem and phloem are reoriented to form an effective bypass around a fire wound. A, burned area; C and D, roots connected by new conducting tissue to surviving bark, B. (*From Kramer and Kozlowski,* 1960; *after Jemison,* 1944.)

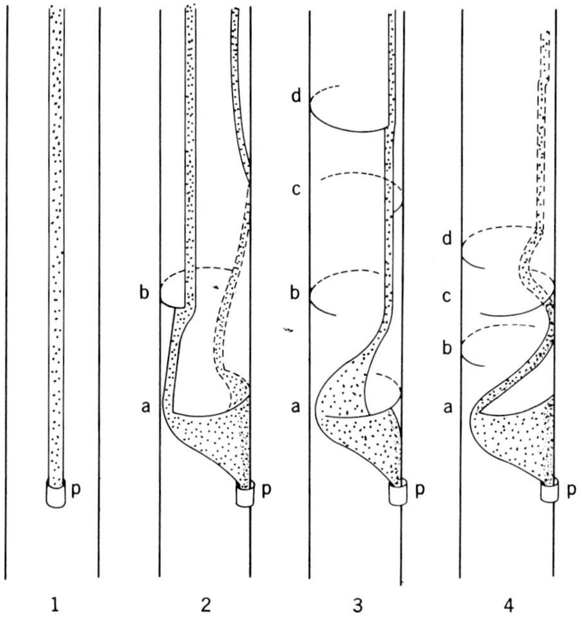

FIG. 8.5 Diagrams showing movement of ^{32}P around cuts in trunks of pine trees. The cuts are designated as *a*, *b*, *c*, and *d*. ^{32}P was supplied at point *P*. The stippled areas indicate the path followed by the isotope. (*From Postlethwait and Rogers,* 1958.)

of several species. Scholander et al. (1957) reported that air entering xylem elements through cuts is confined to the vessels which have been opened. Although the resistance to flow increases, water moves around the blockage caused by cuts in the walls and through the smaller xylem elements which are not plugged by air. Thus, the introduction of air by wounds or otherwise does not prevent the movement of sap; it merely increases the resistance to flow. The vascular bundles in the stems of many herbaceous species are so interconnected at the nodes that many of them can be cut without seriously reducing the flow of water to the leaves. This is true of the large bundles of tomato, but not of the smaller ones (Dimond, 1966).

In spite of the large safety factor in the conducting system of most plants, activities of bacteria, fungi, and insects often result in injury and blockage of the xylem; the result is reduction or prevention of the upward flow of water. There has been much discussion concerning the extent to which so-called wilt diseases are caused by blocking of water flow in the xylem and the extent to which injury is caused by toxins produced by the fungus (Dimond, 1955; Beckman, 1964). It seems unlikely that the mycelium actually blocks the xylem. However, there is adequate evidence that in at least some diseases, oak wilt for example, the vascular system of infected trees is plugged by tyloses and gums which block the movement of water and cause injury by

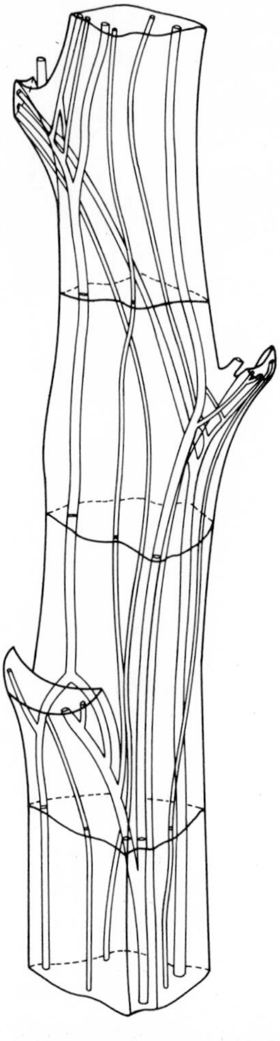

FIG. 8.6 The vascular system of the stem of potato (*Solanum tuberosum*), showing the complex branching at the nodes. Readers are referred to the study by Dimond (1966) of flow relations in the tomato stem, which has a very similar structure. (*From Eames and MacDaniels*, 2d ed., 1947; *after Artschwager*.)

dehydration (Kozlowski et al., 1962; Kuntz and Riker, 1955). Plugging of the vascular system has also been observed in Dutch elm disease, verticillium wilt of elm and maple, and mimosa wilt. It seems possible that toxins produced by the fungus injure the cells adjoining the xylem elements and stimulate the formation of gum and tyloses. On the other hand, it is claimed by Gäumann that fusarium wilt of tomato is caused by a toxic substance or substances released by the fungus (see Dimond, 1955, for references). Dimond (1967) regards toxins as unimportant and attributes the injury largely to blockage of water transport through the xylem. As pointed out by Talboys (see Koslowski, 1968), injuries from vascular or wilt diseases are complex in nature and cannot always be assigned to a single cause.

Efficiency of the conducting system

The fact that the larger part of the xylem can be removed or inactivated without killing the shoot suggests that it forms a very efficient conducting system. Its efficiency can be expressed in terms of specific conductivity under known conditions or as relative conductivity, which is the ratio of conducting surface

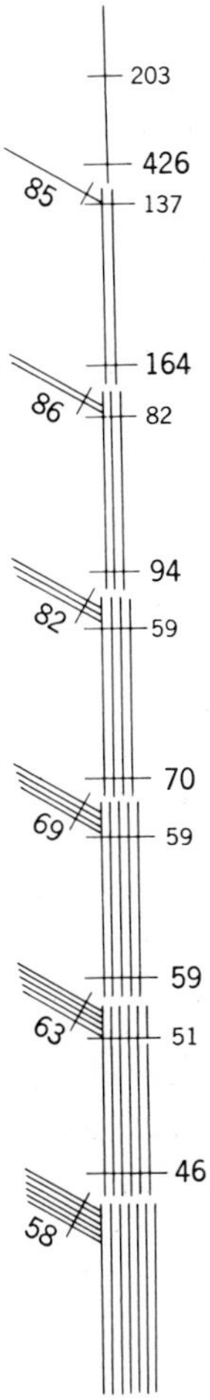

FIG. 8.7 Differences in relative amounts of conducting surface along the main stem and in the side branches of a six-year-old *Abies concolor* seedling, expressed as hundredths of a square millimeter of xylem cross section per gram of needle fresh weight. The relative conductivity is lower at the point of attachment of the whorls of branches (numbers in lightface type) than between nodes (numbers in boldface type), but there is a consistent increase in relative conductivity from base to apex. (*After Huber*, 1928.)

to transpiring surface. Huber (1956*b*) summarized the literature on this subject. He expresses the specific conductivity in terms of volume of water moved per hour under a given pressure in a segment of given length and cross section of conducting tissue, for example, ml/(hr)(cm^2)(bar) over a distance of 1 m. Huber gives relative values of 20 for conifers, 65 to 128 for deciduous broad-leaf trees, 236 to 1,273 for vines, and even higher values for roots. The values for branches and twigs are usually lower than those for the trunk; the specific conductivity of tree trunks is said to decrease from base to top; hence the resistance to water movement is not proportional to the length of the stem.

According to Huber, the relative conductivity or ratio of conducting surface in square millimeters to transpiring surface in grams of fresh weight is about 0.5 for trees, 0.2 for herbaceous shade plants, and 0.10 for desert succulents, but 3.4 for other desert plants and only 0.02 for certain aquatic plants. Data on the relative conductivity of the stem and branches of a white fir tree (*Abies concolor*), given in hundredths of a square millimeter of xylem cross section per gram of leaf fresh weight, are shown in Fig. 8.7. In this species relative conductivity, or the ratio of conducting surface to transpiring surface, increases from bottom to top of the stem and is higher for the branches than for the main stem.

Velocity of sap flow in stems

The differences in specific conductivity of stems of ring-porous and diffuse-porous species result in large differences in velocity of water flow (see Table 8.1). Some measurements have been made by injecting dyes (Greenidge, 1958) or radioactive tracers (Fraser and Mawson, 1953; Moreland, 1950; Kuntz and Riker, 1955), but measurements by this method suffer from the probability that the rate is modified by cutting into the xylem. Huber (1956) discussed the methods of measuring sap flow. Although somewhat qualitative, the most reliable method seems to be the thermoelectric or heat pulse method devised by Huber and his students (Huber and Schmidt, 1937) and used by various others (Bloodworth et al., 1956; Ladefoged, 1960; Skau and Swanson, 1963; Kurtzman, 1966). Heat is supplied to the sap stream by a small heating element attached to the stem; the rate of sap flow is measured by the time required for the warm water to reach a thermocouple placed in or on the wood above the heater. Marshall (1958) discussed the theory and difficulties inherent in this method and concluded that although the sap speed exceeds the speed of the heat pulse, it is a useful method for measuring relative rates of sap flow. The data of Table 8.1 show that the rate is much greater in ring-porous species, where flow is restricted to a single ring, than in conifers and diffuse-porous species, where many annual rings are involved. Differences in rates in various parts of a single tree are shown in Fig. 8.8, with diurnal variations shown in Fig. 8.9.

Table 8.1 Rates of Water Movement in Xylem Measured by Various Methods

Investigator	Method	Material	Rate, m/hr
Bloodworth et al., 1956	Thermoelectric	Cotton	0.8–1.1
Greenidge, 1958	Acid fuchsin	*Acer saccharum*	1.5–4.5
	" "	*Ulmus americana*	4.3–15.5
Huber and Schmidt, 1937	Thermoelectric	Conifers	<0.5
	"	*Liriodendron tulipifera*	2.6
	"	*Quercus pedunculata*	43.6
	"	*Fraxinus excelsior*	25.7
Klemm and Klemm, 1964	^{32}P	*Betula verrucosa*	±3.0
Kuntz and Riker, 1955	^{86}Rb	*Quercus macrocarpa*	27.5–60
Moreland, 1950	^{32}P	*Pinus taeda*	1.2

Bloodworth et al. (1956) and Ladefoged (1960) used the heat pulse method to study the effects of various treatments on rate of sap flow as an indicator of variations in rates of absorption and transpiration. Daum (1967) applied the apparatus to the main stem and the two major branches of a forked ash tree and measured water flow in all three locations. Sap flow started earlier in the morning on the exposed east side of the tree and was more rapid in the morning than in the afternoon in the east side branch. On some occasions there was downward flow in one branch and upward flow in the other. This indicated translocation of water from the slowly to the rapidly transpiring branch. A heavy afternoon rain after a period of rapid transpiration resulted in reversal of the water flow in the tree trunk; downward movement occurred for a time from the wet leaves and branches to the roots.

Direction of sap flow

In this discussion it has been assumed that water movement is upward and outward. This is usually the case because water movement is from regions of higher to regions of lower potential, and loss of water by transpiration lowers the water potential in the leaves. However, water can move in the reverse direction. This was demonstrated in experiments by Hales and other early investigators. In fact, as early as 1669 John Ray described experiments which showed that water would move either way in branches, dispelling the idea

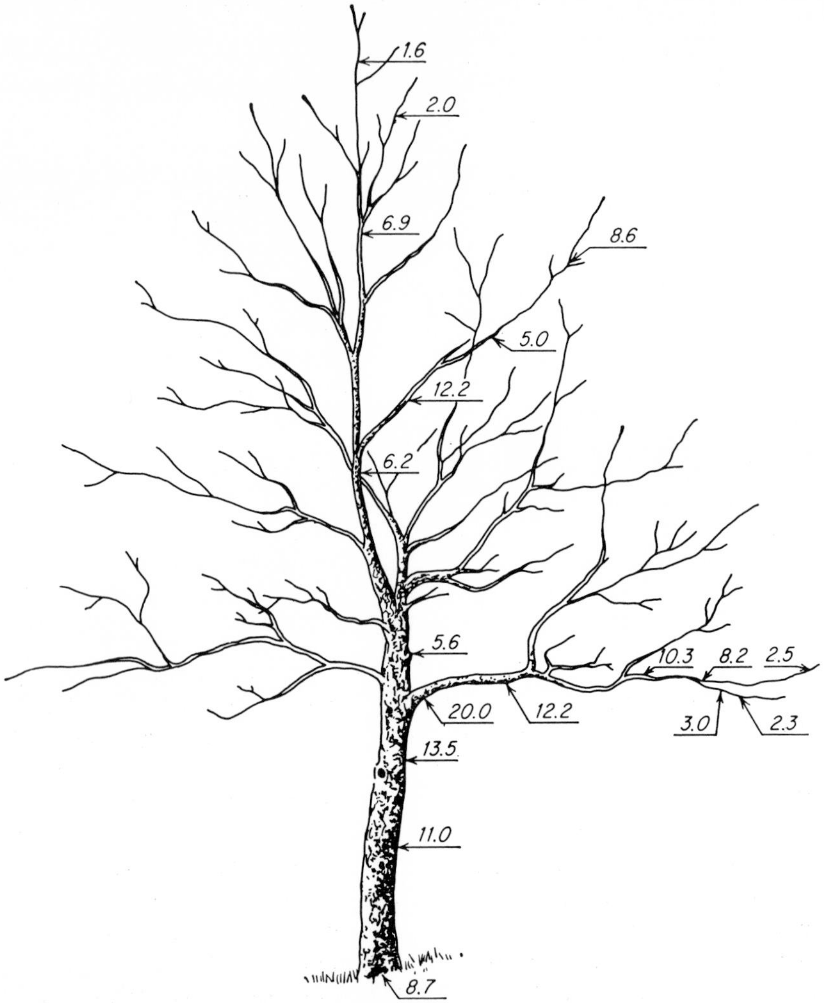

FIG. 8.8 **Rates of water movement in meters per hour in various parts of an oak tree at midday. The rate of flow decreases toward the top because the relative conductivity (see Fig. 8.7) increases toward the top. In birch, relative conductivity decreases and rate of flow increases toward the top. (*From Huber*, 1956*b*; *after Huber and Schmidt*, 1937.)**

that valves exist in the water-conducting system. Williams (1933) reviewed the earlier work and demonstrated enough water movement from leaves immersed in water to keep other leaves in the air turgid for several days. Slatyer (1956) and Jensen et al. (1961) demonstrated that water will flow equally well through plants in both directions. The reverse flow of sap observed by Daum (1967) is mentioned in the preceding section. These and other experiments suggest that there is no more resistance to water movement in the reverse than in the normal direction and that it is free to move in any direction indicated by the prevailing gradient in water potential.

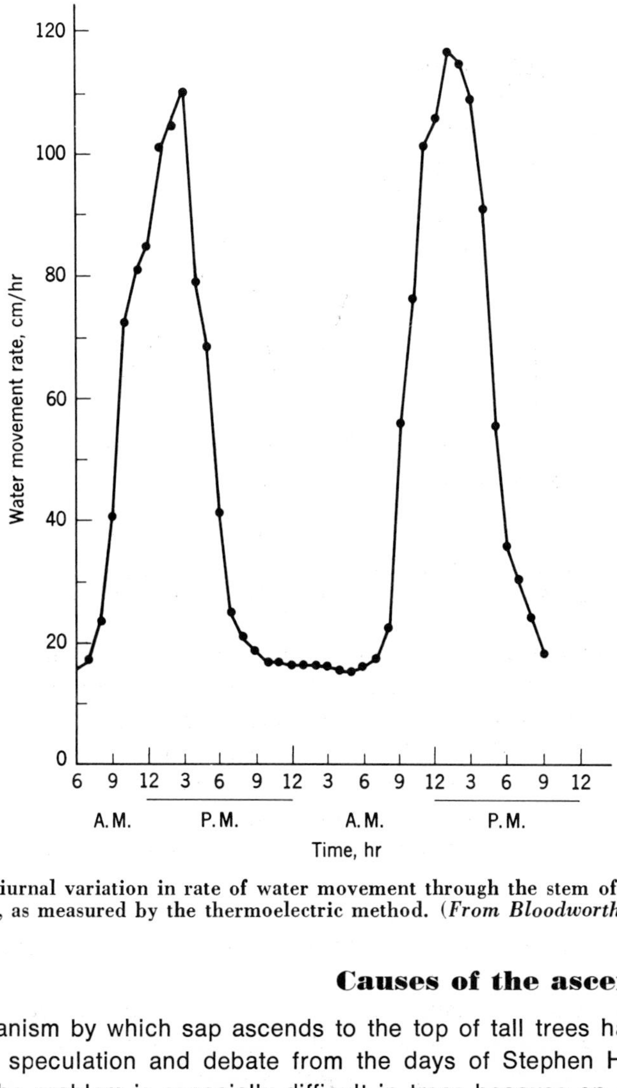

FIG. 8.9 **Diurnal variation in rate of water movement through the stem of a field-grown cotton plant, as measured by the thermoelectric method. (*From Bloodworth et al.,* 1956.)**

Causes of the ascent of sap

The mechanism by which sap ascends to the top of tall trees has been the subject of speculation and debate from the days of Stephen Hales to the present. The problem is especially difficult in trees because on hot summer days hundreds of liters must be transported upward a distance of 10 to 100 m to the transpiring tops. The various theories were reviewed by Kramer (1949); obsolete ones will be discussed only briefly.

Historical review

Root pressure has often been assumed to play an important role in the ascent of sap, and some writers still regard it as important (Rufelt, 1956; White, 1938, 1958). There is no doubt that sometimes sufficient root pressure is developed in certain species of herbaceous plants and in woody plants such as grape

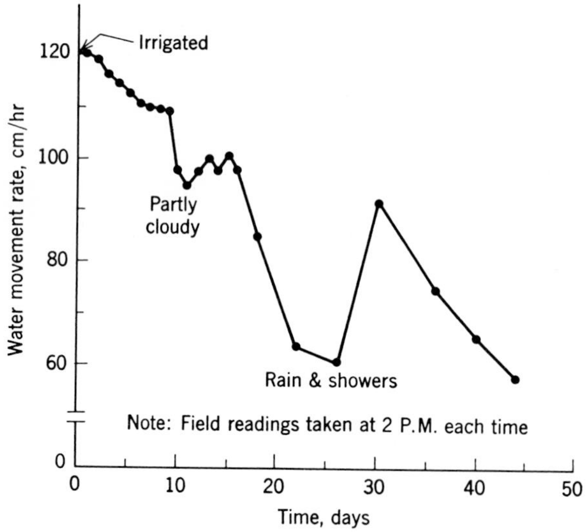

FIG. 8.10 **Effects of changes in soil water content and weather on rate of water movement through stems of field-grown cotton plants.** (*From Bloodworth et al.*, 1956.)

and birch to raise water to the tops when they are transpiring very slowly. This is shown by the occurrence of guttation. However, Hales recognized two and a half centuries ago that root pressure cannot be detected in rapidly transpiring plants and that it does not occur in plants growing in dry soil. Neither does it occur commonly in conifers. In some species it may play a part in refilling xylem elements plugged with air, but it certainly does not play an essential role in the ascent of sap in rapidly transpiring plants.

During the latter half of the nineteenth century, several German physiologists concluded that the living cells of the stems play an essential role in the ascent of sap. It was supposed that the parenchyma cells of the xylem carried on some sort of pumping action which caused the ascent of sap. The view that living cells play an essential role in the ascent of sap has been held by various writers almost to the present time (Handley, 1939; Priestley, 1935; Preston, 1952). Bose (1923) claimed that measurements of electrical potentials on stems showed that rhythmic waves of pulsating activity in stems cause the ascent of sap. However, several investigators, beginning with Boucherie and continuing to the present time (Kurtzman, 1966), have shown that water will rise through stems or stem segments killed by heat or various poisons. It is true that flow through dead stems and stem segments is soon reduced, but this is because gum formation blocks the xylem elements in dead or dying stems rather than because living cells are necessary to cause the flow (Kramer, 1933). However, it is possible that the presence of living cells is essential to prevent infiltration of air and maintain the continuity of the water columns in the xylem.

FIG. 8.11 An example of reversal in the usual direction of translocation produced by inarching one tomato plant to another then cutting it free from its roots. The inked lines show the path followed by water and solutes. (*From Kramer and Kozlowski*, 1960; *after Zimmerman and Connard*, 1934.)

The cohesion theory

The cohesion theory of the ascent of sap was foreshadowed by Hales, and both Sachs and Strasburger concluded that transpiration produces the pull causing the ascent of sap. Boehm (1892) demonstrated that transpiring branches could raise mercury above barometric pressure, but the demonstration by Askenasy (1895) and Dixon and Joly (1895) that water has considerable tensile strength was necessary to make the cohesion theory acceptable.

The following are the essential features of the cohesion theory of the ascent of sap:

(1) Water has high cohesive forces, and when confined in small tubes with wettable walls it can be subjected to a tension of many bars (30 to possibly 300 or more) before the cohering columns rupture.

(2) Water is very strongly bound in the walls of cells such as the mesophyll cells of leaves, from which most water evaporates.
(3) The water in a plant is connected through the water-saturated cell walls and forms a continuous system.
(4) When water evaporates from any part of the plant, such as the cells of the leaves, the reduction in potential at the evaporating surface causes movement of water from the xylem to the evaporating surfaces. This reduces the pressure on the water in the xylem, and if water loss exceeds absorption, the pressure on the water in the xylem may fall below zero and become a tension.
(5) The reduced pressure or tension is transmitted through the hydrodynamic system to the surfaces of the roots, where it reduces the water potential and causes inflow of water from the soil.

Under these conditions in transpiring plants, there is continuous mass flow of water from the soil through the roots, up the stems, and out into the leaves to the evaporating surfaces. Dixon (1914) regarded the water as if it were hanging, suspended from the evaporating surfaces and anchored by the imbibitional forces in the cell walls of the evaporating surfaces.

Objections to the cohesion theory

There have been strong objections to the cohesion theory and a reluctance on the part of some workers to accept it (Greenidge, 1957; Lundegårdh, 1954; Preston, 1957; Priestley, 1935). The objections are concerned chiefly with the adequacy of the cohesive forces of water, the instability of water columns under tension, and the plugging of xylem elements with air.

The theoretical intermolecular attractive forces in water are extremely high, but in practice they are believed to range from 30 to 300 atm. In fact, several workers doubt if they can exceed 30 atm. in the xylem sap (Greenidge, 1957; Loomis, et al. 1960), but this should be sufficient to pull water to the top of a tree 100 m in height. Indirect evidence suggests the occurrence of much higher tensions in the xylem sap. Slatyer (1957) measured water potentials of −70 bars in privet plants and −77 bars in cotton plants near death from desiccation. Arcichovskij and Ossipov (1931) measured water potentials down to −142.9 bars in a desert shrub; it was assumed that equally high tensions must occur in the xylem sap. However, direct measurements were not made until Scholander et al. (1965) introduced the pressure chamber method. In this method a freshly cut twig is enclosed in a pressure chamber with the cut end projecting out through an airtight seal (see Fig. 10.29), and pressure is applied until sap exudes from the cut surface. The pressure required to displace sap is regarded as equal to the tension existing in the water of the xylem before it was cut. Measurements made by

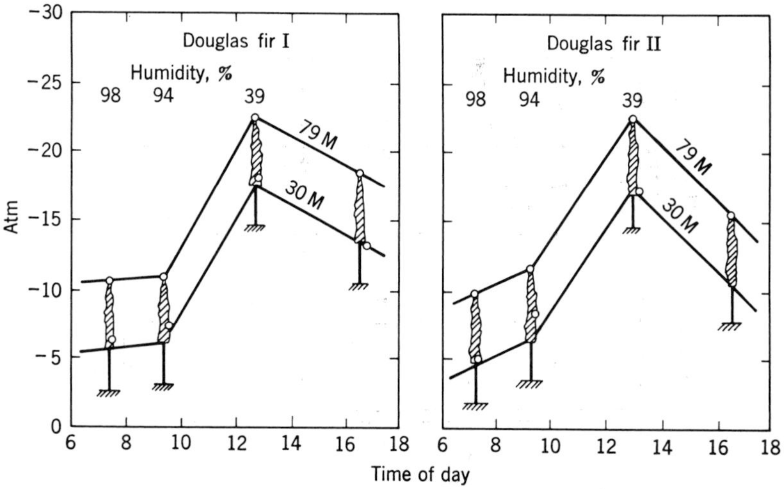

FIG. 8.12 **Differences in hydrostatic pressure in upper and lower parts of crowns of Douglas fir trees at various times of day as measured on excised twigs in a pressure chamber.** (*From Scholander et al.*, 1965.)

this method gave a range of sap tensions from −5 bars in plants of moist habitats to −35 to −60 in halophytes and perhaps −80 bars in the desert shrub creosote bush. The tensions were considerably greater during the day than at night. Measurements of branches taken at various heights from Douglas fir and redwood trees showed increases in tension corresponding to the increase in height of the trees. Although this method is subject to various errors, it seems to provide convincing evidence of the existence of tensions in the xylem of at least −80 bars.

Much has been written about the instability of water columns under tension and the ease with which they break by cavitation in glass capillary tubing. It has been suggested that if they break as easily in the xylem of trees, they would soon become inoperative because of shocks such as those caused by swaying in the wind. However, it seems probable that the nature of the walls of the xylem elements, which are filled with imbibed water, makes the water columns in the stems of plants more stable than those in glass tubes. By use of a vibration detection device, Milburn and Johnson (1966) found evidence of progressive cavitation in individual vessels in the petioles of *Ricinus* leaves under increasing water stress, rather than simultaneous cavitation throughout the xylem. It is true that the water columns in large xylem vessels often break and the lumina of many vessels become filled with air, but the tracheids and smaller vessels remain water-filled. Data of Clark and Gibbs (1957) indicate that up to nearly 50 percent of the water in trunks of some trees may be replaced by air during the summer (see Fig. 8.13). However, experiments by Scholander et al. (1957) indicate that this air is isolated

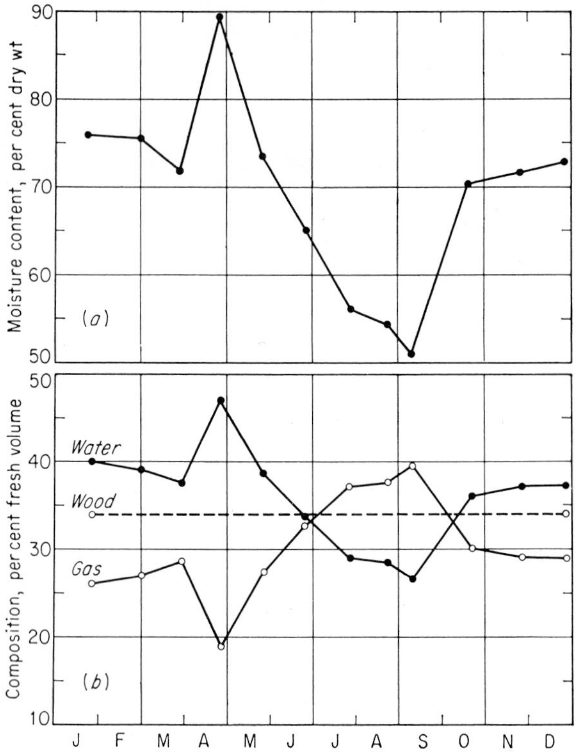

FIG. 8.13 (*a*) Seasonal changes in water content of yellow birch tree trunks calculated from disks cut from the base, middle, and top of the trunks. (*b*) Seasonal changes in gas and water content of yellow birch tree trunks calculated as percentages of total volume. (*From Clark and Gibbs*, 1957.)

in the larger vessels and does not block the entire conducting system. This problem was discussed earlier in connection with injuries to the conducting system.

Another difficulty is the possibility that freezing forces dissolved gas out of solution, the bubbles formed presumably blocking most or all of the xylem elements (Scholander, 1958). However, Hammel (1967) reported that freezing stem segments and twigs of hemlock did not increase resistance to water flow after they were thawed, although it did produce a large increase in resistance to flow in angiosperms. He suggested that in gymnosperms the bordered pits of each tracheid isolate the sap in it, so that pressure develops when the sap freezes and the tiny bubbles of gas formed during freezing are redissolved under this positive pressure. This does not occur in the large vessels of ring-porous angiosperms, where the bubbles are too large and the pressure too low to cause them to dissolve. Zimmerman (1964) suggests that in ring-porous angiosperms, such as oak and ash, the old vessels remain filled with air but new ones are produced early in the spring before the leaves are fully expanded. In another group, including birch and grape, the root pressure which usually occurs in the spring tends to dissolve the gas bubbles in the vessels, causing them to refill with water. Probably most diffuse-porous species behave like gymnosperms, their smaller vessels never being blocked. In view of these observations, blockage of the conductive elements after freezing seems to present less difficulty than formerly supposed.

In spite of some difficulties and questions, the cohesion theory is the only one which is capable of explaining the rise of water in tall trees. It also explains the fact that the absorption of water is usually fairly closely linked to water loss in transpiring plants. The concept of a continuous system of cohering water extending from the absorbing surfaces to the transpiring surfaces is the best explanation of various phenomena which we see in the water relations of plants.

Conduction in leaves

The final step in water conduction through plants is its movement into leaves and distribution to their various tissues. At each node where a leaf is attached, a segment of the vascular system, called a leaf trace, separates from the vascular system of the stem, extends out through the petiole into the leaf blade, and provides a pathway for translocation of water and solutes.

Leaf venation

The arrangement of the vascular system in various kinds of leaves varies widely. In most conifers a single vein extends through the center of the leaf. In grasses numerous veins extend the length of each leaf, parallel to the midvein, and anastomose at the margins near the leaf tips. These veins are connected by small veins extending across the intervening mesophyll tissue. Some dicots have palmate venation, where a few large veins extend outward from the base of the leaf blade and are connected by a complex system of smaller veins. Other dicot leaves are pinnately veined, that is, they have numerous branches extending outward from each side of the midvein. These secondary veins are enclosed in bundle sheaths and contain xylem, cambium, and phloem. The actual distribution of water to the mesophyll occurs chiefly from the smaller veins which branch off from the secondary veins. These branch and rebranch and lose their cambium, then their phloem, and finally their bundle sheaths, ending as single xylem elements buried in the mesophyll. A few xylem elements terminate in epithem tissue near hydathodes. In many species the small veins anastomose and form complex networks. They are so numerous that most cells of a leaf are only a few cells distant from a vein or vein ending. Wylie (1938) found no cell more than 50 μ from a xylem

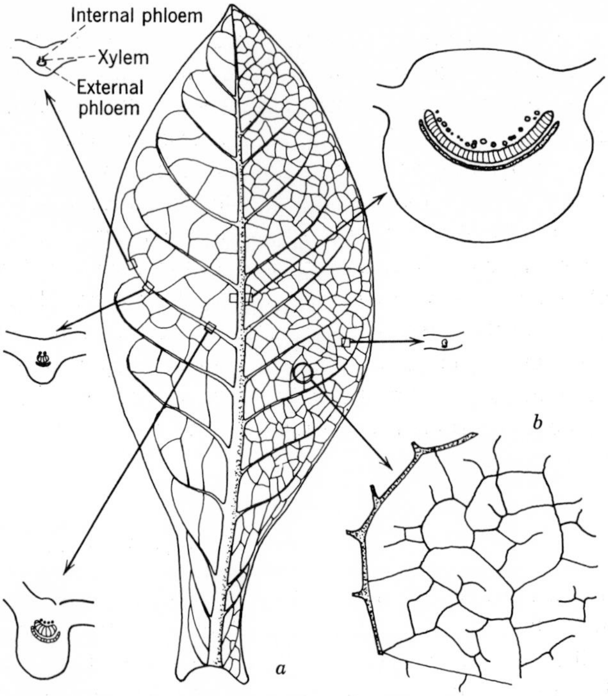

FIG. 8.14 (*a*) **Diagram of venation of a mature tobacco leaf showing midrib and principal lateral veins, also cross sections of midveins and lateral veins of various sizes. Internal phloem is found only in the midrib and principal lateral veins. (*b*) Enlargement of a small section of leaf blade to show the ultimate network of veins. There were 543 mm of veins per square centimeter of leaf blade on this leaf. (*From Avery,* 1933.)**

element in a group of mesomorphic species which he studied. There appears to be so much conductive capacity in most leaves that they often survive the cutting of major veins, as shown in Fig. 8.15.

Conduction outside the veins

There is some difference of opinion concerning the pathway by which water is supplied to the epidermis. Long ago LaRue (1930) reported that the epidermis is normally detached from the mesophyll in *Mitchella repens* and can be detached experimentally over considerable areas in other species without injury. LaRue and later Williams (1950) concluded that the epidermis is supplied with water directly from main veins rather than from the underlying mesophyll.

Many vascular bundles have bundle sheath extensions, which are vertical masses of colorless cells extending outward from the bundles to the upper and lower epidermis and forming partitions which divide the air space of the mesophyll into numerous small chambers. Wylie (1952) reported that bundle sheath extensions are most frequent in deciduous leaves and least common

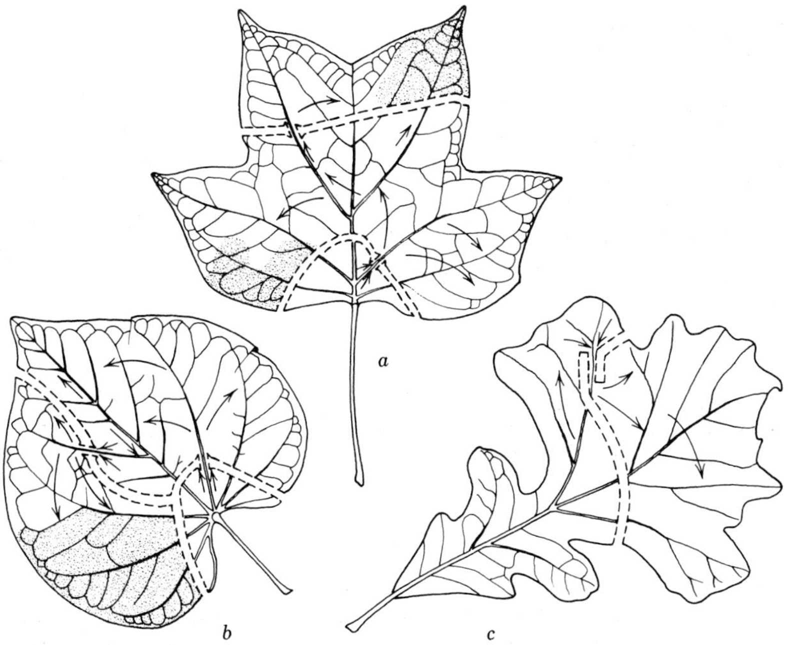

FIG. 8.15 **Ability of leaves to survive severe interruption of normal pathways for water conduction by cutting principal veins. Dotted lines indicate where tissue was cut, and arrows show general direction of water conduction. Shaded areas indicate tissue killed by dehydration. These experiments indicate a large excess of conductive capacity in the small veins and free movement in the reverse direction. Experiments were performed on attached leaves in the open and lasted 20 days. Leaf a is *Liriodendron tulipifera*; leaf b, *Cercis canadensis*; leaf c, *Quercus velutina*. (*From Plymale and Wylie*, 1944.)**

in broad-leaved evergreens. He suggested that these bundle sheath extensions form a pathway for water and solute movement from the xylem of the veins to the epidermal cells, at least in mesomorphic species (Wylie, 1943). It might be easier for water to move out to epidermal cells through the bundle sheath extensions then back to the mesophyll, rather than to move laterally through the mesophyll, because there is limited lateral contact among mesophyll cells.

It has been suggested that there might be a significant amount of water movement to the epidermis as vapor through the intercellular spaces (Russell and Woolley, 1961). According to Williams (1950), this was disproved long ago because the epidermis remained alive even when the inner surface was covered with a waterproof coating. Furthermore, the rate of movement of water as vapor is much slower than its movement as liquid, hence movement as vapor is not likely to be an important source of supply to the epidermis. There is evidence that considerable water movement occurs in the walls of leaf cells rather than across the protoplasts (Weatherley, 1963). This will be discussed in the next section.

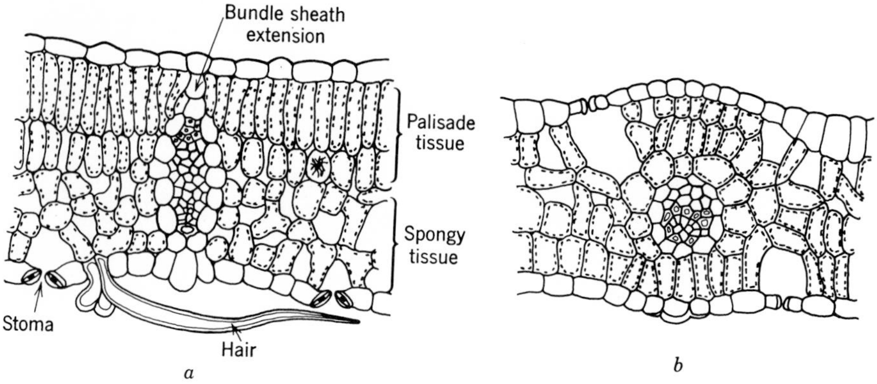

FIG. 8.16 Cross sections of leaves with bundle sheath extensions (*a*) and without (*b*). Section *a* is an oak leaf and shows well-defined palisade and spongy tissue, stomata on the lower surface, and an epidermal hair. Section *b* is an oat leaf with stomata on both surfaces. (*After Eames and MacDaniels*, 1947.)

Conduction outside the vascular system

Although emphasis has been placed on water movement through the xylem, by now it should be apparent to the reader that water movement also occurs outside the xylem. Water entering roots must cross a number of cells before it reaches the xylem, and in leaves it may move across several cells before reaching its ultimate destination in the epidermal or mesophyll cells. If water can move laterally through parenchyma cells, obviously it can also move longitudinally through them, and presumably small amounts of water move longitudinally in the parenchyma cells of the cortex and pith in roots and stems. However, the resistance to flow through living cells is so much higher than resistance to flow through xylem that practically all water movement occurs through the xylem wherever that alternative exists.

This can be illustrated as follows. The average length of tracheids is 5 mm, so water must traverse at least 200 tracheids per meter of xylem. However, if an average parenchyma cell diameter of 50 μ is assumed, there would be 20,000 cells per meter of stem. If there were the same resistance to flow across tracheid and parenchyma walls, the resistance to water flow would be 100 times greater in the parenchyma than through the tracheids. Furthermore, there is the added resistance of the protoplast of each parenchyma cell, which at 10^{-5} cm/(sec)(bar) is over 10,000 times that of tracheid walls (Briggs, 1967, pp. 87–88). Even if water bypasses the protoplasts and moves longitudinally through the walls of the parenchyma cells, the resistance is much higher than for movement through the tracheids. Resistance to movement through xylem-containing vessels is obviously much lower than for movement through tracheids.

There has been considerable discussion concerning the exact pathway followed by extrafascicular water. Does it move across the cells from vacuole to vacuole; does it move through the cytoplasm, bypassing the vacuole; does it move chiefly in the walls; or does it use some combination of these pathways? Strugger (1949) argued from experiments with fluorescent dyes that much water movement occurs in the cell walls. This view is also held by Gaff et al. (1964) and by Weatherley (1963); it is supported by calculations of Russell and Woolley (1961). They estimated that after correction is made for the small volume of wall available as a pathway, the resistance to water movement through the walls is only 5 percent of the resistance to water movement across the protoplasts in the root cortex. Data of Weatherley (1963) indicate that water flow through the walls is 50 or 60 times more rapid than through the protoplasts in the mesophyll cells of leaves. Briggs (1967, pp. 89–91) used another set of values for conductivity in walls and protoplasm and concluded that a pressure difference of 28 bars would be required to move water across the protoplasts in a layer of five root cells at a rate of 1.4×10^{-5} cm/sec. However, a pressure gradient of only 1.4 bars would move water at the same rate through cellulose walls. Since pressures across roots as large as 28 bars are unusual, Briggs concluded that most of the water movement must occur in the walls rather than across the protoplasts.

There are data suggesting that the pathway for mass flow of water in roots may differ from the pathway for movement by diffusion. Raney and Vaadia (1965) reported that cells of roots become equilibrated with tritiated water more slowly in transpiring plants than in detached roots. This suggests that the mass flow of water through roots of transpiring plants bypasses the vacuoles of the cortical cells, but entry by diffusion occurs more uniformly through all parts of the cells, including the vacuoles. Woolley (1965) reported that the half time for movement of tritiated water into roots of transpiring plants is about 160 times that for diffusion of water into detached roots. This seems possible only if the pathway for diffusion is much more extensive and the resistance much lower than for mass flow.

Resistances in various parts of the water-conducting system

One of the most important developments in plant water relations in recent years has been the wide acceptance of the view that there is a continuous flow of water through what is often termed the soil-plant-atmosphere continuum (Philip, 1966). This concept also stresses the analogy between the flow of electricity through a conductor and that of water through the soil-plant-atmosphere system. It assumes that the movement of water through the plant to the air is proportional to the difference in water potential, which is the driving force, and inversely proportional to the resistances in the pathway. This concept seems to have been proposed first by Gradmann

(1928), and it was developed by van den Honert (1948), Slatyer (1960), Rawlins (1963), Philip (1957*a*, 1966), and others. Cowan and Milthorpe (1968) have a detailed discussion of water flow through the plant. Some of the implications of this approach will be examined in the next section. In this section we will discuss the resistances in various parts of the water conducting system.

Root resistance

Considerable attention has been given to the resistance to water movement through roots, because it is generally regarded as an important reason for the lag of absorption behind transpiration which is responsible for the midday water deficits characteristic of rapidly transpiring plants. Also, low temperature and deficient aeration decrease water absorption by increasing root resistance. Early attempts to measure the root resistance involved measuring transpiration from a plant, then cutting off the shoot and measuring the amount of water moved through the same root system under a given pressure gradient. Renner (1929) estimated that pressures of 4 to 11 bars were required, and Kohnlein (1930) 20 to 73 bars, to move water through roots at the rate at which it moves in transpiring plants. However, this assumes no change in permeability, when in fact the permeability of roots tends to increase with increasing pressure gradients (see Chap. 6). This tends to make such estimates of resistance too high. Wilson and Livingston (1937) concluded that most of the resistance to water movement is in the roots and the leaves; Kramer (1938) stated that most of the resistance is in the roots, because the lag of water absorption behind transpiration is decreased by removal of the roots (see Fig. 8.17).

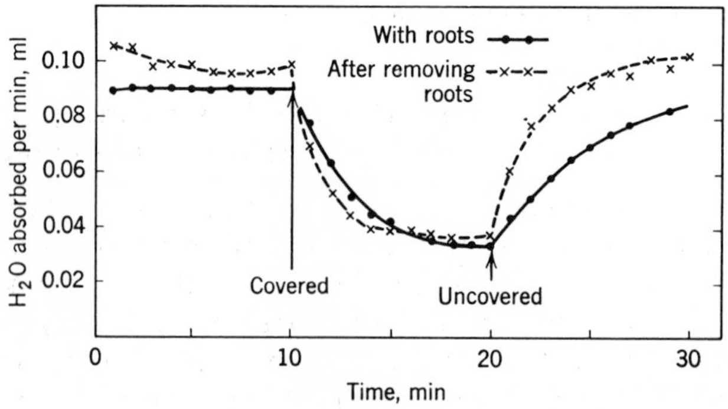

FIG. 8.17 The effect of removing roots on the lag of absorption behind transpiration. Absorption by intact plants was measured for 10 min with shoots exposed. The shoots were then covered for 10 min to reduce transpiration and uncovered for another 10 min. The roots were removed and the procedure repeated. The lag in response of absorption to change in rate of transpiration was much less after removal of roots. (*From Kramer,* 1938.)

Macklon and Weatherley (1965) questioned the importance of root resistance as a cause of leaf water deficits, because the leaf water potential of leaves of *Ricinus communis* growing in water culture were not affected by large changes in rate of transpiration. Similar changes in rate of transpiration produced large changes in leaf water potential in plants rooted in soil. This suggested that the important resistance to water movement is really in the soil rather than in the roots. Root resistance is lower than soil resistance in soils much drier than field capacity because the hydraulic conductivity of soil decreases rapidly with decreasing water content (see Chap. 3). It is often stated that, in soils with a water potential higher than about −1 bar, root resistance is probably greater than soil resistance; but in soils with a lower water potential, the resistance to water movement in the soil may be greater than in the root (Gardner and Ehlig, 1962). However, it is pointed out in Figs. 8.22 and 8.23 that the relative importance of root and soil resistance depends on the amount of root surface in the soil. Increase in soil resistance (decrease in hydraulic conductivity) has much less effect where the amount of root surface per unit of soil is high than where it is low and water must move farther on the average to reach a root surface.

Leaf water deficits sometimes develop even in plants growing in water culture and in soil at field capacity where resistance to water movement is very low, indicating the existence of sufficient resistance in the roots to affect the water supply to the leaves. Skidmore and Stone (1964) reported large changes in root resistance of cotton from midday to midnight. A diurnal change in root resistance was also reported by Barrs and Klepper (1968). The cause of such changes is unknown and deserves further study.

Stem resistance

A careful study of pressure and flow relations in the vascular bundles of tomato by Dimond (1966) indicates that there is relatively low resistance to water flow through the large bundles, also that so much cross transfer occurs that blockage of vessels in a large bundle has little effect on total pressure and flow relationships. Higher pressure was required to move water through the small bundles of the petioles of lower leaves than through the stem to the petioles. Also, more pressure was required to move water to the intermediate leaves than to the uppermost leaves. This indicates that the resistance to water flow is not uniform in the conducting system. Scholander et al. (1965) found that the pressure drop in tall trees and vines is proportional to height, which indicates a fairly uniform resistance to water flow in these systems.

Leaf resistance

It is generally assumed that the resistance to water movement in the leaf veins is quite low. Mer (1940) and Wylie (1938) found that when they cut the main veins of leaves of several species, the area isolated was kept alive by water moving around the cut through the small veins (see Fig. 8.15). However, Rawlins (1963) found evidence of considerable resistance to water movement in tobacco leaves. For example, shaded leaves remained turgid while adjacent leaves in the sun wilted, and one half of a leaf exposed to the sun could be caused to wilt while the shaded half remained unwilted. Large differences in water potential were found in the two halves. Tobacco leaves are reported to wilt first along the margins, which indicates slow movement of water from midveins to margins. Rawlins also found that the resistance to water movement increased in leaves which had been wilted or kept in darkness. Perhaps resistance to water movement in leaves is responsible for the differences in osmotic potential, photosynthesis, and respiration in the base and tip of leaves reported by Slavik (1963). Some differences in resistance may be related to differences in exposure during leaf ontogeny which affect structure and water supply. There should be further investigation of the resistance to water movement in large leaves.

Relative resistances

One of the few attempts to measure the relative resistances in various parts of the conducting system was by Jensen et al. (1961). They arranged plants in a three-compartment chamber (see Fig. 8.18) so that water could be forced through them. By cutting off the roots or the leaves, the conductance in each part could be evaluated. Although the method caused unnatural flow in the sense that the air in the intercellular spaces was first replaced by water under vacuum, the results are interesting and in accord with expectations. They are shown in Table 8.2, where it appears that the root resistance of sunflower is nearly twice as great as the leaf resistance and over three times as great as the stem resistance. The total resistance is lower in tomato than in sunflower, but the root resistance is four times that of the stems and nearly

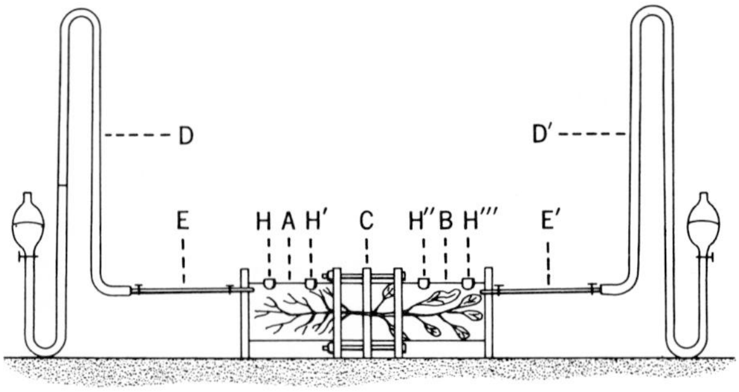

FIG. 8.18 Apparatus for measuring water flow through plants. A and B are closed chambers; C is a partition separating roots from shoots, E and E′ are calibrated capillary tubes attached to mercury manometers D and D′, and H and H′, H″, and H‴ are stoppered openings. (*From Jensen et al.*, 1961.)

Table 8.2 Estimated Relative Resistances to Water Movement through Roots, Stems, and Leaves of Sunflower and Tomato *

	Resistance in sunflower	Resistance in tomato
Whole plant	2.4	1.4
Leaves	0.9	0.6
Stem	0.4	0.24
Roots	1.5	1.00

* *From data of Jensen et al. (1961)*

twice that of the leaves. Water was found to move equally well in either direction in these plants.

Factors affecting water movement

It can be said that in general terms water movement through plants occurs along gradients of decreasing water potential. However, it is impossible to give a neat mathematical summary of the relative magnitudes of the driving forces and resistances at various steps because of the change in state of water from liquid to vapor and the change in magnitude of resistances with change in rate of flow. These facts make it impossible to assume that the rate of movement through the system is proportional to the difference in water potential. In view of this, water movement will first be treated stage by stage, then the application of a unifying theory will be discussed.

Water movement in the soil

Water movement through soils having water contents suitable for plant growth can be assumed to occur as liquid. It usually moves along gradients of decreasing matric potential, which for practical purposes are usually equivalent to differences in total water potential. Although measurable movement as vapor occurs in soil drier than field capacity, this is of minor importance in the water relations of cultivated plants. As pointed out in Chap. 2, the hydraulic conductivity of soil decreases rapidly as the water content decreases. Therefore, as shown in Fig. 2.12, a decrease in soil water potential to about −15 bars is accompanied by a decrease in conductivity to about one-thousandth of the value at field capacity. However, the importance of soil resistance decreases with increase in density of rooting (Andrews and Newman, 1968).

Water movement in the plant

Water movement from the root surfaces to the evaporating surfaces in the leaves occurs in the liquid phase. Over most of this distance it is a mass flow of water and solutes through the xylem elements along gradients of hydrostatic pressure. The greatest resistance to water flow occurs during entrance into the roots where water moves across a mass of living cells. However, it was pointed out earlier that the resistance to water flow across roots decreases as the pressure gradient across them increases (Mees and Weatherley, 1957; Lipushinsky, 1965; also Fig. 6.2*a*). Therefore the rate of water flow through the system cannot be treated as directly proportional to the difference in water potential between root surface and leaf.

It is debatable how much of the water movement across roots is by diffusion and how much is by mass flow. In slowly transpiring plants diffusion of water into the root xylem (active absorption) exceeds water loss, resulting in positive pressure in the xylem. However, in rapidly transpiring plants the rate of water supply diffusion is quite inadequate, and the water in the root cells as well as that in the xylem must be under tension. Under these conditions radial water movement across the root tissues must also occur by mass flow. Probably in rapidly transpiring plants water movement through the entire system from the root surface to the evaporating surfaces of the mesophyll and epidermal cells of leaves is by mass flow. Apparently, in rapidly transpiring plants there is little diffusional movement of water within the plant. However, the difference in hydrostatic pressure which causes this mass flow can be regarded as equivalent to a difference in water potential.

Water movement out of the plant

Water movement from the evaporating surfaces to the bulk air is of course in the form of vapor and occurs by diffusion. The rate is proportional to the driving force, which is the difference in vapor pressure or vapor concentration between the evaporating surfaces and the bulk air and is inversely proportional to the various resistances (cuticular, stomatal, air) in the pathway (see Chap. 9). The change in state from liquid to vapor phase at this point makes a rigorous application of Ohm's law to water movement somewhat misleading.

Factors controlling water movement

As pointed out earlier in this chapter, van den Honert (1948) attempted to develop a unifying theory to identify the factors controlling water movement through the plant, based on an analogy to the flow of current in an electrical conductor. According to Ohm's law

$$\text{Current} = \frac{\text{potential}}{\text{resistance}} \tag{8.1}$$

and van den Honert wrote an expression for steady state flow as follows:

$$\text{Rate of water movement} = \frac{\Psi_{\text{soil}} - \Psi_{\text{root surface}}}{r_{\text{soil}}} = \frac{\Psi_{\text{root surface}} - \Psi_{\text{xylem}}}{r_{\text{root}}} = \frac{\Psi_{\text{xylem}} - \Psi_{\text{leaf}}}{r_{\text{xylem}} + r_{\text{leaf}}} = \frac{\Psi_{\text{leaf}} - \Psi_{\text{air}}}{r_{\text{leaf}} + r_{\text{air}}} \tag{8.2}$$

As Philip (1966) warns, this model is useful but oversimplified. In the first place it represents a steady state situation, something which seldom actually exists in the plant. Attempts are now being made to deal more effectively with the dynamic characteristics of water movement through the soil-plant atmosphere, such as that by Woo et al. (1966). Second, it assumes a constant resistance in each part of the pathway, but the resistances to water movement in the soil, roots, and leaves vary with the rate of water flow. Third, there are difficulties because water moves as liquid through the soil and plant but as vapor from the plant to the air. Thus, in the soil and the plant there are resistances to liquid flow, but in the leaf and the air the resistance is to vapor flow. The basic driving force for both liquid and vapor movement is the difference in water potential, Ψ_w. However, while liquid flow is directly proportional to the gradient in water potential, vapor flow is not, but is proportional to the vapor pressure or vapor concentration gradient. The lack of proportionality between water potential and vapor pressure results from the logarithmic relationship shown in the following equation:

$$\Psi_w = \frac{RT \ln e/e^\circ}{V_w} \tag{1.7}$$

In the original discussions of water movement in the soil-plant-air system by Gradmann and van den Honert, it was emphasized that by far the largest drop in potential occurs from leaf to air. For example, van den Honert (1948) assumed a drop in water potential through the plant of 50 bars, but a drop from leaf to air of 950 bars. Therefore, according to Ohm's law, the resistance in the air pathway should be 19 times the resistance to water movement through the plant; and water movement is controlled in the vapor phase because that is where the highest resistance occurs.

Use of water potential as a measure of gradient in the vapor phase creates a somewhat misleading impression concerning the importance of the resistance to diffusion of water vapor into the air. As stated earlier, water vapor movement is controlled by vapor pressure or vapor concentration gradients, not by water potential gradients. Conversion of vapor pressure gradients into water potential gradients by Eq. (1.7) exaggerates the drop in potential, and the corresponding increase in resistance required by Ohm's law to the movement of water in the liquid phase and the vapor phase produces a somewhat misleading conclusion.

The largest resistance to water movement through the soil-plant-atmosphere continuum is at the leaf-air interface because of the leaf structure, not because a physiological analog of Ohm's law states that it must be there. The total leaf and air resistances to water loss for cotton with open stomata are 4.9 and 1.1 sec/cm at wind speeds of 0.6 and 3.1 cm/sec. The resistances to evaporation from moist filter paper are 3.5 and 1.6 sec/cm at the same wind speeds. However, the cuticular resistance of cotton with closed stomata was 32 sec/cm in experiments of Slatyer and Bierhuizen (1964). There is no doubt that water movement through and out of plants is usually controlled at the leaf-air surface, but this is because of the cuticular and stomatal resistances to water vapor movement. The very large resistance called for by van den Honert's application of Ohm's law to water potentials is largely irrelevant to the control of water vapor loss, which depends chiefly on the vapor pressure gradient and the resistances to vapor diffusion inherent in leaf structure and the air layer surrounding leaves.

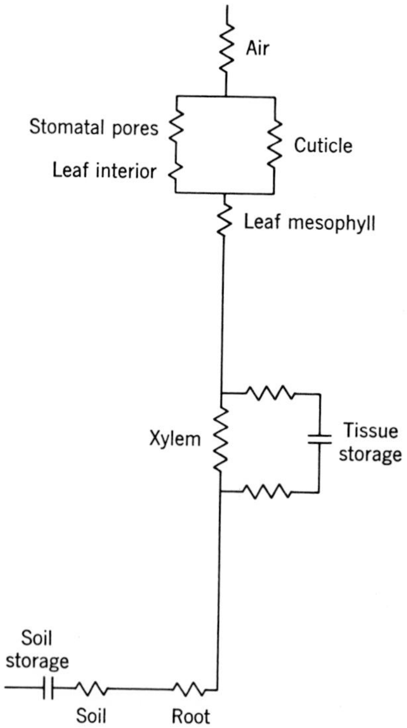

FIG. 8.19 Simplified diagram treating the flow of water through the soil-plant-atmosphere system as analogous to the flow of electrical current through a series of resistances and capacitances. A more detailed diagram is given by Cowan (1965) and a treatment of dynamic flow is described by Woo et al. (1966).

In spite of some inherent weaknesses, this type of approach has stimulated the development of various interesting models, such as that shown in Fig. 8.19. Slatyer (1967) summarized the conclusions from this treatment as follows: (1) Total flow of water is controlled primarily in the vapor phase between the evaporating surfaces in the leaves and the bulk air, (2) the stomata are the chief regulators of water flow through the plant, and (3) development of an increase in resistance to flow elsewhere in the system (e.g., in the roots) can reduce transpiration only indirectly, by reducing leaf turgor sufficiently to cause closure of stomata.

This indirect control was described by Slatyer (1967) and is shown in Fig. 8.20. In this diagram, it is assumed that, at zero time, a plant rooted in

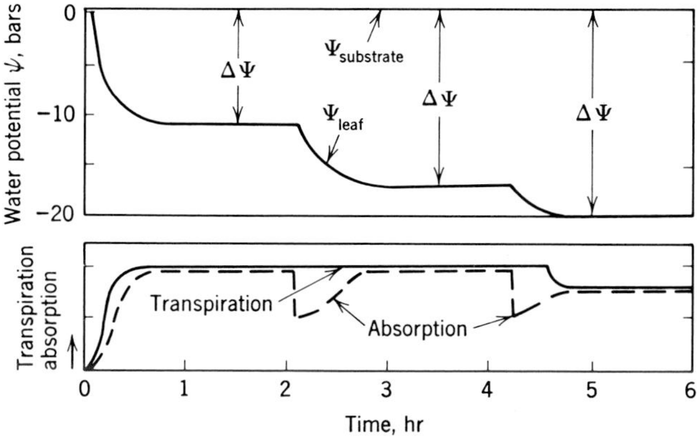

FIG. 8.20 **Diagram showing relationship between rates of absorption and transpiration, leaf water potential Ψ_{leaf}, and the difference between Ψ_{leaf} and $\Psi_{substrate}$ ($\Delta\Psi$) when root resistance is increased in two stages by cooling the substrate.** (*After Slatyer*, 1967.)

water culture in darkness ($\Psi_{root} = 0$, $\Psi_{leaf} = 0$) is illuminated under constant environmental conditions. Transpiration increases to a steady state value in approximately 30 min, and after 1 hr, a steady state situation exists, with transpiration and absorption proceeding at the same rates. The gradient in water potential ($\Delta\Psi = \Psi_{solution} - \Psi_{leaf}$) and the lag of total absorption behind total transpiration are represented by a steady state value of Ψ_{leaf}. Root permeability is then suddenly reduced by cooling the water around the roots, and absorption drops to half its original value. The lag of total absorption behind transpiration increases, even though the rate of absorption commences to increase again, as decreasing Ψ_{leaf} increases $\Delta\Psi$. Finally, the rate of absorption again reaches the transpiration rate, and the steady state system is reestablished. Now, however, a much greater difference in water potential $\Delta\Psi$ is required to maintain the original rate of flow through the roots. It is assumed that transpiration rate is not reduced by the reduction in absorption because the drop in Ψ_{leaf} is insufficient to cause stomatal closure or to change the vapor pressure at the leaf surface.

If it is now assumed that root permeability is reduced a second time, so that rate of absorption is again reduced by one-half, the same sequence of events tends to recur. However, on this occasion, before $\Delta\Psi$ has increased sufficiently to reestablish the rate of absorption at the level of the transpiration rate, Ψ_{leaf} falls to a value which induces partial stomatal closure and causes a reduction in transpiration rate. It is assumed that the degree of closure reduces transpiration sufficiently to prevent further decrease in Ψ_{leaf}, and the rate of transpiration decreases to the level of the prevailing rate of absorption.

This example shows that although the stomata exercise direct control over water movement, the high resistance to liquid water flow through roots is an important factor in controlling the rate of transpiration. In a sense, therefore, the question concerning the location of the controlling mechanism is semantic, since it is clearly valid to state that decreased root permeability causes reduced transpiration, even though direct control over vapor loss may be exerted through change in stomatal opening.

The general question of the main source of control over transpiration was raised again recently by Levitt (1966). However, several workers, including Slatyer and Lake (1967) and Cowan and Milthorpe (1967), strongly reaffirmed the view that final, overall control must lie between the sites of evaporation in the leaf and the free air outside.

Water movement in response to demand

Although stomatal aperture can play an important role in regulating water flow through the soil-plant-atmosphere system, the actual rates of flow are clearly influenced by the atmospheric factors affecting transpiration and by soil water status. Moreover, these factors interact with each other and with the leaf water potential to affect stomatal aperture. For example, under conditions of extremely high potential transpiration, it is possible for Ψ_{leaf} to fall to a level which induces some degree of stomatal closure, even when a plant is rooted in culture solution or in soil at field capacity. However, if potential transpiration rates are extremely low, Ψ_{leaf} may remain above the critical value even in soils having a low water potential. Consequently, the main factors determining actual transpiration rates are (1) those which affect soil water supply, including soil water potential, the hydraulic conductivity of the soil, and root distribution, and (2) those which affect the degree of stomatal closure and control the potential transpiration.

Various attempts have been made to include these factors in an analysis of this problem [see, for example, Philip (1957*a*), Gardner (1960), Visser (1964), Gardner (1965), and Cowan (1965)].

Cowan's model (Fig. 8.19) uses three main functions. The first describes soil water flow to the root in terms of the water potential of the soil mass

(Ψ_{soil}), the water potential at the root surface (Ψ_{root}), the hydraulic conductivity of the soil, and root density (length of root per unit volume of soil). The two remaining functions describe the influence of leaf water potential and resistance to liquid flow in the plant on stomatal aperture and hence on the capacity of the plant to transpire at the potential rate.

The main assumptions are that the rate of water flow, and hence of transpiration, is proportional to $\Psi_{root} - \Psi_{leaf}$, and that a critical value of Ψ_{leaf} ($= \Psi_{crit}$) exists at which stomatal closure reduces transpiration to a degree which prevents further decline in Ψ_{leaf}. At Ψ_{leaf} values higher than Ψ_{crit}, it is assumed that transpiration proceeds at the potential rate.

From the earlier discussion of driving forces and water permeabilities in the liquid pathway, it is apparent that these assumptions are subject to criticism. However, for the present they provide evidence of the interaction between soil, plant, and atmospheric factors in determining the transpiration rate of a hypothetical, but fairly typical, crop.

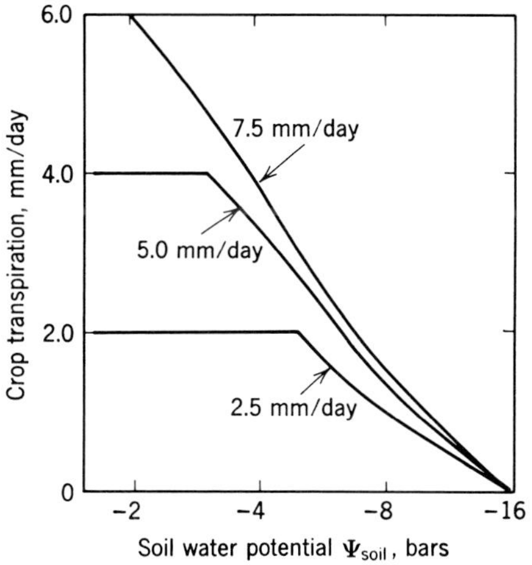

FIG. 8.21 **Estimated relationship between rate of transpiration in millimeters per day and the soil water potential Ψ_{soil} for low, intermediate, and high rates of potential transpiration. At low rates of transpiration the soil dries to a much lower potential before water supply becomes limiting than at high rates of transpiration, where a much steeper gradient of water potential from bulk soil to root surface is required to move water to the roots at the required rate. At very high rates of potential transpiration, actual transpiration may always be below the potential rate. (*After Cowan*, 1965.)**

Typical results from Cowan's analysis are shown in Figs. 8.21 to 8.23. In Fig. 8.21, the expected relationship between transpiration and soil water potential is given for three different levels of potential transpiration. The figure shows that as the soil dries and Ψ_{soil} declines, there is initially no effect on transpiration, but when hydraulic conductivity falls to a level at which stomatal closure commences because of decreasing Ψ_{plant}, a progressive decline in transpiration occurs. This stage is reached at higher values of Ψ_{soil} with higher rates of potential transpiration, since steeper gradients of Ψ_{soil} are required to sustain flow, resulting in lower levels of Ψ_{root}, and hence of Ψ_{leaf}, at any given value of Ψ_{soil}. It can be seen that when very high potential transpiration rates occur, the actual daily transpiration might not at any stage equal the potential daily transpiration.

In Figs. 8.22 and 8.23 the effect of root density on the relationship between transpiration and Ψ_{soil} is depicted. In this case only the intermediate

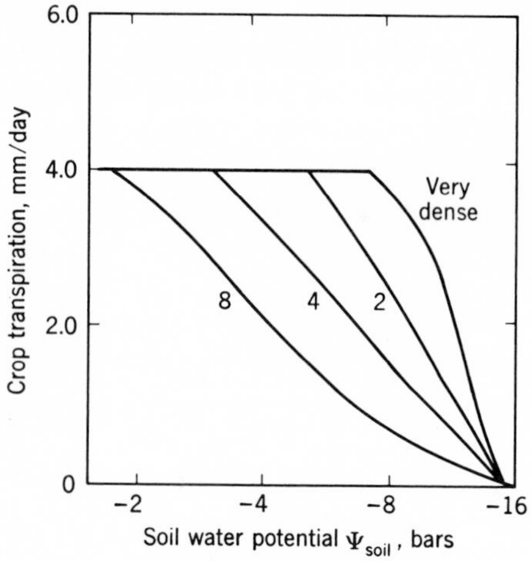

FIG. 8.22 Effect of root density on the expected relationship between daily transpiration of a crop and soil water potential, assuming the intermediate rate of potential transpiration of Fig. 8.21. Soil water potential is less limiting at high root densities than at low densities. Root densities are 8, 4, and 2 cm^3 of soil per centimeter of root length and very dense rooting. These densities are comparatively low compared with those often found under crop plants. (*After Cowan*, 1965.)

level of potential transpiration is used. In Fig. 8.22 it can be seen that as root density increases, higher potential rates of transpiration can be sustained until much lower values of Ψ_{soil} exist, smaller values of $\Psi_{soil} - \Psi_{root}$ being adequate to sustain flow at the desired rate. In Fig. 8.23 the same data are plotted against time, for a soil initially at a water content of 0.2 cm^3 water per cubic centimeter of soil. The marked effect of root density is again apparent, and the response in both figures ranges from a pattern suggesting a progressive decrease in soil water availability for transpiration to a pattern suggesting little effect on soil water availability until low Ψ_{soil} values exist. E. I. Newman (private communication) estimates that if there are 100 cm or more of roots per square centimeter of soil surface, the soil resistance will not become limiting until the matric potential falls to −25 bars. The density of roots probably often exceeds this value, and in such situations soil resistance seldom becomes limiting.

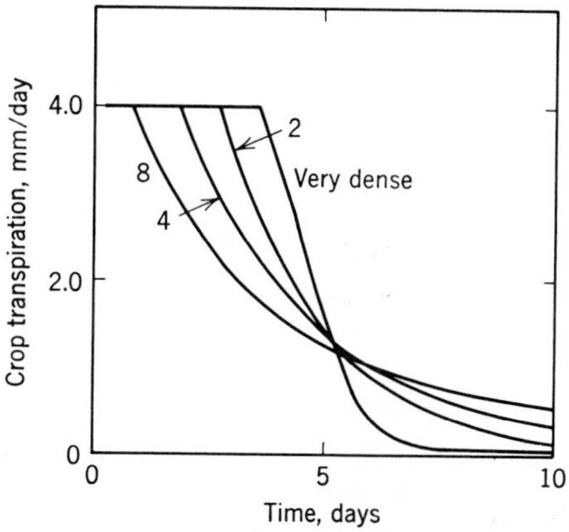

FIG. 8.23 The expected decrease with time in transpiration of a crop having the various root densities specified in Fig. 8.22. At high root densities transpiration is maintained at the potential rate for a longer period, but it then falls more rapidly than for plants with low root density which remove soil water more slowly. (*After Cowan,* 1965.)

The principal conclusions of the Cowan model are confirmed by several field and laboratory studies dealing with the effect of soil water status and evaporative conditions on evapotranspiration (for example, Denmead and Shaw, 1962; Gardner and Ehlig, 1962, 1963; Ehlig and Gardner, 1964). However, differences also occur. For example, Denmead and Shaw (1962) studied the pattern of evapotranspiration from a corn crop, in relation to soil water content under a range of potential transpiration conditions (see Fig. 8.24).

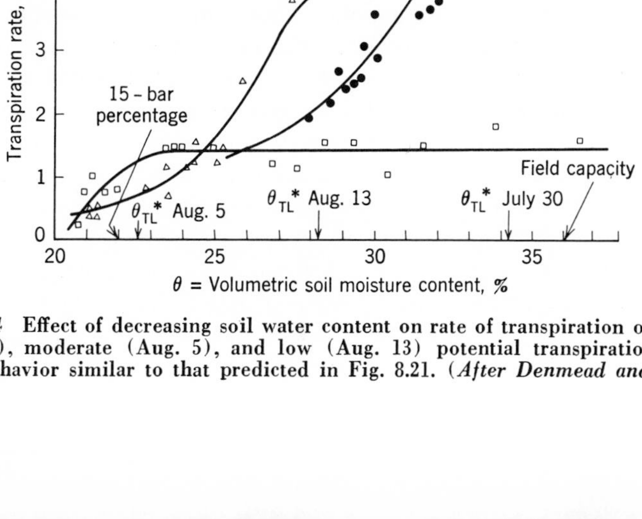

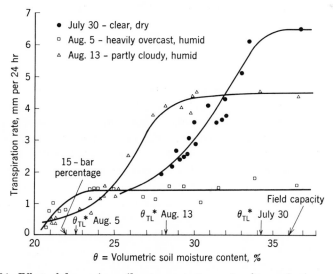

FIG. 8.24 Effect of decreasing soil water content on rate of transpiration on days of high (July 30), moderate (Aug. 5), and low (Aug. 13) potential transpiration. This figure shows behavior similar to that predicted in Fig. 8.21. (*After Denmead and Shaw,* 1962.)

While there is qualitatively good agreement with Fig. 8.21, evapotranspiration of the low- and moderate-potential transpiration treatments can be seen to have been maintained at the potential rate until lower soil water contents developed than those predicted by the model, and the curves tail off as zero transpiration is approached rather than intercepting the abscissa at a value close to $\Psi_{soil} = -15$ bars.

The first of these differences may have been due to the assumption, in the model, of a sharp fluctuation in daytime potential transpiration, whereas, in the field study, the days of low and moderate transpiration were cloudy and humid, and there was probably no pronounced break in the daily evaporative regime. The second is probably due to the oversimplification involved in the assumption that transpiration ceases when Ψ_{soil} reaches a critical value, usually -15 bars. Even so, a great deal of information is to be gained by theoretical studies of the type conducted by Cowan (1965), and, with progressive improvements in experimental techniques, these models will become still more realistic.

Summary

Development of a water-conducting system which permitted rapid mass flow of water from roots to the evaporating surfaces of the exposed shoots was an essential prerequisite to the evolution of large land plants. The resistance to water movement through masses of parenchyma cells is so high that it would be quite impossible for sufficient quantities of water to move from roots to shoots of tall transpiring plants through the pith and cortex. Where water moves through parenchyma tissue, such as roots and leaves, much of the movement probably occurs in the cell walls rather than through the protoplasts.

In addition to providing a pathway of low resistance for rapid movement of water, the continuous water of the xylem provides a control system which tends to keep absorption and transpiration approximately in balance. Loss of water from leaves produces a decrease in water potential which is transmitted to the roots, causing increased absorption. Decrease in water absorption causes a decrease in water potential to be transmitted to the shoots, where decrease in turgor causes closure of stomata and reduction in transpiration.

The movement of water through transpiring plants is often treated as analogous to the flow of electricity through a conducting system containing various resistances. The rigorous application of Ohm's law to the transport of water as a catenary process from soil to air leads to the conclusion that the greatest resistance to the movement of water through plants is in the vapor phase and that this resistance therefore controls water movement through the plant. This is an oversimplification, because it neglects the change in state from liquid (where the movement is proportional to the difference in water potential) to vapor (where the movement is proportional to the difference in vapor pressure).

It seems more satisfactory to deal separately with liquid water movement through the plant, as being driven by the difference in water potential from the soil to the evaporating surfaces of the shoots, and movement as vapor, driven by the difference in vapor pressure. The overall resistance to flow is relatively low, being highest in the roots, intermediate in the leaves, and lowest in the stems, where movement is largely in the vascular system. The driving force for liquid water is generated by the decrease in water potential in the shoots caused by transpiration. Since transpiration is controlled in most plants by the stomatal aperture and the gradient in vapor pressure from leaf to air, the rate of water movement through plants is controlled chiefly in the vapor phase.

When absorption of water is reduced by drying soil or high root resistances caused by low temperature or inadequate aeration, the resulting decrease in water potential causes loss of turgor in the leaf and closure of stomata. Thus increases in soil and root resistance operate indirectly to reduce transpiration by increasing the stomatal resistance.

The velocity of water movement into root surfaces and out of leaves is relatively low, but the velocity of movement through the vascular system of the stems is relatively very high. In corn plants it was estimated that water enters roots and passes out of leaves with a velocity of 0.01 cm/hr, but it passes through the xylem at the base of the stem at the rate of 1,000 cm/hr, or 100,000 times as rapidly. Measurements of the velocity of flow through stems of woody plants indicate rates of 100 to 6,000 cm/hr.

nine
Transpiration

This chapter deals with the nature and importance of transpiration from individual plants and plant communities. Transpiration can be defined as the loss of water from plants in the form of vapor. It is therefore basically an evaporation process. However, unlike evaporation from a water surface, transpiration is modified by plant structure and stomatal behavior operating in conjunction with the physical principles governing evaporation.

Importance of transpiration

Transpiration is the dominant factor in plant water relations because evaporation of water produces the energy gradient which causes movement of water into and through plants. Therefore, it controls the rate of absorption and the ascent of sap. Furthermore, transpiration causes almost daily transient leaf water deficits, and when drying soil causes absorption to lag behind water loss, permanent water deficits develop which cause injury and death by desiccation. In fact, more plants are injured or killed as a result of transpiration exceeding water absorption than by any other cause. If transpiration could be eliminated, crop plants would thrive over large areas which are now semidesert or desert.

The quantitative importance of the process is indicated by measurements which show that a Kansas corn plant lost over 200 liters of water in a summer, or about 100 times its own weight (Miller, 1938), and a field of corn in Illinois transpired over 20 cm of water, or about 80 percent of the precipitation during the growing season. A deciduous forest in southwestern North Carolina transpired over 40 cm, which is equal to about 30 percent of the annual rainfall. A further illustration of the large amount of water lost in transpiration is the fact that from 90 to 500 kg of water are used per kilogram of dry matter produced by crop plants. Of all the water absorbed by plants, approximately 95 percent is lost by transpiration, and 5 percent or less is used in the plant. If it were not for the loss of water by transpiration, a single rain or irrigation would provide enough water for the growth of an entire crop.

Table 9.1 ***Amounts of Water Lost in Various Ways by a North Carolina Watershed Covered with a Deciduous Forest (1940–1941) and the Increase in Runoff Which Followed Cutting of All Woody Vegetation and Elimination of Transpiration (1941–1952)****

	Amount in cm, 1940–1941	Amount in cm, 1941–1942
Precipitation	158.0	158.4
Interception	16.6	9.5
Runoff	53.4	93.0
Soil storage	−0.4	9.7
Evaporation	39.7	46.0
Transpiration	48.7	00.0

* *From Hoover (1944)*

One sometimes reads discussions concerning the importance of transpiration in bringing about the ascent of sap, increasing the absorption of mineral nutrients, and cooling leaves (Clements, 1934; Curtis, 1926, 1936). Of course transpiration speeds up the movement of water through plants, but use of water in growing cells would cause slow ascent of sap even if there were no transpiration. Water moves to the tops of tall trees as they grow upward, and transpiration merely increases the speed and the quantity moved. Much salt absorption occurs independently of water absorption, that is, by active transport (see Chap. 7). However, rapid transpiration does increase the rate of upward movement of salt, and it might be asked whether enough salt would reach the leaves of plants in the absence of transpiration. There are no good data on this problem, although certain kinds of plants thrive in humid environments with very low rates of transpiration. There can be no doubt that the evaporation of water tends to cool the leaves. Leaves which are transpiring very slowly because of wilting, closure of stomata, or other reasons are usually considerably warmer than the air. However, they are rarely injured by overheating. We can agree with Curtis (1926) that transpiration is usually an unavoidable evil, unavoidable because of the structure of leaves and evil because it often results in water deficits and injury by desiccation.

Nature of transpiration

Because of its great importance in the overall water economy of plants, the physical nature of the process of transpiration deserves careful attention.

Driving forces and resistances

The rate of transpiration depends on the supply of energy to vaporize water, the water vapor pressure or concentration gradient which constitutes the driving force, and the resistances to diffusion in the vapor pathway. Most of the water vapor escapes through stomata, but some passes out through the epidermis and its cuticular covering, and in woody species some escapes through lenticels in the bark of the branches and twigs.

The driving force causing liquid water movement through plant tissues is usually the gradient in water potential. In contrast, the driving force for the movement of water vapor in evaporation and transpiration is the gradient in concentration or vapor pressure. Diffusion of water vapor and evaporation (E) from a free water surface can be described by the following equation, which includes a conversion factor from concentration or specific humidity to vapor pressure:

$$E = \frac{c_{\text{water}} - c_{\text{air}}}{r_{\text{air}}} \cong \frac{0.622\rho_{\text{air}}}{P} \frac{e_{\text{water}} - e_{\text{air}}}{r_{\text{air}}} \tag{9.1}$$

E is the evaporation in g/(cm^2)(sec); c_{water} and c_{air} are the water vapor concentrations at the water surface and in the bulk air respectively in g/cm^3; e_{water} and e_{air} are the corresponding water vapor pressures in mm Hg; r_{air} is the surface boundary layer resistance encountered by diffusing molecules in sec/cm; ρ_{air} is the density of air; and P is the atmospheric pressure in mm Hg.

Transpiration differs from evaporation because the escape of water vapor from plants is controlled to a considerable degree by leaf resistances, which are not involved in evaporation from a free water surface. Therefore, the equation for transpiration (T) requires additional terms to include these resistances in the leaf.

$$T = \frac{c_{\mathrm{leaf}} - c_{\mathrm{air}}}{r_{\mathrm{leaf}} + r_{\mathrm{air}}} \cong \frac{0.622\rho_{\mathrm{air}}}{P} \frac{e_{\mathrm{leaf}} - e_{\mathrm{air}}}{r_{\mathrm{leaf}} + r_{\mathrm{air}}} \tag{9.2}$$

where c_{leaf} and e_{leaf} are the water vapor concentration and vapor pressure respectively at the evaporating surfaces within the leaf, and r_{leaf} is the additional diffusive resistance inside the leaf.

Equation (9.2) states that the rate of transpiration in grams per square centimeter per second is proportional to Δc, the difference in concentration of water vapor, or Δe, the difference in vapor pressure between the leaf and the bulk air outside, divided by the sum of the resistances to diffusion ($r_1 + r_a$) in the leaf and in the boundary layer adjacent to it. The situation with respect to resistances is shown diagrammatically in Fig. 9.1. If most of the stomata are on one surface of a leaf, r will differ significantly for the upper and lower surfaces (Holmgren et al., 1965). The term $0.622\rho_{\mathrm{air}}/P$ is a conversion factor to change from Δc to Δe (Byers, 1959, p. 160). It has a value of approximately 10^{-6}, so 1 mm of vapor pressure is equivalent to approximately 1 mg of water vapor per liter of air.

The components of the various terms in Eq. (9.2) will be examined in

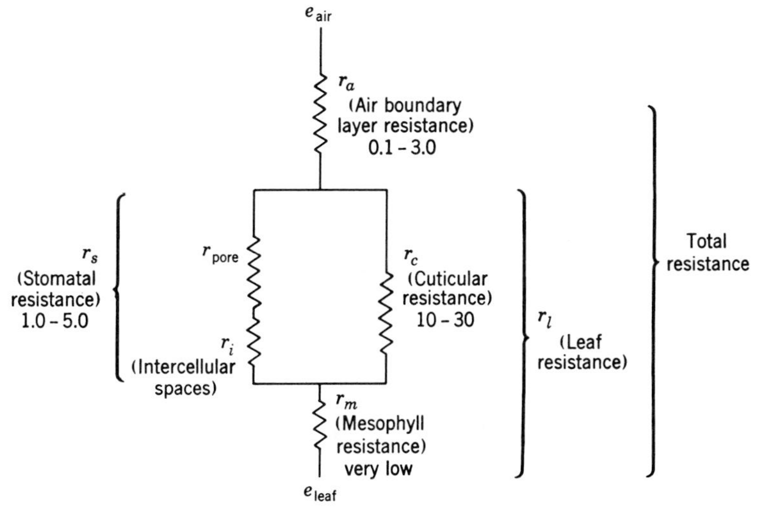

FIG. 9.1 Diagram of resistances to diffusion of water vapor from a leaf. Transpiration is proportional to the steepness of the gradient in water vapor pressure, e_{leaf} to e_{air}, and inversely proportional to the size of the resistances expressed in seconds per centimeter. The numbers are representative of resistances found in mesophytic species, expressed in seconds per centimeter.

more detail in the following sections. First, however, the energy balance of leaves will be discussed.

Energy utilization in transpiration

The energy input to a particular plant or leaf comes from direct solar radiation, radiation reflected and reradiated from surrounding soil and vegetation surfaces, and advective flow of sensible heat from its surroundings. Energy loss is by reradiation and convective flow of sensible heat and by loss of latent heat as water vapor. Several detailed studies of the energy exchange of various types of leaves have been conducted (Raschke, 1956, 1958; Tibbals et al., 1964; Gates et al., 1965; Knoerr and Gay, 1965). General accounts of the energy balance of leaves are given by Gates (1962, 1965) and by Slatyer (1967). Readers can refer to these sources for further details.

It may be helpful to define terms such as convection, advection, latent heat, and sensible heat which are much used in discussions of the energy balance. Convection refers to transfer of energy or heat by movement of the air mass, usually in the vertical direction; advection refers to horizontal transfer. A notable example is the oasis effect, where advective energy transfer from the surroundings to a small, isolated mass of vegetation results in a rate of transpiration exceeding that accounted for by the incident radiation (see Fig. 9.18). Latent heat exchange refers to gain or loss of heat caused by change in state, such as the loss when water evaporates. Sensible heat transfer refers to changes in the heat content of the air.

The energy load of leaves is dissipated by three mechanisms: reradiation, convection of sensible heat, and latent heat transfer by transpiration. According to Idso and Baker (1967), much more heat is dissipated by reradiation than by convection or transpiration. Often more heat is dissipated from individual leaves by convection than by transpiration, but transpiration is the more important mechanism for crops. Convective transfer of sensible heat and water vapor occurs from leaf to air whenever there are gradients of temperature and vapor pressure to provide the driving forces. These gradients usually exist during the day. Occasionally, advective energy increases the supply of heat significantly above that supplied by incident radiation (see Fig. 9.18). Since this results in net transfer of sensible heat toward the leaf, reradiation and transpiration are then the only mechanisms available for energy dissipation. At night gradients of sensible and possibly of latent heat exist toward leaves cooled by radiation to the sky. This often results in deposition of dew.

A small amount of energy is used in photosynthesis, and a small quantity of heat can be stored in the leaf, but these quantities are important only for very precise measurements over short periods of time, such as those by Slatyer and Bierhuizen (1964*a*). Thus, in addition to reradiation, convective transfer of heat (H) and water vapor (lE) are major processes in energy dissipation. Both quantities are expressed in units such as calories per square centimeter per minute; l is the latent heat coefficient for water in calories per gram.

Figure 9.2 shows typical relative quantities of energy dissipated by reradiation, transpiration, and sensible heat transfer for a leaf under given

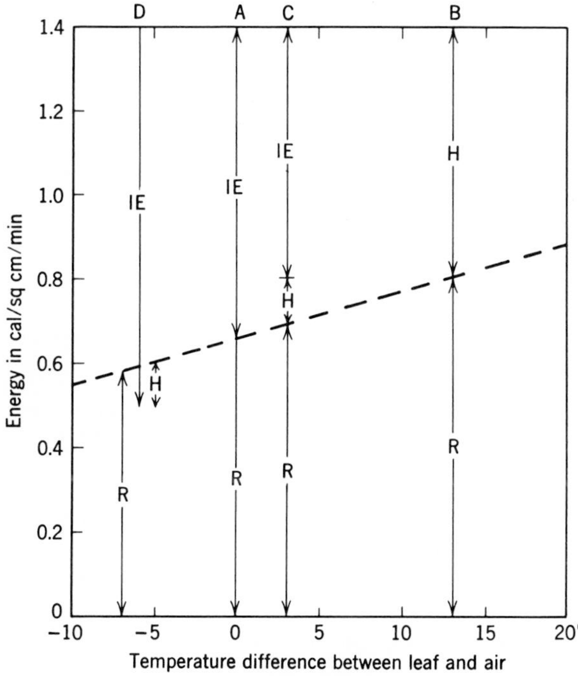

FIG. 9.2 **Diagram showing estimated energy exchange by transpiration (lE) radiation (R), and by sensible heat transfer (H) for a leaf 10 cm wide at 25°C, with wind velocity of 200 cm/sec and energy absorption of 1.4 calories per square centimeter per minute. The small amount of heat used in photosynthesis and released by respiration is disregarded. Data are for one surface only of a leaf exposed to intense radiation. (*After Slatyer,* 1967.)**

conditions. The small fraction of energy used in photosynthesis and stored in the leaf, usually no more than 2 or 3 percent, is not included in the calculations. When air and leaf temperatures are equal, reradiation and transpiration dissipate the entire radiation load (case A). When transpiration becomes virtually zero because of stomatal closure, the load must be dissipated by reradiation and sensible heat transfer (case B). Usually a dynamic equilibrium exists in which all three mechanisms operate and the leaf temperature reaches an intermediate value (case C). In case D there is a flow of heat to a cooler leaf which causes transpiration to increase and dissipate the load. This situation exists in the interior of trees and other large plants where leaves are shaded, but transpiring, and tend to be cooler than the air.

The ratio of the amount of heat transferred as sensible heat to that transferred as latent heat by transpiration is called the Bowen ratio (H/lE). In Fig. 9.2 the Bowen ratio ranges from zero in case A to a value approaching infinity in case B, where transpiration is negligible. It is less than 1 in case C and negative in case D. It averages about 0.1 for crops but has been estimated to be as high as 6.0 for exposed leaves (Idso and Baker, 1967). Thus, the sign and magnitude are determined to a considerable extent by leaf temperature. In fact, leaf temperature affects all three energy dissipation mechanisms. Its absolute value directly determines the outgoing reradiation flux and indirectly, through its effect on saturation vapor pressure, affects the latent heat flux by transpiration; its level relative to the surrounding air directly controls the sensible heat flux by convection. Readers should be reminded that the leaf temperature is not fixed, but rises and falls as a result of continuous interaction involving all of the fluxes.

Water vapor transport

The water vapor pathway

There are two principal sites of evaporation from a leaf, one located in the walls of the mesophyll cells bordering on intercellular spaces, the other in the outer surfaces of the epidermal cells. The outer surfaces of epidermal cells of most plant species are covered with a cuticle which offers considerable resistance to the escape of water vapor. Epidermal cells are supplied with water by diffusion from the leaf veins, apparently directly through the vein extensions in some species (Philpott, 1953; Wylie, 1943). Probably much of this water movement occurs in the cell walls, as is pointed out in Chap. 8. Water which evaporates from the mesophyll cells diffuses as vapor through the intercellular spaces and out through the stomatal pores. Most of the water vapor escapes through the stomata when they are open because the resistance is relatively low in this pathway. When they are closed, the only pathway available is through the epidermal cells and cuticle. Leaf structure (shown in Fig. 9.3) is discussed in more detail later in this chapter.

Since the stomatal and cuticular pathways are in parallel, their resistances can be expressed as follows:

$$\frac{1}{r_l} = \frac{1}{r_c} + \frac{1}{r_s} \tag{9.3}$$

where r_l is the total leaf resistance, r_c the cuticular resistance, and r_s the resistance in the stomatal pathway expressed in seconds per centimeter. The resistance in the stomatal pathway is composed of a number of individual resistances in series. These are in the mesophyll cell wall (r_m), in the

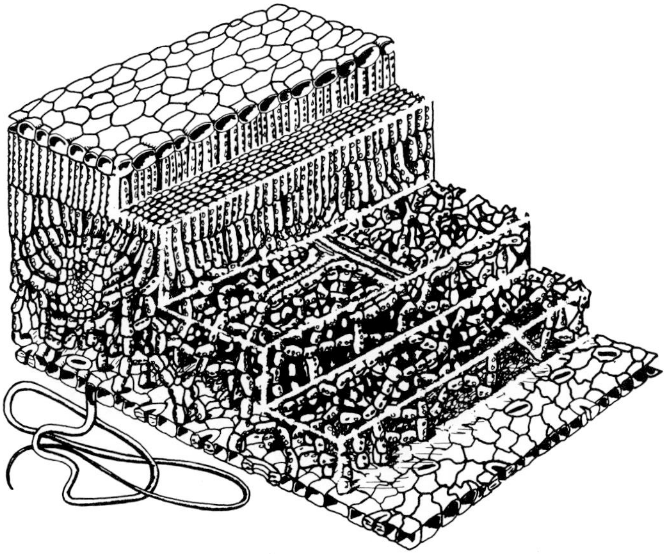

FIG. 9.3 Diagram of internal structure of an apple leaf, showing the large amount of exposed cell surface from which water evaporates and the large volume of internal air space through which water vapor can diffuse to the stomata. Also note the cross section through a large vein and the small vein ending in the spongy mesophyll. (*From Eames and MacDaniels, 2d ed.,* 1947.)

intercellular spaces (r_i), and in the stomatal pores themselves (r_p). Collectively the total stomatal resistance (r_s) can be described by the following equation:

$$r_s = r_m + r_i + r_p \tag{9.4}$$

These specific resistances can also be described by more complicated expressions, such as those used by Milthorpe and Spencer (1957) and Milthorpe (1959), but the simple equations presented are adequate for present purposes.

Relative magnitude of the various diffusion resistances

It can be appreciated from Eqs. (9.2) to (9.4) that the relative magnitudes of the various resistances in the vapor pathway determine the extent to which each can affect transpiration.

EXTERNAL RESISTANCE. Figure 9.4 presents data showing the effect of wind speed and downwind leaf width on leaf boundary layer resistance r_a. Typical values of r_a for a leaf 10 cm wide, such as cotton, range from 3.0 sec/cm at a wind speed of 10 cm/sec to 0.30 sec/cm at 10 m/sec. Comparable values for a grass leaf 1 cm wide would be 1.0 and 0.10 sec/cm. These values agree with those obtained for leaves of a number of plant

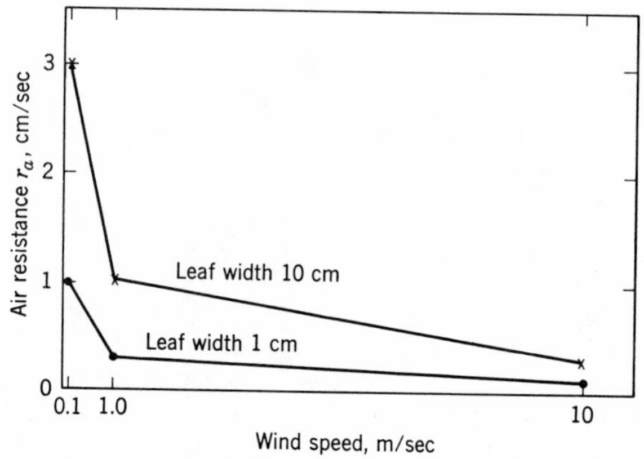

FIG. 9.4 Approximate air resistances at three wind speeds for a cotton leaf 10 cm wide and a grass leaf 1 cm wide. (*From data of Slatyer,* 1967.)

species and for blotting-paper strips by Kuiper (1963), Martin (1943), and Raschke (1956); and, like those given later, they are based on leaf area expressed as the area circumscribed by the leaf outline.

LEAF RESISTANCES. Internal leaf resistances can be evaluated by placing leaves in a wind tunnel or a photosynthesis chamber where air movement and other environmental factors can be measured and controlled. Estimates of cuticular resistance r_c range from 10 sec/cm for shade plants to 100 sec/cm or more for xerophytes; they apparently lie between 10 and 40 sec/cm for many crop plants. Some values for cuticular resistances are given in Tables 9.2 and 9.5. Kuiper (1961) reported values of 9.8 sec/cm for tomato and 9.6 sec/cm for bean; both were measured in darkness, where presumably all transpiration was cuticular. Slatyer and Bierhuizen (1964*a*) obtained a value of 32.3 sec/cm for cotton, while Boyer and Knipling (1965) reported 19.0 sec/cm for cotton and privet in a psychrometer vessel. They also obtained values of 22 sec/cm for tomato, but only 6 sec/cm for geranium. The stomata were assumed to be closed in the dark vessel.

When the stomata are open, stomatal pathway resistance r_s is so much less than resistance to movement through the cuticle that most of the water vapor is lost by the stomatal pathway. Representative values for r_s with open stomata are 0.6 to 2.4 sec/cm for wheat (Penman and Schofield, 1951; Penman and Long, 1960), 1.5 sec/cm for *Zebrina* (Bange, 1953), 1.6 to 1.8 sec/cm for turnip and sugar beet (Gaastra, 1958), and 1.0 sec/cm for cotton (Slatyer and Bierhuizen, 1964*a*; see also Table 9.2). Gates (1966) reported resistances with open stomata ranging from 5 to 10 sec/cm for poplar and birch to 40 or 50 sec/cm for white pine and 200 to 400 sec/cm for *Pteridium aquilinum.* The latter seems excessively high. The most important factor affecting stomatal resistance is stomatal aperture, which, under natural conditions, is

Table 9.2 Resistances to Transfer of Water Vapor and Carbon Dioxide through the Boundary Layer, r_a, the Cuticle, r_c, the Stomata, r_s, and the Mesophyll, r_m, in Leaves of Several Species *

Species	Transfer resistances, sec/cm				
	Water vapor			CO_2	
	r_a	r_s	r_c	r_s	r_m
Betula verrucosa	0.80	0.92	83	1.56	5.8
Quercus robur	0.69	6.7	380	11.3	9.6
Acer platanoides	0.69	4.7	85	8.0	7.3
Circaea lutetiana	0.61	16.1	90	27.4	—
Lamium galeobdolon	0.73	10.6	37	18.0	—
Helianthus annuus	0.55	0.38	—	0.65	2.4

** From Holmgren, Jarvis, and Jarvis (1965)*

influenced primarily by light intensity and leaf water potential—although the light intensity effect and, to a lesser extent, the water potential effect are modified by internal carbon dioxide concentration. An example of the effect of light intensity on stomatal aperture and consequently on stomatal resistance r_s is given in Fig. 9.5.

In addition to the stomatal resistance, it is possible to distinguish resistances in the intercellular spaces and in the mesophyll cells. The resistance in the intercellular spaces measured with an air-flow porometer appears to

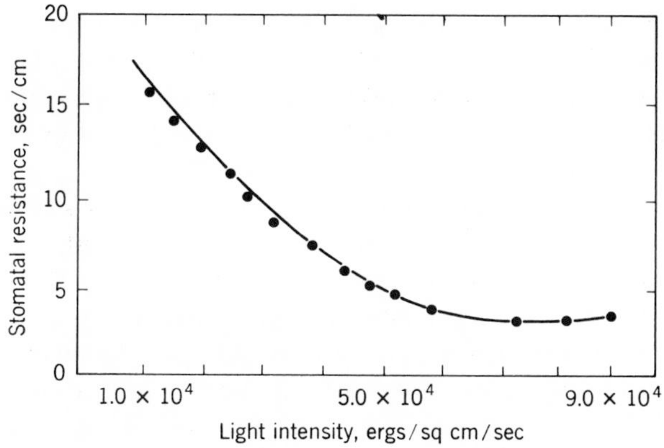

FIG. 9.5 Effect of light intensity in the visible part of the spectrum on stomatal opening of bean, as measured by change in diffusive resistance. (*After Rijtema*, 1965.)

be insignificant in wheat (Milthorpe and Spencer, 1957), but it is measurable in some other species (Bange, 1953; Heath, 1941) and may increase as leaves are dehydrated, possibly because of changes in the internal geometry. Measurements with diffusion porometers (Jarvis and Slatyer, 1967) suggest that resistances in the intercellular spaces may be a significant component of r_s, at least when the stomata are open. It can be expected to be higher in thick leaves and leaves with small intercellular spaces than in thin leaves or those with large intercellular spaces.

The importance of the mesophyll resistance r_m has been the subject of discussion for several decades. It was reported early in this century that the rate of transpiration often decreased with no apparent change in stomatal aperture (Livingston, 1906; Lloyd, 1908; Knight, 1917). This phenomenon was termed "incipient drying" by Livingston and Brown (1912). It was postulated that during periods of rapid transpiration or inadequate water supply the evaporating surfaces retreat into the cell walls, the diffusion pathway is lengthened, the vapor pressure is lowered, and the rate of evaporation decreases. English workers such as Gregory et al. (1950), Milthorpe and his coworkers (1957, 1959, 1960), and Williams and Amer (1957), using air-flow porometers, found no effect of leaf water content on transpiration over a wide range of water contents from fully turgid to wilted, except through stomatal closure. Milthorpe and Spencer (1957) reported that only 2 percent of r_s could be attributed to nonstomatal resistances. However, they found that transpiration after wilting appeared to be as much as 20 percent less than the original rate at the same stomatal aperture, which indicated some kind of internal change.

There is evidence that the exposed walls of the leaf mesophyll cells are cutinized or at least covered with a nonwettable, hydrophobic lipid layer (Frey-Wyssling and Häusermann, 1941; Lewis, 1945, 1948; Scott, 1950, 1959, 1964). This implies that evaporation occurs within pores in the walls rather than from the surfaces, and the result is a lengthened vapor pathway. However, Gaff et al. (1964) introduced gold sol into the transpiration stream and found that it accumulates at the outer surfaces of the mesophyll cells. This suggests that water flows to the cell surfaces in spite of their hydrophobic nature.

Assuming that the largest pores are no more than 10 μ in radius, Slatyer (1966) doubts that any significant retreat of water into the outer mesophyll cell wall surfaces would occur unless Ψ_w is lower than -150 bars. The high permeability of the cell wall water pathway (Russell and Woolley, 1961; Weatherley, 1963) also suggests that water supply should not be limiting. However, if the water conductivity of the cutinized part of the wall is much lower than that of the remainder of the wall, the effective water vapor concentration at the outer surfaces might be significantly lower than the value equivalent to the general level of the leaf cell water potential. If so, transpiration could be expected to vary at constant stomatal aperture with change in e_{leaf} rather than with change in r_{leaf} [see Eq. (9.2)]. Evidence of the occurrence of this phenomenon has been obtained by Shimshi (1963) and by Jarvis and Slatyer (private communication). Weatherspoon (1968) found a negligible amount of mesophyll resistance in mesophytic types of leaves.

Vapor pressure gradient from leaf to air

Having discussed the magnitude of the resistances to movement of water vapor out of leaves, we now turn to the driving force, $c_{\text{leaf}} - c_{\text{air}}$ or $e_{\text{leaf}} - e_{\text{air}}$ of Eq. (9.2). The difference in water vapor pressure or vapor concentration between leaf and air controls the movement of water vapor when the resistances are constant. This difference depends on two variables, the vapor pressure or vapor concentration of the bulk air surrounding leaves and the vapor pressure or concentration at the evaporating surfaces in the leaves.

The vapor pressure of the evaporating surfaces is influenced chiefly by leaf temperature and the water potential at the evaporating surface. If the water potential at the evaporating surface (Ψ_{surface}) is treated as zero, the vapor pressure at the surface is the saturation vapor pressure of water at the existing leaf temperature. The effect of temperature on the saturation vapor pressure of water is shown in Table 9.3. It can be seen that an increase from

Table 9.3 ***Effects of Temperature on Vapor Pressure of Water and on the Vapor Pressure Gradient from Leaf to Air, Δe, Assuming the Leaf Water Potential Is Zero and There Is No Change in Relative Humidity of the Air***

Temperature, °C	Vapor pressure at saturation	Vapor pressure of air at 60% rel. humidity	Vapor pressure gradient, e
0	4.6	2.7	1.9
10	9.2	5.5	3.7
20	17.5	10.5	7.0
30	31.8	19.0	14.8
40	55.3	33.2	22.1

10°C to 30°C more than triples the vapor pressure from 9.2 to 31.8 mm Hg; a further increase to 40° causes another increase to 55.3 mm Hg. From these data, it can be seen that even small changes in leaf temperature can greatly affect the rate of water loss as well as heat loss. In fact, changes in leaf temperature are probably responsible for most short-term changes in transpiration when r_a and r_l are approximately constant.

The assumption that the water potential at the surface of the mesophyll cells is zero is not always true, because it is decreased by the development of water deficits in the cells of transpiring leaves. Furthermore, it is possible that the water potential at the cell surfaces is lower than the water potential of the leaf as a whole. These possibilities will be examined in more detail.

The effect of reduction in Ψ_{leaf} on vapor pressure can be predicted from Eq. (1.7), which relates Ψ to vapor pressure. From this equation, it can be calculated that at values of Ψ_w of −15, −30, and −60 bars, the relative vapor pressures ($e/e°$) at 30°C at the cell surfaces will be approximately 0.99, 0.98, and 0.96. Table 9.4 shows the vapor pressures at the cell surfaces at these

Table 9.4 ***Vapor Pressures at Leaf Cell Surfaces at Three Leaf Water Potentials (Ψ_w) and the Difference in Vapor Pressure between Leaf and Air (Δ_e) at Two Relative Humidities***

(Decrease in leaf Ψ_w to −60 bars reduces Δ_e about 22 percent at 80 percent relative humidity, but the reduction is only about 8 percent at 80 percent relative humidity.)

Ψ_w of mesophyll cells in bars	Δ_e if rel. humidity of air is 80%	Δ_e if rel. humidity of air is 50%
−15	6.37	15.59
−30	5.73	15.27
−60	5.09	14.63

values of Ψ_w and the differences in vapor pressure between leaf and air (Δ_e) at two relative humidities. From these data it can be seen that the reduction in the vapor pressure gradient caused by decrease in Ψ_{leaf} is very small compared with the effects of other factors on Δ_e. Therefore, moderate changes in leaf water potential have little effect on transpiration.

Although it is generally assumed that the water potential at the surfaces of the mesophyll cells is the same as that of the leaf tissue as a whole, this is not necessarily true. It might sometimes be lower. However, because of the difficulty of measuring $\Psi_{surface}$ directly, there has been considerable uncertainty about the actual situation and some question whether the relative humidity of the intercellular spaces really falls much below saturation. Shimshi (1936*b*) reported Ψ_w as low as −90 bars in maize, while Whiteman and Koller (1964) reported potentials of −180 to −320 bars for the desert shrub *Reamuria.* Jarvis and Slatyer (private communication) estimated that the $\Psi_{surface}$ in leaves of well-watered, rapidly transpiring cotton plants approached −100 bars. This suggests that the vapor pressure at the evaporating surfaces might be considerably below saturation.

There seem to be two possible explanations for those low values of $\Psi_{surface}$: (1) limited water supply to the evaporating surface or (2) accumulation of solutes at the surface. The possibility of a limited water supply at the evaporating surface was discussed earlier in the section on leaf resistances. Because of the high permeability of the cell walls to water and the small size of pores in them, Slatyer (1966) doubted if any significant retreat of menisci occurs. However, the mesophyll cell walls appear to be covered with cuticle, and Jarvis and Slatyer (1966*a*) attribute the lowered vapor pressure in rapidly transpiring plants to the low water conductivity of this cuticle layer.

An accumulation of solutes in the unstirred layer at the evaporating surfaces of the leaf cells would also reduce Ψ_{surface} and e_{surface}. Boon-Long (1941) demonstrated that evaporation from a membrane enclosing an unstirred 1.0 *M* sucrose solution was much lower than evaporation from enclosed water alone, presumably because solute accumulation at the evaporative surface reduced Ψ_{surface} and Δ_e. This explanation is supported by unpublished experiments of the author in which Boon-Long's study was repeated with stirring. This enhanced the rate of back diffusion and largely eliminated the reduction in evaporation from the solution. However, it seems doubtful that accumulation of solutes in the unstirred layer at the evaporating surface is of importance in leaves. Jarvis and Slatyer (1967) found no significant increase in estimated Ψ_{surface} after several hours of rapid transpiration, although this should have favored progressive accumulation of solutes at the evaporating surfaces. Also, Slatyer (1966) concluded that the back diffusion of solutes would prevent significant reduction in Ψ_{surface}. It appears that more research is needed on the causes and importance of the reduction in water potential at the evaporating surfaces.

Leaves in relation to transpiration

The size, shape, and surface characteristics of leaves can materially affect their temperature, hence e_{leaf}, and the external resistance r_a, while the internal structure affects the resistance to water vapor movement and hence, r_l.

Leaf area and orientation

The total leaf area and the arrangement or orientation of leaves on plants have significant effects on the water loss from individual leaves and plants. Reduction of leaf surface by pruning usually results in an increase in rate of water loss per unit of surface by the remaining leaves. Kelley (1932) reported that removing half of the leaves from trees increased transpiration of the remaining leaves by 20 to 90 percent. The increase results from increased exposure of the remaining leaves to solar radiation and increased air movement around them. Another reason for transpiration increase is that the increased ratio of roots/shoots may result in a better water supply to individual leaves. Both Bialoglowski (1936) and Parker (1949) found that transpiration was reduced when the ratio of root/leaf surface was decreased. In some species such as *Larrea tridentata,* development of a severe water deficit results in shedding of leaves and reduction in water loss. Curling or rolling of wilting leaves reduces the exposed surface and increases the boundary layer resistance r_a. It is particularly effective for plants which have all or most of the stomata on the lower surface. Stålfelt (1956*a*, p. 336) quotes work by Lemée showing that rolling of leaves reduced transpiration about 35 percent for plants of moist habitats, 55 percent for those of Mediterranean species, and 75 percent in some desert xerophytes.

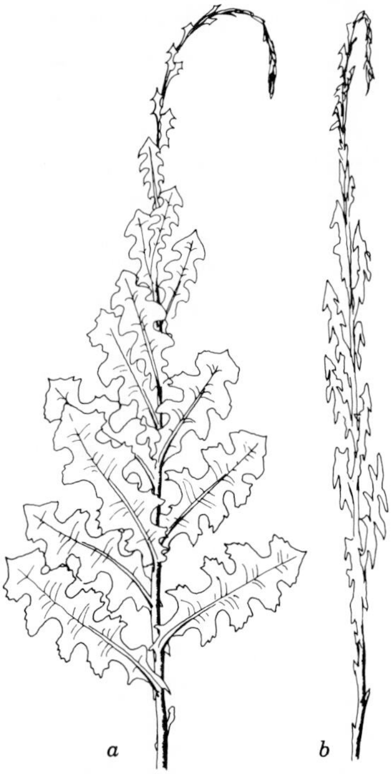

FIG. 9.6 **Leaf orientation of prickly lettuce (*Lactuca scariola*) grown in full sun. *a*, viewed from west; *b*, viewed from south. This leaf orientation considerably reduces the amount of energy absorbed at midday. (*From Transeau*, 1925.)**

Leaf orientation can affect the rate of transpiration because leaves which are at right angles to the sun's rays are warmer than those which are parallel to the incident radiation. The leaves of most plants tend to be oriented more or less perpendicular to the average incident radiation, but the leaves of a few plants, including certain species of *Lactuca* and *Silphium* and seedlings of turkey oak *(Quercus laevis, Walt.),* are oriented approximately parallel to the average incident radiation. It has been claimed that such an orientation aids seedlings to survive in hot dry habitats, but this has never been demonstrated. Pine needles which occur in bundles shade one another. This probably reduces transpiration as well as photosynthesis, but data are available only for the latter (Kramer and Clark, 1947). Wilting and rolling sometimes reduces the amount of radiation received by leaves.

Leaf size and shape also affect the rate of transpiration per unit of leaf surface. Slatyer (1967, p. 261) states that the air resistance r_a is about three times greater for a leaf 10 cm wide than for one only 1 cm wide. This is shown in Fig. 9.4. Small leaves, deeply dissected leaves, and compound leaves with small leaflets all tend to be cooler than large entire leaves because they have thinner boundary layers which permit more rapid sensible heat transfer. The thin boundary layer (lower r_{air}) of small leaves is also more favorable for water vapor loss, so the cooling and the lower r_{air} tend to oppose one another in latent heat transfer (Raschke, 1960). Tibbals et al. (1964) point out that in quiet air broad leaves are considerably warmer than conifer needles. The energy relations of leaves are discussed by Gates (1966), Knoerr (1966), Knoerr and Gay (1965), and Slatyer (1967, pp. 237–247), who cite other work on this interesting topic.

Leaf surfaces

The cuticular resistance varies with the thickness and nature of the cuticular layer, which is a relatively complex structure. It is apparently formed by oxidation of lipid materials on cell wall surfaces exposed to the air and is often covered by deposits of wax. There has been some uncertainty concerning the origin of the wax, but it is usually supposed to be extruded through the wall and cuticle. Hall (1967) and Hall and Donaldson (1962) claim to have found channels through which extrusion might occur. Subcuticular deposits of wax are also said to occur (Schieferstein and Loomis, 1959). The cuticle is fairly permeable to substances applied to leaf surfaces, as is shown by penetration of a wide range of materials applied to leaves (see Chap. 7). Holmgren et al. (1965) reported that the permeability increased as the temperature was increased from 17°C to 22°C. Martin (1943) reported a substantial increase in the cuticular transpiration of sunflowers in darkness at 49°C, suggesting increased permeability at very high temperatures.

The effect of epidermal hairs on transpiration is uncertain. If living, they might increase the evaporating surface, but if dead they should increase the thickness of the boundary layer on the leaf surface and decrease transpiration in moving air. Experiments by Sayre (1920) gave very small increases in transpiration when the hairs were carefully shaved from mullein leaves. It appears that hairs usually have little effect on the energy and water relations of leaves.

The reflective characteristics of the leaf surface may also be important in connection with the proportion of total incident radiation absorbed. Thin shiny leaves reflect a larger proportion of incident radiation and have a higher albedo than dull thick leaves, which absorb more radiation and become warmer. According to Wong and Blevin (1967), the infrared reflectance of most kinds of leaves is very low. Billings and Morris (1951) reported that leaves of desert plants show higher reflectivity than leaves of plants from less exposed situations. The reduction in transpiration sometimes reported to occur after the application of Bordeaux mixture (Miller, 1938) probably results from the lower leaf temperature caused by the light-colored covering it produces on leaves.

Leaf anatomy

The thickness of the cuticle is generally closely related to the cuticular resistance, but Stålfelt (1956*b*) pointed out that there are differences in permeability of the cutin resulting from age and hydration in addition to those resulting from differences in thickness. He states that, in general, cuticular transpiration of shade leaves is higher than that of sun leaves, also that the permeability of the cuticle is higher at night than during the day because it is more hydrated at night. The magnitude of differences among species is illustrated by the data in Tables 9.2 and 9.5.

Table 9.5 ***Cuticular Transpiration of Plants of Several Species under Standard Evaporating Conditions Expressed in Milligrams per Hour per Gram of Fresh Weight*** *

Species	Transpiration
Impatiens noli-tangere	130.0
Caltha palustris	47.0
Fagus sylvatica	25.0
Quercus robur	24.0
Sedum maximum	5.0
Pinus sylvestris	1.53
Opuntia camanchica	0.12

* ***From Pisek and Burger (1938)***

In general, the thicker the cutin layer the lower the rate of cuticular transpiration. However, as cutin grows older, cracks develop and cuticular transpiration from older leaves is said to be more rapid than from younger leaves.

Cutin varies in hydration, and older cutin layers are said to be less hydrophobic than newly formed cutin. Apparently the wax which is extruded through and accumulates on the cuticle has little effect on water loss. The complex nature of cutin layers is brought out by studies such as those of Crafts and Foy (1962), Frey-Wyssling and Mühlethaler (1959), and Schieferstein and Loomis (1959). Permeability varies with water content and is apparently affected by chemicals such as those found in spray materials (van Overbeek, 1956; Hull, 1964).

INTERNAL SURFACE. The area of internal mesophyll surface exposed to intercellular air greatly exceeds the external exposed surface of leaves. Turrell (1936, 1944) found the ratio in several species to range from 6.8 to 31.3. Xeromorphic leaves and sun leaves usually have significantly higher ratios than mesomorphic or shade leaves because of the larger amounts of palisade tissue in them. Swanson (1943) found ratios of 12.9, 7.1, and 4.6 in *Ilex,* tobacco, and *Coleus.* Stålfelt (1956*a*) reported wide variability in the internal volume of air space in leaves compared with their total volume. Volumes ranged from 40 to 70 percent in various summer herbs and 20 to 40 percent in *Tilia* and *Fagus.* Sun leaves had low internal volumes and shade leaves high internal volumes.

STRUCTURE AND ENVIRONMENT. Leaf structure is modified significantly by the environment in which the leaves develop (Stålfelt, 1956*a*). In agricultural practice this is exemplified by growing tobacco under shade to produce the large thin leaves desired for cigar wrappers. Sun leaves usually differ from shade leaves of the same species by having smaller cells and smaller

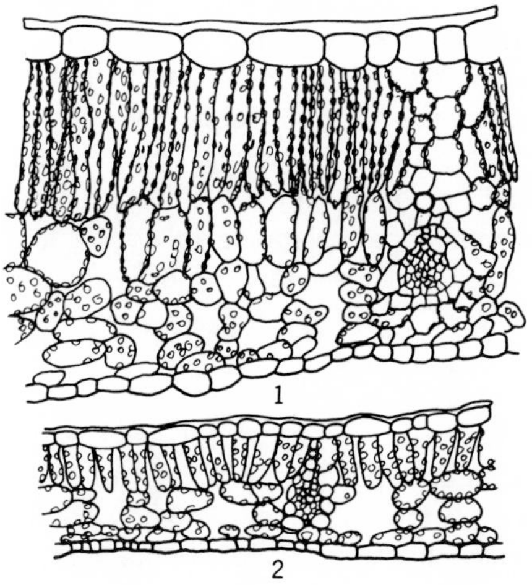

FIG. 9.7 Leaves from the southern edge (1) and the center of the crown (2) of an isolated sugar maple (*Acer saccharum*). (*From Weaver and Clements,* 1938.)

surface area, being thicker and more heavily cutinized, having smaller interveinal areas, and often possessing extra layers of palisade cells. The upper leaves of trees are usually said to be more xeromorphic than the lower ones, which are less exposed to sun and wind. These differences are generally ascribed to the effects of more severe water deficits and greater loss of turgor in exposed leaves, although other less obvious factors such as temperature may be involved (see Stocker, 1960).

Philpott (1956) reported that the leaves of shrubs of the bogs of eastern North Carolina are thicker, contain more palisade tissue and more small veins, and are more heavily cutinized than leaves of related shrubs growing in the mesic forests of the mountains. The presence of more small veins in the leaves of the bog shrubs is probably related to the fact that lateral transfer of water and solutes is more difficult in palisade tissue than in spongy mesophyll (Wylie, 1939).

The xeromorphic characteristics of leaves of bog shrubs are often attributed to the physiological dryness of the poorly aerated bogs, but this is questionable. Mothes (1932) claimed that the xeromorphic structure of leaves of bog plants is caused by deficiency of nitrogen, and Müller-Stoll (1947) held a similar view. Albrecht (1940) claimed that deficiency of calcium was responsible. Neither of these views has been tested experimentally in North American bogs. There is some evidence that leaf succulence is affected differently by various ions—for example, it is increased by chloride and decreased by sulfate ions. It is claimed by van Eijk (1939) that the succulence characteristic of halophytes is caused by an excess of chloride ions. Boyce (1954) also reported that chloride ions increase succulence of leaves. In general, plants grown in a high concentration of salt tend to be more succulent than those grown in solutions of low salt content. Strogonov (1964) discussed the effects of excess salts and specific ions in some detail.

The degree to which these anatomical and morphological features affect intercellular space resistance and consequently r_{leaf} has not been quantitatively assessed, but calculations by Jarvis, Rose, and Begg (1967) suggest that when the stomata are open, intercellular space resistance is approximately equal to stomatal pore resistance in cotton leaves. However, when stomatal closure begins, pore resistance increases rapidly and dominates r_{leaf}.

Stomata

Stomata deserve special attention in any discussion of transpiration, because most of the water escapes through them. We shall briefly discuss their origin, distribution, operating mechanism, and role in the control of leaf transpiration. Some of these topics are discussed in more detail by Heath (1959), Ketellapper (1963), Miller (1938, pp. 417–447), and Stålfelt (1956*c*).

ORIGIN. The development of stomata, or, more accurately, of the two guard cells which surround a stomatal pore, is described by Esau (1965, pp. 163–166). The guard cells originate by transverse division of stomatal mother cells, which differ morphologically and biochemically from their neighbors and apparently exert considerable influence on them (Stebbins and Shah, 1960). The middle lamella disintegrates and the two walls separate, leaving a passageway of variable dimensions between them. Usually the walls of the guard cells adjacent to the pores become considerably thickened, often with odd-shaped projections extending into the pore. The guard cells of many plants are typically bluntly semicircular when closed and crescent- or kidney-shaped when open, but those of grasses are more elongated, and many modifications in

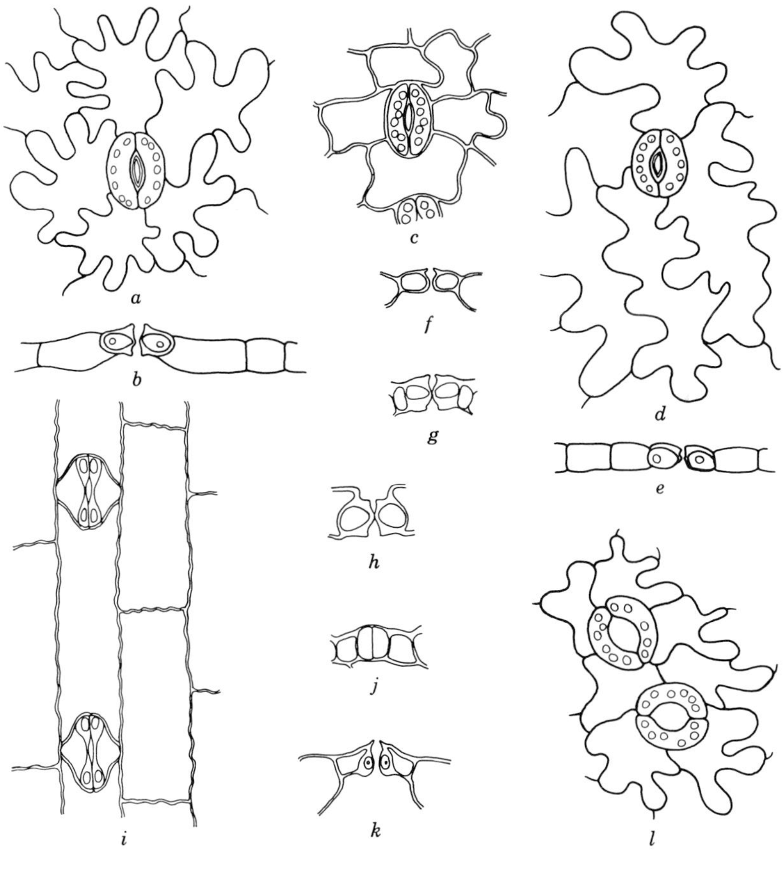

FIG. 9.8 Various types of stomata. *a*, *b*, from *Solanum tuberosum*, in face view and in cross section; *c*, from apple; *d*, *e*, from *Lactuca sativa*; *f*, from *Medeola virginica*; *g*, from *Aplectrum hyemale*; *h*, from *Polygonatum biflorum*; *i*, *j*, *k*, from *Zea mays*. *i* is a face view; *j*, a cross section through ends of guard cells; *k*, a cross section through the center of a stoma. *l* is a face view from *Cucumis sativis*. (*From Eames and MacDaniels, 2d ed.,* 1947.)

shape are found (see Esau, 1965). Several types of stomata are shown in Fig. 9.8.

LOCATION. Stomata occur in the epidermis of all organs of plants except the roots, but the number and manner of occurrence vary widely. In many plants the guard cells are flush with the outer surface of the epidermis or occasionally even elevated above it, but in some species they occur in pits or grooves. In species such as pineapple the grooves are so covered by

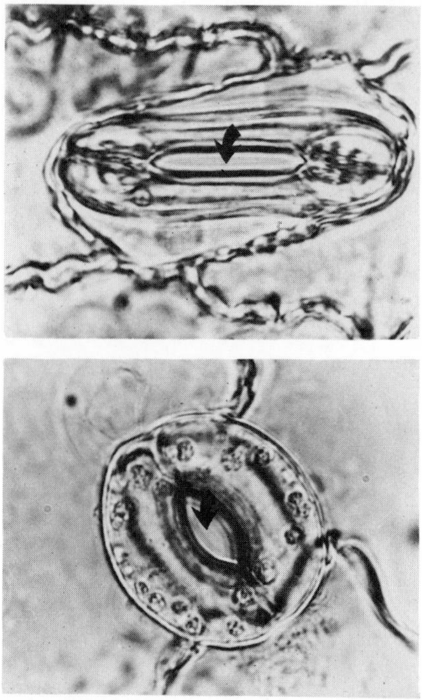

FIG. 9.8a **(Top) Photograph of an open corn stomate magnified about 4,000 times. This is typical of stomata of grasses. (Lower) An open stomate of bean, typical of stomata found on most other types of plants.** (*Courtesy of J. E. Pallas, U.S. Department of Agriculture.*)

scales or hairs that the stomata are practically invisible. Such arrangements lengthen the diffusion pathway, increase the thickness of the boundary air layer, and materially increase the diffusion resistance to loss of water vapor.

Most herbaceous plants have stomata on both the upper and lower leaf surfaces, but many woody species have them only on the lower surface.

NUMBER AND SIZE. The number of stomata per unit of leaf area varies widely, ranging from about 2,000 per square centimeter in oats to 50,000 in black oak, black walnut, and mulberry, and even 100,000 in scarlet oak (Meyer and Anderson, 1952, p. 143). The number varies considerably with environmental conditions. Stomata tend to be smaller and more numerous on leaves developed in full sun and dry habitats than on shade leaves and those in moist habitats. Even when wide open, stomatal pores are very small, ranging from 3 to 12 μ in width and 10 to 30 or occasionally 40 μ in length. The area of the pores when fully open seldom exceeds 3 percent of the total leaf surface.

STOMATAL MECHANISM. It is well known that stomatal closure results from water stress and low light intensity and that the rate of response is affected by the temperature. However, although scores of papers have been written on stomatal opening, there is as yet no fully satisfactory explanation of the biochemical processes involved. It is generally assumed that changes in the carbon dioxide concentration in the intercellular spaces and water stress are the chief factors causing changes in guard cell turgor (Heath, 1959; Ketellapper, 1963; Meidner and Mansfield, 1965; Zelitch, 1965). According to the commonly held hypothesis, a decrease in carbon dioxide concentration in the intercellular spaces results in a decrease in water potential of the guard cells relative to the adjacent cells, followed by water entry, an increase in guard cell turgor and volume, and opening of the stomata. Changes in leaf water content or leaf water potential also affect guard cell turgor directly, often causing midday closure. However, in the absence of water stress, carbon dioxide appears to exercise the primary control over stomatal opening. In fact, Mansfield (1965) reports that stomata are so sensitive to carbon dioxide that breathing on leaves may cause closure.

Since the walls bordering the pores are thicker than those on the opposite side, increase in turgor causes the guard cells to become more crescent-shaped, producing an opening between them. In darkness these steps are reversed; the guard cells lose their turgor, and the stomates are closed. Closure may be aided by pressure from the surrounding epidermal cells, and stomata on some kinds of plants appear transiently to open more widely after a slight loss of leaf turgor than in fully turgid leaves (Stålfelt, 1966).

It is not clear how much of the reaction to light is due directly to photosynthesis, but experimentally induced decreases in carbon dioxide concentration will cause stomatal opening even in darkness (Raschke, 1965; Meidner and Mansfield, 1965).

Zelitch (1963) regards opening and closing as a balance between two opposing processes, opening being accelerated by light. The two processes differ in several respects. He concluded that opening is usually slower than closing, requires oxygen, and is inhibited by low temperature and by substances such as phenylmercuric acetate, chlorogenic acid, and sodium fluoride which do not inhibit closing. Sodium azide inhibits both opening and closing. Walker and Zelitch (1963) suggested that glycolic acid may occupy an important place in the metabolic sequence leading to opening, but this is largely inference at present.

For further details on stomatal physiology the reader is referred to Heath (1959), Meidner and Mansfield (1965), Zelitch (1965), and Raschke (1965).

Stomata vary widely in their normal behavior, which ranges from staying open continuously to remaining closed continuously. Loftfield (1921) classified stomata into three types on the basis of the manner in which they open and close under conditions favorable for growth (see Miller, 1938, pp. 435–436). These are as follows:

(1) The alfalfa type, which normally open in the morning, remain open during the middle of the day, and close slowly during the afternoon. They close at midday if subjected to water stress, and may then open at night. Peas, beans, some cucurbits, apple, peach, pear, and many other thin-leaved mesophytes belong in this group.
(2) The potato type, which are open continuously except for about 3 hr after sunset. Midday closure occurs only when the plants are wilted. The stomata of cabbage, onion, plantain, and pumpkin are said to belong to this group. The stomata of *Equisetum* are said to remain open continuously, even when severely wilted.
(3) The barley type, which is typical of most cereals. The stomata are seldom open for more than an hour or two, and many remain closed during the day. According to Brown and Pratt (1965), the stomata of many grasses native to dry habitats never open perceptibly.

Tal (1966) recently reported that the stomata of certain tomato mutants remain open continuously, probably because of guard cell wall rigidity, resulting in excessive transpiration. Waggoner and Simmonds (1966) reported somewhat similar behavior in a potato mutant.

GASEOUS DIFFUSION THROUGH STOMATAL PORES. Much attention has been given to diffusion of gases through small pores in order to explain the fact that water loss through stomatal pores occupying only a few percent of the leaf area can exceed 50 percent of the rate of evaporation from a free water surface of the same area (Bange, 1953; Brown and Escombe, 1900; Ting and Loomis, 1963). The controversy aroused over this problem is unprofitable, because the importance of resistance in the stomatal pores to diffusion through the whole system depends on the relative magnitudes of the diffusion resistances in the pores and in the boundary layer above the leaves. If the resistance of the boundary layer is reduced to a low value, as by wind, resistance in the pores will control transport, but in still air resistance in the pores may be of little significance. Therefore, in still, humid air transpiration is little affected by partial closure of stomata; but in moving, dry air where

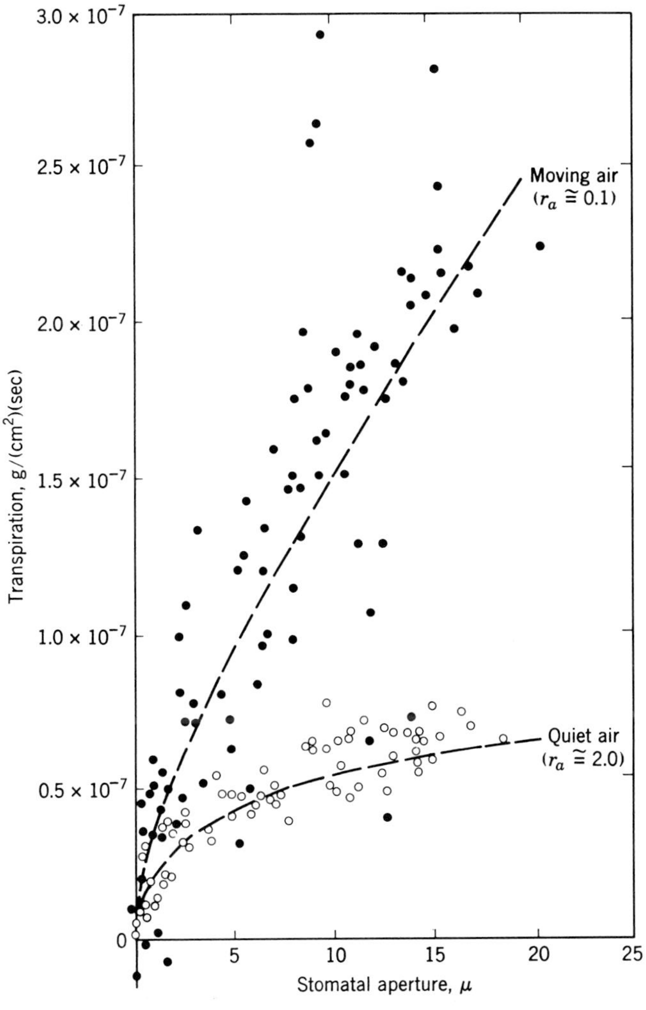

FIG. 9.9 **Effect of increasing stomatal aperture on transpiration rate of *Zebrina* leaves in quiet air ($r_a \cong 2.0$ sec/cm) and in moving air ($r_a \cong 0.1$ sec/cm). The effect of change in stomatal aperture is very large in moving air, but quite small in quiet air, where the air resistance is often as large as the stomatal resistance. (*From Slatyer*, 1967; *after Bange*, 1953.)**

r_a is low, transpiration is closely related to the stomatal aperture (see Fig. 9.9).

MEASUREMENT OF STOMATAL APERTURE. Because of their importance in control of water loss and the entrance of carbon dioxide, there has been much interest in measuring stomatal aperture. Direct observation under the microscope is difficult on attached leaves, and detachment and the handling and illumination necessary for observation results in changes in aperture.

Also, a very small sample is visible in any single field under the microscope. Lloyd (1908) stripped off bits of epidermis and immediately plunged them into absolute alcohol to fix the tissue before changes in aperture could occur. Loftfield (1921) used this method for his classical studies, but it is difficult to strip the epidermis from leaves of some species, and stripping sometimes causes changes in stomatal aperture. A less drastic method involves making impressions of the stomata on collodion film (Clements and Long, 1934) or on a silicone rubber compound (Zelitch, 1961). Impressions made with silicone rubber are being used extensively because they cause little or no injury to leaves. However, they are unsatisfactory on leaves with numerous hairs or where stomata are sunken deeply in pits, as in pine or pineapple, for example. Gloser (1967) discussed some of the difficulties inherent in this method but found it agreed well with measurements made with a gas-flow porometer on sugar beet.

Another method, apparently introduced by Molisch (1912), involves observation of the time required for fluids of various viscosities to infiltrate into leaves. For example, Alvim and Havis (1954) used mixtures of paraffin oil and *n*-dodecane, and workers in Israel used various mixtures of paraffin oil and turpentine or benzol or kerosene alone. The infiltration method involves sampling thousands of stomata, but it may kill the leaf tissue, and it does not work well on hairy leaves or conifer needles. Rutter and Sands (1958) used a saturated solution of crystal violet in ethyl alcohol on pine, and Oppenheimer and Engelberg (1965) made further improvements in this method. Fry and Walker (1967) described a pressure infiltration method which can be used to estimate stomatal apertures of conifers.

Darwin and Pertz (1911) introduced porometers to measure air flow through leaves, and many variations have been developed, including recording porometers (Gregory and Pearse, 1934; Wilson, 1947). Alvim (1965) developed a portable gas-flow porometer of which various modifications have been described (Bierhuizen et al., 1965, for example), some of them being suitable for use in the field. These gas-flow porometers have various disadvantages (Heath, 1959), including the fact that if the pressure is more than about 10 cm of water, stomatal aperture may be affected (Raschke, 1965). This led to the development of diffusion porometers, which measure the diffusion of gas through leaves, as described by Heath (1959). Slatyer and Jarvis (1966) described a diffusion porometer using nitrous oxide which permits continuous recording of stomatal resistance, and Jarvis et al. (1967) compared the behavior of the mass flow and diffusion porometers. Neither flow nor diffusion porometry is practical on leaves which have stomata on only one surface and veins which prevent lateral diffusion (heterobaric type; see Heath, 1959).

The stomatal aperture, or more properly the stomatal resistance, is sometimes estimated from the rate of transpiration by the use of instruments such

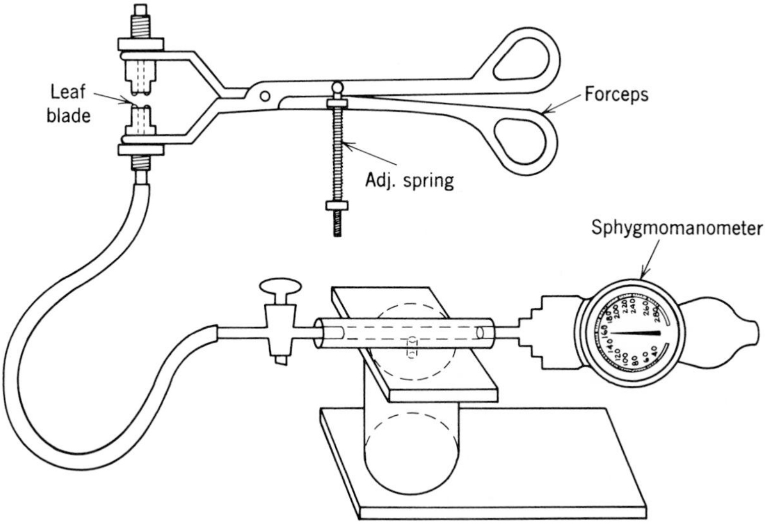

FIG. 9.10 A porometer for measuring stomatal opening by rate of change in air pressure as recorded by the dial of a sphygmomanometer. The leaf blade is inserted between the rubber gaskets and held firmly by a spring-loaded clamp. Pressure is applied by the rubber bulb of the sphygmomanometer, the stopcock is opened, but the time for a given pressure drop on the gauge is measured with a stopwatch. (*Modified from Alvim,* 1966.)

as that described by Wallihan (1964) and modified by van Bavel et al. (1965). Slatyer introduced a fan to stir the air, shortening the response time (see Fig. 9.11). Readings of change in humidity can be made within a minute, decreasing the dangers inherent in methods requiring enclosure of leaves for long periods of time, and the method can be used in the field.

Interaction of factors affecting transpiration

The important environmental factors affecting transpiration are light intensity, atmospheric vapor pressure, and temperature, wind, and water supply to the roots. Plant factors include the extent and efficiency of root systems in absorption, the leaf area, leaf arrangement and structure, and stomatal behavior. There is a relatively complex interaction between the various plant and environmental factors which can be summarized in terms of their effects on various components of Eq. (9.2).

Changes in light intensity cause variation in leaf resistance, r_{leaf}, through its effect on stomatal aperture (see Fig. 9.5), and on the vapor pressure, e_{leaf}, through its effects on leaf temperature. The effects of radiation were discussed earlier in the section on energy utilization. Atmospheric temperature acts through its effects on leaf temperature and hence on leaf water vapor pressure e_{leaf}. Readers are reminded that although an increase in temperature of a given air mass from 20 to 30°C will decrease the relative

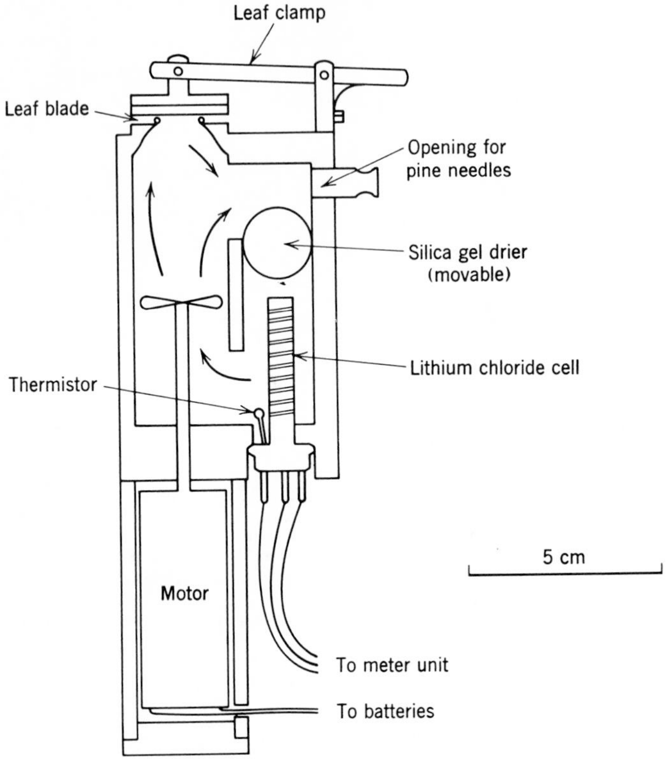

FIG. 9.11 A low-resistance diffusion porometer for measuring leaf diffusion resistance. This instrument follows the design of Slatyer. It uses the principle of the Wallihan–van Bavel instrument, with a motor-driven fan added to improve air circulation and increase the speed of response to change in humidity. (*From Kaufmann*, 1967.)

humidity, it will not change the vapor pressure of the air in an open system. An increase in temperature increases transpiration because it increases the steepness of the vapor pressure gradient from leaf to air, *not* because it is accompanied by a decrease in relative humidity (see Meyer and Anderson, 1952, chap. 11). The magnitude of the effect for any assumed relative humidity can be estimated as shown in Table 9.3. At constant temperature, changes in atmospheric humidity affect transpiration by changing e_{air} and modifying the vapor pressure gradient from leaf to air. Martin (1943) found that at constant temperature in darkness the rate of transpiration is very closely related to the vapor pressure of the atmosphere.

Wind acts directly to increase transpiration by sweeping away the boundary layer of moist air surrounding leaves and reducing r_{air}. It acts indirectly to decrease transpiration by cooling leaves and reducing e_{leaf}. Most of the increase or decrease occurs at a very low velocity (see Fig. 9.12). Knoerr (1966) pointed out that although a breeze should increase transpiration at

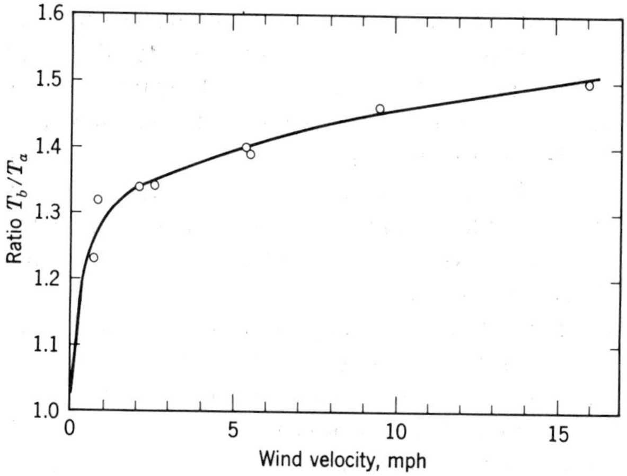

FIG. 9.12 Effect of wind on rate of transpiration of sunflower plants growing in a greenhouse in sun. The ordinate gives the ratio of the rate of plants exposed to wind, T_b, to control plants in quiet air, T_a. The values are averages for 2 to 4 hr of wind. (*From Martin and Clements,* 1936.)

low levels of radiation, at high levels—when leaves tend to be warmer than the air—a breeze should decrease transpiration. His calculations are shown graphically in Fig. 9.13. Other effects of wind, such as ventilation of the intercellular spaces by flexing the leaves and increasing passage of air through amphistomatous leaves, are probably of minor importance (Woolley, 1961).

As mentioned earlier, effects on transpiration of variations in the internal

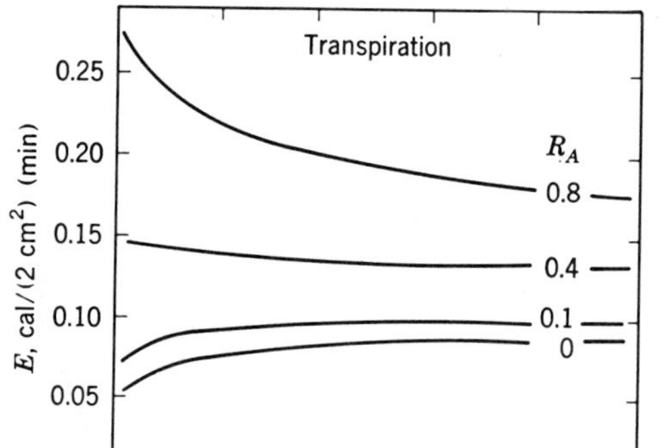

FIG. 9.13 Curves for theoretical latent heat exchange, *E*, by transpiration at various wind speeds and net radiations, R_A. At high levels of radiation wind should decrease transpiration by cooling leaves; at low levels it might increase transpiration by supplying energy. (*From Knoerr,* 1966.)

geometry of leaves, thickness of cutin layers, and degree of stomatal opening are exerted through changes in various components of the leaf resistance which are indicated in Eqs. (9.3) and (9.4). Leaf arrangement affects the exposure of leaves to sun and probably has some effect on the amount of energy received. Upright leaves receive less energy at the hottest time of day than horizontal leaves, and clusters of leaves receive less than those individually attached. Loss of a considerable part of the leaf area materially reduces water loss, and water stress causes development of an abscission layer and shedding of leaves in some species.

It is important to note that a change in one of the factors affecting transpiration does not necessarily produce a proportional change in transpiration rate, because the rate is not controlled by one factor alone. Rather, it is controlled by the difference in vapor pressure between leaf and air ($e_{leaf} - e_{air}$) and by the sum of the resistances in leaf and air pathways ($r_{leaf} + r_{air}$). An example of these interactions is found in Fig. 9.9, where the relationship between stomatal aperture and transpiration is shown. In still air, the resistance r_a is relatively high ($\cong$2.0) compared to stomatal resistance, and large changes in stomatal aperture result in only small changes in transpiration. However, in moving air r_{air} is low ($\cong$0.1), and small decreases in stomatal aperture cause a large decrease in transpiration. Thus wind increases transpiration by reducing r_{air}, but it also cools the leaf and reduces e_{leaf}, which tends to decrease transpiration.

The supply of water to the roots also affects the rate of transpiration. In soil near field capacity, the movement of water toward roots is rapid, and the rate of transpiration is controlled by plant and atmospheric factors. The

Table 9.6 ***Effects of Reduction in Stomatal Aperture on Leaf and Air Resistances and Transpiration Rate in Quiet and Moving Air When Vapor Pressure Gradient ($e_{leaf} - e_{air}$) Is Kept Constant*** *

	r_{leaf} +	r_{air}	Total r	Reduction in transpiration, %
Stomata fully open				
Quiet air	1.0	2.0	3.0	
Moving air	1.0	0.1	1.1	
Stomata 50% open				
Quiet air	1.5	2.0	3.5	15
Moving air	1.5	0.1	1.6	31
Stomata 10% open				
Quiet air	4.5	2.0	6.5	54
Moving air	4.5	0.1	4.6	76

* *From data of Bange (1953)*

midday wilting of rapidly transpiring plants in nutrient solution or soil near field capacity is caused by high root resistance. This is discussed further in Chap. 8. However, as soil water is depleted, the supply of water to the roots may become a limiting factor and cause decrease in the rate of transpiration (Ogata et al., 1960). Inadequate aeration, cold soil, and decrease in the osmotic potential of the root substrate are usually accompanied by large reductions in transpiration. The reduction occurs because restricted absorption results in leaf water stress and closure of stomates, increasing r_{leaf}. As mentioned earlier, change in ratio of root surface/leaf surface may also affect transpiration (Bialoglowski, 1936; Parker, 1949), but the magnitude of the effect is much greater for plants with sparse root systems. Andrews and Newman (1968) reported that removal of 60 percent of the roots from densely rooted wheat seedlings did not reduce transpiration below that of unpruned plants, even in drying soil.

Measurement of transpiration of leaves and plants

Measurement of transpiration has been reviewed by Crafts et al. (1949), Eckardt (1960), Franco and Magalhaes (1965), Slatyer (1967), and Stocker (1956). The usual methods of measuring the rate of transpiration are by change in weight or by the rate of water vapor loss. Water loss from trees has sometimes been estimated from measurements of the rate of sap flow in their stems.

Phytometer method

The majority of measurements of transpiration have been made on plants growing in containers which vary in weight from a few hundred grams to hundreds of kilograms. The soil surface is covered to prevent evaporation, and arrangements are made to replace the water removed from the soil. The pots should also be protected from direct sunlight to prevent overheating. The accuracy of this method depends on the minimum change in weight which can be measured. Because of this limitation, measurements of transpiration over short time intervals are difficult on plants in heavy containers. Correction for change in weight of the plant is necessary for rapidly growing plants over long periods of time. In a few instances, transpiration has been estimated from change in soil water content in the container, measured by a radiation absorption method (Ashton, 1956; also see Chap. 3). Another variation is to grow the plants in nutrient solution and measure the volume removed each day, or to place the root system or cut shoot in a potometer and measure water uptake over short periods of time. Estimates of transpiration over short periods of time, based on rates of absorption, may involve a considerable error caused by lag of absorption behind transpiration in intact plants (see Fig. 10.1).

Extrapolation of transpiration rates measured on potted plants in the greenhouse to estimate rates of plants in the field is unreliable because of the difference in environmental conditions. Extrapolations from individual plants to stands of plants are valid only if the individual plants are measured under field conditions in a stand of similar plants, as is done with well-protected lysimeters (van Bavel, 1961). The use of lysimeters to measure water loss was reviewed by Tanner (1967) and is discussed briefly in Chap. 3. In spite of its obvious limitations, the gravimetric method is useful for comparing species under the same conditions and for measuring the reaction of plants to variations in environmental factors and experimental treatments.

Cut-shoot method

This is sometimes termed the rapid-weighing method and is discussed by Franco and Magalhaes (1965). Numerous measurements of transpiration have been made by weighing cut shoots or single leaves on special balances for a few minutes after detaching them. In spite of its obvious convenience and frequent use, this is a very unreliable method of measuring transpiration for several reasons. Detaching the leaf releases it from whatever tension exists in the water-conducting system, and is often followed by a transient increase in rate of transpiration caused by temporary increase in stomatal aperture (Andersson et al., 1954; Decker and Wien, 1960; Franco and Magalhaes, 1965). Furthermore, in order to weigh the leaf it must be removed from its normal microenvironment and exposed to the abnormal environment of the balance. This causes serious changes in e_{leaf}, e_{air}, and r_{air}. Rates of transpiration measured under such conditions cannot be expected to be similar to those of leaves attached to plants. Weinmann and Le Roux (1946) reported discrepancies of 50 to 100 percent between measurements made by weighing entire plants and cut shoots, and Franco and Magalhaes (1965) report other examples of the discrepancies between the two methods. It has been stated that the errors generally appear to be greatest for leaves from well-watered plants and least for those under severe water stress, because the latter have such a high resistance that changes in the environment have little effect on the rate of transpiration. However, Halevy (1956) reported that on mild days the rates of transpiration obtained by quickly weighing leaves and potted plants were similar, but on very hot, dry days transpiration from potted plants was as much as 50 percent lower. Ringoet (1952) found that, in spite of its defects, he was able to follow successfully the daily course of transpiration of several species using this method.

Water vapor loss method

By this method a leaf, branch, or entire plant is enclosed in a container, and the change in water content of the air moving through the container is measured. The container is usually made of plastic and may vary in size from a container or cuvette to hold a single leaf (Bierhuizen and Slatyer, 1964) to enclosures of flexible, transparent plastic large enough to enclose small trees (Decker et al., 1962). In open systems the difference in water content of the air entering and leaving the chamber is measured (Björkman and Holmgren, 1963; Decker and Wien, 1960; Slatyer and Bierhuizen, 1964*a*), but in closed systems the increase in water vapor concentration or relative humidity is measured (Koller and Samish, 1964; Grieve and Went, 1965).

Many different methods have been used to measure the changes in water content of the air (see Franco and Magalhaes, 1965). These include wet- and dry-bulb thermometers (Decker and Skau, 1964), wet and dry thermocouples (Glover, 1941; Slatyer and Bierhuizen, 1964*c*), the corona hygrometer (Andersson et al., 1954), the microwave hygrometer (Falk, 1966), electrical resistance hygrometers (Grieve and Went, 1965; Wallihan, 1964), and infrared gas analyzers (Decker and Wetzel, 1957). Earlier investigators often absorbed the water vapor from the air in some hygroscopic material such as phosphorus pentoxide (Miller, 1938, pp. 493–494), but this method does not permit continuous measurements. The author's laboratory has found the differential psychrometer described by Slatyer and Bierhuizen (1964*b*) very satisfactory for measuring transpiration (Weatherspoon, 1968). It is much cheaper and simpler than the corona hygrometer, microwave hygrometer, or infrared gas analyzer, yet it has a very rapid response time and is reasonably accurate.

The chief advantage of this method is that it permits continuous measurements of short-term fluctuations in rates accompanying controlled or natural changes in the environment. However, it necessarily imposes an artificial environment on the leaf or plant enclosed in the container. The changes in leaf and air temperature and wind speed caused by enclosure may cause important differences in transpiration rate between the plants measured by these methods and those in the field.

A method developed by Wallihan (1964) and modified by van Bavel et al. (1965) for measuring leaf resistance can also be used to measure transpiration rates. As measurements require only about a minute, enclosure of the leaves probably does not cause serious changes in stomatal opening. However, transpiration is measured in dry air. An example of this apparatus as modified by Slatyer is shown in Fig. 9.11. Ehrler and van Bavel (1968) found good agreement between rates of transpiration measured by weighing and calculated by this method.

One way of controlling the environment in plant enclosures is to provide continuous compensation for changes in humidity and CO_2 concentration and careful control of temperature. Examples of such control can be found in the apparatus described by Koller and Samish (1964) and Moss (1963). Apparatus for control of the environment and continuous simultaneous measurement of water vapor and carbon dioxide exchange and stomatal resistance of leaves was described by Jarvis and Slatyer (1966*a*).

The cobalt chloride method developed by Livingston (see Miller, 1938, p. 491) measures escape of water vapor by the time required to produce a standard change in color on a piece of filter paper impregnated with cobalt chloride and pressed tightly against the leaf. It replaces the normal atmosphere with an artificial environment for a considerable period of time. As a result, rates measured by this method do not agree with rates obtained gravimetrically (Bailey et al., 1952; Milthorpe, 1955).

Velocity of sap flow

Attempts have been made to estimate the rate of transpiration from measurements of the rate of sap flow through the stems. This is usually done by some form of the heat pulse method described by Huber and Schmidt (1937). Ladefoged (1960) used a diathermy apparatus to heat the sap and a 40-junction thermopile to measure the heat pulse. He reported very rapid responses to passage of clouds and gusts of wind, also good seasonal trends (Ladefoged, 1963). He also measured absorption by cutting off trees and placing the base of the trunk in a container of water. From the correlation between velocity of sap flow and rate of absorption, Ladefoged attempted to calculate the rate of transpiration of intact trees. Decker and Skau (1964) also made simultaneous measurements of sap velocity and transpiration and found good agreement. Swanson and Lee (1966) discussed the problems involved in such measurements and cited several other studies. Gale and Poljakoff-Mayber (1965) found discrepancies between the heat pulse velocity and rate of transpiration with decreasing soil water potential, which caused them to question the reliability of hourly measurements of transpiration made by this method. One problem is the lag of absorption behind transpiration, resulting in changes in water content of the plant under observation. Also, it is not certain that the same cross section of the stem is involved in water conduction at different velocities. Although this method certainly indicates major changes in transpiration over considerable periods of time, it appears doubtful that it gives accurate measurements of short-term changes.

Bases for calculations of transpiration rates

After deciding on a suitable method for measuring the rate of water loss from a plant or a leaf, investigators are faced with the problem of selecting a satisfactory method of expressing it. This is particularly important if comparisons are to be made between plants of different species. Shall it be expressed per unit of leaf area or per unit of fresh weight, or even per unit of dry matter produced during the experimental period? If the rate is expressed on a leaf area basis, shall it be for one surface or for two surfaces for amphistomatous leaves? Water loss per unit of land surface is very useful for some purposes. This is discussed in the section on transpiration from plant communities.

Much European work has been expressed as units of water lost per unit of fresh leaf weight. However, the fresh weight of leaves varies from day to day and at various times of day. Furthermore, the ratio of area/weight varies among leaves of various species and even among leaves of the same species grown in different environments, as in sun and shade. The differences in relative rates of transpiration when expressed on fresh weight and leaf area bases are shown for three species in Fig. 9.13*a*. On the whole, leaf

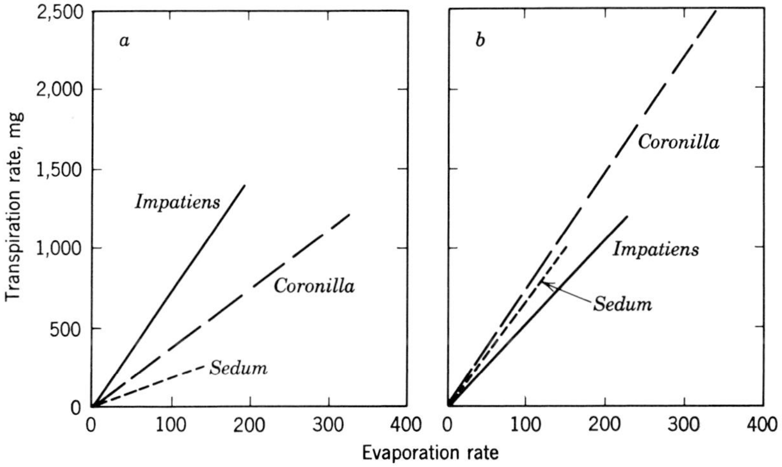

FIG. 9.13a Transpiration rates of three species (*Impatiens noli-tangere, Coronilla varia,* and *Sedum maximum*) expressed in milligrams per gram of leaf fresh weight (left) and in milligrams per square decimeter of leaf surface (right). Transpiration is plotted over rate of evaporation from a filter paper atmometer. The transpiration rate of *Sedum* was affected most by the method used to express it because it has a low ratio of surface to mass. (*Adapted from Pisek and Cartellieri,* 1932.)

surface seems to be the best basis for expressing rates of transpiration and photosynthesis, because the energy received is more closely related to the surface than to the fresh or dry weight. Where measurements of the relative efficiency of water use are of importance, calculation of the transpiration ratio or water loss per unit of dry matter formed is useful. This will be discussed later.

Use of leaf area as the base for calculating water loss introduces the necessity of measuring or calculating the leaf area of the experimental material. This is easy for a few leaves because their outlines can be traced on paper or obtained by holding them against light-sensitive paper (such as blueprint paper) to make a leaf print. The area of the outline can then be obtained by use of a planimeter, or by cutting it out and weighing it, then calculating the area from the weight of a known area of the same kind of paper. Several investigators have used photoelectric devices measuring the amount of light intercepted by leaves of various areas. Another method is based on the good correlation between area and width or length which exists in many kinds of leaves. These methods were reviewed by Miller (1938), and little has been added since then.

Thus far it has been assumed that transpiration from leaves is the only important source of water loss from plants. However, transpiration from the stems of some species of herbaceous plants may constitute a significant fraction of the total water loss. Gračanin (1963*b*) reported that the rate of transpiration per unit of surface from stems of several herbaceous species was considerably higher than that from the leaves. Thus, in spite of the much greater leaf area than stem area, water loss from the stems of a number of species was about half that from the leaves, and in a few species losses from stems were approximately equal to those from leaves. Although the rapid-weighing method used by Gračanin is open to criticism, the results serve to emphasize the fact that measurable amounts of water are lost from stems and that this should be taken into account in calculating rates of transpiration per unit of surface. Huber (1956*a*) cites some data on leaf and stem transpiration. He also gives data on transpiration from the bark of trees which indicate that it is negligible compared with the loss by foliar transpiration.

Rates of transpiration

In view of errors inherent in most of the methods of measuring transpiration, it may seem doubtful if it is worthwhile to give any actual rates. However, differences among species tested under the same conditions and daily and seasonal variations in rates are of considerable interest. Therefore a few data are presented, with a warning concerning their limitations. Furthermore, the actual data are more reliable than might be expected from the criticisms offered earlier. For example, transpiration of potted seedlings of *Liriodendron tulipifera* in full sun in August averaged 10.1 g/(dm^2)(day) at Columbus, Ohio, and 11.7 g/(dm^2)(day) at Durham, North Carolina. Comparisons of the transpiration of six species of trees by two methods and by four different investigators all indicated that spruce had the lowest rate and birch the highest rate (see Kramer and Kozlowski, 1960, p. 299). Measurements of transpiration of orange by quickly weighing detached leaves and potted plants gave similar results on mild days, but not on hot, dry days (Halevy, 1956). It appears that questionable methods sometimes give useful information, provided the sources of error are understood. The difficulty is in attempting to extrapolate measurements made on leaves or isolated plants to stands of vegetation.

Some representative rates of transpiration for potted plants are given in Table 9.7. Curves for diurnal variations in transpiration are given in Fig.

Table 9.7 Midsummer Transpiration Rates of Various Species of Trees Expressed as Grams of Water Lost per Square Decimeter of Leaf Surface per Day
(All seedlings were growing in pots in soil near field capacity.)

Species	Location	Season	Duration, days	No. of plants	Av. transpiration, g/dm^2 per day
Liriodendron tulipifera	Columbus, Ohio	August	1	7	10.11
Liriodendron tulipifera	Durham, N.C.	August	3	4	11.78
Quercus alba	Durham, N.C.	August	3	4	14.21
Quercus rubra	Durham, N.C.	August	3	4	12.02
Quercus rubra	Fayette, Mo.	July	14	6	8.1
Acer saccharum	Fayette, Mo.	July	14	6	12.2
Acer negundo	Fayette, Mo.	July	14	6	6.4
Platanus occidentalis	Fayette, Mo.	July	14	6	8.8
Pinus taeda	Durham, N.C.	August	3	4	4.65
Clethra alnifolia	Durham, N.C.	August	3	4	9.73
Ilex glabra	Durham, N.C.	August	3	4	16.10
Myrica cerifera	Durham, N.C.	August	3	4	10.80
Gordonia lasianthus	Durham, N.C.	July	23	4	17.77
Liriodendron tulipifera	Durham, N.C.	Aug. 26–Sept. 2	12	6	9.76
Quercus rubra	Durham, N.C.	Aug. 26–Sept. 2	12	6	12.45
Pinus taeda	Durham, N.C.	Aug. 26–Sept. 2	12	6	5.08

9.14, and the seasonal course of transpiration for an evergreen and a deciduous species is shown in Fig. 9.15. Daytime rates of transpiration usually range from 0.5 to 2.5 $g/(dm^2)(hr)$, and night rates are likely to be 0.1 $g/(dm^2)(hr)$ or lower. Some plants with crassulacean acid metabolism, such as Opuntia and pineapple, have a lower rate during the day than at night (Joshi et al.,

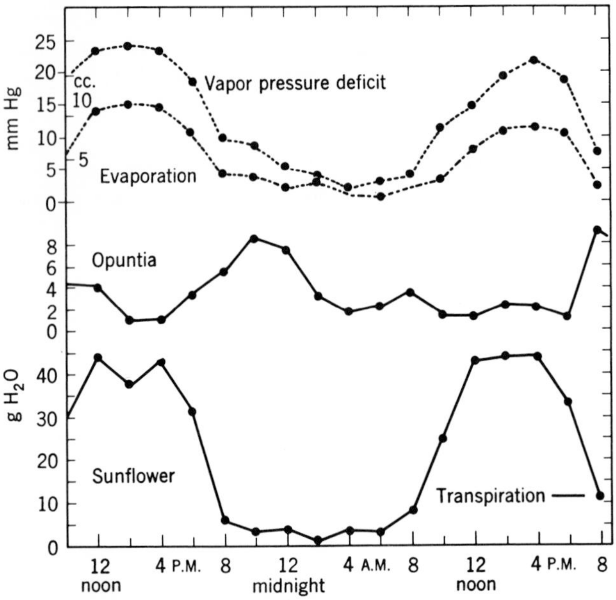

FIG. 9.14 Diurnal course of transpiration of sunflower and Opuntia plants in soil at field capacity on a hot summer day. Note the midday decrease in transpiration of sunflower, probably caused by loss of turgor and partial closure of stomata. Also note that the maximum transpiration of *Opuntia* came at night, a characteristic of plants with crassulacean organic acid metabolism. (*After Kramer*, 1937.)

1965). In terms of individual plants, these rates mean that plants often lose more than their own water content during a clear warm day. The effects of such rapid water loss will be discussed in Chap. 10.

Cyclic variations in transpiration

A number of writers have reported the occurrence of cyclic variations in transpiration which seem to be independent of environmental conditions.

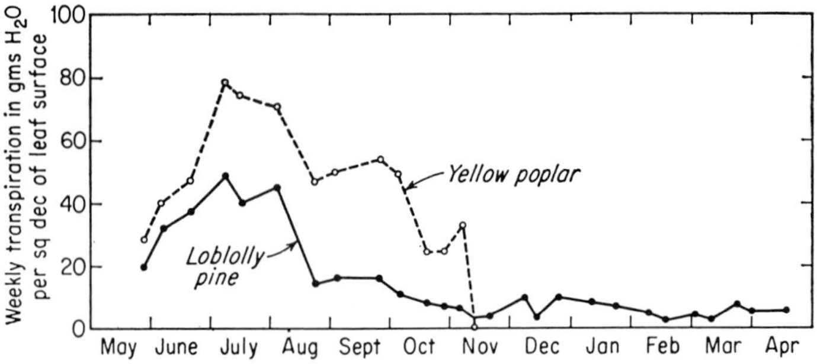

FIG. 9.15 Seasonal course of transpiration of potted seedlings of an evergreen and a deciduous species at Durham, N.C. (*From Kramer and Kozlowski*, 1960.)

Montermoso and Davis (1942) reviewed the earlier literature. They found that *Coleus* leaves and cuttings showed rhythmic fluctuations in transpiration when placed in darkness with constant temperature. Maximum transpiration was about noon and minimum about midnight, but the cycle could be reversed by exposing the plants to an artificially reversed light/dark cycle before placing them in constant darkness. These diurnal variations in transpiration rate have been attributed to rhythmic variations in stomatal opening which continue when plants are kept in constant darkness (Miller, 1938). Perhaps they are also related to the diurnal variations in root resistance mentioned later.

More puzzling are the short-term fluctuations reported by a number of writers (see Barrs and Klepper, 1968) which have a period of about 30 min. Ehrler et al. (1965) reported cyclic changes in transpiration, leaf resistance, leaf thickness, and leaf water deficits of cotton. These were attributed to cyclic changes in leaf turgor which affect stomatal opening. Barrs and Klepper (1968) examined the situation more intensively and found that leaf water potentials also vary cyclically, being lowest when stomatal aperture and transpiration rates are greatest. They concluded that the roots provide the major resistance to water flow through plants and that cycling occurs only when root resistance is high enough to cause leaf water deficits to develop. These results substantiate the report of Skidmore and Stone (1964) that root resistance is lowest toward midday and highest toward midnight. The increase in root resistance toward evening may explain the observation of several investigators that spontaneous cycling only occurs late in the day. It was found that removal of roots prevented cycling, as did reduction in water stress by reducing transpiration.

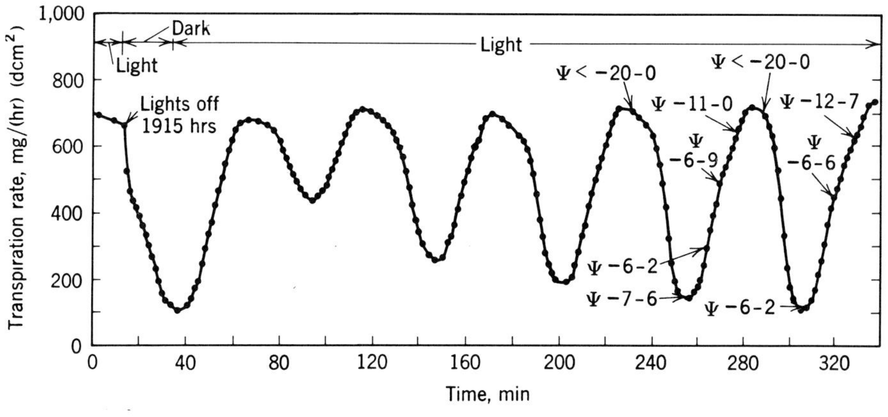

FIG. 9.16 Cyclic variations in transpiration rate and water potential of a cotton leaf induced by a 20-min dark period in the afternoon. Stomatal aperture and transpiration rate were largest when leaf water potential was low. This cycling is attributed to water deficits caused by high root resistance in rapidly transpiring plants. (*From Barrs and Klepper*, 1968.)

Cycling seems to occur because high transpiration combined with high resistance to water flow through roots produces a leaf water deficit, which causes stomatal closure. The resulting reduction in transpiration permits recovery of leaf turgor and reopening of stomata, whereupon a new cycle begins. The often reported midday reduction in transpiration shown by sunflower in Fig. 9.14 is an example of long-term cycling. Begg et al. (1964) reported a midday reduction in transpiration of an entire stand of bulrush millet, caused by stomatal closure.

Transpiration ratio

In agriculture the amount of water used per unit of dry matter produced is useful in evaluating the efficiency of water use. This was formerly known as the water requirement (Shantz and Piemeisel, 1927), but it is now usually termed the transpiration ratio. In experiments in Colorado this was found to range from about 200 to 300 for sorghum and millet to 700 to 1,000 for alfalfa (Shantz and Piemeisel, 1927). The lowest transpiration ratio reported for a cultivated crop appears to be about 50 for pineapple (Ekern, 1965). Slatyer (1967) points out that it may be as high as 5,000 for desert plant communities where plant cover is scattered and much of the soil water is lost by evaporation. It is well established that where sufficient rainfall occurs to support plant cover over the entire soil surface, the efficiency of water use is increased by

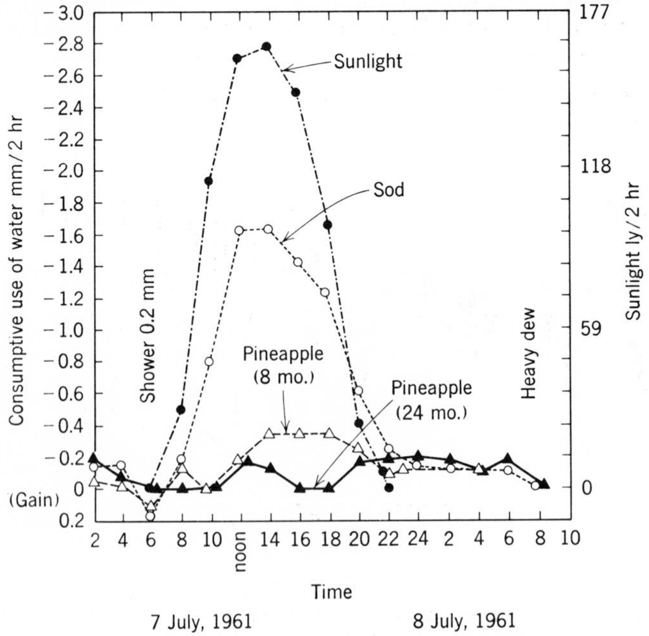

FIG. 9.17 Evapotranspiration from a Bermudagrass sod and from pineapple 8 and 24 months after planting, growing in lysimeters in Hawaii. (*From Ekern*, 1965.)

increasing the density of the cover, so that water loss by evaporation from the soil surface is kept to a minimum. The extreme example is pineapple, which, as shown in Fig. 9.17, has such a low rate of transpiration that when it completely shades the soil, evapotranspiration is reduced below the level that occurs in a field of young plants where the soil is partly exposed and much water evaporates from it (Ekern, 1965). Fertilization and other cultural treatments which increase yields usually tend to decrease the transpiration ratio and increase the efficiency of water use. An example is given in Table 10.2, which shows differences among species and varieties as well as the effects of supplying nitrogen. Koch (1957) points out that factors such as sulfur dioxide injury which decrease photosynthesis more than transpiration decrease the efficiency of water use. Attacks on leaves by fungi sometimes have the same effect. The use of transpiration-suppressing chemicals is based on the expectation that transpiration can be reduced more than photosynthesis, thus increasing the efficiency of water use.

Reduction of transpiration

Agriculturists and horticulturists have long been interested in reducing the rate of transpiration. Reduction following transplanting would enable plants to maintain an adequate water balance until their root systems are reestablished. Reduction of transpiration during droughts would enable crop plants to survive with a minimum of injury, and reduction of transpiration from the plant cover on watersheds might increase the yield of water. One possibility of accomplishing this is to coat plants with a waterproof film to reduce the escape of water vapor; another is to apply substances which will cause closure of stomata.

A variety of fungicides, herbicides, metabolic inhibitors, and growth regulators have been reported to reduce transpiration, often by causing closure of stomata. Various kinds of films, such as emulsions of latex, polyvinyl waxes, polyethylene, and vinyl-acrylate, and higher alcohols such as hexadecanol have been applied with varying results. Their effectiveness seems to depend on the species, stage of development, and atmospheric conditions during the test period. The literature on this subject was reviewed by Gale and Hagan (1966).

Use of films is not very promising, because they are relatively impermeable to carbon dioxide (Woolley, 1967) and reduce photosynthesis. Furthermore, repeated applications are necessary to keep the transpiring surface of rapidly growing plants covered. Application of substances which bring about closure of stomata seems more promising, because partial closure of stomata should reduce transpiration more than it reduces photosynthesis (Shimshi, 1963*a*; Slatyer and Bierhuizen, 1964*b*; Zelitch and Waggoner, 1962). This assumption is based on the claim originally made by Gaastra (1959) that the movement of carbon dioxide into the mesophyll cells and the chloroplasts introduces an additional relatively large resistance to the entrance of carbon dioxide which does not affect the exit of water vapor. Gaastra (1959) claims this mesophyll resistance for entrance of carbon dioxide is considerably higher than the air and stomatal resistances which affect exit of water vapor. Thus a large change in stomatal resistance would have less effect on photosynthesis than on transpiration. Slatyer and Bierhuizen (1964*b*) reported that they were able to reduce transpiration more than photosynthesis by closing stomata with phenylmercuric acetate. On the other hand Barrs (1968) measured photosynthesis and transpiration of cotton during cyclic opening and closing of stomata. The ratio of net photosynthesis to transpiration was constant over a wide range of gas exchange rates and even in wilting leaves. It seemed clear that exchange of both water vapor and carbon dioxide was limited by stomatal aperture over a wide range of leaf water stress. This led Barrs to conclude that the importance of the mesophyll resistance to entrance of carbon dioxide has been overestimated. Brix (1962) also found that photosynthesis and transpiration were decreased to the same extent in loblolly pine and tomato by increasing water stress.

One of the most promising compounds for causing closure of stomata is phenylmercuric acetate. It has been shown to reduce transpiration significantly in several species. Waggoner and Bravdo (1967) reported that it significantly reduced water loss from a pine stand. Unfortunately there have been some instances of toxicity from the use of phenylmercuric acetate. Keller (1966) reported that both the film-forming polyvinyl compound VL-600 and phenylmercuric acetate reduced transpiration of spruce seedlings, but they also reduced photosynthesis and root growth, and phenylmercuric acetate increased respiration and caused injury to spruce needles. Another method of reducing transpiration might be to increase the concentration of carbon dioxide high enough to close the stomata. Moss et al. (1961) reported that increasing the carbon dioxide concentration around corn leaves from 0.031 to 0.057 percent decreased transpiration by 23 percent and increased net photosynthesis by 30 percent. Unfortunately this approach is practical only for enclosed spaces or limited areas. The possibility of reducing transpiration obviously requires more study before it will become feasible on a large scale, if indeed this can ever be done.

Transpiration from plant communities

Under field conditions leaves vary widely in size and in orientation with respect to incident energy and air movement. Moreover, on any one plant and in any plant community there is an array of leaves of different ages and structure, located in a range of microenvironments. As a consequence, the behavior of a single leaf in a specified or controlled environment may bear little relationship to behavior of one in a field or forest. It is important to emphasize the limited applicability of single-leaf or single-plant studies to conditions different from those in which the studies were made. Isolated measurements of transpiration rate without accompanying measurements of the microenvironment are of very limited value.

To have physiological significance, measurements must be put into context with the terms of Eq. (9.2); and to have ecological significance, they must be put into context with the microenvironmental relationships of plant communities.

Evapotranspiration from plant communities

Evaporation and transpiration from plant communities is frequently termed evapotranspiration, to describe the combined plant-soil nature of the evaporating surfaces. Evaporation from soil and transpiration from plants involve basically similar processes, and soil and plants may be thought of as alternative paths through which water moves on its way to the surface, from which it finally diffuses into the bulk air. The heterogeneity of the evaporating surface and the biologically controlled variability of the resistance to internal flow merely modify the movement of water through vegetation. Slatyer (1967, pp. 46–56) discussed this complex problem in detail.

The characteristics of the evapotranspiration process from a plant stand are sometimes represented by an equation of the same form as Eq. (9.2) (Slatyer, 1967, p. 48):

$$E = \frac{c_{\mathrm{int}} - c_{\mathrm{air}}}{r_{\mathrm{int}} + r_{\mathrm{air}}} = \frac{0.622\rho_{\mathrm{air}}}{P} \, \frac{e_{\mathrm{int}} - e_{\mathrm{air}}}{r_{\mathrm{int}} + r_{\mathrm{air}}} \tag{9.5}$$

The subscript $_{\mathrm{int}}$ refers to the internal part of the pathway between the evaporating surface and the soil or plant surface.

In soils which are wet to the surface, r_{int} can be neglected, and c_{int} is effectively the vapor concentration at the soil surface. As the soil dries, however, the evaporating surface retreats into the soil, lengthening and increasing the tortuosity of the water vapor diffusion pathway. This is largely responsible for the sudden drop in evaporation so frequently observed in drying soils.

In single plants the evaporating surfaces are always within the leaf surfaces, except when the leaves are wet. However, in stands of plants the evaporating surfaces are the leaves within the canopy. Equation (9.5) follows Monteith (1963) in treating the surface of a stand of plants as the evaporating surface. This is an oversimplification, because both air and leaf conditions vary from top to bottom of the canopy, resulting in different rates of transpiration at various levels. Examples of vertical differences in leaf properties and transpiration rate within a crop are given by Begg et al. (1964). Waggoner and Reifsnyder (1968) developed a model to deal with the complex interaction of factors which affect evaporation within a plant canopy. It consists of a vertical series of resistors and conductors to represent various levels or strata in the canopy through which downward flow of energy and upward flow of water vapor occur along gradients of potential. This approach permits estimation of evaporation in each layer or stratum of leaves and, from these data, of total evaporation from the crown, if the radiation, resistances, and leaf areas of the various strata are known. The procedures and some applications are described by Waggoner and Reifsnyder (1968).

The leaf area of stands of plants is often three to six times the area of the land on which the plants grow (leaf-area index of 3 to 6). As most leaves lose water from both surfaces, the total external surface should be regarded as twice that amount. Nevertheless, evaporation from a plant community

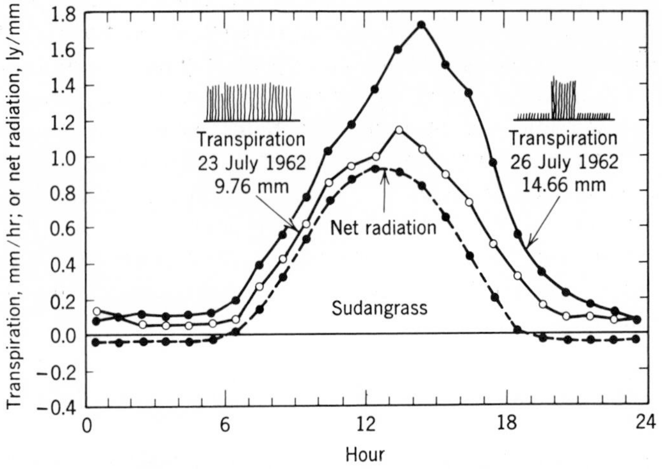

FIG. 9.18 **Transpiration from a lysimeter in a closed stand of Sudangrass and from the same stand three days later after it had been isolated by removing the surrounding crop. Radiation was essentially the same on the two days, but additional energy supplied by advection caused much higher transpiration from the isolated plot. (*From van Bavel et al.*, 1963.)**

never exceeds that from a similar area of wet soil or water with the same exposure. Sometimes incident energy on an evaporating surface is supplemented by advection, as in the case of cultivated fields surrounded by desert. This situation is shown in Fig. 9.18, where the energy dissipated by evapotranspiration from a well-watered crop of Sudangrass growing in a lysimeter exceeded that which it received as net radiation. The discrepancy was exaggerated when the exposure was increased by removing the surrounding vegetation.

Agriculturists and foresters are particularly interested in the effects of changes in density and height of stands of plants on the rate of evapotranspiration. For example, under similar conditions open forests or orchards or stands of irregular height might have higher rates than dense stands of plants of uniform height, because of greater turbulence in the open, irregular stands. Furthermore, thinning a stand may or may not decrease evapotranspiration. Very light thinning may merely open it up and increase turbulence, decreasing the air resistance and increasing evapotranspiration. Only when the degree of thinning becomes large enough to bring about a great reduction in leaf area can significant reductions in community evapotranspiration be expected. Nevertheless, increase in water yield was approximately proportional to the reduction in basal area of vegetation on a watershed (Hewlett and Hibbert, 1961). Part of the increased water yield probably resulted from elimination of water loss by interception in the forest canopy. Substantial leaf shedding, such as occurs in deciduous species and in such desert perennials as *Euphorbia* and *Larrea* when exposed to water stress, clearly can reduce water loss materially. Seasonal and ontogenetic changes in transpiring surfaces of a variety of species are discussed by Killian and Lemée (1956).

Relative losses by evaporation and transpiration

The relative amounts of water removed from soil by evaporation and transpiration are of great interest, particularly in regions of limited rainfall, and a few data are presented to illustrate the magnitude of these losses. Over the United States as a whole about one-fourth of the total precipitation escapes as stream flow, and about three-fourths is returned to the atmosphere by evaporation from the soil and vegetation (Ackerman and Loff, chap. 3, 1959). However, wide local variations occur in the relative amounts of water disposed of in these two ways. It is of particular interest to differentiate between losses by evaporation and transpiration on watersheds in order to determine whether changes in the vegetation cover might increase the water available for stream flow. Those who calculate water losses from the energy balance have tended toward the view that the nature of the plant cover is less important than the energy supply and that the water yield will be affected very little by the kind of species growing on the watershed, as long as conditions favor active growth. However, Rider (1957) found that different crops covering areas of several acres or more modify the environment over them differently, resulting in differences in water loss. Also, under conditions where advection occurs, communities with an open, well-ventilated structure can be expected to transpire more than dense, even communities. Furthermore, differences in depth of rooting, length of growing season, morphology of a stand, and degree of stomatal control over transpiration may result in differences in water loss among species.

Extensive experiments at the Coweeta Hydrologic Laboratory in humid southwestern North Carolina indicate that conversion of mature forest to low-growing vegetation increased stream flow by the equivalent of 12.5 to 40 cm/year the first year after cutting, the increase being much greater on north-facing than on south-facing slopes (Hewlett and Hibbert, 1961). When vegetation was allowed to resprout and grow, stream flow began to decrease. It is said that experiments in various parts of the temperate zone other than Coweeta indicate that clear-cutting watersheds (removal of woody vegetation) increases stream flow up to 25 cm/year, with an average increase of about half of this. Clear-cutting lodgepole pine in Colorado increased stream flow from 25.7 to 33.7 cm, or about 30 percent (Wilm and Dunford, 1948). It is claimed that in both the Central Valley of California (Biswell and Schultz, 1957) and southern California (Pillsbury et al., 1961), conversion of chaparral to grass increases and even doubles runoff. The effects of cutting on stream flow will vary with the amount and seasonal distribution of precipitation and the amount per storm (Hewlett and Hibbert, 1967). Various aspects of this complex problem are reviewed in the symposium on forest hydrology edited by Sopper and Lull (1967).

Several instances have been reported where clear-cutting of forests resulted in measurable rise in a shallow water table. According to Wilde et al. (1953), removal of aspen caused a rise of 35 cm in the water table and converted a fairly well-drained soil into a semiswamp. Trousdell and Hoover (1955) reported that the water table was lower in an uncut stand of loblolly pine than in the adjacent selectively cut stand on similar soil. However, the water table became higher than in the selectively cut stand after the uncut stand was clear-cut. In some wheat-growing areas of western Australia, the hydrologic balance has been changed completely by replacing deep-rooting, evergreen *Eucalyptus* woodlands with shallow-rooted annual crops (Burvill, 1947).

It is difficult to separate water loss by evaporation from that by transpiration, but the difference between subsurface drainage before and after removal of forest certainly represents water which would have been lost by transpiration, although losses by evaporation are probably higher after cutting than before. Thus, the actual transpiration of forest at Coweeta was probably greater than the increased runoff of 12.5 to 40 cm/year. Patric et al. (1965) estimated from soil water depletion data that water absorption from April through October by a 21-year-old loblolly pine stand was 41 cm, and that for a mountain stand of oak and hickory it was 37.2 cm. The soil on these plots was covered with plastic film to prevent evaporation, so the absorption data can be regarded as equivalent to the losses by transpiration.

Experiments with corn plots at Urbana, Illinois, in which the soil of some plots was covered with plastic to prevent evaporation, indicate that 50 percent or more of the total evapotranspiration loss was by evaporation from the soil surface (Peters and Russell, 1959). Total evapotranspiration from the uncovered plots exceeded the rainfall during the same period by 9.7 to 13.7

Table 9.8 ***Relative Water Losses by Transpiration and Evaporation from an Illinois Cornfield during the Period from mid-June to Early September*** *

Year	Total evapotranspiration from uncovered plot, cm	Transpiration from covered plot, cm	Transpiration as percentage of evapotranspiration	Total precipitation, cm	Excess of evapotranspiration over rainfall, cm
1954	32.25	16.5	51%	18.5	13.75
1955	34.50	17.5	51%	23.0	11.50
1957	33.75	15.1	45%	24.0	9.75

* *From Peters and Russell (1959)*

cm. Reimann, Van Doren, and Stauffer (1946) studied the water relations of a corn crop on a silt loam soil near Urbana, Illinois, during a summer with a total rainfall of 25.5 cm. Of this, 5.0 cm was intercepted by the corn plants and lost by evaporation, and 20.5 cm reached the soil. Transpiration dissipated 20.5 cm and evaporation 13.2 cm, so the total water loss exceeded the water reaching the soil by 13.2 cm. The difference was supplied by depletion of soil moisture, the amount of available water in the upper 2.3 m of soil being reduced from 31.0 to 17.7 cm.

It is fairly common for water loss to exceed precipitation for one or more growing seasons in regions of limited rainfall. Wiggans (1936, 1937, 1938) reported that in an apple orchard 18 to 20 years old in eastern Nebraska, evapotranspiration removed 25 to 38 cm/year more water than was replaced by precipitation. In one year 94.5 cm of water was removed but only 56.5 supplied as precipitation, and the deficit of 38 cm came from the reserve deep in the soil. Wiggans estimated that in three more years all available water under this orchard would be exhausted to the depth of over 9 m. Presumably, these trees would then have died unless they were irrigated. Unfortunately they were killed by a severe autumn freeze so the accuracy of the estimate was not determined.

A similar situation exists with respect to deep-rooted perennial herbaceous plants in areas of limited rainfall. In certain areas of Nebraska, alfalfa made excellent growth the first three years after seeding, quite independently of current rainfall, after which the yield declined and became closely correlated with current rainfall. The decline was caused by depletion of available water in the subsoil during the first three years, after which the yield was dependent on the current rainfall. Alfalfa may absorb water from depths of at least 9 m in well-aerated soils, and this is replaced very slowly under rainfall conditions existing in Nebraska. In fact, after 15 years of cropping with cereals following alfalfa, very little water had been replaced below the 2.0-m level (Kiesselbach et al., 1929).

Trees can sometimes be established in areas of limited rainfall and grow for a number of years on reserve moisture, but they die when it is exhausted. Bunger and Thomson (1938) reported this situation in the panhandle of Oklahoma, where rainfall is inadequate for indefinite growth of trees. Many plantations of forest trees were established in the Prairie and Plains states in the 1880s and 1890s. These grew well for a number of years, but eventually the trees began to die, probably because they finally exhausted all the reserve soil water available to their roots. A similar explanation has been advanced to explain the failure of trees and grasses established on sand dunes in the Rajasthan desert of India.

Measurement of evapotranspiration

Measurement of the amount of water lost by evapotranspiration from areas of land is of great interest to agriculturists and foresters in connection with studies of water use by various kinds of plant cover, prediction of yield from watersheds, and timing of irrigation. There are four general methods of measuring water loss from land areas. These are (1) determination of the evaporation term in the water balance equation, (2) determination from the energy consumed in evaporating water, (3) determination of the net upward flow of water vapor in the air layers near the ground, and (4) estimation from meteorological data or rates of pan evaporation. The various methods will be discussed very briefly in this chapter. Readers are referred to Tanner (1967) for a review of the advantages and disadvantages of the various methods.

WATER BALANCE METHOD. This method involves determination of the evaporation term E in the water balance equation, Eq. (3.1). It can be used for entire watersheds and provides useful information over long periods of time where leakage is negligible and runoff can be measured. Measurement of soil water depletion can also be used to provide an estimate of evapotranspiration over periods of a week or more. However, in humid regions with frequent rainfall serious errors may result from drainage. The most precise results are obtained with lysimeters, where a volume of soil is enclosed in a container, permitting measurement of water gained and lost. Tanner (1967) discussed the problems involved in the operation of lysimeters. Measurement by the water balance method is also briefly discussed in Chap. 3.

THE ENERGY BALANCE METHOD. The energy balance method described by Tanner (1960), Slatyer and McIlroy (1961), Lemon (1963), and Begg et al. (1964) involves measuring or estimating the radiant energy entering and leaving the community and measuring or estimating the proportions of the net radiation absorbed by the community which are involved in sensible and latent heat (evapotranspiration) transfer between the community and the air. These relationships can be expressed in the following equation

$$R_n = H + lE + G + aA \tag{9.1}$$

where R_n represents net radiation, or the difference between incoming and outgoing radiation, H is sensible heat exchange with the atmosphere, lE is latent heat used in evapotranspiration, G is the heat exchange with the soil and vegetation, and aA is the energy used in photosynthesis and released by respiration. Over periods of time as long as days, the only important changes are in H *and* lE, and if H is measured, the amount of evapotranspiration can be calculated from the lE term. Slatyer (1967, p. 35) discusses this method in more detail. It is widely used, even though it is restricted to areas of relatively homogeneous vegetation such as crops and some forests. Advection does not invalidate the method, but it makes measurement considerably more complex.

AERODYNAMIC OR VAPOR FLOW METHOD. The most direct measurement, the determination of the upward vapor flow above the surface being studied, requires very sensitive and delicate instrumentation, and few measurements of this type have been made as yet (Swinbank, 1951; Dyer, 1961). However, estimates of the vapor flux can be obtained from measurements of the humidity gradient above a crop using an appropriate transfer coefficient, and procedures of this type have also given satisfactory results. Slatyer and McIlroy (1961), and Slatyer (1967) provide general information about methods, and specific details are given by Pasquill (1949), Harbeck et al. (1958), and Webb (1960).

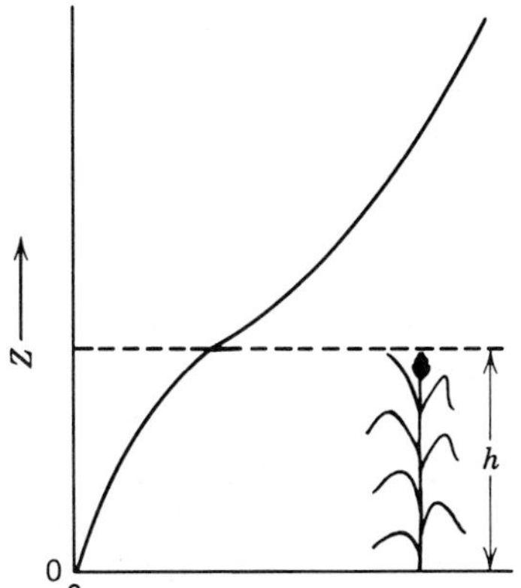

FIG. 9.19 **Diagram showing wind velocity *U* in and above a plant community of height *h*. *Z* refers to height above the soil surface. (*After Lemon*, 1967.)**

Measurements of the change in water content of an airstream passing through a large chamber mounted over a whole plant or a portion of a plant community are also, in a sense, vapor flow procedures (Decker et al., 1962; Moss et al., 1961). However, the same cautionary remarks made earlier with respect to single-leaf and single-shoot enclosures must also be applied to this procedure if it is desired to estimate, from the results obtained, evapotranspiration rates of the unenclosed plots or areas. The most useful applications of this procedure are for studies of the transpiration process under known rather than natural environmental conditions, or to compare relative evapotranspiration rates from different plant communities under similar conditions.

Empirical formulas

Several methods have been developed to provide estimates of evapotranspiration from commonly measured meteorological elements such as bulk air temperature and humidity, wind speed, cloudiness, and length of daylight (Thornthwaite, 1948, 1954; Penman, 1948, 1956; Blaney and Criddle, 1950). Most of the formulas were applied originally to evaporation from free water and have been modified for use for vegetation by the introduction of empirical weighting coefficients. They were designed to be used in calculating the water balance of plant communities or to predict irrigation schedules, drought frequency, stream flow, etc. Accounts of most of them were given by Dzerdzeevski (1958) and Penman (1963).

The Penman method is probably the most soundly based. It has had the most successful application to biological and hydrological problems. It combines an aerodynamic and energy balance approach and, although more complicated than most other evaporation formulas, requires only meteorological data. In use, an estimate of free water evaporation is first obtained from air temperature, humidity, and wind speed data. If available, net radiation and surface temperature data can be used, but they can be estimated (Penman, 1948). The potential evapotranspiration from an extensive area of green crop, physiologically active and well supplied with water, is then assumed to be from 0.6 to 0.8 of the free water evaporation, depending on season and length of day.

Other refinements have been incorporated from time to time to allow for the effect of soil water availability (Penman, 1949, 1956), and the method has been used extensively for a wide range of water balance and related studies (see for example, van Bavel and Verlinden, 1956; Slatyer, 1960*b*; Fitzgerald and Rickard, 1960). Although it gives invalid estimates in the presence of advection, when the observed ratios of evapotranspiration to evaporation from free water may exceed 1.0, it was not intended for use in such situations. In situations where it is applicable, it generally compares favorably with other methods developed for similar purposes (Tanner and Pelton, 1960; Stanhill, 1931). Tanner (1967) discussed these methods in some detail.

Summary

Transpiration is the escape of water from plants in the form of vapor. Although basically an evaporation process, it is affected by plant structure and stomatal behavior as well as by the physical factors which control evaporation. Transpiration is the dominant factor in plant water relations because it produces the energy gradient which causes the movement of water into and through plants. In addition it often produces midday leaf water deficits, and when drying soil causes absorption to lag behind water loss, permanent water deficits develop which cause injury and even death. In fact, more plants are injured or killed by excessive transpiration than by any other single factor.

The rate of transpiration depends on the supply of energy available to vaporize water, the water vapor pressure or concentration gradient which constitutes the driving force, and the resistances in the vapor pathway. The chief leaf resistances are the cuticle and the stomata. Another important resistance is the air layer surrounding the leaf.

There is a complex interaction among the factors affecting the rate of transpiration. The chief environmental factors are light intensity, vapor pressure and temperature of the air, wind, and water supply to the root. Plant factors include the extent and efficiency of the root system as an absorbing surface, leaf area, leaf arrangement and structure, and stomatal behavior. If any one of these factors is altered enough to cause a change in rate of transpiration, other factors will be changed, causing further adjustments in the transpiration rate.

Measurements of transpiration rates of individual plants, twigs, or leaves can be made in a number of ways. Measurement of the water loss from potted plants by weighing the plant and container is most reliable. Measurement of the water vapor lost from a plant or group of plants enclosed in some kind of transparent container is subject to various errors because the environment in the container is different from that in the open. Measurement of water loss from cut twigs or leaves is often subject to very large errors. Great caution must be used in extending measurements of the transpiration rate of single plants or plant parts to plants or stands of plants growing in the open under very different conditions.

The loss of water by evapotranspiration from plant communities can be measured in several ways. One method involves estimating the evaporation term in the water balance equation, a second requires determination of the fraction of the energy in the energy balance used to evaporate water, and a third depends on determining the net upward flow of water in the air near the ground. More often evapotranspiration is estimated from easily measured meteorological factors such as bulk air temperature and humidity, wind speed, and amount of sunlight.

ten

Water Stress & Plant Growth

Introduction

This chapter discusses the nature and causes of plant water stress, the effects of water stress on the principal physiological processes, measurement of plant water stress, differences in efficiency of water use, and the nature of drought resistance. It was stated in Chap. 1 that environmental water deficits reduce plant growth by modifying the physiological processes and conditions which control growth. Plant growth is therefore controlled directly by plant water stress and only indirectly by atmospheric and soil water stress.

Plant water stress or water deficit refers to situations where cells and tissues are less than fully turgid. Water stress can vary in degree from a small decrease in water potential detectable only by instrumental measurements, through the transient midday wilting often observed in hot sunny weather, to permanent wilting and death by desiccation. In simplest terms, water deficit or water stress occurs whenever the loss of water in transpiration exceeds the rate of absorption. It is characterized by decrease in water content, osmotic potential, and total water potential, accompanied by loss of turgor, closure of stomata, and decrease in growth. If severe, water stress results in drastic reduction in photosynthesis and disturbance of many other physiological processes, cessation of growth, and finally in death by desiccation.

Causes of plant water stress

Plant water stress is caused by either excessive loss of water or inadequate absorption, or a combination of the two.

The absorption lag

Midday water deficits occur because absorption tends to lag behind transpiration, as shown in Fig. 10.1. The midday absorption lag results from

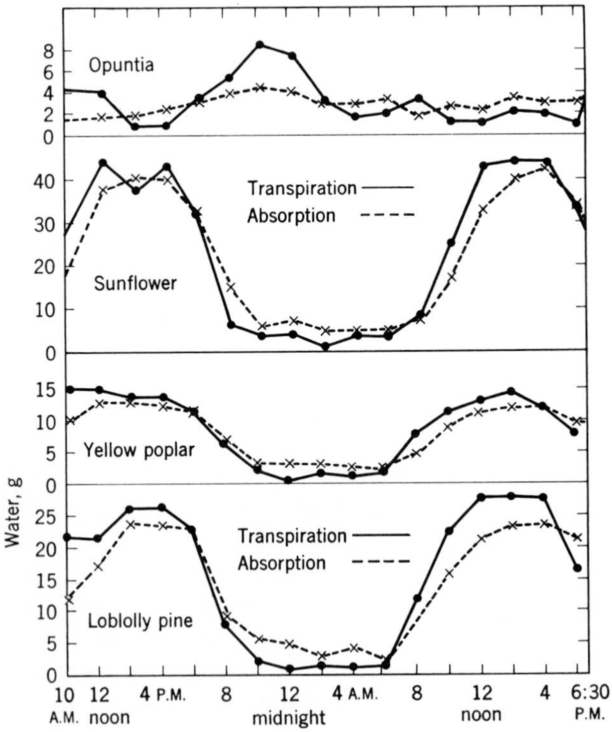

FIG. 10.1 Lag of absorption behind transpiration in four very dissimilar species of plants. Plants were growing in soil supplied with water by an autoirrigation system which permitted measurement of water absorption. Note the midday drop in transpiration of sunflower, presumably caused by loss of turgor and midday closure of stomata; also the evening maximum in transpiration of *Opuntia*. (*After Kramer*, 1937.)

resistance to water flow through plants (see Chap. 8) plus the fact that the rates of water absorption and transpiration are controlled by different sets of factors. The transpiration rate is controlled by (1) leaf area and structure, (2) stomatal opening, and (3) those factors which affect the steepness of the vapor pressure gradient from plant to air. Absorption, on the other hand, is controlled by (1) rate of water loss, (2) extent and efficiency of root systems, and (3) the water potential and hydraulic conductivity of the soil. It is not surprising that processes controlled by different sets of factors are not perfectly synchronized, even though they are partly interdependent and linked together by the continuous water columns extending from roots to leaves.

Since water is inelastic, it might be expected that a change in rate either of water loss or of water absorption would be instantly transmitted to the other process. However, there is considerable resistance to movement of water through the roots. There is also a buffer in the system in the form of the parenchyma tissue, which functions as a reservoir, losing water during periods when transpiration exceeds absorption and gaining water when the reverse occurs. As a result, the first effect of a high rate of transpiration is decrease in water content and loss of turgidity by leaf cells, which often culminates in wilting.

Transpiration versus absorption

According to this view, water loss by transpiration is the primary cause of plant water deficits. This certainly is true with respect to temporary midday deficits, because even plants growing in moist soil or aerated culture solutions develop water stress and sometimes wilt on hot sunny days. The development of midday water stress is also shown by the fact that during hot weather plants often grow more at night than during the day. For example, Loomis (1934) and Thut and Loomis (1944) found that in Iowa extension growth of corn and plants of other species is reduced more often by excessive transpiration than by deficient soil moisture. In North Carolina young pine trees made nearly twice as much growth at night as during the day in late June and early July when transpiration rates were high (Reed, 1939). Other examples are given by Miller (1938). In contrast to this view, Orchard (1967) found no evidence that growth of kale at Rothamsted Experiment Station in England was affected by plant water deficits other than those caused by deficient soil moisture. Perhaps in mild English summers transpiration rates are too low to produce midday water deficits in plants in moist soil, but this certainly is not true in climates with bright sun and high temperatures. Examples of short-term changes in leaf water potential caused by changes in rate of transpiration are shown in Fig. 9.16.

Although excessive transpiration is responsible for temporary midday daily water deficits, decreased absorption caused by decreasing availability

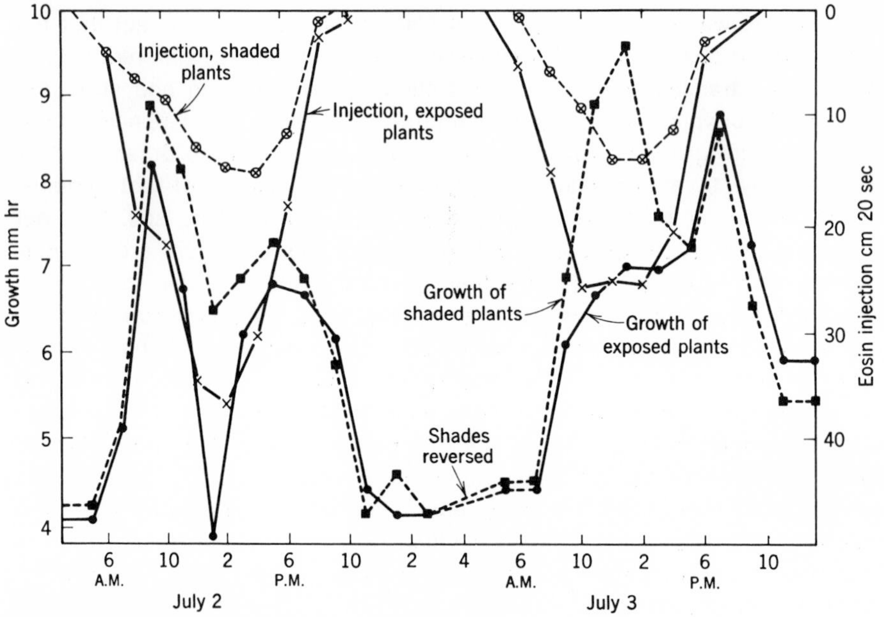

FIG. 10.2 Midday reduction in growth of shaded and unshaded corn plants. The reduction is attributed to a midday water deficit, as indicated by the rapid injection of dye into cut leaves. (*From Thut and Loomis,* 1944.)

of soil water is responsible for the long severe periods of water stress which cause the largest reductions in plant growth. The level of soil water potential sets the maximum possible level of plant water potential, and on this ceiling is imposed the daily decrease in potential caused by transpiration. The development of water stress is described in detail in a later section.

Soil water and plant growth

The literature of agriculture, horticulture, and forestry is full of papers in many languages describing the effects of drought on growth and yield of all kinds of plants. Much work on the relationship between soil water and plant growth was summarized by Richards and Wadleigh (1952) and by Stanhill (1957). In general there is decreased growth with decreasing soil water, but exceptions occur, as will be noted later. It might be useful if environmental conditions could be characterized in terms of the number of days when plant water stress is severe enough to limit growth. Such an attempt was made by Denmead and Shaw (1962), who related growth of corn to the number of days the soil water content was below the estimated wilting point (Fig. 10.3). Some progress has also been made in characterizing areas in terms of the probability of droughts and soil water deficits severe enough to reduce crop growth (van Bavel, 1953; van Bavel and Verlinden, 1956).

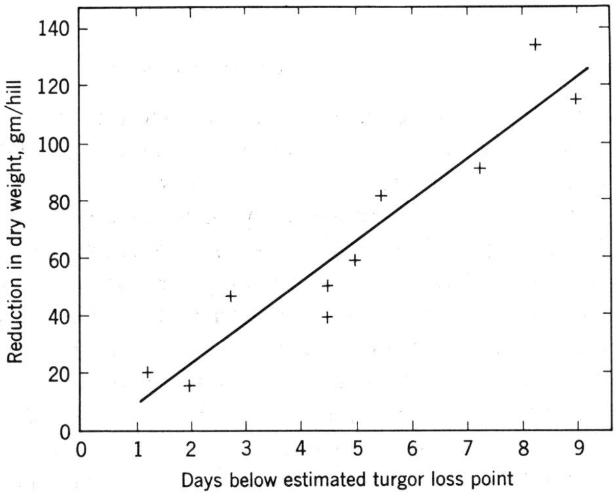

FIG. 10.3 **Reduction in dry weight of corn plants subjected to various numbers of days of soil water stress severe enough to cause loss of turgor. (*After Denmead and Shaw,* 1962.)**

Often there is a fairly good correlation between soil water stress and plant growth. For example, Bassett (1964) calculated available water in the root zone from rainfall and evapotranspiration data, then converted it to soil water potential in bars for every day of the growing season for 20 years. He found that actual basal area and volume growth rates of a mixed pine stand were very nearly the same as the potential growth rates calculated from the data for available soil water. Trees have such a long growing season in the southeastern United States, where this study was conducted, that short periods of high transpiration or soil water deficit will have less effect on the season's growth than in the case of annual crop plants.

Even with trees there is sometimes poor correlation between rainfall, soil moisture, and tree growth (Glock and Agerter, 1962, for example). Because of the effect of high temperature on transpiration, more growth may occur in a summer of below average rainfall with temperatures below average than in a summer with high rainfall but also with high temperatures, which cause excessive water loss (Coile, 1936). Also, trees on a wet site may grow more in a dry year than in a wet year, when saturated soil and poor aeration reduce absorption (Fraser, 1962). Furthermore, the rainfall of one season sometimes has measurable effects on diameter growth the following year. The complex relationships between water supply and tree growth are discussed by Kramer (1964) and Zahner (1968). The discussion later in this chapter of various specific effects of water stress on plant processes is equally applicable to woody and herbaceous plants.

Relative availability of soil water

Unfortunately, there has been controversy concerning the point at which soil water begins to limit plant growth. Veihmeyer and Hendrickson (1950) argued that soil water is equally available from field capacity to permanent wilting, while others, whose views are expressed by Richards and Wadleigh (1959), claimed that decreasing availability of soil water affects growth before wilting occurs. Hagan et al. (1959) attempted to bring together and explain these opposing claims. Today it is generally agreed that as the soil content decreases, water becomes progressively less available, and there is no definite point at which it becomes unavailable to plants. The steady decrease in growth with decreasing soil water potential is shown in Fig. 10.21. However, it should be clear by now that measurements of soil water stress cannot always be expected to show good correlations with yield, because plant growth is controlled directly by plant water stress and only indirectly by soil water stress. For example, Denmead and Shaw (1962) found that transpiration and growth of corn were limited by soil water content at a higher level on sunny days with high transpiration than on cloudy days when transpiration rates were low.

How water stress develops

Fairly large water deficits and low leaf water potentials can develop within less than an hour when transpiration is rapid (see Barrs and Klepper, 1968; also Fig. 9.16). However, most injury to plants results from water stress which has developed over a period of several days because of decreasing soil water supply.

Progressive development of stress

Slatyer (1967) carefully analyzed the changes which occur in a transpiring plant as it progressively reduces the water potential of the soil over a period of several days. His treatment is illustrated in Fig. 10.4. This diagram shows a daily cycle in plant water potential which occurs because absorption lags behind transpiration. It also shows that in the absence of added soil water both plant and soil water potential decrease over a period of days until $\Psi_{\text{plant}} = \Psi_{\text{soil}}$ and the plant ceases to absorb water because there no longer is a gradient in water potential from soil to roots. At first, while the soil water potential is relatively high, Ψ_{plant} returns to a value equal to Ψ_{soil} at night, but as Ψ_{soil} and the water conductivity of the soil decrease, this becomes impossible, because the rate of water movement toward the roots is too slow to replace daytime losses. At this point permanent wilting occurs, as shown on days 4 and 5 in Fig. 10.4.

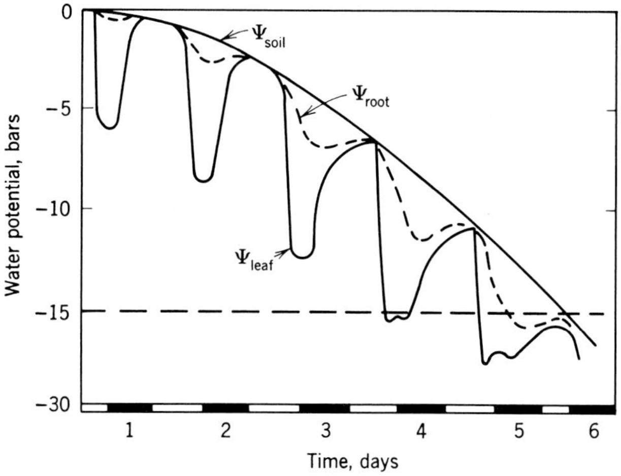

FIG. 10.4 Diagram showing probable changes in leaf water potential Ψ_{leaf} and root water potential Ψ_{root} of a transpiring plant rooted in soil allowed to dry from a water potential Ψ_{soil} near zero (wet soil) to a water potential at which wilting occurs. The dark bars indicate darkness. (*From Slatyer*, 1967.)

The daily cycle in water stress is controlled chiefly by the rate of transpiration, but the long-term decrease in plant water potential and related values is controlled chiefly by soil water potential and soil water conductivity. In soil near field capacity (Ψ_{soil} −1 bar) the hydraulic conductivity is relatively high, and although there is probably some lag in water supply to the roots at midday, only a small gradient in water potential will maintain adequate water flow to plants with well-branched root systems. As a result, recovery to the condition $\Psi_{plant} = \Psi_{soil}$ occurs overnight. However, as soil water content and Ψ_{soil} decrease, the hydraulic conductivity decreases even more rapidly (see Fig. 2.12), so a much larger difference in water potential between root and soil is required to move enough water from soil to roots to replace transpiration losses. Finally, plant water potential falls as low as the osmotic potential, and the loss of turgor is accompanied by wilting and stomatal closure. The latter will reduce transpiration. However, water movement toward the roots has now become so slow that recovery overnight is impossible, and permanent wilting occurs. This analysis indicates that the permanent wilting percentage is controlled by the osmotic potential of the leaves, rather than by soil characteristics (see also Chap. 3, and Slatyer, 1957, 1967). The actual values observed by Gardner and Nieman (1964) resemble the hypothetical values of Fig. 10.4 and support Slatyer's interpretation of the development of water stress and permanent wilting. The close relationship between the values for leaf water potential, root water potential, and soil water potential shown in Table 10.1 also supports this view.

Variations in water content

There are wide variations of water content in various parts of a plant as well as among plants of different species and stages of development. Some data on water content of various tissues and organs were given in Table 1.1. The variations in water content in various parts of woody plants are discussed in detail by Kramer and Kozlowski (1960, pp. 342–359) and by Kozlowski (1964). The sapwood has a higher water content than the heartwood in most species, though exceptions to this are found. According to Stewart (1967), the average water contents for sapwood of ring-porous, diffuse-porous, and nonporous species as percentages of dry weight are less than 75 percent, about 100 percent, and greater than 130 percent respectively. In general, at least in deciduous species, the water content of sapwood approaches a maximum in early summer, decreases to a minimum in late summer, and rises in late autumn and early winter.

Table 10.1 ***Comparison of Root and Needle Water Potential with Decreasing Soil Water Potentials for Loblolly Pine Seedlings Sampled at 11:30 A.M.****

Soil water potential	Root water potential	Leaf water potential
−1	−4	−4
−4	−5	−9
−8	−8	−13
−12	−12	−14
−16	−16	−18

* *From Kaufmann (1968)*

The diurnal changes in water content of tree trunks and other structures cause measurable changes in diameter (see Fig. 10.5), and droughts may cause prolonged periods of no growth or even of shrinkage in the tree trunks (Fig. 10.6). Kozlowski (1967) discussed variations in stem diameter of tree seedlings in some detail.

Leaves and other herbaceous structures may show even wider variations in water content. However, identification of diurnal and seasonal changes in water content of leaves is often obscured by changes in dry weight (Halevy and Monselise, 1963; Pharis, 1967). Because of seasonal increase in dry weight, the water content of leaves generally decreases as the season progresses (see Ackley, 1954). An example of the simultaneous variation in water content in the various organs of sunflower plants is shown in Fig. 10.7. It is suggested

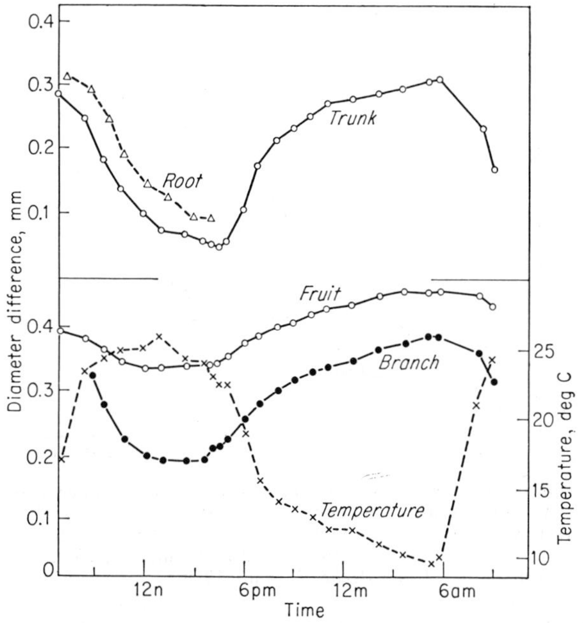

FIG. 10.5 **Midday shrinkage in various parts of an avocado tree caused by water deficit resulting from rapid transpiration. (*From Schroeder and Wieland,* 1956.)**

that because of their relatively large volume, the stems function as a water reservoir on which the leaves draw during periods of rapid transpiration. Tree trunks must be very effective water reservoirs, as indicated by their large diurnal and seasonal variations in water content.

Internal competition for water

The basic factor controlling water movement within plants is the loss of water from the shoots, which produces a gradient of decreasing water potential from roots to leaves. However, this gradient is modified in many ways. Because of differences in exposure, various parts of the shoot lose water at different rates and develop different levels of water deficit and water potential. Also various stages of growth are associated with differences in ability to compete for water. Young leaves and fruits usually obtain water at the expense of older leaves, and the latter usually die first when plants are subjected to severe water stress. Stem tips of tomato continue to elongate even when the leaves are wilting (Slatyer, 1957; Wilson, 1948), but stem elongation of cotton (Balls, 1908) and corn (Loomis, 1934) is stopped. There is also active competition for water between leaves and fruits (Bartholomew, 1926; Furr and Taylor, 1939; Hendrickson and Veihmeyer, 1941; Magness et al., 1935; Tukey, 1964; Schroeder and Wieland, 1956). The enlargement of various

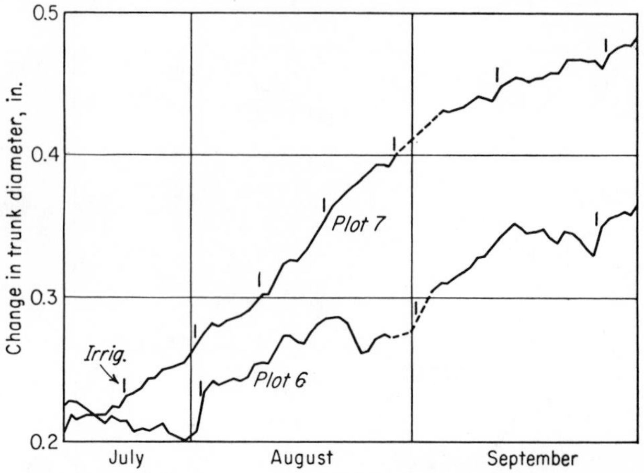

FIG. 10.6 The relationship between soil water stress and diameter growth of avocado trees. Trees in plot 7 were watered whenever the soil matric potential fell to −0.5 bar. Trees in plot 6 were watered when the matric potential fell to −10 bars. The short vertical bars indicate irrigation. Note cessation of growth and shrinkage of trunks before irrigation of trees on dry plot. (*From Richards et al.*, 1958.)

kinds of fruits decreases during the hours when transpiration is rapid and is often more rapid at night than during the day (see Fig. 10.5).

Decrease in fruit enlargement is said to be a good indicator of the development of water stress in lemon trees (Furr and Taylor, 1939). According to Rokach (1953) very young orange fruits do not lose water to leaves, and such loss begins only after the fruits are about 35 mm in diameter. Anderson and Kerr (1943) found a similar situation in young cotton bolls (see Fig. 10.8).

The variations in water content and water potential in various parts of a transpiring plant and the accompanying competition for water create important problems in sampling and measuring the degree of water stress in plants. These will be discussed in the section on measurement of water stress.

Effects of water stress on plant growth

Water stress affects practically every aspect of plant growth, modifying the anatomy, morphology, physiology, and biochemistry. Some of the effects are related to decrease in turgor, some to the decrease in water potential, and perhaps—as claimed by Huber (1965)—some are caused by decrease in osmotic potential. Readers are referred to Crafts (1968), Evenari (1960), Gates (1968), Kozlowski (1964, 1968), Slatyer (1967), Slavik (1965), Stocker (1960), and Vaadia et al. (1961) for surveys of the extensive literature on this topic. Only a few important effects on growth and related processes will be discussed in this section. Many other effects of water stress are discussed elsewhere in this book.

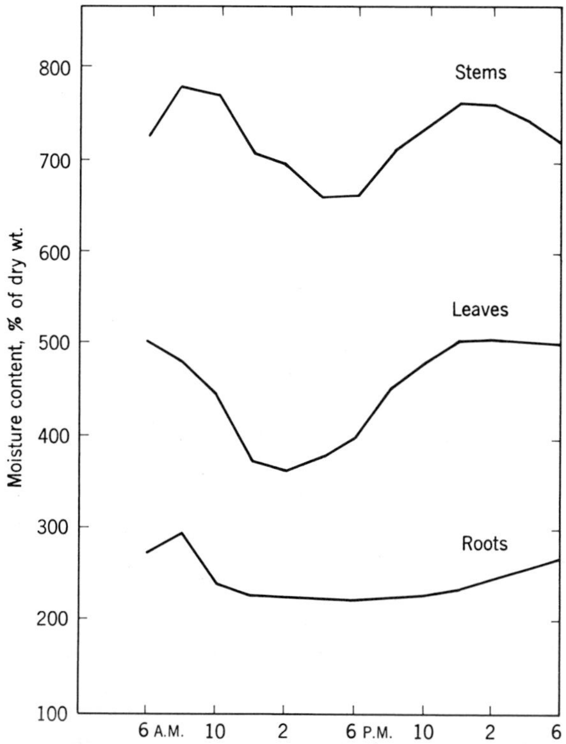

FIG. 10.7 Diurnal variation in water content of roots, stems, and leaves of sunflower plants growing in moist soil on a hot summer day. (*Wilson et al.*, 1953.)

General effects of water stress

Everyone is aware of the general reduction in size of plants subjected to prolonged water stress. Although reduced cell turgor is the most important reason for reduced plant size, water stress affects nearly every process in a plant, and other factors in addition to turgor are involved. Turgor pressure is low in enlarging cells, but some minimum level of turgor is necessary for cell expansion. The relationship of turgor to cell enlargement is discussed later on in the section on water stress and cell enlargement. Turgor is also important in relation to the opening and closing of stomata, expansion of leaves and flowers, and various movements of plant parts. On the other hand enzyme-mediated processes are presumably controlled more directly by the water potential. It is sometimes questioned whether a reduction of a few bars in water potential can cause enough change in structure of proteins to modify enzyme action (Gale et al., 1967). However, it will be seen later that both carbohydrate and nitrogen metabolism are modified by water deficits of only a few bars, and the most obvious explanation seems to be that changes in structure of proteins caused by decrease in water potential affect enzyme

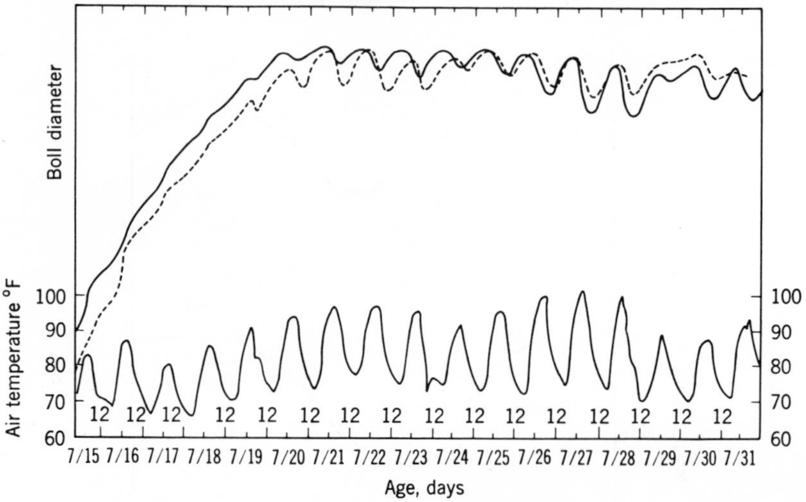

FIG. 10.8 **Variation in diameter of two cotton bolls during the period 11 to 27 days after flowering. Young bolls do not shrink during the midday water deficit, but older bolls show the midday shrinkage often observed in other kinds of fruits. Rain prevented shrinkage on July 29. (*From Anderson and Kerr,* 1943.)**

activity. Other possibilities involve disturbance of the fine structure in cells and diversion of food from normal metabolic pathways. Walter (1963, 1965) has emphasized the importance of osmotic potential, pointing out that the protoplasm in cells must be in osmotic equilibrium with the solutes. Thus, protoplasm will be less hydrated in cells with an osmotic potential of −20 bars than in cells with an osmotic potential of −10 bars, although the water potential of both is zero if they are fully turgid. It seems impossible at present to evaluate the relative importance of water potential and osmotic potential in plant metabolism. Slavik (1965) cites Russian and German work indicating that the decrease in yield of several crops is quantitatively related to the decrease in osmotic potential of the plant sap. Kreeb (1957) reported that the grain yield of barley decreased about 60 kg/hectare and the yield of straw about 1,000 kg/hectare per bar decrease in mean osmotic potential. Vegetables showed even greater decreases in yield per unit decrease in osmotic potential. However, a small decrease in osmotic potential may be accompanied by a large decrease in turgor and water potential (see Fig. 1.11), and the latter might be the controlling factor.

Water stress usually has multiple effects on plant growth. For example, photosynthesis is reduced by closure of stomata, which decreases the supply of carbon dioxide, but water stress also reduces the capacity of the protoplasm to carry on photosynthesis, and reduced translocation might hinder it by accumulation of end product. The reduction in photosynthesis, decreased translocation of carbohydrates and growth regulators, and disturbance of nitrogen metabolism all add to the effects of reduced turgor in reducing

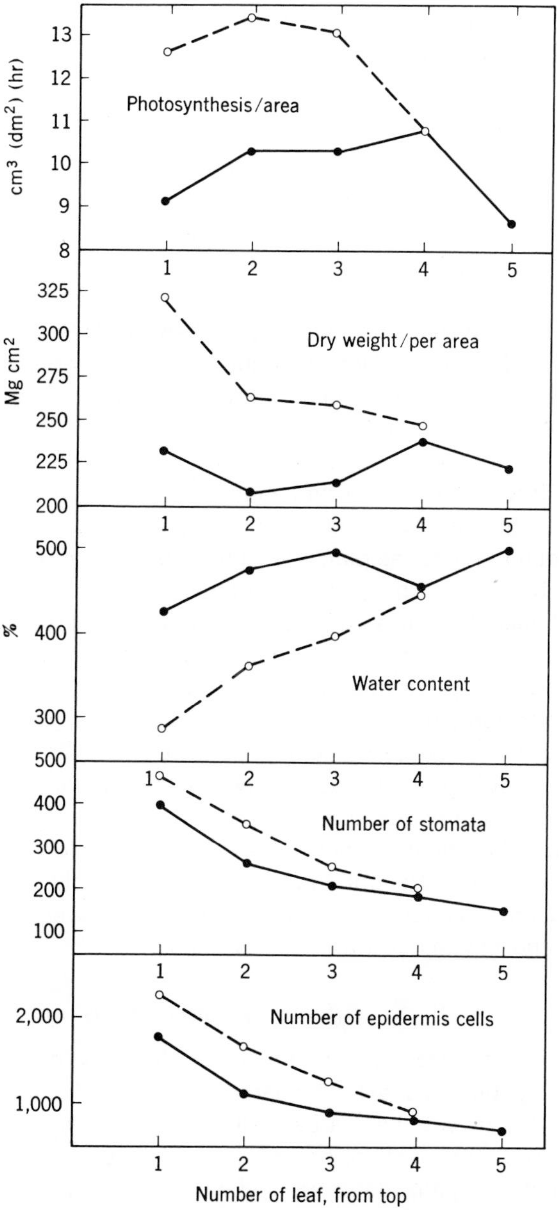

FIG. 10.9 Effects of water stress on leaf characteristics of *Trifolium incarnatum* grown in soil kept near 80 percent (—●—●—) and 30 percent (—o—o—) of field capacity. (*From Stocker,* 1960; *after Simonis.*)

growth. In turn, reduced growth reduces the photosynthetic surface, further decreasing the relative amount of carbohydrate available for growth, as compared with unstressed plants.

The stage in plant growth at which water stress occurs

There is general agreement that water stress at certain critical stages in plant growth causes more injury than at other stages. The critical period usually comes at the time reproductive organs are formed and pollination and fertilization occur. The shedding of bolls by water-stressed cotton is well known, and severe water stress at the time of silking and tasseling of corn can drastically reduce yields (Denmead and Shaw, 1962; Robins and Domingo, 1953). Slavik (1965) cites Russian work showing the injurious effects of water stress on yield of grain when it occurs during tillering, heading, and anthesis. According to W. V. Brown (1952), water stress at the time of flower initiation greatly increases the percentage of cleistogamous flowers in *Stipa leucotricha*. D. S. Brown (1952) reported that water stress in the late summer reduces the number of flower buds formed by apricots. On the other hand, Alvim (1960) found that coffee plants must be subjected to water stress before flowering can be induced by rain or irrigation.

In view of these facts it is obvious that water stress produces quite different effects at different stages in the growth cycle. For example, irrigation of sugar beets at Davis, California, did not increase yield and even resulted in decreased sugar content. However, irrigation at Logan, Utah, caused large increases in yield and sugar content (Hagan et al., 1959). The difference in results probably occurred at least partly because beets are planted much earlier at Davis than at Logan and therefore have time to develop a deep root system and achieve much of their growth before hot, dry weather occurs. Effects of irrigation and drought can be evaluated reliably only if plant water stress is measured sufficiently often during the growing season to make certain when and how much water stress occurs.

Effects of water stress on plant structure

It is well known that plants subjected to water stress not only show a general reduction in size but also exhibit characteristic modifications in structure, particularly of the leaves. Leaf area, cell size, and intercellular volume are usually decreased. Cutinization, hairiness, vein density, stomatal frequency, and thickness of both palisade layers and entire leaves are usually increased. This often results in the relatively thick, leathery, highly cutinized type of foliage generally described as xeromorphic. Nearly all kinds of plants show at least some of these modifications when subjected to water stress. Shading of tobacco to produce large thin leaves for cigar wrappers is an example of decreasing water stress to the minimum in order to produce a desired type of leaf structure. So sensitive is leaf growth to water stress that it has been recommended for use as an indicator of the need for irrigation (Higgins et al., 1964).

The differences in leaf structure from the lower to the upper part of a tree and between sun and shade leaves are often attributed to differences in leaf water stress (Farkas and Rajhathy, 1955; Yapp, 1912). However, not all xeromorphic types of leaves are developed in dry habitats. Stocker (1960), for example, cites instances where an excess of soil water caused xeromorphic modifications in leaf structure. This might occur because an excess of soil water reduced absorption and caused leaf water stress. The xeromorphic structure of leaves of plants growing in bogs is often attributed to physiological drought, but this is questioned by Caughey (1945). This type of xeromorphism is more probably related to deficiency of nitrogen or other nutrients (Albrecht, 1940; Mothes, 1932; Stocker, 1960). More research is needed on this problem. Readers are referred to the section on leaf structure in Chap. 9 for further discussion.

Killian and Lemée (1956) and Stocker (1960) summarized considerable literature on the interrelationships between environment and plant structure.

Water stress at the cellular level

The water status of plants as a whole is controlled by cell water stress. The osmotic aspects and terminology of cell water relations were discussed in Chap. 1. Here we shall concentrate our attention on effects of water stress on cell division, cell turgor, and protoplasmic properties.

WATER STRESS AND CELL DIVISION. In meristematic tissues the principal effects of water deficits are presumably on synthetic activities, e.g., synthesis of DNA, RNA, and cell wall materials. However, some minimum level of turgor is essential for cell enlargement. The susceptibility of meristematic regions to water stress appears to vary among species. Balls (1908), working with cotton, Loomis (1934) with corn, and Thut and Loomis (1944) with corn and other species, all found shoot elongation completely inhibited during periods of high transpiration. In contrast, Wilson (1948) and Slatyer (1957) found elongation of tomato stem tips continuing even when the leaf tissue was wilted. The reasons for these differences in behavior deserve further investigation.

Cell division seems to be affected less by water deficit than cell elongation. According to Gardner and Niemann (1964), the DNA content of cotyledonary leaves of radish is reduced to about 40 percent of the control leaves at a leaf water potential of −2 bars and to 20 percent at −8 bars. Further decrease in leaf water potential did not reduce DNA appreciably more. As amount of DNA is related to cell number, these observations indicate that

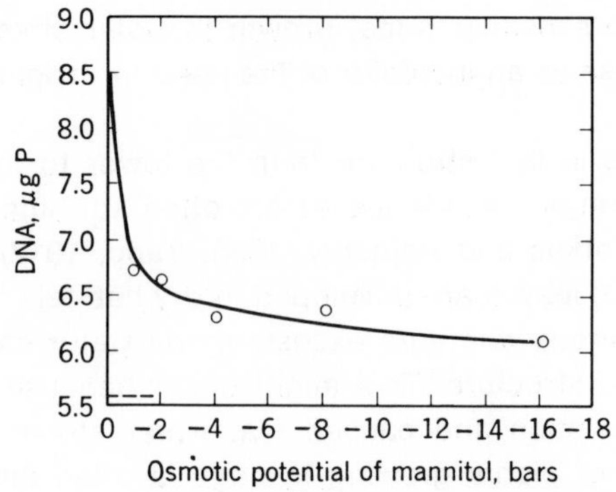

FIG. 10.10 **DNA content of detached cotyledonary radish leaves grown for 28 hr on media of various osmotic potentials. The dashed line indicates DNA content at the beginning of the experiment. Although DNA synthesis was drastically reduced at −1 bar, some synthesis occurred at −16 bars, indicating a limited amount of cell division. (*From Gardner and Niemann*, 1964.)**

cell division did not cease even when cell turgor fell to zero. Gates and Bonner (1959) found that RNA synthesis continued in tomato plants subjected to water stress, further supporting the idea that cell division continued.

WATER STRESS AND CELL ENLARGEMENT. Reduction in turgor causes reduction in cell enlargement, which in turn decreases shoot and root elongation and leaf enlargement. It also interferes with other processes dependent on cell turgor, such as stomatal opening. The relationship between turgor and cell enlargement has been studied by many workers (Brouwer, 1963; Burström, 1953; Ordin, 1958, 1960; Ordin et al., 1956; Probine and Preston, 1962). An example of the effect of decrease in turgor on expansion of bean leaves is shown in Fig. 10.11. In this experiment leaf osmotic potential de-

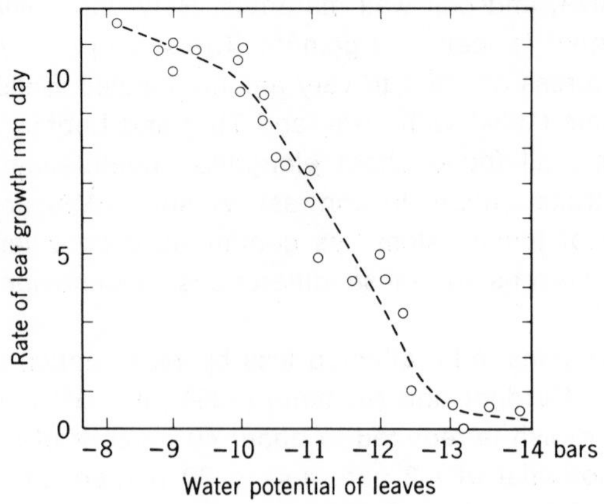

FIG. 10.11 **Rate of growth of bean leaves plotted over leaf water potential. (*From Brouwer*, 1963.)**

creased 1 bar, water potential decreased from −8 to −14 bars, and the change in turgor pressure over the range of water potentials used was from 5 at −8 to zero at −14 bars. Wadleigh and Gauch (1948) and others have demonstrated close relationships between decreasing turgor and decreasing enlargement of plant organs. Boyer (1968) reported that sunflower leaf enlargement ceased when the leaf water potential fell below −3.8 bars and the turgor pressure below 6.8 bars.

Ordin (1958, 1960) attempted to separate the effects of reduced turgor from the effects of decreased osmotic potential on enlargement of oat coleoptiles. He imposed the same water potential with different internal values of osmotic potential and turgor by producing water stress with mannitol, which is little accumulated, and sodium chloride, which is accumulated freely. As the greatest reduction in cell enlargement occurred with mannitol, which caused the greatest reduction in turgor, but little change in osmotic potential, it appeared that turgor was more important than osmotic potential. Ordin (1960) thought that reduced turgor pressure affected both enlargement and cell wall metabolism, possibly through cellulose synthesis. According to Whitmore and Zahner (1967), the ability of young xylem tissue to incorporate ^{14}C-labeled glucose is greatly decreased by decreasing water potential.

PROTOPLASMIC EFFECTS OF WATER STRESS. It is probable that most of the effects of water stress, other than those caused directly by loss of turgor, can be attributed to dehydration of the protoplasm. Removal of part of the water surrounding protein molecules may cause changes in configuration affecting permeability, hydration, viscosity, and enzyme activity (Klotz, 1958; Tanford, 1964). Gaff (1966) reported that dehydration of protein from cabbage leaves caused changes in amount of reactive sulfhydryl which he attributed to changes in configuration of the protein. According to Chen et al. (1964), relatively small changes in osmotic potential cause marked changes in protein structure and enzyme activity.

The literature in this field was reviewed by Stocker (1960), who presents the divergent views on this rather confused subject. Stocker treats dehydration as occurring in two stages—the reaction phase, when plants are first subjected to water stress, and the restitution and hardening phase, which occurs if water stress lasts several days. These two phases are said to be exhibited both by protoplasmic structure and by physiological processes such as respiration. The reaction phase is characterized by decrease in viscosity of protoplasm, increased permeability to water, urea, and glycerine, increased proteolysis, and increased respiration. If water stress continues, there is a restitution stage involving increase in viscosity above the original value, decrease in permeability to water and urea, and decrease in physiological processes such as respiration. If the plants are rewatered before permanent injury occurs, these processes are reversible, and conditions often return toward normal. However, in some instances the original condition is not recovered. Occurrence of the course of events outlined by Stocker depends on gradual drying and recovery, and the two stages are not always clearly defined. An example of protoplasmic changes in plants subjected to water stress is shown in Fig. 10.12. According to Henckel (1964), Russian physiologists support Stocker's concept of a two-stage response to water stress.

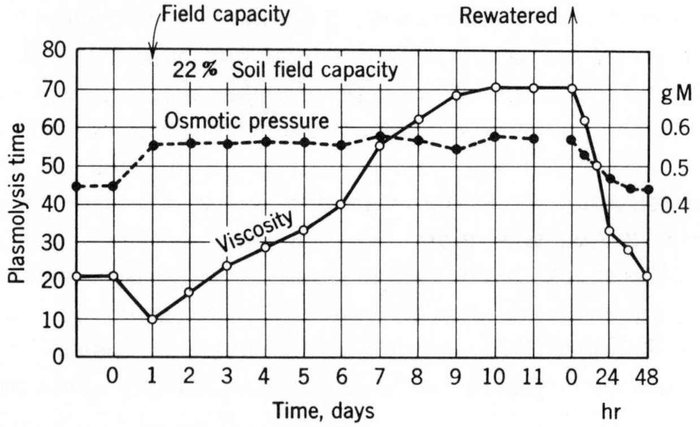

FIG. 10.12 Changes in viscosity of protoplasm of *Lamium maculatum* during dehydration (reaction phase) and rehydration (restitution phase). At day 0, the soil began to dry and was said to have been kept at 22 percent of field capacity until day 11, when it was rewatered. (*After Stocker and Ross*, 1956.)

Some investigators claim that injuries from drought, heat, and freezing are similar in nature but this is questioned by others (Larcher, 1963). Levitt (1956) summarized considerable evidence in support of this view. Later he suggested that tolerance of heat, cold, and dehydration is related to resistance against oxidation of sulfhydryl (—SH) to disulfide (—SS) (Levitt, 1962). Gaff (1966) reported that there is a considerable decrease in reactive —SH in the soluble protein from cabbage leaves subjected to water stress. There is also considerable conversion of —SH to —SS at the point where death from desiccation occurs. Gaff attributed injury and death to degradation of lipoprotein membranes caused by changes in conformation of structural protein molecules. Russian investigators likewise regard heat and drought resistance as closely related and claim that dehydrated plants suffer from the effects of high temperatures because of their reduced transpiration. They also regard resistance as closely related to protoplasmic properties such as viscosity, hydration, and permeability and emphasize the possibility of increasing protoplasmic resistance to dehydration by prior treatments such as soaking of seeds in concentrated salt solutions. The Russian literature was reviewed by Henckel (1964). Protoplasmic effects of water stress are also discussed later in this chapter in the section on hardening.

Effects of water stress on photosynthesis and respiration

Water stress can reduce photosynthesis by reduction in leaf area, closure of stomata, and reduction in activity of the dehydrated protoplasmic machinery. Some investigators claim that the most serious effect of drought is to reduce the photosynthetic surface and the production of dry matter. However, reduction in rate per unit of surface is also important. The large decrease in photosynthesis per unit of leaf area which occurs in plants subjected to water stress is usually attributed to stomatal closure. This view is supported by the fact that transpiration and photosynthesis are often reduced to about the

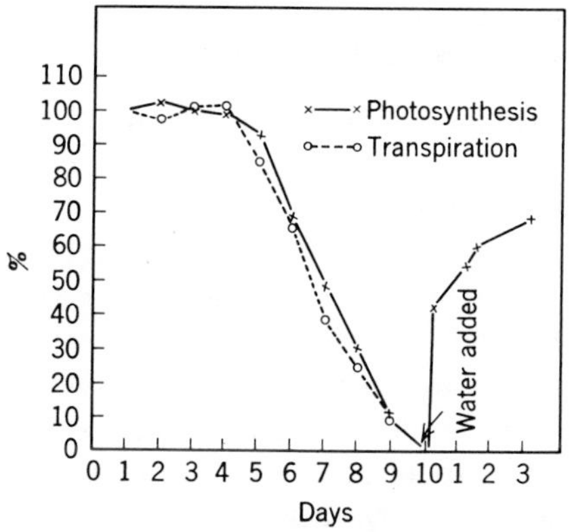

FIG. 10.13 **Simultaneous decrease in rates of photosynthesis and transpiration of tomato plants subjected to increasing water stress by withholding water, and the return to normal when the soil was rewatered. (*From Brix,* 1962.)**

same extent (see Figs. 10.13 and 10.14). However, as Slatyer (1967, pp. 293–294) and others point out, there is supposed to be an additional mesophyll resistance in the carbon dioxide pathway, so a given degree of stomatal closure should reduce transpiration more than carbon dioxide uptake. Barrs (1968) questioned this, because he found that photosynthesis and transpiration varied to the same extent over a wide range of rates of gas exchange (see Figs. 10.13 and 10.14).

Experiments with aquatic species in water and with plants such as lichens and mosses, which lack stomata, also show reduction in photosynthesis with increasing water stress (Ensgraber, 1954; Greenfield, 1942; Slavik, 1965; Stocker and Holdheide, 1937). Boyer (1965) found a substantial decrease in net photosynthesis of cotton subjected to high water stress under conditions such that the stomatal resistance was not increased (see Fig. 10.15). It therefore seems clear that decreased hydration of the protoplasm directly reduces photosynthesis.

In general, the rate of apparent photosynthesis begins to decrease at a

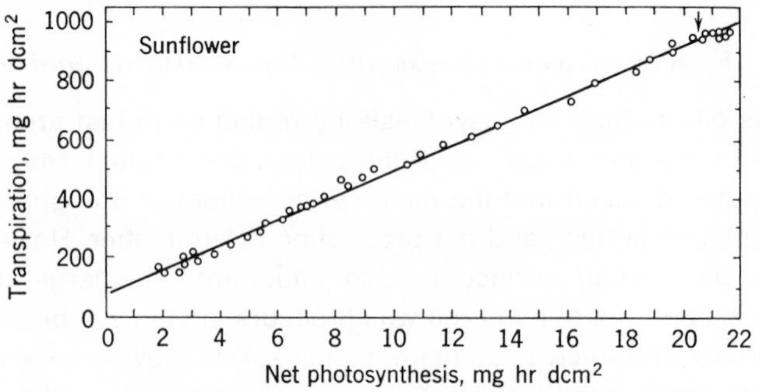

FIG. 10.14 **Relation between transpiration and photosynthesis of a drying sunflower leaf. Points to the left of the arrow were obtained after excision of the leaf, which had then begun to dry. These data suggest that both photosynthesis and transpiration decrease at the same rate, probably because of stomatal closure. (*From Barrs,* 1968.)**

water deficit of a few bars and ceases at approximately zero turgor (Bourdeau, 1954; Brix, 1962; Loustalot, 1945), or even falls below the compensation point if respiration exceeds photosynthesis. Occasionally a small increase in photosynthesis occurs between full turgor and the point at which the rate begins to decrease (see Fig. 10.16). Stålfelt (1935) attributed this to increased stomatal aperture following a small loss of turgor by the leaf cells. There is usually more delay in the return to normal of photosynthesis than of transpiration when stressed plants are watered (see Fig. 10.17). This suggests that considerable time is required for the protoplasm to regain full photosynthetic capacity after it is dehydrated.

Stocker (1960) cited work indicating that plants subjected to repeated wilting or prolonged water stress have a higher rate of photosynthesis per

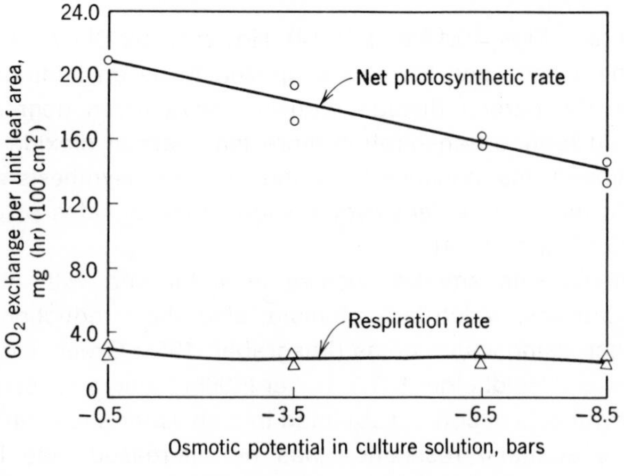

FIG. 10.15 **Photosynthesis and respiration of cotton plants growing in solutions of various osmotic potentials. The stomata remained open in all plants, and the reduction in photosynthesis at low water potentials was therefore attributed to protoplasmic effects of the reduced water potential. (*From Boyer,* 1965.)**

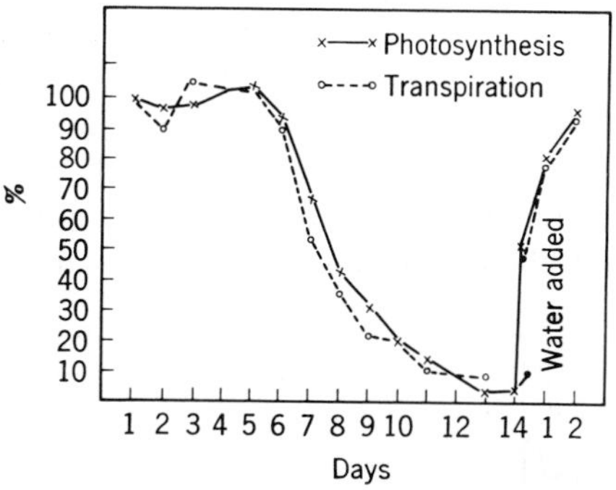

FIG. 10.16 Changes in rate of transpiration and photosynthesis of loblolly pine with decreasing soil water content and after rewatering the soil. Note the small increase in rates before the decrease began. This is attributed by Stålfelt to wider opening of stomata accompanying a slight loss in turgor. Rates are expressed as percentages of rates with soil at field capacity. (*From Brix,* 1962.)

unit of leaf area than plants not subjected to water stress. He regarded the higher rate of photosynthesis in leaves produced under water stress as evidence of the restitution stage of response to stress. It seems more likely that the difference is caused by modification of leaf structure or stomatal behavior than by protoplasmic modifications.

The effects of water deficit on respiration are variable. In some experiments there has been a temporary increase in respiration, followed by a

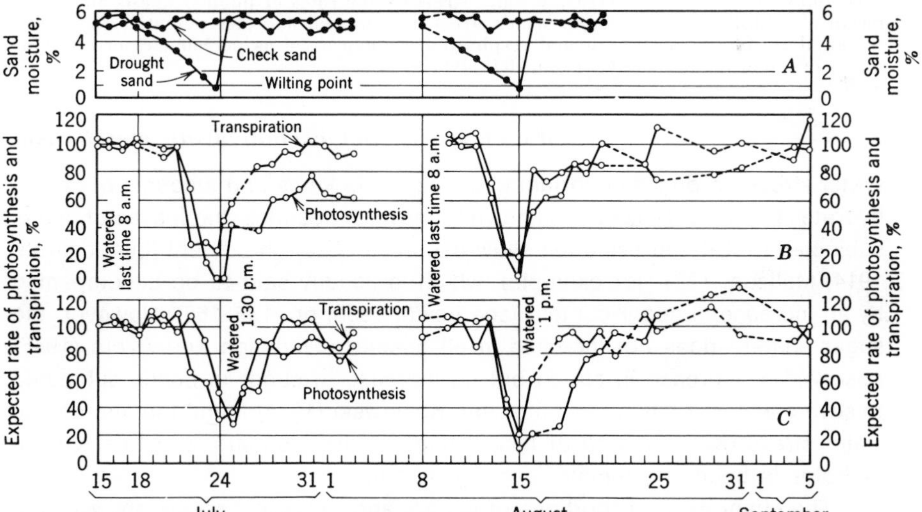

FIG. 10.17 Effects of decreasing soil water content on photosynthesis and transpiration of pecan seedlings growing in sand: A, water content of the sand; B, rates of photosynthesis and transpiration during the afternoon, expressed as percentages of rates of well-watered plants; C, similar rates for the morning. Effects of dry soil were greater in the afternoon, because there was a greater leaf water deficit in the afternoon. (*From Loustalot,* 1945.)

decrease as more severe water stress developed (Brix, 1962; Parker, 1952). This may result from an increase in substrate caused by the hydrolysis of starch to sugar sometimes observed in plants subjected to water stress. In most experiments respiration decreases steadily with increasing water stress. However, it often decreases more slowly than photosynthesis, leading to further decrease in net photosynthesis under water stress.

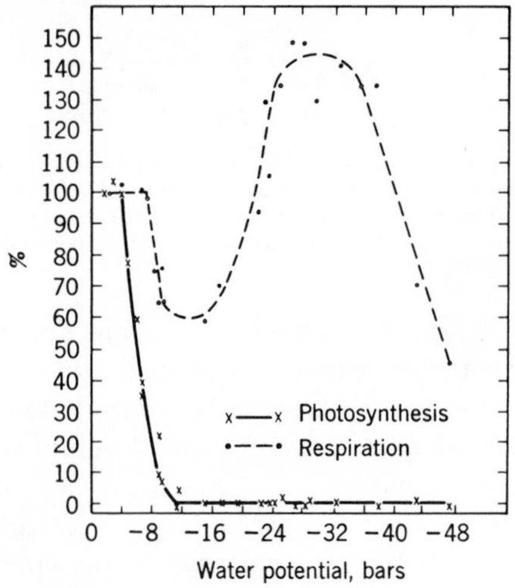

FIG. 10.18 **The effect of increasing water stress on the rates of photosynthesis and respiration of loblolly pine seedlings. Water stress is expressed as water potential Ψ_w of the leaves. Rates of photosynthesis and respiration are expressed as percentages of rates for plants in soil at field capacity. (*From Brix,* 1962.)**

Water stress and carbohydrate metabolism

Water stress produces important changes in the kinds and amounts of carbohydrates in plants. It has been known for several decades that leaves of plants subjected to water stress often show decrease in starch content (Lundegårdh, 1914; Molisch, 1921, for example), which is usually said to be accompanied by an increase in sugar content (see Levitt, 1956, pp. 166–168). However, the sugar content does not increase in all species. Wadleigh and Ayers (1945) observed a decrease in starch but no increase in sugar in beans subjected to severe water stress. Woodhams and Kozlowski (1954) found that increasing water stress over a period of time reduced starch, sugars, and total carbohydrates in bean and tomato (see Fig. 10.19).

The changes in proportions of sugars and polysaccharides are presumably related to changes in enzyme activity. Spoehr and Milner (1939) reported that the amylase activity increased in leaves subjected to water stress. Eaton and Ergle (1948) reported that cotton leaves allowed to wilt daily contained four times as much amylase activity but only one-third as much starch and

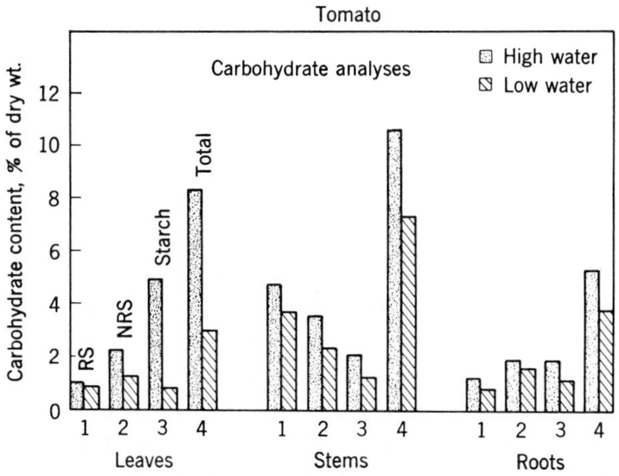

FIG. 10.19 **Changes in carbohydrate content of two groups of tomato plants after 8 weeks in soil maintained near field capacity (high water) and in soil subjected to four periods without irrigation. RS, reducing sugars; NRS, nonreducing sugars.** (*From Woodhams and Kozlowski,* 1954.)

one-half as much total carbohydrate as well-watered controls. According to Spoehr (1919), decreasing water content causes accumulation of polysaccharides in cacti, and stress is said to cause accumulation of pentosans in some kinds of plants (Rosa, 1921). Evidently, there are important differences among species in the effects of water stress on carbohydrate metabolism. The reaction is complicated by the fact that respiration often decreases more slowly than photosynthesis, causing depletion of food reserves and change in the proportions of various carbohydrates.

Water stress and nitrogen metabolism

Several investigators have reported that considerable hydrolysis of proteins occurs in wilted plants and is accompanied by increase in amino acids (Barnett and Naylor, 1966; Chen et al., 1964; Chibnall, 1954; Mothes, 1956; Petrie and Wood, 1938). Chen et al. (1964) observed three stages in citrus seedlings subjected to increasing soil water stress over a period of 10 days. In the first stage, prior to wilting, there was a rapid drop in water content and a small increase in protein, possibly because RNA synthesis continued. During wilting there was much hydrolysis of protein. Finally an apparent increase in protein occurred, although this may have been an increase in peptides rather than protein. Barnett and Naylor (1966) found a general decrease in soluble protein in water-stressed Bermudagrass, also a decrease in protein-bound arginine. The amount of proline increased greatly and appeared to function as a storage compound in stressed plants. Kudrev (1967) reported large increases in free proline content of wilted pumpkin plants and a smaller increase in free glutamic acid.

Shah and Loomis (1965) reported that synthesis of RNA and protein are decreased in sugar beet before visible wilting occurs. Water stress and aging both produce considerable changes in the course of protein synthesis, and it seems possible that water stress greatly accelerates the changes normally associated with senescence. Gates and Bonner (1959) reported that RNA synthesis continued in water-stressed tomatoes, but there was a decrease in total RNA caused by rapid breakdown. According to Dove (1967) an increase in ribonuclease activity occurs in tomato leaves subjected to water stress. West (1962) reported that water stress caused an increase in RNA content of germinating corn seedlings subjected to water stress. However, very young seedlings are more resistant to water stress than older plants. According to Todd and Yoo (1964), there is a decrease in protein content and in the activity of several enzymes in detached wheat leaves subjected to water stress. More research is needed on the effects of water stress on various aspects of nitrogen metabolism.

Water stress and growth regulators

Little information is available concerning the effects of water stress on the synthesis or transport of growth regulators. However, in view of its effects on carbohydrate and nitrogen metabolism, it seems likely that the synthesis of growth regulators is also affected. Larson (1964) suggested that water stress inhibits auxin formation in the stem tips of trees, reducing or cutting off the supply to the cambium and modifying cambial activity. Itai and Vaadia (1965) reported that there is significantly less cytokinin activity in the xylem exudate from roots of sunflowers subjected to water stress than in exudate from unstressed control plants. Ben-Zioni et al. (1967) found that the capacity of disks of tobacco leaf tissue to fix *l*-leucine labeled with ^{14}C into proteins was decreased about 50 percent if the plants from which the disks were cut had been subjected to water stress. If plants were allowed to recover from water stress for 72 hr, the disks recovered the ability to incorporate leucine. Also, if disks from stressed leaves were treated with kinetin before being incubated with leucine, they recovered part of their ability to incorporate leucine. These results suggest that cytokinins from roots are important in leaf metabolism and that water stress reduces the supply. There is also a possibility that roots supply gibberellins to the shoots (Jones and Lacey, 1968; Skene, 1967).

The possibility that water deficit reduces the supply of growth-regulating substances supplied to the shoot by the roots provides plausible explanations of certain general effects of water stress. For example, Meyer and Gingrich (1964) found that when only a part of the root system of a wheat plant was subjected to a mild water deficit (−1 bar), growth and metabolism were materially affected. The reduction in growth which occurs when the water potential of the root substrate is decreased by addition of salt cannot be explained solely in terms of reduced water absorption (Bernstein, 1961; Slatyer, 1961). Perhaps reduced synthesis of growth regulators such as cytokinins and gibberellins in the roots is an important factor in the reduction of growth observed in plants subjected to water stress. It might also be a factor in the rapid senescence of leaves on plants subjected to water stress. This problem deserves further investigation.

Water stress and translocation

There appears to be considerable reduction in translocation of organic compounds in plants subjected to water stress. There are also changes in the pattern of translocation. The literature on this subject was reviewed by Roberts (1964) and Wiebe and Wihrheim (1962). Unfortunately, the degree of water stress was measured quantitatively in only a few of these experiments. Wiebe and Wihrheim (1962) reported that translocation of ^{14}C-labeled photosynthate out of sunflower leaves was reduced about one-third as the water potential was decreased to −10 or −12 bars. Hartt (1967) found that water stress reduced translocation of ^{14}C-labeled photosynthate out of leaves more than it reduced formation of labeled photosynthate. Translocation of ^{14}C-labeled 2,4-D out of bean leaves was reduced greatly as the relative water content was reduced below 80 percent. According to Roberts (1964), translocation of ^{14}C-labeled photosynthate out of leaves of yellow-poplar is

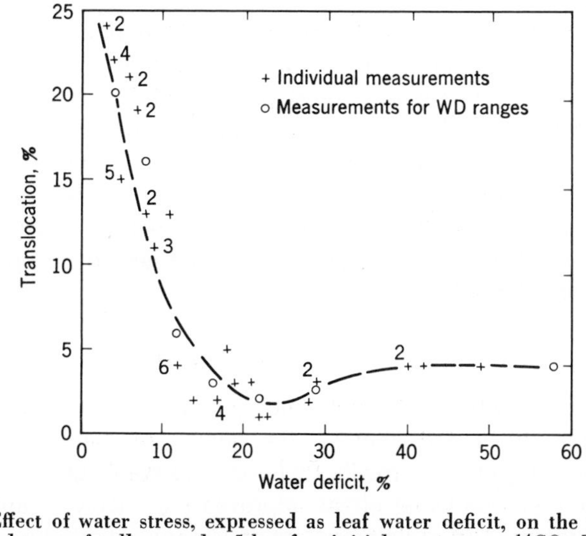

FIG. 10.20 **Effect of water stress, expressed as leaf water deficit, on the translocation of ^{14}C out of the leaves of yellow-poplar 5 hr after initial exposure to $^{14}CO_2$. The curve shows the percentage of total ^{14}C in counts per minute which occurred outside the exposed leaf. Crosses represent the average of the number of observations indicated. Circles indicate water deficit measurements. (*From Roberts,* 1964.)**

reduced to a very low level at a water deficit of 16 to 20 percent (see Fig. 10.20). Gates (1955) reported disturbance of normal patterns of translocation for both organic and inorganic constituents in wilted tomato plants.

It seems possible that the reduction in translocation of photosynthate out of leaves might be a factor in the reduction of photosynthesis observed in plants subjected to water stress. Reduction in translocation of auxin and other growth regulators may be as important as reduction in translocation of carbohydrates, but this has not been investigated. As mentioned earlier, water stress causes reduction in translocation of herbicides (Basler et al., 1961; Pallas and Williams, 1962).

Water stress in relation to disease and insect resistance

There is evidence that resistance to diseases and insect attacks is sometimes related to the level of plant water stress. In fact, J. Parker (1965) treats water stress itself as a physiological disease and lists various symptoms of drought injury. Bier (1959) reported that the fungus causing bark cankers in willow invades the bark only when the relative water content (relative turgidity) is below 80 percent. He found a similar relationship with respect to bark canker of poplar (*Populus trichocarpa).* A. F. Parker (1961) reviewed considerable literature on the relationship between water content and disease and concluded that development of bark cankers is usually correlated with decreased water content of the bark. The incidence of blossom-end rot of tomato fruits is said to be higher on plants subjected to severe water stress (Carolus et al., 1965).

There is also evidence that attacks on trees by boring insects which live in the inner bark and outer wood are more severe in dry seasons than in seasons when little water stress develops (Vité, 1961). According to Vité, infestation of ponderosa pine by beetles is much more severe in trees suffering from water deficits than in well-watered trees. Lorio and Hodges (private communication) reported that loblolly pine trees on dry sites are more subject to beetle attacks than trees on moist sites. Trees with low water stress have high oleoresin exudation pressures, which seem to be unfavorable to the establishment of beetles.

Beneficial effects of water stress

Water stress is not always entirely injurious. Under some circumstances moderate water stress can improve the quality of plant products, even though it reduces vegetative growth. Richards and Wadleigh (1952) cite work indicating that moderate water stress improves the quality of apples, pears, peaches, and prunes. The protein content of wheat is said to be increased by water stress during maturation. Although water stress decreases total vegetative growth, it generally increases the rubber content of guayule, as shown

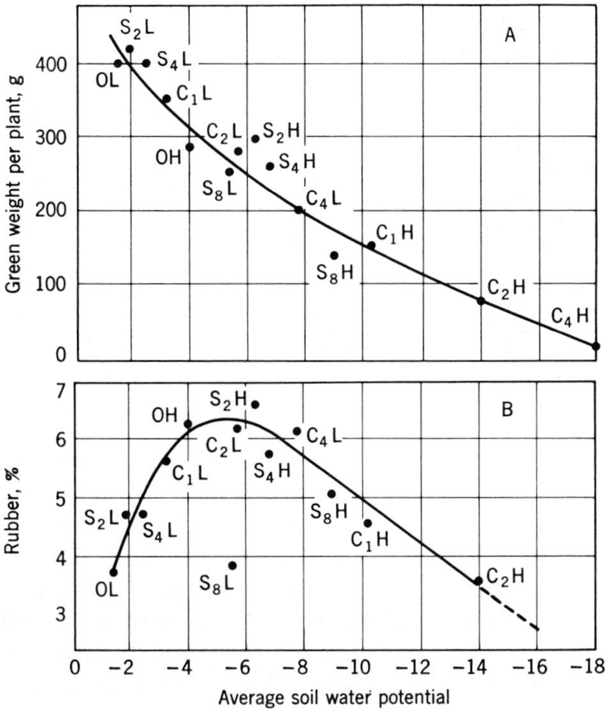

FIG. 10.21 Relationship between average moisture stress developed between irrigations and (A) the fresh weight of guayule plants and (B) the percentage of rubber in the plants. (*From Wadleigh et al.*, 1946.) The symbols refer to various treatments as follows:

O = no salt added	S_2 = 0.2% Na_2SO_4 added
C_1 = 0.1% NaCl added	S_4 = 0.4% Na_2SO_4 added
C_2 = 0.2% NaCl added	S_8 = 0.8% Na_2SO_4 added
C_4 = 0.4% NaCl added	

L = Low-tension series. Water was added to bring soil back to field capacity when a matric potential of 300 to 400 cm of water was indicated by tensiometers.

H = High-tension series. Water was added when the average water content fell to the permanent wilting percentage.

in Fig. 10.21. This situation explains why early attempts to cultivate guayule with heavy irrigation resulted in failure because of low rubber content.

Water stress increases the desirable aromatic properties of Turkish tobacco (Wolf, 1962). However, water stress increases the nitrogen and nicotine content of cigarette tobacco, which is undesirable (van Bavel, 1953). Evenari (1960) cites claims that water stress increases the alkaloid content of *Atropa belladonna, Datura,* and *Hyoscyamus muticus.* However, Loustalot et al. (1947) reported that water stress decreased the alkaloid content of *Cinchona ledgeriana.* It is also said that water stress increases the essential oil content of mint and the oil content of olive fruits (Evenari, 1960). It also increases the percentage of oil in soybean seed but decreases the yield of oil per acre (Miller and Beard, 1967).

Drought resistance

The meaning of the term drought resistance is so debatable that Evenari (1960) refused to use it. However, we shall define it as referring to the various means by which plants survive periods of environmental water stress. Basically, plants are drought-resistant either because their protoplasm is able to endure dehydration without permanent injury or because they possess structural or physiological characteristics which result in avoidance or postponement of a lethal level of desiccation. Parker (1968) described the various mechanisms which contribute to drought resistance in plants.

Desiccation tolerance

Many mosses and lichens and a few ferns and seed plants possess protoplasm which can be dehydrated to or nearly to the air-dry condition without being killed. Examples of seed plants are various desert grasses and shrubs such as sagebrush (*Artemisia* spp.) and creosote bush (*Larrea tridentata*). There are also differences in tolerance of dehydration among various species of grasses, shrubs, and trees of less arid habitats, differences which are very important where survival is more important than yield. Perhaps an outstanding example is olive, which thrives where it is too dry for any other kind of fruit tree because its leaves tolerate extreme dehydration.

Differences in tolerance of desiccation are of minor importance for crop plants because by the time they have been dehydrated to the survival level the crop is likely to be a total loss. Interest centers on the yield of crop plants rather than on their survival, and yield begins to be decreased by soil water potentials of only -1 or -2 bars, or long before they are in danger of death by desiccation. However, the ability to resume growth after a period of water stress possessed by sorghum and certain other plants is a valuable characteristic. The reasons for this ability deserve further study.

Desiccation avoidance or postponement

Much of the rather small degree of drought resistance found in most mesophytes probably results from physiological or morphological characteristics which result in avoidance or postponement of plant water stress. There are three types of this kind of resistance.

ADJUSTMENT OF GROWING SEASON. Many desert annuals germinate, grow, and flower within a few weeks after rains have wetted the surface soil. Such plants complete their life cycle before severe water stress develops. Winter annuals also avoid summer drought, and some species such as certain Mediterranean grasses become dormant during the hot, dry season (McWilliam, 1968). Varieties of crop plants which can be planted early enough to mature before summer droughts develop usually yield much better than later-maturing varieties.

EXTENSIVE ROOT SYSTEMS. One of the most effective safeguards against drought injury is a deep, wide-spreading, much-branched root system, such as that of sorghum. Plants with shallow, sparsely branched root systems, such as potatoes, onions, and lettuce, suffer sooner than deeper-rooted species like alfalfa, maize, and tomato. A combination of a potentially deep-rooting plant and soil conditions favorable for deep rooting is likely to provide conditions advantageous for drought avoidance.

CONTROL OF TRANSPIRATION. The third way of postponing plant water stress is by reduction in transpiration. Some plants such as *Larrea* react to water stress by shedding their leaves. Many plants react by closure of stomata. Both reactions reduce water loss, and plants showing them survive longer than those which do not. Responsive stomata which close promptly at the onset of water stress, combined with heavily cutinized leaves, result in very effective control of transpiration. Tal (1966) described "wilty" tomato mutants which are difficult to grow even in a humid greenhouse because their stomata do not close. Waggoner and Simmonds (1966) found a similar mutant in potato. An outstanding example of combined stomatal and cuticular control of transpiration occurs in pineapple (*Ananas comosus*) (Joshi et al., 1965). More information about the mechanisms controlling water loss from crop plants is needed to provide a sound basis for selection and breeding programs to increase drought resistance.

Efficiency of water use

Efficiency of water use in terms of units of water used per unit of dry matter produced is important, especially where there is a limited supply of water. According to Slatyer (1964), the efficiency varies from 200 to 500 for high-yielding crops to 2,000 or more for sparsely vegetated arid lands. In general, the higher the yield of dry matter the higher the efficiency, because dry matter production increases more rapidly than water loss. Therefore, the efficiency of water use is increased by use of high-yielding, deep-rooted varieties grown at optimum density with adequate fertilization. However, even under optimum conditions 200 to 500 units of water are used to produce 1 unit of dry matter. This is because efficient photosynthetic structures permitting entrance of large amounts of carbon dioxide permit the exit of large amounts of water vapor. The only exception among cultivated crops is pineapple, which is said to produce as much dry matter per year as sugarcane, but often uses only 10 or 12 percent as much water (Ekern, 1965). The low water loss occurs because the stomata of pineapple are closed most of the day. Pineapple has crassulacean acid metabolism, which enables it to store carbon dioxide as organic acids at night and convert it into carbohydrate during the day, resulting in efficient production of dry matter.

It would be helpful if other useful crop plants with crassulacean metabolism could be found, but this seems unlikely. The use of transpiration-reducing chemicals does not seem practical as yet (see Chap. 9). The most promising method of obtaining increased efficiency of water use, apparently, is not decreasing the use of water but encouraging the production of dry matter. Viets (1962) presented a discussion of the problems involved in bringing about increased efficiency of water use. He points out that maximum efficiency of water use is not always practical because it requires production of the maximum possible yield, which often is unprofitable. For example, the yield may increase with increasing frequency of irrigation, but the yield per unit of water applied may decrease at high rates of irrigation. In some instances fertilization greatly increases efficiency of water use (Viets, 1962). Examples of differences among species and varieties and of the effects of fertilization on efficiency of water use are shown in Table 10.2.

Table 10.2 Effects of Drought and Nitrogen Supply on Efficiency of Water Use by Various Grasses *

(The summer of 1953 had above average rainfall, while 1954 was extremely dry. The figures given are for pounds of water used per pound dry matter produced.)

	50 lb N per acre		200 lb N per acre	
	1953	1954	1953	1954
Common Bermudagrass	6,812	9,738	1,546	4,336
Coastal Bermudagrass	2,478	1,547	803	641
Suwannee Bermudagrass	1,923	1,107	692	452
Pensacola Bahiagrass	2,200	3,103	870	1,293
Pangolagrass	2,249	2,843	2,240	3,016

* *Burton et al. (1957)*

Hardening

It is believed by most writers that sudden development of severe water stress causes more injury than gradual development over a long period of time. Plants subjected to one or more periods of moderate water stress are said to be "hardened," because they usually survive drought with less injury than plants not previously stressed. Some European and Russian investigators attribute hardening chiefly to such protoplasmic changes as increased water-binding capacity and viscosity along with decreased permeability (Henckel, 1964). This view led Russian workers to attempt to increase drought resistance by treatment of seeds before sowing. Before sowing, the seeds are soaked in water then air dried, or soaked in salt solution. Henckel claims good results from this treatment, but attempts to reproduce the results have been disappointing (May et al., 1962). The writer feels that the protoplasmic changes observed in plants subjected to water stress are more often merely results of the stress than adaptations which increase tolerance. However, the protoplasmic basis for tolerance of water stress needs more study, because the protoplasm of some seeds and plants can be dehydrated to the air-dry condition without injury.

The increased ratio of roots to shoots, smaller leaves, thicker cutin, and denser venation found in plants subjected to water stress are probably beneficial when the plants are again subjected to atmospheric water stress. Such characteristics probably provide a better water supply to the leaf tissue and lower transpiration per unit of leaf surface when the stomata are closed by water stress. Thus, plants previously subjected to water stress might have better control over water loss than those not previously stressed. An example is soybean, which transpires more slowly after being subjected to water stress because stressed plants have more lipids on the leaf surfaces (Clark and Levitt, 1956).

Kelley et al. (1945) found that guayule plants subjected to high water stress resumed growth sooner and survived better than plants given an abundance of water. Seedlings of herbaceous species are often "hardened" by reducing the water supply for some days before transplanting. The physiological basis for this treatment has not been adequately studied. Orchard (1967) reported that leaves of kale (*Brassica oleracea* var. *fruticosa* Metz) which expanded during a drought survived longer than leaves which expanded while the plants were being watered daily. It is not clear how much of the advantage possessed by stressed plants is structural and how much protoplasmic. More research on the response of plants to drought is needed to separate structural and protoplasmic changes and evaluate their relative importance.

Although water stress reduces growth, it has been observed that plants subjected to moderate stress sometimes grow more rapidly for a time after rewatering than similar plants not subjected to water stress. Among such reports are those by Gates (1955) for tomato, Owen and Watson (1956) for sugar beet, and Petrie and Arthur (1943) for tobacco. Miller (1965) observed prolonged stimulation of stem elongation of loblolly pine seedlings when stress was removed by rewatering. Perhaps carbohydrates and nitrogen compounds are accumulated in the stressed plants and are available to stimulate growth when water becomes available.

Measurement of plant water stress

The essential feature in plant water relations is the internal water balance or degree of water stress, because this controls the physiological processes and conditions which determine the quantity and quality of plant growth. Attempts to estimate plant water stress from measurements of soil water content or rates of evapotranspiration are useful for some purposes. However, they do not supply information reliable enough to evaluate the effects of water supply on plant processes and plant growth. The only reliable indicators of plant water stress are direct measurements made on the plants. Many of the inconclusive and contradictory results from experiments on the relationship between water supply and plant growth exist because of failure to measure plant water stress.

What should be measured

The importance of measuring plant water stress was recognized by ecologists and physiologists early in this century, as indicated by the numerous measurements of osmotic potential in the early literature (Fitting, 1911; Harris, 1934; Korstian, 1924; Miller, 1938, pp. 39–45). However, interest shifted from measurements of osmotic potential to measurements of what is now termed the water potential (see Chap. 1). Unfortunately, the difficulties attending reliable measurements of water potential have discouraged many investigators. As a result, numerous attempts were made to evaluate water stress in terms of water content, relative turgidity or relative water content, saturation deficit, and even stomatal aperture. Recent improvements in methods of measuring the water potential of plant tissue are resulting in renewed interest in its measurement. The important methods will be reviewed briefly. Readers can find more details in Slatyer (1967, pp. 150–160), Slatyer and Shmueli (1967), and in a review by Barrs (1968).

Perhaps one reason for failure to make routine measurements of plant water stress is the uncertainty concerning what should be measured. Obviously that characteristic should be measured which is most closely related to the essential physiological processes of plants. In addition to the water content, there are three important characteristics of cells—the osmotic potential, the turgor or pressure potential, and the water potential.

Walter (1955, 1965) has urged the importance of measurements of osmotic potential as indicators of cell water status, because the osmotic potential of cells affects the hydration of the protoplasm. This is discussed in the section on hydratur in Chap. 1. However, the normal range of osmotic potentials varies widely among different kinds of plants, and a value normal for halophytes would be much lower than that characteristic of mesophytes.

There is no question about the importance of turgor potential in cell enlargement and stomatal opening, but turgor pressure is difficult to measure directly. It is sometimes estimated from the resonance frequency of pieces of tissue by the method described by Falk et al. (1958) and Burström et al. (1967). However, it is usually calculated as the difference between cell water potential and osmotic potential from Eq. (1.10).

Sampling problems

Whatever method is employed, the use of representative and comparable samples is very important. Most measurements are made on leaves because they are accessible and easily sampled and very important physiologically. Because of their exposure they are sensitive indicators of water stress. However, the water status of leaves varies widely. For example, old and young leaves often differ in water content (see Fig. 10.22) and osmotic potential, as do leaves in the sun and in the shade, or those from the top and the bottom of a plant. Large errors can be introduced by comparing samples of different ages, those obtained from locations with different exposures, or those obtained at different times of day. Great care must also be used to avoid changes in water content of samples during and after collection. They should be placed in tight containers and kept in the shade, and measurements should be made as quickly as possible. Preparative procedures such as cutting leaf samples to fit psychrometer chambers or test tubes should be done in a humid chamber (Kreeb, 1960).

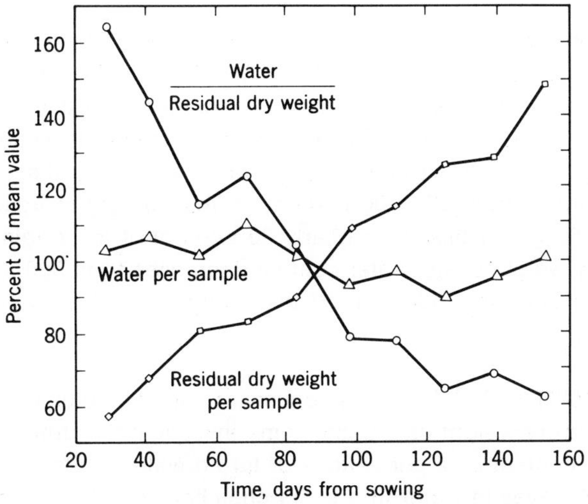

FIG. 10.22 Change in water content of cotton leaves based on residual dry weight. An increase in dry weight creates the impression that the water content is decreasing, although the water content per leaf remains practically unchanged. (*After Weatherley*, 1950.)

Direct measurement of water content

The oldest method of measuring the water status of plants is in terms of water content as a percentage of fresh or dry weight.

WATER CONTENT ON A DRY WEIGHT BASIS. To calculate water content on a dry weight basis, the material is usually oven-dried at 85°C, because higher temperatures sometimes result in loss of dry weight. This procedure is unsatisfactory, especially for leaves, which often show a large increase in dry weight during their life. For example, the dry weight of pear leaves increased 250 percent from May to August, causing the water content on a percentage basis to decrease proportionately, although the amount of water per leaf remained unchanged during most of this period (Ackley, 1954). Important changes in dry weight have been reported by many other investigators, including Kozlowski and Clausen (1965), Miller (1917), and Weatherley (1950). The effects of such changes are shown in Fig. 10.22.

Photosynthesis, respiration, and translocation cause measurable changes in the amount of solutes in leaves and therefore cause diurnal changes in leaf dry weight. Denny (1932) and Mason and Maskell (1928) attempted to reduce such fluctuations by extracting the easily hydrolyzable constituents with dilute hydrochloric acid, leaving only the inert cell walls and other inert materials. This decreases short-term variations in dry weight, but does not eliminate the long-term changes caused by increase in cell wall material.

WATER CONTENT ON A FRESH WEIGHT BASIS. Water content is often expressed as a percentage of fresh weight, but this is very unsatisfactory because of the wide fluctuations in water content. Furthermore, the fresh weight basis is relatively insensitive to small changes in water content, as shown in Fig. 10.23. Curtis and Clark (1950, p. 259) note that a decrease in leaf water content from 85 to 80 percent of fresh weight represents a loss of 25 g water/100 g leaf tissue, or nearly 30 percent of the original water content. On a dry weight basis, water content decreases from 566 to 400 percent, giving a more accurate indication of the change than is provided on a fresh weight basis.

RELATIVE WATER CONTENT. Because of the difficulties experienced in using fresh or dry weight as a base, some investigators turned to expression of leaf water content as a percentage of turgid water content. Stocker (1929) placed intact leaves in water in a moist chamber until they reached constant weight and calculated what he termed the "water deficit" as follows:

$$\text{Water deficit} = \frac{\text{turgid wt} - \text{field wt}}{\text{turgid wt} - \text{oven-dry wt}} \times 100 \qquad (10.1)$$

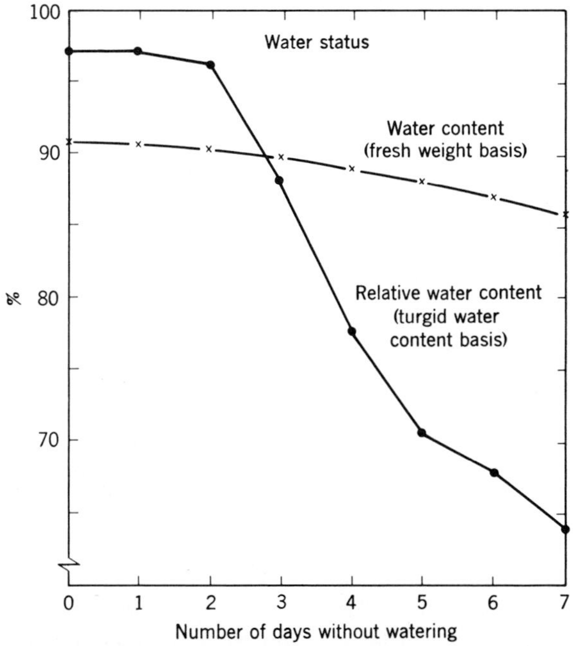

FIG. 10.23 Change in relative water content of corn leaves compared with change in water content expressed on a fresh weight basis. The fresh weight basis for calculating changes in water content is not sufficiently sensitive. (*From Barrs,* 1968; *after Mattas and Pauli,* 1965.)

This is sometimes called the saturation deficit. Although Hewlett and Kramer (1963) obtained good results with entire leaves, there have been complaints that intact leaves require too long to attain equilibrium. Weatherley (1950, 1951) proposed the use of disks of leaf tissue, which can be floated on water, but this introduced errors from infiltration along the cut edges. To eliminate this, Čatsky (1965) placed the disks on pieces of water-saturated polyurethane foam in a moist chamber. It has been shown that water uptake can be divided into two phases, the first associated chiefly with elimination of the water deficit, the second associated with growth. The first phase requires only a few hours, whereas the second can continue for days, as shown in Fig. 10.24. Barrs and Weatherley (1962) reduced the water uptake time to 4 hr, eliminating the need for duplicate samples to correct for loss of dry weight by respiration. The shorter period also minimizes uptake of water by growth. Millar (1966) reported that considerable errors can be caused by measuring water uptake at a temperature very different from the temperature at which the tissue was growing.

Weatherley calculated what he termed relative turgidity—probably better termed the relative water content—by the following equation:

$$\text{Relative water content} = \frac{\text{field wt} - \text{oven-dry wt}}{\text{turgid wt} - \text{oven-dry wt}} \times 100 \qquad (10.2)$$

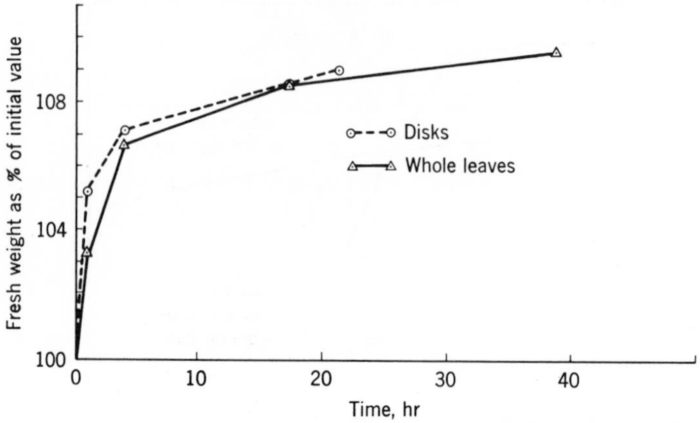

FIG. 10.24 Uptake of water by disks cut from leaves and by whole leaves. The initial rapid increase is associated with elimination of leaf water deficits; the long, slow continued uptake is associated with growth. (*From Barrs and Weatherley*, 1962.)

Water deficit and relative water content are complementary; relative water content = 100 − water deficit.

The exact procedure most likely to give a reliable measure of full turgidity varies with the species. Clausen and Kozlowski (1965) and Harms and McGregor (1962) found the use of entire needles satisfactory for several species of conifers. Kramer and Hewlett (1963) found entire leaves more satisfactory than disks for some species. In all procedures care in sampling and in handling samples is necessary. One source of error is from infiltration of intercellular spaces with water; another is in drying the leaf surfaces before weighing. Measurements should be started as soon after collection as possible.

Measurements of relative water content or water deficit form a convenient method for following changes in water content without errors caused by changing dry weight. Unfortunately a given water deficit or relative water content does not represent the same level of water potential in leaves of different species or ages, or from different environments. This is shown in Figs. 10.27 and 10.28. Comparisons of values for different kinds of tissue are therefore questionable.

Indirect measurement of water content

The development of so-called beta gauges has provided a useful indirect method of measuring changes in leaf water content. A source of beta radiation is placed on one side of the leaf and a radiation detector on the other side. The amount of radiation absorbed by the intervening leaf tissue changes with change in mass per unit of area. Over short periods of time changes in mass (really leaf thickness) closely parallel changes in water content (Mederski, 1961; Mederski and Alles, 1968; Nakayama and Ehrler, 1964). Unfortunately, leaves differ in thickness, and there are even differences in thickness from place to place on a single leaf, making calibration difficult. Jarvis and Slatyer (1966) describe a method of calibration which minimizes these difficulties, but Mederski and Alles (1968) state that there is no substitute for direct calibration in leaves which shrink during loss of water.

Qualitative estimates of water stress

As mentioned in Chap. 3, qualitative estimates of the development of plant water stress can be made from the appearance of the leaves of those plants which wilt with a small decrease in water content. A more sensitive indicator is premature closure of stomata, because the guard cells are very responsive to decreases in leaf turgor. Furthermore, closure of stomata is important because it reduces the supply of carbon dioxide required for photosynthesis. A variety of methods for measuring stomatal aperture are listed by Slatyer and Shmueli (1967). The simplest field technique is the infiltration method, but porometers and surface film impressions are useful for some purposes.

Another indirect method of following changes in water status used successfully on pine trees is measurement of the oleoresin exudation pressure (Lorio and Hodges, 1968). This appears to be well correlated with the water status and can be used on large trees to indicate the water status in the trunk. Apparently the water status of plants with a well-developed system of latex ducts can be estimated from the latex pressure (Buttery and Boatman, 1966). However, such methods are useful only as indicators of the occurrence of water stress and give no quantitative measure of its severity.

Measurement of water potential

The most useful single value for characterization of plant water stress is probably the water potential. Several methods of measurement will be described.

LIQUID EQUILIBRATION. The oldest method of measuring water potential is to immerse tissue in a series of solutions covering a range of osmotic potentials and determine the osmotic potential at which it neither gains nor loses water. Sometimes this is determined by measuring changes in size of individual cells, but more often change in length of strips of tissue is measured. The method works fairly well on thin-walled tissue containing no large veins. In the gravimetric form of this procedure pieces of tissue are weighed, immersed in a series of solutions for a time, then removed, blotted dry, and reweighed. Errors arise from infiltration of the tissue by the solution in which it is immersed, from failure to dry the surfaces uniformly, and from loss of weight during the handling and weighing. It is impossible to make reliable measurements on tissue dehydrated to the point of plasmolysis (Slatyer, 1958). The gravimetric method is suitable only for relatively massive tissue, because it is impossible to obtain a sufficient number of comparable samples from thin leaves and other small plant parts.

A more useful method of liquid equilibration which avoids some of the problems inherent in the gravimetric method is to measure the change in concentration of the solution in which the tissue is immersed. In one procedure similar samples of tissue are placed in a series of sucrose solutions with osmotic potentials covering the range of water potentials expected in the tissue under study. The refractometric index of each solution is measured initially and again after the tissue has been immersed in it for a few hours. The tissue loses water to solutions with a lower potential and gains water from those with a higher potential, making it easy to estimate the approximate water potential. The method seems to have been developed in Russia and was first described in English by Ashby and Wolf (1947). Additional information can be found in papers by Barrs (1968), Knipling (1970*a*), Kramer and Brix (1965), Rehder and Kreeb (1961), and others.

The most frequently used version of this procedure is the dye method, sometimes called the Shardakov method after the Russian who originally described it (Shardakov, 1948). This method eliminates the need for an expensive refractometer. Test tubes containing duplicate series of sucrose solutions are prepared with osmotic potentials covering the range of water potentials expected in the tissue. Representative samples of the leaf tissue are immersed in each test solution while the control series is kept for comparison. After an appropriate period of time, usually 2 to 8 hr, the tissue is removed from the tubes and enough of a dye such as methylene blue or methyl orange is added to each test solution to color it slightly. A drop of colored solution is transferred with a medicine dropper from each test solution to the corresponding tube of control solution. The drop will rise if the test solution has been diluted by absorption from the plant tissue. It will fall if it has been concentrated by loss of water to the plant tissue. The solution in which the drop of test solution neither rises nor falls has a potential equal to the water potential of the leaf. The procedure is shown diagrammatically in Fig. 10.25.

This method is convenient for use in the field, but some species show large errors (Brix, 1966; Knipling and Kramer, 1967). These result chiefly from contamination of the test solutions by solutes escaping from the cut surfaces of the leaf disks. Partial compensation for this error can be attained by increased immersion times. An equilibration period of 2 hr seems adequate for cut leaves of tomato, tobacco, and yellow-poplar, but 4 to 8 hr are needed for dogwood and some species of oak and maple. Less time is required for whole leaves of dogwood and tomato than for cut leaves, because there is

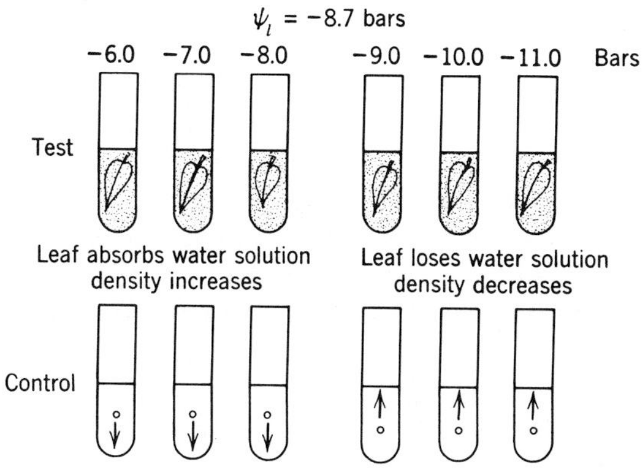

FIG. 10.25 **Diagram showing operation of dye method for measuring leaf water potential. The water potential is assumed to lie between that of the solution in which the drop rises and that of the solution in which it sinks, i.e., −8.5 ± 0.5 bar. (*From Knipling, 1967a.*)**

less contamination. However, 24 hr are required for loblolly pine needles, and reliable measurements appear to be difficult or impossible on needles of some species of conifers. Whole leaves of some species seem to provide more reliable values than cut leaves, but the opposite is true of other species (Knipling, 1967*a*).

VAPOR EQUILIBRATION. Measurement of leaf water potential by vapor equilibration avoids some of the errors associated with liquid methods. The oldest method is to place weighed samples of tissue over solutions of known osmotic potential in airtight containers and immerse them in a water bath for some hours, then reweigh them. The solution over which no change in weight occurs is assumed to have the same potential as the tissue. Errors can occur because of loss of weight by respiration during the equilibration period. This method is described by Slatyer (1958), who used it effectively for both leaves and soil.

In recent years thermocouple psychrometers have been used extensively to measure the water potential of soil and of plant tissue. The psychrometer method determines the relative vapor pressure of the air in equilibration with the tissue. Relative vapor pressure is related to water potential by Eq. (1.7). The tissue is enclosed in small containers which are immersed in a water bath kept at a constant temperature ± 0.001°C. One type of psychrometer, proposed by Spanner (1951) and modified by Monteith and Owen (1958), uses the Peltier effect to condense water on a thermocouple. When the current is turned off, the condensed water starts to evaporate, cooling the thermocouple junction. The resulting current is measured with a galvanometer. In the Richards and Ogata (1958) version, a drop of water is placed on the wet junction, and the current is measured when a steady state of evaporation is attained. Both types are calibrated by obtaining readings with solutions or pieces of filter paper wetted with solutions of known vapor pressure. The Richards and Ogata psychrometer reaches equilibrium more rapidly than the Spanner instrument and also covers a wider range of water potentials. Rawlins (1964) reported considerable errors in psychrometric measurements caused by the resistance of leaf tissue to diffusion of water vapor, but Barrs (1965) did not find this important. Boyer and Knipling (1965) found only moderate errors due to leaf resistance and proposed the use of an isopiestic method which eliminates this problem. Boyer (1966) reported very accurate measurements

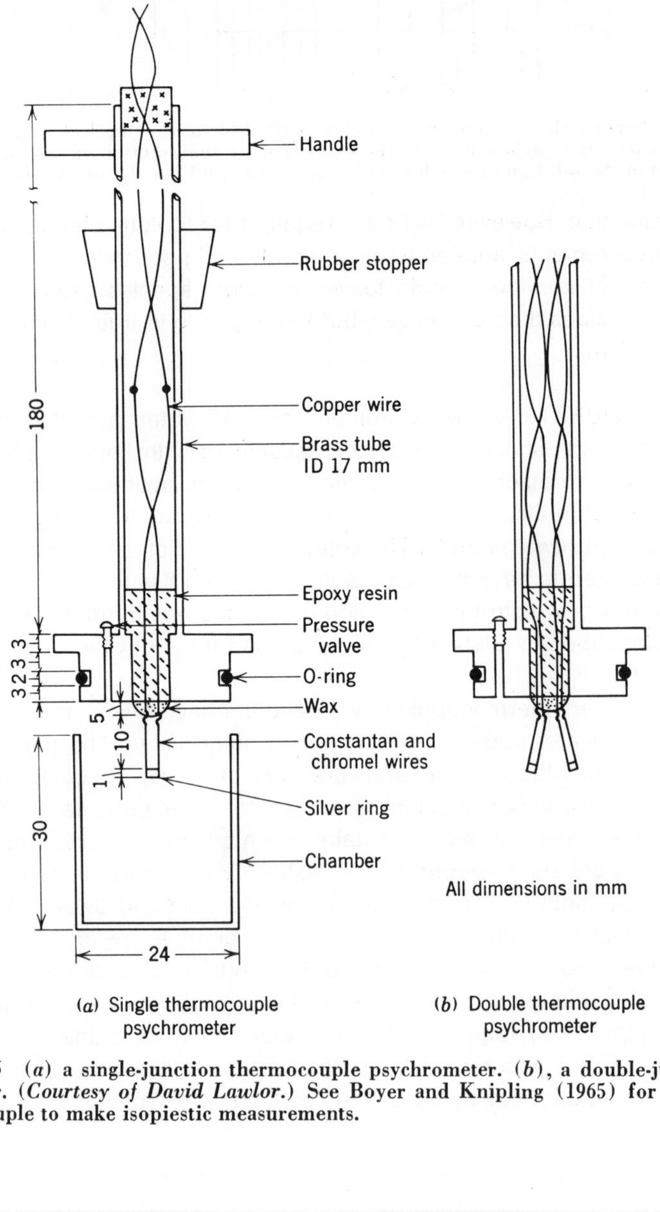

FIG. 10.26 (*a*) a single-junction thermocouple psychrometer. (*b*), a double-junction psychrometer. (*Courtesy of David Lawlor.*) See Boyer and Knipling (1965) for details of a thermocouple to make isopiestic measurements.

of the water potential of sunflower leaves by this method. Barrs (1964) discovered that increase in temperature caused by respiration of the tissue in the psychrometer chamber can cause significant errors. He later suggested methods of correcting for this by use of a second dry-junction thermocouple to measure the temperature in the chamber. Wet- and dry-junction thermocouples have been used extensively in the author's laboratory. Salt or other osmotically active solutes on the leaf surfaces also cause inaccurate measurements.

Thermocouple psychrometers can also be used to measure the osmotic potential. After measuring the water potential the container is immersed in a freezing bath to kill the tissue and then returned to the water bath for a second measurement. Since turgor pressure is eliminated by death, the new measurement gives the osmotic potential. Some of the problems involved in operating thermocouple psychrometers are discussed by Barrs (1968*a*), Ehlig, (1962), and Boyer and Knipling (1965). Lang and Barrs (1965) and Lambert and van Schilfgaarde (1965) described techniques which enable measurements to be made on attached leaves.

INDIRECT ESTIMATION OF WATER POTENTIAL. Weatherley and Slatyer (1957) suggested that if a relative water content/water potential curve could be constructed for a given kind of tissue, a relationship might be estimated from the relative water content. Unfortunately, this relationship changes with age and habitat (Knipling, 1967*b*), as shown in Fig. 10.27, and among species (Jarvis and Jarvis, 1963; Slatyer, 1960). The different relationships of water potential and osmotic potential to relative water content for three different species are shown in Fig. 10.28. In view of the variability in this relationship, attempts to estimate water potential from relative water content or water deficit appear to have limited usefulness.

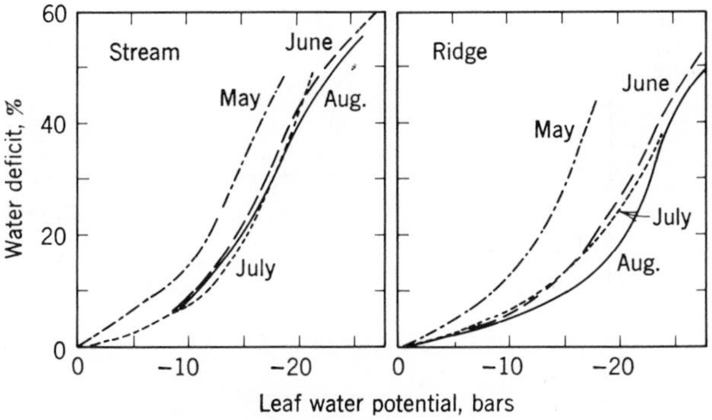

FIG. 10.27 **Comparisons of the relationship between water deficit and water potential for dogwood leaves of increasing age from two habitats. The relationship changed more with age for exposed leaves from the ridge than for the leaves from the better-watered trees beside a stream. (*From Knipling,* 1967*b*.)**

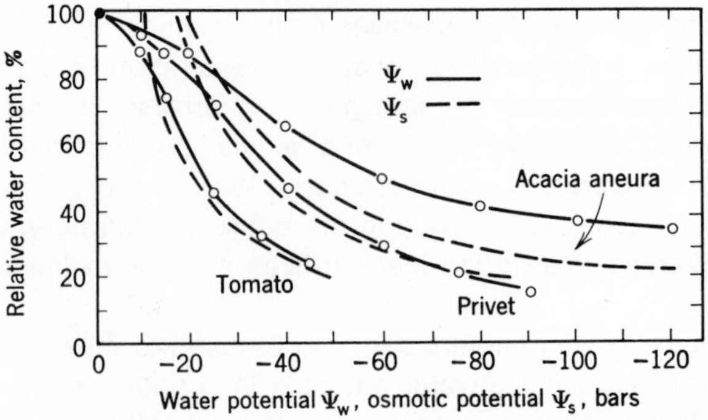

FIG. 10.28 **Relationship between leaf water potential, osmotic potential, and relative water content of three dissimilar species. (*From Slatyer*, 1967; *after Slatyer*, 1960*a*.)**

PRESSURE EQUILIBRATION. Scholander and his colleagues (Scholander et al., 1964, 1965) reintroduced a method first described by Dixon (1914) which has wide applicability. A leafy shoot is sealed in a pressure chamber with the cut surface protruding, as shown in Fig. 10.29. Pressure is applied to the shoot until xylem sap appears at the cut surface. The amount of pressure which must be applied to force water from the leaf cells back into the xylem is regarded as equal to the water potential of the leaf cells. Boyer

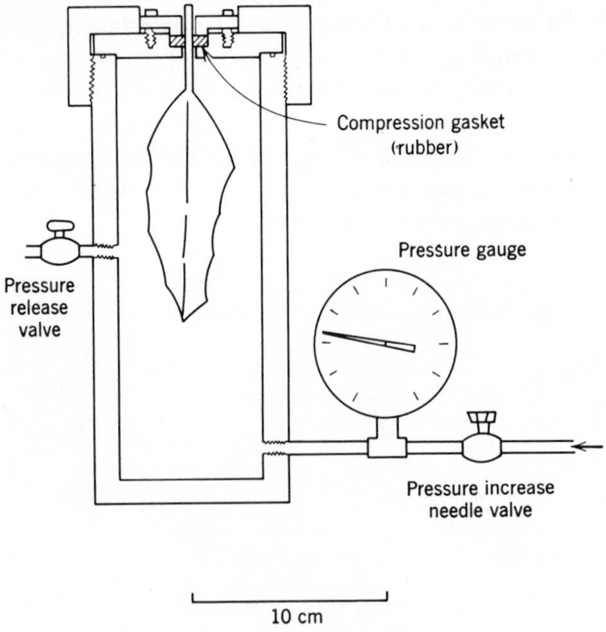

FIG. 10.29 **Cross section through a pressure chamber used to measure water potential by pressure equilibration. (*From Kaufmann*, 1937.)**

(1967) found that measurements of the leaf water potentials of yew and sunflower in a pressure chamber and by the psychrometer method agreed reasonably well, but measurements of rhododendron by the two methods did not agree. Apparently the discrepancies in rhododendron occur because sap is forced into intercellular spaces. Deformation of the stem under pressure can also cause errors.

Measurements can be made very rapidly with the pressure chamber, and the equipment can be carried into the field (Waring and Cleary, 1967, for example). The method will apparently be useful for a wide variety of purposes. The pressure chamber method will certainly provide approximate measurements of water potential, but the variability will be greater than with a thermocouple psychrometer. The reliability of the method for any new kind of plant tissue ought to be tested by comparison with psychrometer measurements. It requires stems or petioles firm enough to be anchored in the pressure chamber.

Summary

Plant growth is controlled directly by plant water stress and only indirectly by soil and atmospheric water stress. Plant water stress or water deficits develop when water loss exceeds water absorption. Temporary midday water deficits occur in rapidly transpiring plants because the resistance to water movement through roots causes absorption to lag behind transpiration even in moist soil. Longer-term and more severe water deficits develop when decreasing soil water potential and hydraulic conductivity cause decreased absorption of water. Thus, daily cycles in water stress are controlled chiefly by transpiration, but long-term, severe water deficits develop chiefly because of decreasing availability of soil water.

Water stress affects practically every aspect of plant growth, modifying the anatomy, morphology, physiology, and biochemistry. It is uncertain how much of the reduction in growth is caused by decreased turgor, how much by decreased water potential, and how much by decrease in osmotic potential. Under average conditions midday water deficits reduce cell enlargement and stem elongation somewhat, and there is often a midday decrease in photosynthesis. Decreased cell enlargement results in smaller plants, thicker leaves which are more heavily cutinized, more dry matter, and a higher ratio of roots to shoots. As the soil water content decreases, water deficits become more severe, and breakdown of RNA, DNA, and proteins increases, photosynthesis decreases, and respiration increases. Finally, photosynthesis and growth cease. As stress increases, synthesis and translocation of growth regulators is probably inhibited, and translocation of other compounds is hindered. Carbohydrates and proteins are hydrolyzed, and there is often an increase in soluble sugars and nitrogen compounds. Finally, protoplasmic dehydration becomes so severe that cells and tissues begin to die.

It is essential that the degree of plant water stress be measured in all research concerning the effects of water on plant growth, because it is impossible to evaluate plant water stress reliably from soil water data. Measurements of water content on either a fresh or dry weight basis are unsatisfactory because water contents vary widely with age and kind of tissue. Relative water content, that is, water content expressed as a percentage of water content at full turgidity, is more satisfactory. However, the degree of water stress is probably best expressed in terms of water potential, because this appears to be most closely related to the physiological and biochemical processes which control growth. Use of water potential also allows soil and plant water stress to be expressed in the same units.

Plant water potential can be measured by both liquid and vapor equilibration methods. It appears that liquid equilibration determination by the dye method is fairly satisfactory for field measurements, but the thermocouple psychrometer method is more accurate. The pressure equilibration method is rapid and can be used in the field as well as in the laboratory, but it is unreliable for some species.

Bibliography

Abell, C. A., and C. R. Hursh, 1931. Positive gas and water pressures in oaks. *Science* **73:**449.

Ackerman, E., and G. Loff, 1959. "Technology in American Water Development." The Johns Hopkins Press, Baltimore.

Ackley, W. B., 1954. Seasonal and diurnal changes in the water contents and deficits of Bartlett pear leaves. *Plant Physiol.* **29:**445–448.

Addoms, R. M., 1937. Nutritional studies on loblolly pine. *Plant Physiol.* **12:**199–205.

Aitchison, G. D., P. F. Butler, and C. G. Gurr, 1951. Techniques associated with the use of gypsum block soil moisture meters. *Aust. J. Appl. Sci.* **2:**56–75.

Alberda, Th., 1948. The influence of some external factors on growth and phosphate uptake of maize plants of different salt conditions. *Rec. trav. bot. Neer.* **41:**541–602.

———, 1953. Growth and root development of lowland rice and its relation to oxygen supply. *Plant and Soil* **5:**1–28.

Albertson, F. W., and J. E. Weaver, 1945. Injury and death or recovery of trees in prairie climate. *Ecol. Monogr.* **15:**393–433.

Albrecht, W. A., 1940. Calcium-potassium-phosphorus relation as a possible factor in ecological array of plants. *J. Amer. Soc. Agron.* **32:**411–418.

Aldrich, W. W., R. A. Work, and M. R. Lewis, 1935. Pear root concentration in relation to soil-moisture extraction in heavy clay soil. *J. Agr. Res.* **50:**975–988.

Aljibury, F. K., W. M. Tomlinson, and C. E. Houston, 1965. Tensiometers, automatic timing for sprinkler control. *Calif. Agr.* **19**(5):2–4.

Allen, R. M., 1967. Influence of the root system on height growth of three southern pines. *Forest Sci.* **13:**253–257.

Alvim, P. de T., 1959. Stomatal opening as a practical indicator of water stress in plants. *Proc. 9th Int. Bot. Congr.* **4:**5.

———, 1960. Moisture stress as a requirement for flowering of coffee. *Science* **132:**354.

———, 1965. A new type of porometer for measuring stomatal opening and its use in irrigation studies. *Arid Zone Res.* **25:**325–329.

——— and J. R. Havis, 1954. An improved infiltration series for studying stomatal opening as illustrated with coffee. *Plant Physiol.* **29:**97–98.

van Andel, O. M., 1953. The influence of salts on the exudation of tomato plants. *Acta Bot. Neer.* **2:**445–521.

Anderson, A. B. C., and N. E. Edlefsen, 1942. The electrical capacity of the 2-electrode plaster of paris block as an indicator of soil-moisture content. *Soil Sci.* **54:**35–46.

Anderson, D. B., and T. Kerr, 1943. A note on the growth behavior of cotton bolls. *Plant Physiol.* **18:**261–269.

Anderson, W. P., and C. R. House, 1967. A correlation between structure and function in the root of *Zea mays*. *J. Exp. Bot.* **18:**544–555.

Anderssen, F. G., 1929. Some seasonal changes in the tracheal sap of pear and apricot. *Plant Physiol.* **4:**459–476.

Andersson, N. E., D. H. Hertz, and H. Rufelt, 1954. A new fast recording hygrometer for plant transpiration measurements. *Physiol. Plant.* **7:**753–767.

Andrews, R. E., and E. I. Newman, 1962. The influence of root pruning on the growth and transpiration of wheat under different soil moisture conditions. *New Phytol.* **67:**617–630.

Arcichovskij, V., and A. Ossipov, 1931. Die Saugkraft der baumartigen Pflanzen der zentralasiastischen Wüsten nebst Transpirationsmessungen am Saxaul (*Arthrophytum haloxylon* Litw.). *Planta* **14:**552–565.

Arikado, H., 1955. The ventilating pressure of rice plants growing under paddy field conditions. *Bull. Fac. Agr. Mie Univ.* no. 11.

Arisz, W. H., 1956. Significance of the symplasm theory for transport across the root. *Protoplasma* **46:**5–62.

———, 1964. Influx and efflux of electrolytes. II. Leakage out of cells and tissues. *Acta Bot. Neer.* **13:**1–58.

———, I. J. Camphius, H. Heikens, and A. J. van Tooren, 1955. The secretion of the salt glands of *Limonium lactifolium* Ktze. *Acta Bot. Neer.* **4:**322–338.

Army, T. J., and T. T. Kozlowski, 1951. Availability of soil moisture for active absorption in drying soil. *Plant Physiol.* **26:**353–362.

Arndt, C. H., 1937. Water absorption in the cotton plant as affected by soil and water temperatures. *Plant Physiol.* **12:**703–720.

———, 1945. Temperature-growth relations of the roots and hypocotyls of cotton seedlings. *Plant Physiol.* **20:**200–220.

Arnold, A., 1952. Über den Functionsmechanismus der Endodermiszellen der Wurzeln. *Protoplasma* **41:**189–211.

Arnon, D. I., and D. R. Hoagland, 1940. Crop production in artificial culture solutions and in soils with special reference to factors influencing yields and absorption of inorganic nutrients. *Soil Sci.* **50:**463–484.

——— and C. M. Johnson, 1942. Influence of hydrogen ion concentration on growth of higher plants under controlled conditions. *Plant Physiol.* **17:**525–539.

Ashby, E., and R. Wolf, 1947. A critical examination of the gravimetric method of determining suction force. *Ann. Bot.* **11**:261–268.

Ashton, F. M., 1956. Measurement of soil moisture. *Plant Physiol.* **31**:266–274.

Askenasy, E., 1895. Ueber das Saftsteigen. *Bot. Centralbl.* **62**:237–238.

Aslyng, H. C., 1963. Soil physics terminology. *Int. Soc. Soil Sci. Bull.* **23**:1–4.

Atkins, W. R. G., 1916. Some recent researches in plant physiology. Whitaker and Co., London.

Atkinson, M. R., G. P. Findlay, A. B. Hope, M. G. Pitman, H. D. W. Saddler, and K. R. West, 1967. Salt regulation in the mangroves *Rhizophora mucronata* Lam. and *Aegialitis annulata* R. BR. *Aust. J. Biol. Sci.* **20**:589–599.

Avery, G. S., Jr., 1933. Structure and development of the tobacco leaf. *Amer. J. Bot.* **20**:565–592.

Ayers, A. D., C. H. Wadleigh, and O. C. Magistad, 1943. The interrelationships of salt concentration and soil moisture content with the growth of beans. *J. Amer. Soc. Agron.* **35**:796–810.

Ayers, H. D., and V. E. A. Wikramanayake, 1958. The effect of the water storage capacity of the soil on mass infiltration. *Can. J. Soil. Sci.* **38**:44–48.

Bailey, L. F., J. S. Rothacher, and W. H. Cummings, 1952. A critical study of the cobalt chloride method of measuring transpiration. *Plant Physiol.* **27**:562–574.

Baker, K. F., and W. C. Snyder (eds.), 1965. Ecology of soil-borne plant pathogens, prelude to biological control. Univ. of California Press, Berkeley.

Bakke, A. L., and N. L. Noecker, 1933. The relation of moisture to respiration and heating in stored oats. *Iowa Agr. Exp. Sta. Res. Bull.* 165.

Bald, J. G., 1952. Stomatal droplets and the penetration of leaves by plant pathogens. *Amer. J. Bot.* **39**:97–99.

Balls, W. L., 1908. The cotton plant in Egypt. The Macmillan Company, New York.

Bange, G. G. J., 1953. On the quantitative explanation of stomatal transpiration. *Acta Bot. Neer.* **2**:255–297.

Barber, D. A., M. Ebert, and N. T. S. Evans, 1962. The movement of ^{15}O through barley and rice plants. *J. Exp. Bot.* **13**:397–403.

Barber, S. A., 1962. A diffusion and mass-flow concept of soil nutrient availability. *Soil Sci.* **93**:39–49.

Barley, K. P., 1962. The effects of mechanical stress on the growth of roots. *J. Exp. Bot.* **13**:95–110.

Barnes, R. L., and A. W. Naylor, 1959. In vitro culture of pine roots and the use of *Pinus serotina* roots in metabolic studies. *Forest Sci.* **5**:158–168.

Barnett, N. M., and A. W. Naylor, 1966. Amino acid and protein metabolism in Bermuda grass during water stress. *Plant Physiol.* **41**:1222–1230.

Barney, C. W., 1951. Effects of soil temperature and light intensity on root growth of loblolly pine seedlings. *Plant Physiol.* **26**:146–163.

Barrs, H. D., 1964. Heat of respiration as a possible cause of error in the estimation by psychrometric methods of water potential in plant tissue. *Nature* **203**:1136–1137.

———, 1965. Psychrometric measurement of leaf water potential: lack of error attributable to leaf permeability. *Science* **149**:63–65.

———, 1966. Root pressure and leaf water potential. *Science* **152**:1266–1268.

———, 1968*a*. Determination of water deficits in plant tissues. In T. T. Kozlowski (ed.), "Water deficits and plant growth, I," Academic Press Inc., New York.

———, 1968*b*. Effect of cyclic variations in gas exchange under constant environmental conditions on the ratio of transpiration to net photosynthesis. *Physiol. Plant.* **21**:918–929.

——— and B. Klepper, 1968. Cyclic variations in plant properties under constant environmental conditions. *Physiol. Plant.* **21**:711–730.

——— and P. E. Weatherley, 1962. A reexamination of the relative turgidity technique for estimating water deficits in leaves. *Aust. J. Biol. Sci.* **15**:413–428.

Bartholomew, E. T., 1926. Internal decline of lemons. III. Water deficit in lemon fruits caused by excessive leaf evaporation. *Amer. J. Bot.* **13**:102–117.

Basler, E., G. W. Todd, and R. E. Meyer, 1961. Effects of moisture stress on absorption, translocation, and distribution of 2,4-dichlorophenoxyacetic acid in bean plants. *Plant Physiol.* **36**:573–576.

Bassett, J. R., 1964. Tree growth as affected by soil-moisture availability. *Soil Sci. Soc. Amer. Proc.* **28**:436–438.

Bates, C. G., 1924. Relative resistance of tree seedlings to excessive heat. *U.S. Dept. Agr. Bull.* 1263.

Batjer, L. P., J. P. Magness, and L. O. Regeimbal, 1939. The effect of root temperature on growth and nitrogen intake of apple trees. *Amer. Soc. Hort. Sci. Proc.* **37**:11–18.

van Bavel, C. H. M., 1953. Chemical composition of tobacco leaves as affected by soil moisture conditions. *Agron. J.* **45**:611–614.

———, E. E. Hood, and N. Underwood, 1954. Vertical resolution in the neutron method for measuring soil moisture. *Trans. Amer. Geol. Union* **35**(4):595–600.

——— and F. J. Verlinden, 1956. Agricultural drought in North Carolina. *N. C. Agr. Exp. Sta. Tech. Bull.* 122.

———, 1961. Lysimetric measurements of evapotranspiration in the Eastern United States. *Soil Sci. Soc. Amer. Proc.* **25**:138–141.

——— and L. E. Myers, 1962. An automatic weighing lysimeter. *Agr. Eng.* **43**:580–583, 586–588.

———, L. J. Fritschen, and W. E. Lewis, 1963. Transpiration by Sudangrass as an externally controlled process. *Science* **141**:269–270.

———, F. S. Nakayama, and W. L. Ehrler, 1965. Measuring transpiration resistance of leaves. *Plant Physiol.* **40**:535–540.

Baver, L. D., 1956. Soil physics. John Wiley & Sons, Inc., New York.

Bayliss, W. M., 1924. "Principles of General Physiology," 4th ed. Longmans, Green & Co., Inc., New York.

Beament, J. W. L., 1965. The active transport of water: evidence, models and mechanisms. *Symp. Soc. Exp. Biol.* **19**:273–298.

Beckman, C. A., 1964. Host responses to vascular infection. *Ann. Rev. Phytopathol.* **2**:231–252.

Begg, J. E., J. F. Bierhuizen, E. R. Lemon, D. K. Misra, R. O. Slatyer, and W. R. Stern, 1964. Diurnal energy and water exchanges in bulrush millet in an area of high solar radiation. *Agr. Meteorol.* **1**:294–312.

Belcher, D. J., T. R. Cuykendall, and H. S. Sack, 1950. The measurement of soil moisture and density by neutron and gamma-ray scattering. *Tech. Rep.* 127, U.S. Civil Aeronaut. Admin.

Bell, C. W., and O. Biddulph, 1963. Translocation of calcium. Exchange versus mass flow. *Plant Physiol.* **38**:610–614.

Bennett, J. P., F. G. Anderssen, and Y. Milad, 1927. Methods of obtaining tracheal sap from woody plants. *New Phytol.* **26**:316–323.

Bennet-Clark, T. A., 1959. Water relations of cells. In F. C. Steward (ed.), "Plant Physiology." Academic Press Inc., New York, vol. 2, pp. 105–191.

———, A. D. Greenwood, and J. W. Barker, 1936. Water relations and osmotic pressures of plant cells. *New Phytol.* **35**:277–291.

Ben-Zioni, A., C. Itai, and Y. Vaadia, 1967. Water and salt stresses, kinetin and protein synthesis in tobacco leaves. *Plant Physiol.* **42**:361–365.

Bernal, J. D., 1965. The structure of water and its biological implications. *Symp. Soc. Exp. Biol.* **19:**17–32. Cambridge University Press, New York.

Bernstein, L., 1961. Osmotic adjustment of plants to saline media. I. Steady state. *Amer. J. Bot.* **48:**909–918.

———, 1963. Salt tolerance of plants and the potential use of saline waters for irrigation. *Desalination Res. Conf. Proc.,* pp. 273–283. *Nat. Acad. Sci.–Nat. Res. Counc. Publ.* 942.

———, J. W. Brown, and H. E. Hayward, 1956. The influence of rootstock on growth and salt accumulation in stone-fruit trees and almonds. *Proc. Amer. Soc. Hort. Sci.* **68:**66–95.

———, W. R. Gardner, and L. A. Richards, 1959. Is there a vapor gap around roots? *Science* **129:**1750, 1753.

——— and R. H. Nieman, 1960. Apparent free space of plant roots. *Plant Physiol.* **35:**589–598.

Bialoglowski, J., 1936. Effect of extent and temperature of roots on transpiration of rooted lemon cuttings. *Proc. Amer. Soc. Hort. Sci.* **34:**96–102.

Biddulph, A., 1959. Translocation of inorganic solutes. In F. C. Steward (ed.), "Plant Physiology," vol. 2, pp. 553–603. Academic Press Inc., New York.

——— and J. Markle, 1944. Translocation of radioactive phosphorus in the phloem of the cotton plant. *Amer. J. Bot.* **31:**551–555.

———, F. S. Nakayama, and R. Cory, 1961. Transpiration and ascension of calcium. *Plant Physiol.* **36:**429–436.

Bier, J. E., 1959. The relation of bark moisture to the development of canker diseases caused by native, facultative parasites. I. Cryptodiaporthe canker on willow. *Can. J. Bot.* **37:**229–238.

Bierhuizen, J. F., and R. O. Slatyer, 1964. An apparatus for the continuous and simultaneous measurements of photosynthesis and transpiration under controlled environmental conditions. *C.S.I.R.O. Aust., Div. Land Res. Tech. Paper* 24.

———, R. O. Slatyer, and C. W. Rose, 1965. A porometer for laboratory and field operation. *J. Exp. Bot.* **16:**182–191.

Billings, W. D., and P. J. Godfrey, 1967. Photosynthetic utilization of internal carbon dioxide by hollow-stemmed plants. *Science* **158:**121–123.

——— and R. J. Morris, 1951. Reflection of visible and infrared radiation from leaves of different ecological groups. *Amer. J. Bot.* **38:**327–331.

Bingham, E. C., and R. F. Jackson, 1918. Standard substances for the calibration of viscometers. *Bull. Bur. Stand.* **14:**59–86.

Biswell, H. H., and A. M. Schultz, 1957. Spring flow affected by brush. *Calif. Agr.* **11**(10): 3–4, 10.

——— and J. E. Weaver, 1933. Effect of frequent clipping on the development of roots and tops of grasses in prairie sod. *Ecology* **14:**368–390.

Björkman, E., 1942. Über die Bedingungen der Mykorrhizabildung bei Kiefer und Fichte. *Symb. Bot. Upsal.* **6**(2):1–190.

———, 1949. The ecological significance of the ectotrophic mycorrhizal association in forest trees. *Svensk. Bot. Tidskr.* **43:**223–262.

——— and P. Holmgren, 1963. Adaptability of the photosynthetic apparatus to light intensity in ecotypes from exposed and shaded habitats. *Physiol. Plant.* **16:**889–914.

Black, R. F., 1956. Effect of NaCl in water culture on the ion uptake and growth of *Atriplex hastata* L. *Aust. J. Biol. Sci.* **9:**67–80.

Blaney, H. F., and W. D. Criddle, 1950. Determining water requirements in irrigated areas from climatological and irrigation data. *U.S. Dept. Agr. Soil. Conserv. Ser. Tech. Paper* 96.

Blinks, L. R., and R. L. Airth, 1951. The role of electroosmosis in living cells. *Science* **113**:474–475.

Bloodworth, M. E., and J. B. Page, 1957. Use of thermistors for the measurement of soil moisture and temperature. *Soil Sci. Soc. Amer. Proc.* **21**:11–15.

———, ———, and W. R. Cowley, 1956. Some applications of the thermoelectric method for measuring water flow rates in plants. *Agron. J.* **48**:222–228.

Bode, H. R., 1923. Beiträge zur dynamik der wasserbewegung in den gefässpflanzen. *Jahrb. Wiss. Bot.* **62**:92–127.

Bodman, G. B., and E. A. Colman, 1944. Moisture and energy conditions during downward entry of water into soils. *Soil Sci. Soc. Amer. Proc.* **8**:116–122.

Boehm, J., 1892. Ueber einen eigenthümlichen Stammdruck. *Ber. Deut. Bot. Ges.* **10**:539–544.

Bogen, H. J., 1940. Untersuchungen über den Quellungseffekt permeierender Anelektrolyte. I. Ionenwirkung auf die Permeabilität von *Rhoeo discolor. Z. Bot.* **36**:65–106.

———, 1953. Beitrage zur Physiologie der nichtosmotischen Wasseraufnahme. *Planta* **42**:140–155.

———, 1956. Nichtosmotische Aufnahme von Wasser und gelösten Anelektrolyten. *Ber. Deut. Bot. Ges.* **69**:209–222.

——— and H. Prell, 1953. Messung nichtosmotischer Wasseraufnahme an plasmolysierten Protoplasten. *Planta* **41**:459–479.

Böhning, R. H., and B. Lusanandana, 1952. A comparative study of gradual and abrupt changes in root temperature on water absorption. *Plant Physiol.* **27**:475–488.

Bollard, E. G., 1953. The use of tracheal sap in the study of apple-tree nutrition. *J. Exp. Bot.* **4**:363–368.

———, 1958. Nitrogenous compounds in tree xylem sap. In K. V. Thimann (ed.), "The Physiology of Forest Trees," pp. 83–93. The Ronald Press Company, New York.

———, 1960. Transport in the xylem. *Ann. Rev. Plant Physiol.* **11**:141–166.

——— and G. W. Butler, 1966. Mineral nutrition of plants. *Ann. Rev. Plant Physiol.* **17**:77–112.

Bolt, G. H., and M. J. Frissel, 1960. Thermodynamics of soil moisture. *Neth. J. Agr. Sci.* **8**:57–78.

Bolz, C., 1927. Lassen sich bei Wurzeln Nutationsbewegungen feststellen under welcher Art sind sie? *Bot. Arch.* **19**:450–459.

Bonner, J., 1946. Relation of toxic substances to growth of guayule in soil. *Bot. Gaz.* **107**:343–351.

———, R. S. Bandurski, and A. Millerd, 1953. Linkage of respiration to auxin-inducea water uptake. *Physiol. Plant.* **6**:511–522.

Boon-Long, T. S., 1941. Transpiration as influenced by osmotic concentration and cell permeability. *Amer. J. Bot.* **28**:333–343.

Bormann, F. H., 1957. Moisture transfer between plants through intertwined root systems. *Plant Physiol.* **32**:48–55.

——— and B. F. Graham, Jr., 1959. The occurrence of natural root grafting in eastern white pine, *Pinus strobus* L., and its ecological implications. *Ecology* **40**:677–691.

——— and ———, 1960. Translocation of silvicides through root grafts. *J. Forest.* **58**:402–403.

Bose, J. C., 1923. "Physiology of the Ascent of Sap." Longmans, Green & Co., Inc., New York.

Böszörmenyi, Z., 1965. A comparison between the bromide absorption by excised roots and intact plants. *Acta Agr. Hung.* **14**:219–234.

——— and E. Cseh, 1964. Studies of ion-uptake by using halide ions changes in the relationships between ions depending on concentration. *Physiol. Plant.* **17**:81–90.

Bourdeau, P. F., 1954. Oak seedling ecology determining segregation of species in Piedmont oak-hickory forests. *Ecol. Monogr.* **24**:297–320.

Bouyoucos, G. J., 1931. The alcohol method for determining moisture content of soils. *Soil Sci.* **32**:173–179.

———, 1949. Nylon electrical resistance unit for continuous measurement of soil moisture in the field. *Soil Sci.* **67**:319–330.

———, 1951. Effect of fertilizers on the plaster of paris electrical resistance method of measuring soil moisture in the field. *J. Amer. Soc. Agron.* **43**:508–511.

———, 1953*a*. An improved type of soil hydrometer. *Soil Sci.* **76**:377–378.

———, 1953*b*. More durable plaster of paris moisture blocks. *Soil Sci.* **76**:447–451.

———, 1954. New type electrode for plaster of paris moisture blocks. *Soil Sci.* **78**:339–342.

——— and A. H. Mick, 1940. An electrical resistance method for the continuous measurement of soil moisture under field conditions. *Mich. Agr. Exp. Sta. Tech. Bull.* 172.

——— and ———, 1948. A comparison of electric resistance units for making a continuous measurement of soil moisture under field conditions. *Plant Physiol.* **23**:532–543.

Bowen, G. D., 1965. Mycorrhiza inoculation in forestry practice. *Aust. Forest.* **29**:231–237.

———, 1968. Phosphate uptake by mycorrhizas and uninfected roots of *Pinus radiata* in relation to root distribution. *Trans. Ninth Int. Cong. Soil Sci.* **2**:219–228.

——— and C. Theodorou, 1967. Studies of phosphate uptake by mycorrhizas. *I.U.F.R.O. Proc. 14th Congr.* **5**:116–138.

Bowling, D. J. F., A. E. S. Macklon, and R. M. Spanswick, 1966. Active and passive transport of the major nutrient ions across the root of *Ricinus communis*. *J. Exp. Bot.* **17**:410–425.

——— and P. E. Weatherley, 1965. The relationship between transpiration and potassium uptake in *Ricinus communis*. *J. Exp. Bot.* **16**:732–741.

Boyce, S. G., 1954. The salt spray community. *Ecol. Monogr.* **24**:29–67.

Boyer, J. S., 1965. Effects of osmotic water stress on metabolic rates of cotton plants with open stomata. *Plant Physiol.* **40**:229–234.

———, 1966. Isopiestic technique: measurement of accurate leaf water potentials. *Science* **154**:1459–1460.

———,1967. Leaf water potentials measured with a pressure chamber. *Plant Physiol.* **42**:133–137.

——— and E. B. Knipling, 1965. Isopiestic technique for measuring leaf water potentials with a thermocouple psychrometer. *Proc. Nat. Acad. Sci.* **54**:1044–1051.

———, 1968. Relationship of water potential to growth of leaves. *Plant Physiol.* **43**:1056–1962.

Boyko, H., 1966. Salinity and aridity; new approaches to old problems. W. Junk, The Hague.

Boynton, D., and O. C. Compton, 1944. Normal seasonal changes of oxygen and carbon dioxide percentages in gas from the larger pores of three orchard subsoils. *Soil Sci.* **57**:107–117.

Branton, D., and L. Jacobson, 1962. Iron transport in pea plants. *Plant Physiol.* **37**:539–545.

Brauner, L., 1930. Über polare Permeabilität. *Ber. Deut. Bot. Ges.* **49**:109–118.

———, 1956. Die Permeabilität der Zellwand. *Encycl. Plant Physiol.* **2**:337–357.

——— and M. Hasman, 1946. Untersuchungen über die anomale Komponente des osmotischen Potentials Ubender Pflanzenzellen. *Rev. Fac. Sci. Univ. Istanbul, Ser.* **B11**:1–37.

———, 1952. Weitere Untersuchungen über den Wirkungsmechanismus des Heteroauxins bei der Wasseraufnahme von Pflanzenparenchymen. *Protoplasma* **41**:302–326.

Bray, J. R., 1963. Root production and the estimation of net productivity. *Can. J. Bot.* **41**:65–72.

Bray, R. H., 1954. A nutrient mobility concept of soil-plant relationships. *Soil Sci.* **78**:9–22.

Breazeale, E. L., and W. T. McGeorge, 1953. Exudation pressure in roots of tomato plants under humid conditions. *Soil Sci.* **75**:293–298.

Breazeale, J. F., and F. J. Crider, 1934. Plant association and survival, and the build-up of moisture in semi-arid soils. *Ariz. Agr. Exp. Sta. Tech. Bull.* 53.

Briggs, G. E., 1967. "Movement of Water in Plants." F. A. Davis Company, Philadelphia.

———, A. B. Hope, and R. N. Robertson, 1961. "Electrolytes and Plant Cells." Blackwell Scientific Publications, Ltd., Oxford.

——— and R. N. Robertson, 1957. Apparent free space. *Ann. Rev. Plant Physiol.* **8**:11–30.

Briggs, L. J., and H. L. Shantz, 1911. A wax seal method for determining the lower limit of available soil moisture. *Bot. Gaz.* **51**:210–219.

——— and ———, 1912. The relative wilting coefficients for different plants. *Bot. Gaz.* **53**:229–235.

Brix, H., 1962. The effect of water stress on the rates of photosynthesis and respiration in tomato plant and loblolly pine seedlings. *Physiol. Plant.* **15**:10–20.

———, 1966. Errors in measurement of leaf water potential of some woody plants with the Schardakow dye method. *Forest. Branch Dept. Publ. no.* 1164, *Can. Dept. Forest. Rural Develop.*

Brooks, S. C., and M. M. Brooks, 1941. The permeability of living cells. *Protoplasma Monogr.* 19.

Brouwer, R., 1953. Water absorption by the roots of *Vicia faba* at various transpiration strengths. I, II. *Proc. Kon. Ned. Akad. Wet.* **C56**:106–115, 129–136.

———, 1954*a*. The regulating influence of transpiration and suction tension on the water and salt uptake by the roots of intact *Vicia faba* plants. *Acta Bot. Neer.* **3**:264–312.

———, 1954*b*. Water absorption by the roots of *Vicia faba* at various transpiration strengths. III. Changes in water conductivity artificially obtained. *Proc. Kon. Ned. Akad. Wet.* **C57**:68–80.

———, 1963. The influence of the suction tension of the nutrient solutions on growth, transpiration and diffusion pressure deficit of bean leaves (*Phaseolus vulgaris*). *Acta Bot. Neer.* **12**:248–261.

———, 1964. Responses of bean plants to root temperatures. I. Root temperatures and growth in the vegetative stage. *Jaarb. I.B.S.* 11–22.

———, 1965*a*. Water movement across the root. *Symp. Soc. Exp. Biol.* **19**:131–149.

———, 1965*b*. Ion absorption and transport in plants. *Ann. Rev. Plant Physiol.* **16**:241–266.

——— and A. Hoogland, 1964. Responses of bean plants to root temperatures. II. Anatomical aspects. *Jaarb. I.B.S.* 23–31.

Brown, D. S., 1952. Relation of irrigation practice to the differentiation and development of apricot flower buds. *Bot. Gaz.* **114**:95–102.

Brown, E. M., 1939. Some effects of temperature on the growth and chemical composition of certain pasture grasses. *Mo. Agr. Exp. Sta. Res. Bull.* 299.

Brown, H. T., and F. Escombe, 1900. Static diffusion of gases and liquids in relation to the assimilation of carbon and translocation of plants. *Phil. Trans. Roy. Soc. (London)* **B193:**223–291.

Brown, J. C., 1963. Iron chlorosis in soybeans as related to the genotype of root stock. 4. Sorption and translocation as two separable phases of iron nutrition. *Soil Sci.* **96:**387–394.

———, L. O. Tiffin, R. S. Holmes, A. W. Specht, and J. W. Resnicky, 1959. Internal inactivation of iron in soybeans as affected by root growth medium. *Soil Sci.* **87:**89–94.

Brown, R., 1947. The gaseous exchange between the root and the shoot of the seedling of *Cucurbita pepo. Ann. Bot.* **11:**417–437.

Brown, W. V., 1952. The relation of soil moisture to cleistogamy in *Stipa leucotricha. Bot. Gaz.* **113:**438–444.

——— and G. A. Pratt, 1965. Stomatal inactivity in grasses. *Southwest. Natur.* **10**(1):48–56.

Broyer, T. C., 1947. The movement of materials into plants. Part I. Osmosis and the movement of water into plants. *Bot. Rev.* **13:**1–58.

———, 1950. Further observations on the absorption and translocation of inorganic solutes using radioactive isotopes with plants. *Plant Physiol.* **25:**367–377.

Brumfield, R. T., 1942. Cell growth and division in living root meristems. *Amer. J. Bot.* **29:**533–543.

Bryant, A. E., 1934. Comparison of anatomical and histological differences between roots of barley grown in aerated and nonaerated culture solutions. *Plant Physiol.* **9:**389–391.

Buckingham, E. A., 1907. Studies of the movement of soil moisture. *U.S. Dept. Agr. Bull.* 38.

Bukovac, M. J., and S. H. Wittwer, 1951. Absorption and mobility of foliar applied nutrients. *Plant Physiol.* **32:**429–435.

Bunger, M. T., and H. J. Thomson, 1938. Root development as a factor in the success or failure of windbreak trees in the southern high plains. *J. Forest.* **36:**790–803.

Burgerstein, A., 1920. "Die Transpiration der Pflanzen." II. G. Fischer, Jena.

Burström, H., 1947. A preliminary study on mineral nutrition and cell elongation of roots. *Kgl. Fysiogr. Sällsk. Lund Förh.* **17**(1):1–11.

———, 1953. Studies on metabolism of roots. IX. Cell elongation and water absorption. *Physiol. Plant.* **6:**262–276.

———, 1956. Temperature and root cell elongation. *Physiol. Plant.* **9:**682–692.

———, 1959. Growth and formation of intercellularies in root meristems. *Physiol. Plant.* **12:**371–385.

———, 1962. Influence of azide on the permeability of Rhoeo cells. *Indian J. Plant Physiol.* **5:**88–96.

———, 1965. The physiology of plant roots. In K. F. Baker, W. C. Snyder, et al. (eds.), "Ecology of Soil-borne Plant Pathogens," pp. 154–169. Univ. of California Press, Berkeley.

———, I. Uhrström, and R. Wurscher, 1967. Growth, turgor, water potential, and Young's modulus in pea internodes. *Physiol. Plant.* **20:**213–231.

Burton, G. W., G. M. Prine, and J. E. Jackson, 1957. Studies of drouth tolerance and water use of several southern grasses. *Agron. J.* **49:**498–503.

———, E. H. DeVane, and R. L. Carter, 1954. Root penetration, distribution and activity in southern grasses measured by yields, drought symptoms, and P^{32} uptake. *Agron. J.* **46:**229–233.

Burvill, G. H., 1947. Soil salinity in the agricultural area of Western Australia. *J. Aust. Inst. Agr. Sci.* **13:**9–19.

Büsgen, M., and E. Münch, 1926. "The Structure and Life of Forest Trees," 3d ed., Eng. trans. by Thomson. John Wiley & Sons, Inc., New York.

Bushnell, J., 1941. Exploratory tests of subsoil treatments inducing deeper rooting of potatoes on Wooster silt loam. *J. Amer. Soc. Agron.* **33:**823–828.

Buswell, A. M., and W. H. Rodebush, 1956. Water. *Sci. Amer.* **194**(4):77–89.

Butler, G. W., 1953. Ion uptake by young wheat plants. II. The "apparent free space" of wheat roots. *Physiol Plant.* **6:**617–635.

Butler, P. F., and J. A. Prescott, 1955. Evapotranspiration from wheat and pasture in relation to available moisture. *Aust. J. Agr. Res.* **6:**52–61.

Buttery, B. R., and S. G. Boatman, 1966. Manometric measurement of turgor pressures in laticiferous phloem tissues. *J. Exp. Bot.* **17:**283.

Buvat, R., 1963. Electron microscopy of plant protoplasm. *Int. Rev. Cytol.* **14:**41–155.

Byers, R. B., 1959. "General Meteorology," 3d ed. McGraw-Hill Book Company, New York.

Cameron, S. H., 1941. The influence of soil temperature on the rate of transpiration of young orange trees. *Proc. Amer. Soc. Hort. Sci.* **38:**75–79.

Campbell, R. B., C. A. Bower, and L. A. Richards, 1949. Change of electrical conductivity with temperature and the relation of osmotic pressure to electrical conductivity and ion concentration for soil extracts. *Soil Sci. Soc. Amer. Proc.* **13:**66–69.

Campbell, W. A., and O. L. Copeland, 1954. Littleleaf disease of shortleaf and loblolly pines. *U.S. Dept. Agr. Circ.* 940.

Canning, R. E., and P. J. Kramer, 1958. Salt absorption and accumulation in various regions of roots. *Amer. J. Bot.* **45:**378–382.

Cannon, W. A., 1911. Root habits of desert plants. *Carnegie Inst. Wash. Publ.* 131.

———, 1932. Absorption of oxygen by roots when the shoot is in darkness or in light. *Plant Physiol.* **7:**673–684.

Carolus, R. L., A. E. Erickson, E. H. Kidder, and Z. R. Wheaton, 1965. The interaction of climate and soil moisture on water use, growth, and development of tomatoes. *Mich. State Univ. Agr. Exp. Sta. Quart. Bull.* **47:**542.

Carter, J. C., 1945. Wet wood of elms. *Ill. Natur. Hist. Surv.* **23**(4):401–448.

Carter, M. C., and H. S. Larsen, 1965. Soil nutrients and loblolly pine xylem sap composition. *Forest Sci.* **11:**216–220.

Cary, J. W., and S. A. Taylor, 1967. The dynamics of soil water. Part II. Temperature and solute effects. In R. M. Hagan et al. (eds.), "Irrigation of Agricultural Lands," pp. 245–253. *Amer. Soc. Agron.,* Madison, Wis.

Čatsky, J., 1965. Leaf-disc method for determining water saturation deficit. *Arid Zone Res.* **25:**353–360, UNESCO, Paris.

Caughey, M. G., 1945. Water relations of pocosins or bog shrubs. *Plant Physiol.* **20:**671–689.

Chang, C. W., and R. S. Bandurski, 1964. Exocellular enzymes of corn roots. *Plant Physiol.* **39:**60–64.

Chang, H. T., and W. E. Loomis, 1945. Effect of carbon dioxide on absorption of water and nutrients by roots. *Plant Physiol.* **20:**221–232.

Chang, J., 1961. Microclimate of sugar cane. *Hawaii. Plant. Rec.* **56**(3):195–225.

Chapman, H. D., and E. R. Parker, 1942. Weekly absorption of nitrate by young bearing orange trees growing out of doors in solution cultures. *Plant Physiol.* **17:**366–376.

Chen, D., B. Kessler, and S. P. Monselise, 1964. Studies on water regime and nitrogen metabolism of citrus seedlings grown under water stress. *Plant Physiol.* **39:**379–386.

Chibnall, A. C., 1939. "Protein Metabolism in the Plant," pp. 265–266. Yale Univ. Press, New Haven.

———, 1954. Protein metabolism in rooted runner bean leaves. *New Phytol.* **53**:31–38.

Childers, N. F., and D. G. White, 1942. Influence of submersion of the roots on transpiration, apparent photosynthesis, and respiration of young apple trees. *Plant Physiol.* **17**:603–618.

Clark, J., and R. D. Gibbs, 1957. Studies in tree physiology. IV. Further investigations of seasonal changes in moisture content of certain Canadian forest trees. *Can. J. Bot.* **35**:219–253.

Clark, J. A., and J. Levitt, 1956. The basis of drought resistance in the soybean plant. *Physiol. Plant.* **9**:598–606.

Clark, W. S., 1874. The circulation of sap in plants. *Mass. State Board Agr. Ann. Rep.* **21**:159–204.

———, 1875. Observations upon the phenomena of plant life. *Mass. State Board Agr. Ann. Rep.* **22**:204–312.

Clausen, J. J., and T. T. Kozlowski, 1965. Use of the relative turgidity technique for measurement of water stresses in gymnosperm leaves. *Can. J. Bot.* **43**:305–316.

Clements, F. E., 1921. Aeration and air content. *Carnegie Inst. Wash. Publ.* 315.

——— and F. L. Long, 1934. The method of collodion films for stomata. *Amer. J. Bot.* **21**:7–17.

——— and E. V. Martin, 1934. Effect of soil temperature on transpiration in *Helianthus annuus. Plant Physiol.* **9**:619–630.

Clements, H. F., 1934. Significance of transpiration. *Plant Physiol.* **9**:165–172.

——— and T. Kubota, 1942. Internal moisture relations of sugar cane—The selection of a moisture index. *Hawaii. Plant. Rec.* **46**:17–36.

Cline, J. F., 1953. Absorption and metabolism of tritium and tritium gas by bean plants. *Plant Physiol.* **28**:717–723.

Coile, T. S., 1936. Soil samplers. *Soil Sci.* **42**:139–142.

———, 1937. Distribution of forest tree roots in North Carolina Piedmont soils. *J. Forest.* **35**:247–257.

———, 1940. Soil changes associated with loblolly pine succession on abandoned agricultural land of the Piedmont plateau. *Duke Univ. Sch. Forest. Bull.* 5.

Collander, R., 1941. Selective absorption of cations by higher plants. *Plant Physiol.* **16**:691–720.

———, 1957. Permeability of plant cells. *Ann. Rev. Plant Physiol.* **8**:335–348.

———, 1959. Cell membranes: Their resistance to penetration and their capacity for transport. In F. C. Steward (ed.), "Plant Physiology," vol. 2, pp. 4–102. Academic Press Inc., New York.

Colman, E. A., 1947. A laboratory procedure for determining the field capacity of soils. *Soil Sci.* **63**:277–283.

——— and T. M. Hendrix, 1949. The fiberglas electrical soil-moisture instrument. *Soil Sci.* **67**:425–438.

——— and G. B. Bodman, 1945. Moisture and energy conditions during downward entry of water into moist and layered soils. *Proc. Soil Sci. Soc. Amer.* **9**:3–11.

Conway, V. M., 1940. Aeration and plant growth in wet soils. *Bot. Rev.* **6**:149–163.

Cooil, B. J., R. K. de la Fuente, and R. S. de la Pena, 1965. Absorption and transport of sodium and potassium in squash. *Plant. Physiol.* **40**:625–632.

Cooper, A. J., 1958. Observations on growth trends of the tomato plant throughout the whole of the growing season. *J. Hort. Sci.* **33**:43–48.

Cooper, W. C., B. S. Gorton, and E. O. Olson, 1952. Ionic accumulation in citrus as influenced by rootstock and scion and concentration of salts and boron in the substrate. *Plant Physiol.* **27**:191–203.

Cormack, R. G. H., 1944. The effect of environmental factors on the development of root hairs in *Phleum pratense* and *Sporobolus cryptandrus. Amer. J. Bot.* **31:**443–449.

———, 1945. Cell elongation and the development of root hairs in tomato roots. *Amer. J. Bot.* **32:**490–496.

———, 1962. Development of root hairs in angiosperms. II. *Bot. Rev.* **28:**446–464.

Coult, D. A., 1964. Observations on gas movement in the rhizome of *Menyanthes trifoliata* L. with comments on the role of the endodermis. *J. Exp. Bot.* **15:**205–218.

Cowan, I. R., 1965. Transport of water in the soil-plant-atmosphere system. *J. Appl. Ecol.* **2:**221:239.

——— and F. L. Milthorpe, 1967. Resistance to water transport in plants—a misconception misconceived. *Nature* **213:**740–741.

——— and ———, 1968 Plant factors influencing the water status of plant tissues. In T. T. Kozlowski (ed.), "Water Deficits and Plant Growth," vol. 1, 137–193. Academic Press Inc., New York.

Crafts, A. S., 1936. Further studies on exudation in cucurbits. *Plant Physiol.* **11:**63–79.

———, 1931. Movement of organic materials in plants. *Plant Physiol.* **6:**1–41.

———, 1961. "The Chemistry and Mode of Action of Herbicides." Interscience Publishers, Inc., New York. 269 pp.

———, 1968. Water deficits and physiological processes. In T. T. Kozlowski (ed.), "Water Deficits and Plant Growth," vol. 2, pp. 85–133. Academic Press Inc., New York.

——— and T. C. Broyer, 1938. Migration of salts and water into xylem of the roots of higher plants. *Amer. J. Bot.* **25:**529–535.

———, H. B. Currier, and C. R. Stocking, 1949. "Water in the Physiology of the Plant." Chronica Botanica Co., Waltham, Mass.

——— and C. L. Foy, 1962. The chemical and physical nature of plant surfaces in relation to the use of pesticides and their residues. *Residue Rev.* **1:**112–139. Academic Press Inc., New York.

——— and S. Yamaguchi, 1960. Absorption of herbicides by roots. *Amer. J. Bot.* **47:**248–255.

Crawford, R. M. M., 1967. Alcohol dehydrogenase activity in relation to flooding tolerance in roots. *J. Exp. Bot.* **18:**458–464.

Crider, F. J., 1933. Selective absorption of ions not confined to young roots. *Science* **78:**169.

Crowdy, S. H., 1959. Uptake and translocation of organic chemicals by higher plants. In C. S. Holton et al. (eds.), "Plant Pathology: Problems and Progress." Univ. of Wisconsin Press, Madison, Wis.

Currier, H. B., 1944. Water relations of root cells of *Beta vulgaris. Amer. J. Bot.* **31:**378–387.

Curtis, L. C., 1943. Deleterious effects of guttated fluids on foliage. *Amer. J. Bot.* **30:**778–781.

———, 1944. The exudation of glutamine from lawn grass. *Plant Physiol.* **19:**1–5.

Curtis, O. F., 1926. What is the significance of transpiration? *Science* **63:**267–271.

———, 1936. Leaf temperatures and the cooling of leaves by radiation. *Plant Physiol.* **11:**343–364.

———, 1937. Vapor pressure gradients, water distribution in fruits and so-called infra-red injury. *Amer. J. Bot.* **24:**705–710.

——— and D. G. Clark, 1950. "An Introduction to Plant Physiology." McGraw-Hill Book Company, New York.

Dainty, J., 1963*a*. Water relations of plant cells. In R. D. Preston (ed.), "Advances in Botanical Research," vol. 1, pp. 279–326. Academic Press Inc., New York.

———, 1963*b*. The polar permeability of plant cell membranes to water. *Protoplasma* **57**:220–228.

———, 1965. Osmotic flow. *Soc. Exp. Biol. Symp.* **19**:75–85. Academic Press Inc., New York.

——— and B. Z. Ginzburg, 1964*a*. The measurement of hydraulic conductivity (osmotic permeability to water) of internodal Characean cells by means of transcellular osmosis. *Biochim. Biophys. Acta* **79**:102–111.

——— and ———, 1964*b*. The permeability of the cell membranes of *Nitello translucens* to urea, and the effects of high concentrations of sucrose on this permeability. *Biochim. Biophys. Acta* **79**:112–121.

——— and A. B. Hope, 1959. The water permeability of cells of *Chara Australis* R. BR. *Aust. J. Biol. Sci.* **12**:136–146.

Daniels, F., and R. A. Alberty, 1963. "Physical Chemistry," 2d ed. John Wiley & Sons, Inc., New York.

Danielson, R. E., 1967. Root systems in relation to irrigation. In R. M. Hagan et al. (eds.), "Irrigation of Agricultural Lands," pp. 390–424. *Amer. Soc. Agron.*, Madison, Wis.

Darwin, F., and D. F. M. Pertz, 1911. On a new method of estimating the aperture of stomata. *Proc. Roy. Soc.* (London) **B84**:136–154.

Daum, C. R., 1967. A method for determining water transport in trees. *Ecology* **48**:425–431.

Davis, C. H., 1940. Absorption of soil moisture by maize roots. *Bot. Gaz.* **101**:791–805.

Davis, R. M., and J. C. Lingle, 1961. Basis of shoot response to root temperature in tomato. *Plant Physiol.* **36**:153–161.

Davson, H., and J. F. Danielli, 1952. "The Permeability of Natural Membranes." Cambridge Univ. Press. New York.

Day, P. R., G. H. Bolt, and D. M. Anderson, 1967. Nature of soil water. In R. M. Hagan et al. (eds.), "Irrigation of Agricultural Lands," pp. 193–208. *Amer. Soc. Agron.*, Madison, Wis.

Decker, J. P., W. G. Gaylor, and F. D. Cole, 1962. Measuring transpiration of undisturbed tamarisk shrubs. *Plant Physiol.* **37**:393–397.

——— and C. M. Skau, 1964. Simultaneous studies of transpiration rate and sap velocity in trees. *Plant Physiol.* **39**:213–215.

——— and B. F. Wetzel, 1957. A method for measuring transpiration of intact plants under controlled light, humidity, and temperature. *Forest Sci.* **3**:350–354.

——— and J. D. Wien, 1960. Transpirational surges in tamarix and eucalyptus as measured with an infrared gas analyzer. *Plant Physiol.* **35**:340–343.

Denaeyer-DeSmet, S., 1967. Contribution a l'étude chimique de la sève du bois de *Corylus avellana* L. *Bull. Soc. Roy. Bot. Belg.* **100**:353-371.

Denmead, O. T., and R. H. Shaw, 1962. Availability of soil water to plants as affected by soil moisture content and meteorological conditions. *Agron. J.* **54**:385–390.

Denny, F. E., 1917. Permeability of certain membranes to water. *Bot. Gaz.* **63**:373–397.

———, 1932. Changes in leaves during the night. *Contrib. Boyce Thompson Inst.* **4**:65–83.

De Plater, C. V., 1955. Portable capacitance-type soil moisture meter. *Soil Sci.* **80**:391–395.

De Roo, H. C., 1957. Root growth in Connecticut tobacco soils. *Conn. Agr. Exp. Sta. Bull.* 608.

———, 1961. Deep tillage and root growth. *Conn. Agr. Exp. Sta. Bull.* 644.

Diamond, J. M., 1965. The mechanism of isotonic water absorption and secretion. *Soc. Exp. Biol. Symp.* **19**:329-348.

Dick, D. A. T., 1966. "Cell Water." Butterworths Inc., Washington, D.C.

Van Die, J., 1958. On the occurrence of α-keto acids and organic nitrogen compounds in xylem exudates of cucumber and tomato plants. *Kon. Ned. Akad. Wetensch. Proc.* **C61:**572–578.

Diebold, C. H., 1954. Effect of tillage practices upon intake rates, run off and soil losses of dry farm land soils. *Proc. Soil Sci. Soc. Amer.* **18:**88–91.

Dimbleby, G. W., 1952. The root sap of birch on a podzol. *Plant and Soil* **4:**141–153.

Dimond, A. E., 1955. Pathogenesis in the wilt diseases. *Ann. Rev. Plant Physiol.* **6:**329–350.

———, 1966. Pressure and flow relations in vascular bundles of the tomato plant. *Plant Physiol.* **41:**119–131.

———, 1967. Physiology of wilt diseases. In C. J. Mirocha and I. Uritanic (eds.), "The Dynamic Role of Molecular Constituents in Plant-parasite Interaction." *Amer. Phytopathol. Soc.,* St. Paul, Minn.

Dittmer, H. J., 1937. A quantitative study of the roots and root hairs of a winter rye plant (*Secale cereale*). *Amer. J. Bot.* **24:**417–420.

Dixon, H. H., 1914. Transpiration and the ascent of sap in plants. The Macmillan Company, New York.

——— and J. Joly, 1895. The path of the transpiration current. *Ann. Bot.* **9:**416–419.

Doneen, L. D., and J. H. MacGillivray, 1946. Suggestions on irrigating commercial truck crops. Calif. Agr. Exp. Sta. Lithoprint.

Döring, B., 1935. Die Temperaturabhängigkeit der Wasseraufnahme und ihre ökologische Bedeutung. *Z. Bot.* **28:**305–383.

Dove, L. D., 1967. Ribonuclease activity of stressed tomato leaflets. *Plant Physiol.* **42:**1176–1178.

Drost-Hansen, W., 1965. The effects on biologic systems of higher-order phase transitions in water. *Ann. N.Y. Acad. Sci.* **125** (art. 2):471–501.

van Duin, R. H. A., 1955. Tillage in relation to rainfall intensity and infiltration capacity of soils. *Neth. J. Agr. Sci.* **3:**189–191.

Duncan, H. F., and D. A. Cooke, 1932. A preliminary investigation on the effect of temperature on root absorption of the sugar cane. *Hawaii. Plant. Rec.* **36:**31–39.

Durbin, R. P., and F. G. Moody, 1965. Water movement through a transporting epithelial membrane: the gastric mucosa. *Soc. Exp. Biol. Symp.* **19:**299–306.

Duvdevani, A., 1964. Dew in Israel and its effect on plants. *Soil Sci.* **98:**14–21.

Dybing, C. D., and H. B. Currier, 1961. Foliar penetration by chemicals. *Plant Physiol.* **36:**169–174.

Dyer, A. J., 1961. Measurements of evaporation and heat transfer in the lower atmosphere by an automatic eddy correlation technique. *Quart. J. Roy. Meteorol. Soc.* **87:**401–412.

Dzerdzeevskii, B. L., 1958. On some climatological problems and microclimatological studies of arid and semi-arid regions in U.S.S.R. *Arid Zone Res.* **16:**315–325. UNESCO, Paris.

Eames, A. J., and L. H. MacDaniels, 1947. "An Introduction to Plant Anatomy," 2d ed. McGraw-Hill Book Company, New York.

Eaton, F. M., 1931. Root development as related to character of growth and fruitfulness of the cotton plant. *J. Agr. Res.* **43:**875–883.

———, 1941. Water uptake and root growth as influenced by inequalities in the concentration of the substrate. *Plant Physiol.* **16:**545–564.

———, 1942. Toxicity and accumulation of chloride and sulfate salts in plants. *J. Agr. Res.* **64:**357–399.

———, 1943. The osmotic and vitalistic interpretations of exudation. *Amer. J. Bot.* **30:**663–674.

——— and D. R. Ergle, 1948. Carbohydrate accumulation in the cotton plant at low moisture levels. *Plant Physiol.* **23**:169–187.

——— and H. E. Joham, 1944. Sugar movement to roots, mineral uptake, and the growth cycle of the cotton plant. *Plant Physiol.* **19**:507–518.

Eckardt, F. E., 1960. Eco-physiological measuring techniques applied to research on water relations of plants in arid and semi-arid regions. *Arid Zone Res.* **15**:139–171. UNESCO, Paris.

Edlefsen, N. E., 1941. Some thermodynamic aspects of the use of soil-moisture by plants. *Trans. Amer. Geophys. Union* **22**:917–940.

——— and G. B. Bodman, 1941. Field measurements of water movement through a silt loam soil. *Amer. Soc. Agron. J.* **33**:713–731.

Ehlig, C. F., 1960. Effect of salinity on four varieties of table grapes grown in sand culture. *Proc. Amer. Soc. Hort. Sci.* **76**:323–335.

———, 1962. Measurement of energy status of water in plants with a thermocouple psychrometer. *Plant Physiol.* **37**:288–290.

——— and W. R. Gardner, 1964. Relationship between transpiration and the internal water relations of plants. *Agron. J.* **56**:127–130.

Ehrler, W. L., 1963. Water absorption of alfalfa as affected by low root temperature and other factors of a controlled environment. *Agron. J.* **55**:363.

——— and C. H. M. van Bavel, 1968. Leaf diffusion resistance, illuminance, and transpiration. *Plant Physiol.* **43**:208–214.

———, C. H. M. van Bavel, and F. S. Nakayama, 1966. Transpiration, water absorption, and internal water balance of cotton plants as affected by light and changes in saturation deficit. *Plant Physiol.* **41**:71–74.

——— and L. Bernstein, 1958. Effects of root temperature, mineral nutrition and salinity on the growth and composition of rice. *Bot. Gaz.* **120**:67–74.

———, F. S. Nakayama, and C. H. M. van Bavel, 1965. Cyclic changes in water balance and transpiration of cotton leaves in a steady environment. *Physiol. Plant.* **18**:766–775.

van Eijk, M., 1939. Analyze der Wirkung des NaCl auf die Entwicklung, Sukkulenz und Transpiration bei *Salicornia herbacea,* sowie Untersuchungen über den Einfluss der Salzaufnahme auf die Wurzelatmung bei *Aster Tripolium. Rec. Trav. Bot. Neer.* **36**:559–657.

Ekern, P. C., 1965. Evapotranspiration of pineapple in Hawaii. *Plant Physiol.* **40**:736–739.

Elazari-Volcani, T., 1936. The influence of a partial interruption of the transpiration stream by root pruning and stem incisions on the turgor of citrus trees. *Palestine J. Bot. Hort. Sci.* **1**:94–96.

Elrick, D. E., and C. B. Tanner, 1955. Influence of sample pretreatment on soil moisture retention. *Soil Sci. Soc. Amer. Proc.* **19**:279–282.

Emmert, F. H., 1961. Evidence of a barrier to lateral penetration of P-32 across roots of intact transpiring plants, based on measurements of xylem stream composition. *Physiol. Plant.* **14**:478–487.

Emmert, E. M., and F. K. Ball, 1933. The effect of soil moisture on the availability of nitrate, phosphate and potassium to the tomato plant. *Soil Sci.* **35**:295–306.

Engel, H., and I. Friederichsen, 1952. Weitere Untersuchungen über periodische Guttation etiolierter Haferkeimlinge. *Planta* **40**:529–549.

England, C. B., 1965. Changes in fiber-glass soil-moisture electrical-resistance elements in long-term installations. *Soil Sci. Soc. Amer. Proc.* **29**:229–231.

———, and E. H. Lesesne, 1962. Evapotranspiration research in Western North Carolina. *Agr. Eng.* **43**:526–528.

Ensgraber, A., 1954. Uber den Einfluss der Antrocknung auf die Assimilation und Atmung von Moosen und Flechten. *Flora* **141**:432–475.

Epstein, E., 1955. Passive permeation and active transport of ions in plant roots. *Plant Physiol.* **30:**529–535.

———, 1956. Passive passage and active transport of ions in plant roots. *U.S. At. Energy Comm. Rep.* TID–7512:297–301.

———, 1960. Spaces, barriers, and ion carriers: ion absorption in plants. *Amer. J. Bot.* **47:**393–399.

———, 1961. The essential role of calcium in selective cation absorption. *Plant Physiol.* **36:**437–444.

———, 1965. Mineral metabolism. In J. Bonner and J. E. Varner (eds.), "Plant Biochemistry," pp. 438–466. Academic Press Inc., New York.

———, 1966. Dual pattern of ion absorption by plant cells and by plants. *Nature* **212:**1324–1327.

——— and R. L. Jefferies, 1964. The genetic basis of selective ion transport in plants. *Ann. Rev. Plant Physiol.* **15:**169–184.

——— and J. L. Leggett, 1954. The absorption of alkaline earth cations by barley roots: kinetics and mechanism. *Amer. J. Bot.* **41:**785–792.

———, D. W. Rains, and W. E. Schmid, 1962. Course of cation absorption by plant tissue. *Science* **136:**1051–1052.

Erickson, A. E., 1965. Short-term oxygen deficiencies and plant responses. In Amer. Soc. Agr. Eng., Conference on Drainage for Efficient Crop Production, pp. 11–12, 23.

Erickson, L. C., 1946. Growth of tomato roots as influenced by oxygen in the nutrient solution. *Amer. J. Bot.* **33:**551–561.

Esau, K., 1941. Phloem anatomy of tobacco affected with curly top and mosaic. *Hilgardia* **13:**437–470.

———, 1943. Vascular differentiation in the pear root. *Hilgardia* **15:**299–324.

———, 1965. "Plant Anatomy," 2d ed. John Wiley & Sons, Inc., New York.

Evenari, M., 1960. Plant physiology and arid zone research. *Arid Zone Res.* **18:**175–195. UNESCO, Paris.

———, 1961. Chemical influences of other plants (allelopathy). In W. Ruhland (ed.), "Encyclopedia of Plant Physiology," vol. 16, pp. 691–736. Springer-Verlag OHG, Berlin.

Ewart, G. Y., 1951. The mechanics of field irrigation scheduling utilizing Bouyoucos blocks. *Agr. Eng.* **32:**148–151, 154.

Fadeel, A. A., 1963. Assimilation of carbon dioxide by chlorophyllous roots. *Physiol. Plant.* **16:**870–888.

Falk, M., and G. S. Kell, 1966. Thermal properties of water: discontinuities questioned. *Science* **154:**1013–1014.

Falk, S., C. H. Hertz, and H. I. Virgin, 1958. On the relation between turgor pressure and tissue rigidity. I. Experiments on resonance frequency and tissue rigidity. *Physiol. Plant.* **11:**802–818.

Falk, S. O., 1966. A microwave hygrometer for measuring plant transpiration. *Z. Pflanzenphysiol.* **55:**31–37.

Farkas, G. L., and T. Rajhathy, 1955. Untersuchungen über die xeromorphischen Gradienten einiger Kulturpflanzen. *Planta* **45:**535–548.

Fensom, D. S., 1957. The bioelectric potentials of plants and their functional significance. *Can. J. Bot.* **35:**573–582.

———, 1958. The bioelectric potentials of plants. II. The patterns of bio-electric potential and exudation rate in excised sunflower roots and stems. *Can. J. Bot.* **36:**367–383.

Ferguson, H., and W. H. Gardner, 1962. Water content measurement in soil columns by gamma ray absorption. *Soil Sci. Soc. Amer. Proc.* **26**:11–14.

Figdor, W., 1898. Untersuchungen über die Erscheinung des Blutungsdruckes in den Tropen. *Sitzungsber. Wien. Akad. Abt.* 1, **107**:639–669.

Firbas, F., 1931. Untersuchungen über den Wasserhaushalt der Hochmoorpflanzen. *Jahrb. wiss. Bot.* **74**:457–696.

Fishback, P. E., and F. L. Duley, 1950. Intake of water by claypan soils. *Proc. Soil Sci. Soc. Amer.* **15**:404–408.

Fisher, J. E., 1964. Evidence of circumnutational growth movements of rhizomes of *Poa pratensis* L. that aid in soil penetration. *Can. J. Bot.* **35**:339–347.

Fitting, H., 1911. Die Wasserversorgung und die osmotischen Druckverhältnisse der Wüstenpflanzen. *Z. Bot.* **3**:209–275.

Fitzgerald, P. D., and D. S. Rickard, 1960. A comparison of Penman's and Thornthwaite's method of determining soil moisture deficits. *N. Z. J. Agr. Res.* **3**:106–112.

Fletcher, J. E., 1939. A dielectric method for determining soil moisture. *Soil Sci. Soc. Amer. Proc.* **4**:84–88.

Fox, D. G., 1933. Carbon dioxide narcosis. *J. Cell. Comp. Physiol.* **3**:75–100.

Foy, C. D., and S. A. Barber, 1958. Magnesium absorption and utilization by two inbred lines of corn. *Proc. Soil Sci. Soc. Amer.* **22**:57–62.

Franck, J., and J. E. Mayer, 1947. An osmotic diffusion pump. *Arch. Biochem.* **14**:297–313.

Franco, C. M., and A. C. Magalhaes, 1965. Techniques for the measurement of transpiration of individual plants. *Arid Zone Res.* **25**:211–224. UNESCO, Paris.

Franke, W., 1967. Mechanisms of foliar penetration of solutions. *Ann. Rev. Plant Physiol.* **18**:281–300.

Fraser, D. A., 1957. Annual and seasonal march of soil temperature on several sites under a hardwood stand. *Can. Dept. N. Aff. Natur. Resour. Forest Res. Div. Tech.* Note 56.

———, 1962. Tree growth in relation to soil moisture. In T. T. Kozlowski (ed.), "Tree Growth." The Ronald Press Company, New York.

——— and C. A. Mawson, 1953. Movement of radioactive isotopes in yellow birch and white pine as detected with a portable scintillation counter. *Can. J. Bot.* **31**:324–333.

Frey-Wyssling, A., 1941. Die Guttation als allgemeine Erscheinung. *Ber. Schweiz. Bot. Ges.* **51**:321.

———, 1953. "Submicroscopic Morphology of Protoplasm." American Elsevier Publishing Company, New York.

——— and E. Häusermann, 1941. Über die Ausgleidung der Mesophyllinterzellularen. *Ber. Schweiz. Bot. Ges.* **51**:430–43.

——— and K. Mühlethaler, 1959. Über das submikroskopische Geschehen bei der Kutinisierung pflanzlicher Zellwande. *Vierteljahressch. Naturforsch. Ges. Zürich* **104**:294–299.

——— and ———, 1965. "Ultrastructural Plant Cytology." American Elsevier Publishing Company, New York.

Fried, M., and H. Broeshart, 1967. "The Soil-Plant System in Relation to Inorganic Nutrition." Academic Press Inc., New York.

——— and R. E. Shapiro, 1961. Soil-plant relationships in ion uptake. *Ann. Rev. Plant Physiol.* **12**:91–112.

Friesner, R. C., 1920. Daily rhythms of elongation and cell division in certain roots. *Amer. J. Bot.* **7**:380–406.

———, 1940. An observation on the effectiveness of root pressure in the ascent of sap. *Butler Univ. Bot. Stud.* **4**:226–227.

Fry, K. E., and R. B. Walker, 1967. A pressure-infiltration method for estimating stomatal opening in conifers. *Ecology* **48**:155–157.

Fulton, J. M., and A. E. Erickson, 1964. Relation between soil aeration and ethyl alcohol accumulation in xylem exudates of tomatoes. *Soil Sci. Soc. Amer. Proc.* **28**:610–614.

Furkova, N. S., 1944. Growth reactions in plants under excessive watering. *C. R.* (Doklady) *Acad. Sci.* **42**:87–90.

Furr, J. R., and W. W. Aldrich, 1943. Oxygen and carbon dioxide changes in the soil atmosphere of an irrigated date garden on calcareous very fine sandy loam soil. *Proc. Amer. Soc. Hort. Sci.* **42**:46–52.

——— and J. O. Reeve, 1945. The range of soil-moisture percentages through which plants undergo permanent wilting in some soils from semiarid irrigated areas. *J. Agr. Res.* **71**:149–170.

——— and C. A. Taylor, 1939. Growth of lemon fruits in relation to moisture content of the soil. *U.S. Dept. Agr. Tech. Bull.* 640.

Gaastra, P., 1959. Photosynthesis of crop plants as influenced by light, carbon dioxide, temperature, and stomatal diffusion resistances. *Meded. Landbouwhogesch. Wageningen* **59**:1–68.

Gaff, D. F., 1966. The sulfhydryl-disulphide hypothesis in relation to desiccation injury of cabbage leaves. *Aust. J. Biol. Sci.* **19**:291–299.

———, T. C. Chambers, and K. Markus, 1964. Studies of extrafascicular movement of water in the leaf. *Aust. J. Biol. Sci.* **17**:581–586.

Gale, J., and R. M. Hagan, 1966. Plant antitranspirants. *Ann. Rev. Plant Physiol.* **17**: 269–282.

———, H. C. Kohl, and R. M. Hagan, 1967. Changes in the water balance and photosynthesis of onion, bean and cotton plants under saline conditions. *Physiol. Plant.* **20**:408–420.

——— and A. Poljakoff-Mayber, 1965. Antitranspirants as a research tool for the study of the effects of water stress on plant behaviour. *Arid Zone Res.* **25**:269–274. UNESCO, Paris.

Gardner, W. R., 1958. Some steady state solutions of the unsaturated moisture flow equation with application to evaporation from a water table. *Soil Sci.* **85**:228–232.

———, 1960. Dynamic aspects of water availability to plants. *Soil Sci.* **89**:63–73.

———, 1964. Relation of root distribution to water uptake and availability. *Agron. J.* **56**:41–45.

——— and C. F. Ehlig, 1962. Some observations on the movement of water to plant roots. *Agron. J.* **54**:453–456.

——— and ———, 1963. The influence of soil water on transpiration by plants. *J. Geophys. Res.* **68**:5719–5724.

——— and ———, 1965. Physical aspects of the internal water relations of plant leaves. *Plant Physiol.* **40**:705–710.

——— and M. Fireman, 1958. Laboratory studies of evaporation from soil columns in the presence of a water table. *Soil Sci.* **85**:244–249.

——— and R. H. Nieman, 1964. Lower limit of water availability to plants. *Science* **143**:1460–1462.

Gast, P. R., 1937. Studies on the development of conifers in raw humus. III. The growth of Scots pine seedlings in pot cultures of different soils under varied radiation intensities. *Medd. Statens Skog.* **29**:587–682.

Gates, C. T., 1955. The response of the young tomato plant to a brief period of water shortage. I. The whole plant and its principal parts. *Aust. J. Biol. Sci.* **8**:196–214.

———, 1968. Water deficits and growth of herbaceous plants. In T. T. Kozlowski (ed.), "Water Deficits and Plant Growth," vol. 2, pp. 135–190. Academic Press Inc., New York.

——— and J. Bonner, 1959. The response of young tomato plants to a brief period of water shortage. IV. Effects of water stress on the ribonucleic acid metabolism of tomato leaves. *Plant Physiol.* **34**:49–55.

Gates, D. M., 1962. "Energy Exchange in the Biosphere." Harper & Row, Publishers, Incorporated, New York.

———, 1965. Energy, plants, and ecology. *Ecology* **46**:1–13.

———, 1966. Transpiration and energy exchange. *Quart. Rev. Biol.* **41**:353–364.

———, E. C. Tibbals, and L. Kreith, 1965. Radiation and convection for Ponderosa pine. *Amer. J. Bot.* **52**:66–71.

Geisler, G., 1963. Morphogenetic influence of (CO_2 + − HCO_3) on roots. *Plant Physiol.* **38**:77–80.

———, 1965. The morphogenetic effect of oxygen on roots. *Plant Physiol.* **40**:85–88.

Gill, W. R., and G. H. Bolt, 1955. Pfeffer's studies of the root growth pressures exerted by plants. *Agron. J.* **47**:166–168.

Girton, R. E., 1927. The growth of citrus seedlings as influenced by environmental factors. *Univ. Calif. Publ. Agr. Sci.* **5**:83–117.

Glasstone, V. F. C., 1942. Passage of air through plants and its relation to measurement of respiration and assimilation. *Amer. J. Bot.* **29**:156–159.

Glinka, Z., and Leonora Reinhold, 1964. Reversible changes in the hydraulic permeability of plant cell membranes. *Plant Physiol.* **39**:1043–1050.

Glock, W. S., and S. R. Agerter, 1962. Rainfall and tree growth. In T. T. Kozlowski (ed.), "Tree Growth," pp. 23–56. The Ronald Press Company, New York.

Gloser, J., 1967. Some problems of the determination of stomatal aperture by the microrelief method. *Biol. Plant.* (*Praha*) **9**:28–33.

Glover, J., 1941. A method for the continuous measurement of transpiration of single leaves under natural conditions. *Ann. Bot.* **5**:25–34.

Goatley, J. I., and R. W. Lewis, 1966. Composition of guttation fluid from rye, wheat, and barley seedlings. *Plant Physiol.* **41**:373–375.

González-Bernáldez, F., J. F. López-Sáez, and G. Garciá-Ferrero, 1968. Effect of osmotic pressure on root growth, cell cycle and cell elongation. *Protoplasma* **65**:255–262.

Goodwin, R. H., and W. Stepka, 1945. Growth and differentiation in the root tip of *Phleum pratense. Amer. J. Bot.* **32**:36–46.

Gortner, R. A., 1938. "Outlines of Biochemistry," 2d ed. John Wiley & Sons, Inc., New York.

Grable, A. R., 1966. Soil aeration and plant growth. *Advan. Agron.* **18**:58–106.

——— and R. E. Danielson, 1965. Influence of CO_2 on growth of corn and soybean seedlings. *Soil Sci. Soc. Amer. Proc.* **29**:233–238.

Gračanin, M., 1963*a*. Die kritsche Bodenfeuchtigkeit für die Guttation. *Ber. Deut. Bot. Ges.* **75**:445–473.

———, 1963*b*. Über Unterschiede in der Transpiration von Blattspreite und Stamm. *Phyton* **10**:216–224.

———, 1964. Zur Rolle osmotischer Kräft bei Guttation und Exudation. *Flora* **154**:21–35.

Gradmann, H., 1928. Untersuchungen über die Wasserverhältnisse des Bodens als Grundlage des Pflanzenwachstums. *Jahrb. Wiss. Bot.* **69**:1–100.

Greenfield, S. S., 1942. Inhibitory effects of inorganic compounds on photosynthesis in *Chlorella. Amer. J. Bot.* **29:**121–131.

Greenidge, K. N. H., 1952. An approach to the study of vessel length in hardwood species. *Amer. J. Bot.* **39:**570–574.

———, 1955. Studies in the physiology of forest trees. II. Experimental studies of fracture of stretched water columns in transpiring trees. *Amer. J. Bot.* **42:**28–37.

———, 1957. Ascent of sap. *Ann. Rev. Plant Physiol.* **8:**237–256.

———, 1958. A note on the rates of upward travel of moisture in trees under differing experimental conditions. *Can. J. Bot.* **36:**357–361.

Greenway, H., 1962. Plant responses to saline substrates. I. Growth and ion uptake of several varieties of Hordeum during and after sodium chloride treatment. *Aust. J. Biol. Sci.* **15:**16–38.

Gregory, F. G., F. L. Milthorpe, H. L. Pearse, and H. J. Spencer, 1950. Experimental studies of the factors controlling transpiration. II. The relation between transpiration rate and leaf water content. *J. Exp. Bot.* **1:**15–28.

——— and H. L. Pearse, 1934. The resistance porometer and its application to the study of stomatal movement. *Proc. Roy. Soc.* **B114:**477–493.

Gries, G. A., 1943. Juglone (5-hydroxy-1;4-naphthoquinone)—a promising fungicide. *Phytopathol.* **33:**1112.

Grieve, B. J., and F. W. Went, 1965. An electric hygrometer apparatus for measuring water vapour loss from plants in the field. *Arid Zone Res.* **25:**247–256. UNESCO, Paris.

Groenewegen, H., and J. A. Mills, 1960. Uptake of mannitol into the shoots of intact barley plants. *Aust. J. Biol. Sci.* **13:**1–4.

Grossenbacher, K. A., 1938. Diurnal fluctuation in root pressure. *Plant Physiol.* **13:**669–676.

Guest, P. L., and H. D. Chapman, 1944. Some effects of pH on growth of citrus in sand and solution cultures. *Soil Sci.* **58:**455–465.

Guilliermond, A., 1941. The cytoplasm of the plant cell. Chronica Botanica Co., Waltham, Mass.

Gurr, C. G., 1962. Use of gamma rays in measuring water content and permeability in unsaturated columns of soil. *Soil Sci.* **94:**224–229.

———, T. J. Marshall, and J. T. Hutton, 1952. Movement of water in soil due to a temperature gradient. *Soil Sci.* **74:**335–345.

Gutknecht, J., 1967. Membranes of *Valonia ventricosa:* Apparent absence of water-filled pores. *Science* **158:**787–788.

Guttenberg, H. V., and G. Meinl, 1952. Über den Einfluss von Wirkstoffen auf die Wasserpermeabilität des Protoplasmas. II. Über den Einfluss des pH-Wertes und der Temperatur auf die durch Heteroauxin bedingten Veränderungen der Wasserpermeabilität. *Planta* **40:**431–442.

Haas, A. R. C., 1936. Growth and water losses in citrus affected by soil temperature. *Calif. Citrogr.* **21:**467, 469.

———, 1945. Influence of chlorine on plants. *Soil Sci.* **60:**53–61.

———, 1948. Effect of the rootstock on the composition of citrus trees and fruit. *Plant Physiol.* **23:**309–330.

Haberlandt, G., 1914. "Physiological Plant Anatomy," Eng. trans. by M. Drummond. The Macmillan Company, New York.

Hackett, D. P., 1961. Effects of salt on DPNH oxidase activity and structure of sweet potato mitochondria. *Plant Physiol.* **36:**445–452.

——— and K. V. Thimann, 1952. The nature of the auxin-induced water uptake by potato tissue. *Amer. J. Bot.* **39:**553–560.

Hacskaylo, E., and J. G. Palmer, 1957. Effects of several biocides on growth of seedling pines and incidence of mycorrhizae in field plots. *Plant Dis. Rep.* **41**:354–358.

Hagan, R. M., 1949. Autonomic diurnal cycles in the water relations of nonexuding detopped root systems. *Plant Physiol.* **24**:441–454.

———, 1950. Soil aeration as a factor in water absorption by the roots of transpiring plants. *Plant Physiol.* **25**:748–762.

———, 1955. Factors affecting soil moisture–plant growth relations. *Rept. 14th Int. Hort. Congr.* 82–102.

———, H. R. Haise, and T. W. Edminster (eds.), 1967. "Irrigation of Agricultural Lands." *Amer. Soc. Agron.*, Madison, Wis.

——— and J. F. Laborde, 1966. Plants as indicators of need for irrigation. *Proc. 8th Int. Congr. Soil Sci.* (Bucharest, Rumania).

———, Y. Vaadia, and M. B. Russell, 1959. Interpretation of plant responses to soil moisture regimes. *Advan. Agron.* **11**:77–98.

Haise, H. R., and R. M. Hagan, 1967. Soil, plant, and evaporative measurements as criteria for scheduling irrigation. In R. M. Hagan, H. R. Haise, and T. W. Edminster (eds.), "Irrigation of Agricultural Lands, " pp. 577–604. *Amer. Soc. Agron.*, Madison, Wis.

——— and O. J. Kelley, 1946. Relation of moisture tension to heat transfer and electrical resistance in plaster of paris blocks. *Soil Sci.* **61**:411–422.

Hales, S., 1727. Vegetable staticks. W. & J. Innys and T. Woodward, London.

Halevy, A., 1956. Orange leaf transpiration under orchard conditions. IV. A contribution to the methodology of transpiration measurements in citrus leaves. *Res. Counc. Israel Bull.* **5D**:155–164.

———, and S. P. Monselise, 1963. Meaning of apparent midnight decrease in water content of leaves. *Bot. Gaz.* **124**:343–346.

Hall, D. M., 1967. Wax microchannels in the epidermis of white clover. *Science* **158**:505–506.

——— and L. A. Donaldson, 1962. Secretion from pores of surface wax on plant leaves. *Nature* **194**:1196.

Hall, N. S., W. F. Chandler, C. H. M. van Bavel, P. H. Reid, and J. H. Anderson, 1953. A tracer technique to measure growth and activity of plant root systems. *N.C. Agr. Exp. Sta. Tech. Bull.* 101.

Hall, W. C., 1949. Effects of photoperiod and nitrogen supply on growth and reproduction in the gherkin. *Plant Physiol.* **24**:753–769.

Hamilton, K. C., and K. P. Buchholz, 1955. Effect of rhizomes of quack grass (*Agropyron repens*) and shading on the seedling development of weedy species. *Ecology* **36**:304–308.

Hammel, H. T., 1967. Freezing of xylem sap without cavitation. *Plant Physiol.* **42**:55–66.

Handley, W. R. C., 1939. The effect of prolonged chilling on water movement and radial growth in trees. *Ann. Bot.* **3**:803–813.

Hansen, C., 1926. The water-retaining power of the soil. *J. Ecol.* **14**:111–119.

Hanson, J. B., and O. Biddulph, 1953. The diurnal variation in the translocation of minerals across bean roots. *Plant Physiol.* **28**:356–370.

Harbeck, G. E., M. A. Kohler, and G. E. Coberg, 1958. Water loss investigations: Lake Mead studies. *U.S. Geol. Surv. Prof. Paper* 298.

Harley, J. L., 1956. The mycorrhiza of forest trees. *Endeavour* **15**:43–48.

———, 1959. "Biology of Mycorrhiza." Leonard Hill Books, London.

———, 1965. Mycorrhiza. In K. F. Baker, W. C. Snyder, et al. (eds.), "Ecology of Soil-borne Plant Pathogens," pp. 218–230. Univ. of California Press, Berkeley.

Harms, W. R., and W. H. D. McGregor, 1962. A method for measuring the water balance of pine needles. *Ecology* **43**:531–532.

Hartt, C. E., 1967. Effect of moisture supply upon translocation and storage of ^{14}C in sugarcane. *Plant Physiol.* **42**:338–346.

Harris, J. A., 1934. "The Physico-chemical Properties of Plant Saps in Relation to Phytogeography." Univ. of Minnesota Press, Minneapolis.

Hatch, A. B., 1937. The physical basis of mycotrophy in Pinus. *Bull* 6, *Black Rock Forest.*

——— and K. D. Doak, 1933. Mycorrhizal and other features of the root systems of Pinus. *J. Arnold Arboretum* **14**:85–99.

Haynes, J. L., and W. R. Robbins, 1948. Calcium and boron as essential factors in the root environment. *J. Amer. Soc. Agron.* **40**:795–803.

Hayward, H. E., and W. M. Blair, 1942. Some responses of Valencia orange seedlings to varying concentrations of chloride and hydrogen ions. *Amer. J. Bot.* **29**:148–155.

———, W. M. Blair, and P. E. Skaling, 1942. Device for measuring entry of water into roots. *Bot. Gaz.* **104**:152–160.

——— and E. M. Long, 1942. The anatomy of the seedling and roots of the Valencia orange. *U.S. Dept. Agr. Tech. Bull.* 786.

———, 1943. Some effects of sodium salts on the growth of the tomato. *Plant Physiol.* **18**:556–569.

———, E. M. Long, and R. Uhvits, 1946. Effect of chloride and sulfate salts on the growth and development of the Elberta peach on Shalil and Lovell rootstocks. *U.S. Dept. Agr. Tech. Bull.* 922.

——— and O. C. Magistad, 1946. The salt problem in irrigation agriculture. *U.S. Dept. Agr. Misc. Publ.* 607.

——— and W. B. Spurr, 1943. Effects of osmotic concentration of substrate on the entry of water into corn roots. *Bot. Gaz.* **105**:152–164.

———, 1944. Effects of isosmotic concentrations of inorganic and organic substrates on entry of water into corn roots. *Bot. Gaz.* **106**:131–139.

Head, G. C., 1964. A study of 'exudation' from the root hairs of apple roots by time-lapse cine-photomicrography. *Ann. Bot. N.S.* **28**:495–498.

———, 1965. Studies of diurnal change in cherry root growth and nutational movements of apple root tips by time-lapse cinematography. *Ann. Bot. N.S.* **29**:219–224.

———, 1966. Estimating seasonal changes in the quantity of white unsuberized root on fruit trees. *J. Hort. Sci.* **41**:197–206.

———, 1967. Effects of seasonal changes in shoot growth on the amount of unsuberized root on apple and plum trees. *J. Hort. Sci.* **42**:169–180.

Heath, O. V. S., 1941. Experimental studies of the relation between carbon assimilation and stomatal movement. II. The use of the resistance porometer in estimating stomatal aperture and diffusive resistance. *Ann. Bot. N.S.* **5**:455–500.

———, 1959. The water relations of stomatal cells and the mechanisms of stomatal movement. In F. C. Steward (ed.), "Plant Physiology," vol. 2, pp. 193–250. Academic Press Inc., New York.

Helder, R. J., 1952. Analysis of the process of anion uptake of intact maize plants. *Acta Bot. Neer.* **1**:361–434.

———, 1956. The loss of substance by cells and tissues (salt glands). "Encyclopedia of Plant Physiology," vol. 2, pp. 468–488. Springer Verlag OHG, Berlin.

Hellebust, J. A., and D. F. Forward, 1962. The invertase of the corn radicle and its activity in successive stages of growth. *Can. J. Bot.* **40**:113–126.

Hellmers, H., 1963. Some temperature and light effects in the growth of Jeffrey pine seedlings. *Forest Sci.* **9**:189–201.

———, J. S. Horton, G. Juhren, and J. O'Keefe, 1955. Root systems of some chaparral plants in southern California. *Ecology* **36**:667–678.

Henckel, P. A., 1964. Physiology of plants under drought. *Ann. Rev. Plant Physiol.* **15**:363–386.

Henderson, L., 1934. Relation between root respiration and absorption. *Plant Physiol.* **9**:283–300.

Henderson, L. J., 1913. "The Fitness of the Environment." The Macmillan Company, New York.

Hendrickson, A. H., and F. J. Veihmeyer, 1929. Irrigation experiments with peaches in California. *Calif. Agr. Exp. Sta. Bull.* 479.

——— and ———, 1941. Some factors affecting the growth rate of pears. *Proc. Amer. Soc. Hort. Sci.* **39**:1–7.

——— and ———, 1942. Irrigation experiments with pears and apples. *Calif. Agr. Exp. Sta. Bull.* 667.

——— and ———, 1945. Permanent wilting percentages of soils obtained from field and laboratory trials. *Plant Physiol.* **20**:517–539.

Hewitt, W. B., and M. E. Gardiner, 1956. Some studies of the absorption of zinc sulfate in Thompson seedless grape canes. *Plant Physiol.* **31**:393–399.

Hewlett, J. D., and J. E. Douglass, 1961. A method for calculating error of soil moisture volumes in gravimetric sampling. *Forest. Sci.* **7**:265–272.

———, J. E. Douglass, and J. L. Clutter, 1964. Instrumental and soil moisture variance using the neutron-scattering method. *Soil Sci.* **97**:19–24.

——— and A. R. Hibbert, 1961. Increases in water yield after several types of forest cutting. *Int. Assoc. Sci. Hydrol. Bull.* **6**:5–17.

——— and ———, 1967. Factors affecting the response of small watersheds to precipitation in humid areas. In W. E. Sopper and H. W. Lull (eds.), "Forest Hydrology," pp. 275–290. Pergamon Press, New York.

——— and P. J. Kramer, 1963. The measurement of water deficits in broadleaf plants. *Protoplasma* **57**:381–391.

Heyl, J. G., 1933. Der Einfluss von Aussenfaktoren auf das Bluten der Pflanzen. *Planta* **20**:294–353.

Hiatt, A. J., 1967. Relationship of cell sap pH to organic acid change during ion uptake. *Plant Physiol.* **42**:294–298.

Higgins, J. J., J. R. Haun, and E. J. Koch, 1964. Leaf development: index of plant response to environmental factors. *Agron. J.* **56**:489–492.

Hoagland, D. R., 1944. Lectures on the inorganic nutrition of plants. Chronica Botanica Co., Waltham, Mass.

——— and T. C. Broyer, 1936. General nature of the process of salt accumulation by roots with description of experimental methods. *Plant Physiol.* **11**:471–507.

Hodges, T. K., and Y. Vaadia, 1964. Chloride uptake and transport in roots of different salt status. *Plant Physiol.* **39**:109–114.

Hodgson, R. H., 1954. A study of the physiology of mycorrhizal roots on *Pinus taeda* L. M.A. thesis, Duke Univ., N.C.

Höfler, K., 1920. Ein Schema für die osmotische Leistung der Pflanzenzellen. *Ber. deut. bot. Ges.* **38**:288–298.

Hofmeister, W., 1862. Ueber Spannung, Ausflussmenge und Ausflussgeschwindigkeit von Säften lebender Pflanzen. *Flora* **45**:97–108, 113–120, 138–144, 145–152, 170–175.

Hofstra, J. J., 1964. Amino acids in the bleeding sap of fruiting tomato plants. *Acta Bot. Neer.* **13**:148–158.

Hohn, K., 1951. Beziehungen zwischen Blutung und Guttation bei *Zea mays. Planta* **39**:65–74.

Holch, A. E., 1931. Development of roots and shoots of certain deciduous tree seedlings in different forest sites. *Ecology* **12**:259–298.

Holdridge, L. R., 1962. The determination of atmospheric water movements. *Ecology* **43**:1–9.

Hollaender, A. (ed.), 1956. "Radiation Biology," vol. 3. McGraw-Hill Book Company, New York.

Holmes, J. W., 1956. Calibration and field use of the neutron scattering method of measuring soil water content. *Aust. J. Appl. Sci.* **7**:45–58.

———, 1960. Water balance and the water table in deep sandy soils of the upper south-east, South Australia. *Aust. J. Agr. Res.* **11**:970–988.

——— and A. F. Jenkinson, 1959. Techniques for using the neutron moisture meter. *J. Agr. Eng. Res.* **4**:100–109.

———, S. A. Taylor, and S. J. Richards, 1967. Measurement of soil water. In R. M. Hagan et al. (eds.), "Irrigation of Agricultural Lands," pp. 275–303. *Amer. Soc. Agron.,* Madison, Wis.

Holmgren, P., P. G. Jarvis, and M. S. Jarvis, 1965. Resistances to carbon dioxide and water vapour transfer in leaves of different plant species. *Physiol. Plant.* **18**:557–573.

Honda, S. I., R. N. Robertson, and J. M. Gregory, 1958. Studies in metabolism of plant cells. XII. Ionic effects on oxidation of reduced diphosphopyridine nucleotide and cytochrome c by plant mitochondria. *Aust. J. Biol. Sci.* **11**:1–15.

van den Honert, T. H., 1948. Water transport as a catenary process. *Faraday Soc. Discuss.* no. 3, 146–153.

Hoover, M. D., 1949. Hydrologic characteristics of South Carolina Piedmont forest soils. *Proc. Soil Sci. Soc. Amer.* **14**:353–358.

Hope, A. B., and P. G. Stevens, 1952. Electrical potential differences in bean roots and their relation to salt uptake. *Aust. J. Sci. Res.* **B5**:335–343.

Hough, W. A., F. W. Woods, and M. L. McCormack, 1965. Root extension of individual trees in surface soils of a natural longleaf pine–turkey oak stand. *Forest Sci.* **11**:223–242.

House, C. R., and N. Findlay, 1966. Water transport in isolated maize roots. *J. Exp. Bot.* **17**:344–354.

Howard, A., 1925. The effect of grass on trees. *Proc. Roy. Soc. Proc.* (London). **B97**:284–321.

Howell, J., 1932. Relation of western yellow pine seedlings to the reaction of the culture solution. *Plant Physiol.* **7**:657–671.

Huber, B., 1928. Weitere quantitative Untersuchungen über das Wasserleitungssystem der Pflanzen. *Jahrb. wiss. Bot.* **67**:877–959.

———, 1935. Die physiologische Bedeutung der Ring- und Zerstreutporigkeit. *Ber. deut. bot. Ges.* **53**:711–719.

———, 1956*a*. Die Transpiration von Sprossachsen und anderen nicht foliosen Organen. "Encyclopedia of Plant Physiology," vol. 3, pp. 427–435. Springer-Verlag OHG, Berlin.

———, 1956*b*. Die Gefässleitung. "Encyclopedia of Plant Physiology," vol. 3, pp. 541–582. Springer-Verlag OHG, Berlin.

——— and K. Höfler, 1930. Die Wasserpermeabilität des Protoplasmas. *Jahrb. wiss. Bot.* **73**:351–511.

——— and E. Schmidt, 1937. Eine Kompensationsmethode zur thermoelektrischen Messung Langsamer Saftstrome. *Ber. deut. bot. Ges.* **50**:514–529.

Hudson, J. P., 1960. Relations between root and shoot growth in tomatoes. *Sci. Hort.* **14**:49–54.

Huffaker, R. C., and A. Wallace, 1959. Effect of potassium and sodium levels on sodium distribution in some plant species. *Soil Sci.* **88**:80–82.

Hull, H. M., 1964. Leaf structure as related to penetration of organic substances. Symposium on Absorption and Translocation of Organic Substances in Plants, southern section, Amer. Soc. Plant Physiol.

Hunt, F. M., 1951. Effects of flooded soil on growth of pine seedlings. *Plant Physiol.* **26:**363–368.

Hunter, A. S., and O. J. Kelley, 1946. A new technique for studying the absorption of moisture and nutrients from soil by plant roots. *Soil Sci.* **62:**441–450.

Hunter, C., and E. M. Rich, 1925. The effect of artificial aeration of the soil on *Impatiens balsamina* L. *New Phytol.* **24:**257–271.

Hutchinson, G. E., 1957. "A Treatise on Limnology." John Wiley & Sons, Inc., New York.

Hylmö, B., 1953. Transpiration and ion absorption. *Physiol. Plant.* **6:**333–405.

———, 1955. Passive components in the ion absorption of the plant. I. The zonal ion and water absorption in Brouwer's experiments. *Physiol. Plant.* **8:**433–449.

———, 1958. Passive components in the ion absorption of the plant. II. The zonal water flow, ion passage, and pore size in roots of *Vicia faba. Physiol. Plant.* **11:**382–400.

Idso, S. B., and D. G. Baker, 1967. Relative importance of reradiation, convection, and transpiration in heat transfer from plants. *Plant Physiol.* **42:**631–640.

Ingelsten, B., 1966. Absorption and transport of sulfate by wheat at varying mannitol concentration in the medium. *Physiol. Plant.* **19:**563–579.

——— and B. Hylmö, 1961. Apparent free space and surface film determined by a centrifugation method. *Physiol. Plant.* **14:**157–170.

Itai, C., and Y. Vaadia, 1965. Kinetin-like activity in root exudate of water-stressed sunflower plants. *Physiol. Plant.* **18:**941–944.

Ivanoff, S. S., 1944. Guttation-salt injury on leaves of cantaloupe, pepper, and onion. *Phytopathol.* **34:**436–437.

Jackson, J. E., and P. E. Weatherley, 1962. The effect of hydrostatic pressure gradients on the movement of potassium across the root cortex. *J. Exp. Bot.* **13:**128–143.

Jackson, W. T., 1955. The role of adventitious roots in recovery of shoots following flooding of the original root systems. *Amer. J. Bot.* **42:**816–819.

———, 1956. The relative importance of factors causing injury to shoots of flooded tomato plants. *Amer. J. Bot.* **43:**637–639.

———, 1962. Use of carbowaxes (polyethylene glycols) as osmotic agents. *Plant Physiol.* **37:**513–519.

Jacoby, B., 1965. Sodium retention in excised bean stems. *Physiol. Plant.* **18:**730–739.

———, 1964. Function of the root and stems in sodium retention. *Plant Physiol.* **39:**445–449.

Jamison, V. C., 1946. The penetration of irrigation and rain water into sandy soils of central Florida. *Soil Sci. Soc. Amer. Proc.* **10:**25–29.

Janes, B. E., 1955. Vegetable rotation studies in Connecticut. III. Effect of sweet corn and vetch on the growth of several crops which follow. *Proc. Amer. Soc. Hort. Sci.* **65:**324–330.

Jantti, A., and P. J. Kramer, 1956. Regrowth of pastures in relation to soil moisture and defoliation. *Proc. 7th Int. Grasslands Congr.*

Jarvis, P. G., and M. S. Jarvis, 1963. Water relations of tree seedlings. I. Growth and water use in relation to soil water potential. *Physiol. Plant.* **16:**215–235.

———, C. W. Rose, and J. E. Begg, 1967. An experimental and theoretical comparison of viscous and diffusive resistance to gas flow through amphistomatous leaves. *Agr. Metab.* **4:**103–117.

——— and R. O. Slatyer, 1966*a*. A controlled environment chamber for studies of gas exchange by each surface of a leaf. *C.S.I.R.O. Aust., Div. Land Res. Tech. Paper* 29.

———, 1966*b*. Calibration of beta gauges for determining leaf water status. *Science* **153:**78–79.

Jemison, G. H., 1944. The effect of basal wounding by forest fires on the diameter growth of some southern Appalachian hardwoods. *Duke Univ. Sch. Forest. Bull.* 9.

Jennings, D. H., 1963. "The Absorption of Solutes by Plant Cells." Iowa State Univ. Press, Ames, Iowa.

Jenny, H., and K. Grossenbacher, 1963. Root-soil boundary zones as seen in the electron microscope. *Soil Sci. Soc. Amer. Proc.* **27:**273–277.

——— and R. Overstreet, 1939. Cation interchange between plant roots and soil colloids. *Soil Sci.* **47:**257–272.

Jensen, C. R., J. Letey, and L. H. Stolzy, 1964. Labeled oxygen: Transport through growing corn roots. *Science* **144:**550–552.

———, L. H. Stolzy, and J. Letey, 1967. Tracer studies of oxygen diffusion through roots of barley, corn and rice. *Soil Sci.* **103:**23–29.

Jensen, G., 1962. Relationship between water and nitrate uptake in excised tomato root systems. *Physiol. Plant.* **15:**791–803.

Jensen, R. D., S. A. Taylor, and H. H. Wiebe, 1961. Negative transport and resistance to water flow through plants. *Plant Physiol.* **36:**633–638.

Jeschke, W. D., 1967. Die Cyclische und die Nichtcyclische Photophosphorylierung als Energiequellen der Lichtabhängigen Chloridionenaufnahme bei *Elodea. Planta* **73:**161–174.

Johnson, J., 1936. Relation of root pressure to plant disease. *Science* **84:**135–136.

Johnson, L. P. V., 1944. Sugar production by white and yellow birches. *Can. J. Res. Sec.* **C22:**1–6.

———, 1945. Physiological studies on sap flow in the sugar maple, *Acer saccharum* Marsh. *Can. J. Res. Sec.* **C23:**192–197.

Johnston, C. N., 1942. Water-permeable jacketed thermal radiators as indicators of field capacity and permanent wilting percentage in soils. *Soil Sci.* **54:**123–126.

Jones, C. G., A. W. Edson, and W. J. Morse, 1903. The maple sap flow. *Vt. Agr. Exp. Sta. Bull.* 103.

Jones, O. P., and H. J. Lacey, 1968. Gibberellin-like substances in the transpiration stream of apple and pear trees. *J. Expt. Bot.* **19:**526–531.

Joshi, M. C., J. S. Boyer, and P. J. Kramer, 1965. Growth, carbon dioxide exchange, transpiration, and transpiration ratio of pineapple. *Bot. Gaz.* **126:**174–179.

Kalela, E., 1954. Mantysiemenpuiden japuustojen juuroisuhteista (On root relations of pine seed trees). *Acta Forest. Fennica* **61:**1–17.

Kamiya, N., and M. Tazawa, 1956. Studies of water permeability of a single plant cell by means of transcellular osmosis. *Protoplasma* **46:**394–422.

———, ———, and T. Takata, 1962. Water permeability of the cell wall in *Nitella. Plant and Cell Physiol.* **3:**285–292.

———, ———, and ———, 1963. The relation of turgor pressure to cell volume in *Nitella* with special reference to the mechanical properties of the cell wall. *Protoplasma* **57:**501–521.

Katznelson, H., J. W. Rouatt, and E. A. Peterson, 1962. The rhizosphere effect of mycorrhizal and nonmycorrhizal roots of yellow birch seedlings. *Can. J. Bot.* **40:**377–382.

Kaufman, M. R., 1967. Water relations of pine seedlings in relation to root and shoot growth. Ph.D. dissertation, Duke Univ., N.C.

———, 1968. Water relations of pine seedlings in relation to root and shoot growth. *Plant Physiol.* **43:**281–288.

Kavanau, J. L., 1964. "Water and Solute-water Interactions." Holden-Day, Inc., San Francisco.

Keever, Catherine, 1950. Causes of succession on old fields of the Piedmont, North Carolina. *Ecol. Monogr.* **20:**229–250.

Keller, R., 1930. Der elektrische Faktor des Wassertransports in Lichte der Vitalfarbung. *Ergeb. Physiol.* **30:**294–407.

Keller, Th., 1966. Über den Einfluss von transpirationshemmenden Chemikalien (Antitranspirantien) auf Transpiration, CO_2-Aufnahme und Wurzel. *Forstw. Cbl.* **85:**65–79.

Kelley, A. P., 1950. Mycotrophy in plants. Chronica Botanica Co., Waltham, Mass.

Kelley, O. J., 1944. A rapid method of calibrating various instruments for measuring soil moisture in situ. *Soil Sci.* **58:**433–440.

———, A. S. Hunter, and C. H. Hobbs, 1945. The effect of moisture stress on nursery-grown guayule with respect to the amount and type of growth and growth response on transplanting. *J. Amer. Soc. Agron.* **37:**194–216.

———, ———, H. R. Haise, and C. H. Hobbs, 1946. A comparison of methods of measuring soil moisture under field conditions. *J. Amer. Soc. Agron.* **38:**759–784.

Kelley, U. W., 1932. The effect of pruning of excised shoots on the transpiration rate of some deciduous fruit species. *Proc. Amer. Soc. Hort. Sci.* **29:**71–73.

Kelly, S., 1947. The relations between respiration and water uptake in the oat coleoptile. *Amer. J. Bot.* **34:**521–526.

Kende, H., 1965. Kinetinlike factors in the root exudate of sunflowers. *Proc. Nat. Acad. Sci.* **53:**1302–1307.

Kenefick, D. G., 1962. Formation and elimination of ethanol in sugar beet roots. *Plant Physiol.* **37:**434–439.

Ketellapper, H. J., 1963. Stomatal physiology. *Ann. Rev. Plant Physiol.* **14:**249–270.

Kiesselbach, T. A., J. C. Russel, and A. Anderson, 1929. The significance of subsoil moisture in alfalfa production. *J. Amer. Soc. Agron.* **21:**241–268.

Killian, Ch., and G. Lemée, 1956. Les xérophytes: leur économie d'eau. "Encyclopedia of Plant Physiology," vol. 3, pp. 787–824. Springer-Verlag OHG, Berlin.

Kinman, C. F., 1932. A preliminary report on root growth studies with some orchard trees. *Proc. Amer. Soc. Hort. Sci.* **29:**220–224.

Klemm, M., and W. Klemm, 1963. Die Verwendung von Radioisotopen zur kontinuierlichen Bestimmung des Tagesverlaufes der Transpirationsstromgeschwindigkeit bei Bäumen. *Flora* **154:**89–93.

Klepper, B., 1967. Effects of osmotic pressure on exudation from corn roots. *Aust. J. Biol. Sci.* **20:**723–735.

——— and M. R. Kaufmann, 1966. Removal of salt from xylem sap by leaves and stems of guttating plants. *Plant Physiol.* **41:**1743–1747.

Klotz, I. M., 1958. Protein hydration and behaviour. *Science* **128:**815–822.

Knapp, R., H. F. Linskens, H. Lieth, and R. Wolf, 1952. Untersuchungen über die Bodenfeuchtigkeit in verschiedenen Pflanzengesellschaften nach neuren Methoden. *Ber. deut. bot. Ges.* **65:**113-132.

Knight, R. C., 1917. The interrelations of stomatal aperture, leaf water-content and transpiration rate. *Ann. Bot.* **31:**221–240.

Knipling, E. B., 1967*a*. Measurement of leaf water potential by the dye method. *Ecology* **48:**1038–1041.

———, 1967*b*. Effect of leaf aging on water deficit–water potential relationship of dogwood leaves growing in two environments. *Physiol. Plant.* **20:**65–72.

——— and P. J. Kramer, 1967. Comparison of the dye method with the thermocouple psychrometer for measuring leaf water potentials. *Plant Physiol.* **42:**1315–1320.

Knoerr, K., 1966. Contrasts in energy balances between individual leaves and vegetated surfaces. In Sopper and Lull (eds.), "International Symposium on Forest Hydrology," pp. 391–401. Pergamon Press, New York.

Knoerr, K. R., and L. W. Gay, 1965. Tree leaf energy balance. *Ecology* **46:**17–24.

Knoll, H. A., D. J. Lathwell, and N. C. Brady, 1964. The influence of root zone temperature on the growth and contents of phosphorus and anthocyanin of corn. *Soil. Sci. Soc. Amer. Proc.* **28:**400–403.

Koch, W., 1957. Der Tagesgang der "Produktivität der Transpiration." *Planta* **48:**418–452.

Kohn, P. G., 1965. Tables of some physical and chemical properties of water. *Soc. Exp. Biol. Symp.* **19:**3–16.

Köhnlein, E., 1930. Untersuchungen über die Höhe des Wurzelwiderstandes und die Bedeutung aktiver Wurzeltätigkeit für die Wasserversorgung der Pflanzen. *Planta* **10:**381–423.

Koller, D., and Y. Samish, 1964. A null-point compensating system for simultaneous and continuous measurement of net photosynthesis and transpiration by controlled gas-stream analysis. *Bot. Gaz.* **125:**81–88.

Korn, E. D., 1966. Structure of biological membranes. *Science* **153:**1491–1498.

Korstian, C. F., 1924. Density of cell sap in relation to environmental conditions in the Wasatch Mountains of Utah. *J. Agr. Res.* **28:**845–909.

——— and N. J. Fetherolf, 1921. Control of stem girdle of spruce transplants caused by excessive heat. *Phytopathol.* **11:**485–490.

Kosaroff, P., 1897. Einfluss verschiedener äusserer Factoren auf die Wasseraufnahme der Pflanzen. Inaugural Dissertation, Univ. of Leipzig.

Kozinka, V., and S. Klenovska, 1965. The uptake of mannitol by higher plants. *Biol. Plant.* **7:**285–292.

Kozlowski, T. T., 1943. Transpiration rates of some forest tree species during the dormant season. *Plant Physiol.* **18:**252–260.

———, 1949. Light and water in relation to growth and competition of Piedmont forest trees. *Ecol. Monogr.* **19:**207–231.

———, 1961. The movement of water in trees. *Forest Sci.* **7:**177–192.

———, 1964. "Water Metabolism in Plants." Harper & Row, Publishers, Incorporated, New York.

———, 1967. Diurnal variations in stem diameters of small trees. *Bot. Gaz.* **128:**60–68.

——— (ed.), 1968. "Water Deficits and Plant Growth." Academic Press, Inc., New York.

——— and J. J. Clausen, 1965. Changes in moisture contents and dry weights of buds and leaves of forest trees. *Bot. Gaz.* **126:**20–26.

———, J. F. Hughes, and E. Leyton, 1966. Patterns of water movement in dormant gymnosperm seedlings. *Biorheol.* **3:**77–85.

——— and W. H. Scholtes, 1948. Growth of roots and root hairs of pine and hardwood seedlings in the Piedmont. *J. Forest.* **46:**750–754.

———, C. H. Winget, and J. H. Torrie, 1962. Daily radial growth of oak in relation to maximum and minimum temperature. *Bot. Gaz.* **124:**9–17.

——— and C. H. Winget, 1963. Patterns of water movement in forest trees. *Bot. Gaz.* **124:**301–311.

Kramer, P. J., 1932. The absorption of water by root systems of plants. *Amer. J. Bot.* **19:**148–164.

———, 1933. The intake of water through dead root systems and its relation to the problem of absorption by transpiring plants. *Amer. J. Bot.* **20:**481–492.

———, 1937. The relation between rate of transpiration and rate of absorption of water in plants. *Amer. J. Bot.* **24:**10–15.
———, 1938. Root resistance as the cause of the absorption lag. *Amer. J. Bot.* **25:**110–113.
———, 1939. The forces concerned in the intake of water by transpiring plants. *Amer. J. Bot.* **26:**784–791.
———, 1940*a*. Root resistance as a cause of decreased water absorption by plants at low temperatures. *Plant Physiol.* **15:**63–79.
———, 1940*b*. Causes of decreased absorption of water by plants in poorly aerated media. *Amer. J. Bot.* **27:**216–220.
———, 1941. Soil moisture as a limiting factor for active absorption and root pressure. *Amer. J. Bot.* **28:**446–451.
———, 1942. Species differences with respect to water absorption at low soil temperatures. *Amer. J. Bot.* **29:**828–832.
———, 1946. Absorption of water through suberized roots of trees. *Plant Physiol.* **21:**37–41.
——— and W. S. Clark, 1947. A comparison of photosynthesis in individual pine needles and entire seedlings at various light intensities. *Plant Physiol.* **22:**51–57.
———, 1949. "Plant and Soil Water Relationships." McGraw-Hill Book Company, New York.
——— and K. M. Wilbur, 1949. Absorption of radioactive phosphorus by mycorrhizal roots of pine. *Science* **110:**8–9.
———, 1950. Effects of wilting on the subsequent intake of water by plants. *Amer. J. Bot.* **37:**280–284.
———, 1951. Causes of injury to plants resulting from flooding of the soil. *Plant Physiol.* **26:**722–736.
———, W. S. Riley, and T. T. Bannister, 1952. Gas exchange of cypress (*Taxodium distichum*) knees. *Ecology* **33:**117–121.
——— and H. H. Wiebe, 1952. Longitudinal gradients of P^{32} absorption in roots. *Plant Physiol.* **27:**661–674.
——— and W. T. Jackson, 1954. Causes of injury to flooded tobacco plants. *Plant Physiol.* **29:**241–245.
———, 1955*a*. Water content and water turnover in plant cells. "Encyclopedia of Plant Physiology," vol. 1, pp. 194–222. Springer-Verlag OHG, Berlin.
———, 1955*b*. Bound water. "Encyclopedia of Plant Physiology," vol. 1, pp. 223–242. Springer-Verlag OHG, Berlin.
———, 1955*c*. Physical chemistry of the vacuoles. "Encyclopedia of Plant Physiology," vol. 1, pp. 649–660. Springer-Verlag OHG, Berlin.
———, 1956*a*. The uptake of salts by plant cells. "Encyclopedia of Plant Physiology," vol. 2, pp. 290–315. Springer-Verlag OHG, Berlin.
———, 1956*b*. The uptake of water by plant cells. "Encyclopedia of Plant Physiology," vol. 2, pp. 316–336. Springer-Verlag OHG, Berlin.
———, 1956*c*. Permeability in relation to respiration. "Encyclopedia of Plant Physiology," vol. 2, pp. 358–368. Springer-Verlag OHG, Berlin.
———, 1956*d*. Physical and physiological aspects of water absorption. "Encyclopedia of Plant Physiology," vol. 3, pp. 124–159. Springer-Verlag OHG, Berlin.
———, 1956*e*. Roots as absorbing organs. "Encyclopedia of Plant Physiology," vol. 3, pp. 188–214. Springer-Verlag OHG, Berlin.
———, 1957. Outer space in plants. *Science* **125:**633–635.
——— and T. T. Kozlowski, 1960. "Physiology of Trees." McGraw-Hill Book Company, New York.

———, 1963. Water stress and plant growth. *Agron. J.* **55:**31–35.
———, 1964. The role of water in wood formation. In M. H. Zimmerman (ed.), "The Formation of Wood in Forest Trees," pp. 519–532. Academic Press Inc., New York.
——— and H. Brix, 1965. Measurement of water stress in plants. *Arid Zone Res.* **25:**343–351. UNESCO, Paris.
——— and H. C. Bullock, 1966. Seasonal variations in the proportions of suberized and unsuberized roots of trees in relation to the absorption of water. *Amer. J. Bot.* **53:**200–204.
———, E. B. Knipling, and L. N. Miller, 1966. Terminolgy of cell-water relations. *Science* **153:**889–890.
Kreeb, K., 1957. Hydratur und Ertrag. *Ber. deut. bot. Ges.* **70:**121–136.
———, Über die gravimetrische Methode zur Bestimmung der Sangspannung und das Problem des negativen Turgors. I. Mitt. *Planta* **55:**274–282.
———, 1963. Hydrature and plant production. In A. J. Rutter and F. H. Whitehead (eds.), "The Water Relations of Plants," pp. 272–288. John Wiley & Sons, Inc., New York.
———, 1965. Die ökologische Bedeutung der Bodenversalzung. *Angew. Bot.* **34:**1–15.
———, 1967. Entgegnung an R. O. Slatyer. *Z. Pflanzenphysiol.* **56:**95–97.
Kudrev, T. G., 1967. The effect of drought on the amount of free and bound glutamic acid and proline in pumpkins. *C. R. Acad. Bulg. Sci.* **20:**61–63.
Kuiper, P. J. C., 1961. The effects of environmental factors on the transpiration of leaves, with special reference to stomatal light response. *Meded. Landbouw hogesch. Wageningen* **61:**1–49.
———, 1963. Some considerations on water transport across living cell membranes. In I. Zelitch (ed.), "Stomata and Water Relations in Plants," pp. 59–68. *Conn. Agr. Exp. Sta. Bull.* 664, New Haven, Conn.
———, 1964*a*. Water transport across root cell membranes: effects of alkenylsuccinic acids. *Science* **143:**690–691.
———, 1964*b*. Inducing resistance to freezing and desiccation in plants by decenylsuccinic acid. *Science* **146:**544–546.
———, 1964*c*. Water uptake of higher plants as affected by root temperature. *Meded. Landbouw hogesch. Wageningen* **63:**1–11.
Kuntz, J. E., and A. J. Riker, 1955. The use of radioactive isotopes to ascertain the role of root grafting in the translocation of water, nutrients, and disease-inducing organisms. *Int. Cont. Peaceful Uses At. Energy* (Geneva, Switzerland) *Proc.* **12:**144–148.
Kurtzman, R. H., Jr., 1966. Xylem sap flow as affected by metabolic inhibitors and girdling. *Plant Physiol.* **41:**641–646.
Kylin, A., 1966. Uptake and loss of Na^+, Rb^+, and Cs^+ in relation to an active mechanism for extrusion of Na^+ in *Scenedesmus. Plant Physiol.* **41:**579–584.
——— and B. Hylmö, 1957. Uptake and transport of sulphate in wheat. Active and passive components. *Physiol. Plant.* **10:**467–483.

Ladefoged, K., 1948. Analysis of the root sap of the birch. *Plant and Soil* **1:**127–134.
———, 1960. A method for measuring the water consumption of larger intact trees. *Physiol. Plant.* **13:**648–658.
———, 1963. Transpiration of forest trees. *Physiol. Plant.* **16:**378–414.
Lagerwerff, J. V., and H. E. Eagle, 1961. Osmotic and specific effects of excess salts on beans. *Plant Physiol.* **36:**472–477.
———, 1962. Transpiration related to ion uptake by beans from saline substrates. *Soil Sci.* **93:**420–430.

———, G. Ogata, and H. E. Eagle, 1961. Control of osmotic pressure of culture solutions with polyethylene glycol. *Science* **133:**1486–1487.

Laing, H. E., 1940*a*. The composition of the internal atmosphere of *Nuphar advenum* and other water plants. *Amer. J. Bot.* **27:**861–868.

———, 1940*b*. Respiration of the rhizomes of *Nuphar advenum* and other water plants. *Amer. J. Bot.* **27:**574–581.

Lamb, C. A., 1936. Tensile strength, extensibility, and other characteristics of wheat roots in relation to winter injury. *Ohio Agr. Exp. Sta. Bull.* 568.

Lambert, J. R., and J. van Schilfgaarde, 1965. A method of determining the water potential of intact plants. *Soil Sci.* **100:**1–9.

Lamport, T. D. A., 1965. The protein component of primary cell walls. *Advan. Bot. Res.* **2:**151–218.

Lang, A. R. G., and H. D. Barrs, 1965. An apparatus for measuring water potentials in the xylem of intact plants. *Aust. J. Biol. Sci.* **18:**487–497.

Larcher, W., 1963. Zur Frage des Zusammenhanges zwischen Austrocknungsresistenz und Frosthärte bei Immergrünen. *Protoplasma* **57:**569–587.

Larson, P. R., 1964. Some indirect effects of environment on wood formation. In M. H. Zimmerman (ed.), "The Formation of Wood in Trees," pp. 345–365. Academic Press Inc., New York.

LaRue, C. D., 1930. The water supply of the epidermis of leaves. *Papers Mich. Acad. Sci.* **13:**131–139.

Laties, G. G., 1954. The osmotic inactivation in situ of plant mitochondrial enzymes. *J. Exp. Bot.* **5:**49–70.

———, 1959. Active transport of salt into plant tissue. *Ann. Rev. Plant Physiol.* **10:**87–112.

———, 1962. Controlling influence of thickness on development and type of respiratory activity in potato slices. *Plant Physiol.* **37:**679–690.

——— and K. Budd, 1964. The development of differential permeability in isolated steles of corn roots. *Proc. Nat. Acad. Sci.* **52:**462-469.

Latimer, W. M., and W. H. Rodebush, 1920. Polarity and ionization from the standpoint of the Lewis theory of valence. *J. Amer. Chem. Soc.* **42:**1419–1433.

Lawton, K., 1946. The influence of soil aeration on the growth and absorption of nutrients by corn plants. *Proc. Soil Sci. Soc. Amer.* **10:**263–268.

Lebedeff, A. F., 1928. The movement of ground and soil waters. *Proc. 1st Int. Congr. Soil Sci.* **1:**459–494.

Lee, J. A., 1960. A study of plant competition in relation to development. *Evolution* **14:**18–28.

Leggett, J. E., and W. A. Gilbert, 1967. Localization of the Ca-mediated apparent ion selectivity in the cross-sectional volume of soybean roots. *Plant Physiol.* **42:**1658–1664.

Lemon, E. R., 1963. Energy and water balance of plant communities. In L. T. Evans (ed.), "Environmental Control of Plant Growth," pp. 55–78. Academic Press Inc., New York.

———, 1967. Aerodynamic studies of CO_2 exchange between the atmosphere and the plant. In A. San Petro, F. A. Greer, and T. J. Army (eds.), "Harvesting the Sun," pp. 263–290. Academic Press Inc., New York.

——— and A. E. Erickson, 1952. The measurement of oxygen diffusion in the soil with a platinum microelectrode. *Soil Sci. Soc. Amer. Proc.* **16:**160–163.

Leonard, E. R., and G. C. Head, 1958. Technique and preliminary observations on growth of the roots of glasshouse tomatoes in relation to that of the tops. *J. Hort. Sci.* **33:**171–185.

Leonard, O. A., 1944. Use of root pressures in determining injury to roots by cultivation. *Plant Physiol.* **19**:157–163.

——— and J. A. Pinckard, 1946. Effect of various oxygen and carbon dioxide concentrations on cotton root development. *Plant Physiol.* **21**:18–36.

Lepeschkin, W. W., 1923. Über aktive und passive Wasserdrusen und Wasserspalten. *Ber. deut. bot. Ges.* **41**:298–300.

Leshem, B., 1965. The annual activity of intermediary roots of the Aleppo pine. *Forest Sci.* **11**:291–298.

Letey, J., 1966. Measuring aeration. Proc. conf. on drainage for efficient crop production, pp. 6–10. Amer. Soc. Agr. Eng., St. Joseph, Mich.

——— and L. H. Stolzy, 1964. Measurement of oxygen diffusion rates with the platinum microelectrode. I. Theory and equipment. *Hilgardia* **35**:545–554.

———, ———, and W. D. Kemper, 1967. Soil aeration. In R. M. Hagan et al. (eds.), "Irrigation of Agricultural Lands," pp. 941–949. *Amer. Soc. Agron.,* Madison, Wis.

———, ———, N. Valoras, and T. E. Szuszkiewics, 1962. Influence of oxygen diffusion rate on sunflower growth at various soil and air temperatures. *Agron. J.* **54**:316–319.

———, N. Welch, R. E. Pelishek, and J. Osborn, 1962. Effects of wetting agents on irrigation of water repellent soils. *Calif. Agr.* **16**:213.

Levitt, J., and G. W. Scarth, 1936. Frost-hardening studies with living cells. II. Permeability in relation to frost resistance and the seasonal cycle. *Can. J. Res.* **C14**:285–305.

———, ———, and R. D. Gibbs, 1936. Water permeability of isolated protoplasts in relation to volume change. *Protoplasma* **26**:237–248.

———, 1947. The thermodynamics of active (non-osmotic) water absorption. *Plant Physiol.* **22**:514–525.

———, 1948. The role of active water absorption in auxin-induced uptake by aerated potato discs. *Plant Physiol.* **23**:505–515.

———, 1951. Frost, drought, and heat resistance. *Ann. Rev. Plant Physiol.* **2**:245–268.

———, 1956. "The Hardiness of Plants." Academic Press, Inc., New York.

———, 1957. The significance of "apparent free space" (A.F.S.) in ion absorption. *Physiol. Plant.* **10**:882–888.

———, 1962. A sulfhydryl-disulfide hypothesis of frost injury and resistance in plants. *J. Theoret. Biol.* **3**:355–391.

———, 1966. Resistance to water transport in plants—a misconception. *Nature* **212**:527.

———, 1967. Active water transport once more: a reply to J. J. Oertli. *Physiol. Plant.* **20**:263–264.

Lewis, D. G., and J. P. Quirk, 1967. Phosphate diffusion in soil and uptake by plants. III. P^{31} movement and uptake by plants as indicated by P^{32} autoradiography. *Plant and Soil* **26**:445–453.

Lewis, F. J., 1945. Physical condition of the surface of the mesophyll cell walls of the leaf. *Nature* **156**:407–490.

———, 1948. Water movement in leaves. *Disc. Faraday Soc.* **3**:159–162.

Lewis, R. W., 1962. Guttation fluid: effects on growth of *Claviceps purpurea* in vitro. *Science* **138**:690–691.

Leyton, L., and L. Z. Rousseau, 1958. Root growth of tree seedlings in relation to aeration. In K. V. Thimann (ed.), "The Physiology of Forest Trees," pp. 467–475. The Ronald Press Company, New York.

Liming, F. G., 1934. A preliminary study of the lengths of the open vessels in the branches of the American elm. *Ohio J. Sci.* **34**:415–419.

Lingle, J. C., and R. M. Davis, 1959. The influence of soil temperature and phosphorus fertilization on the growth and mineral absorption of tomato seedlings. *Proc. Amer. Soc. Hort. Sci.* **73**:312–322.

Livingston, B. E., 1906. The relation of desert plants to soil moisture and to evaporation. *Carnegie Inst. Washington Publ.* 50.

———, 1908. A method for controlling plant moisture. *Plant World* **11**:39–40.

———, 1918. Porous clay cones for the auto-irrigation of potted plants. *Plant World* **21**:202–208.

——— and W. H. Brown, 1912. Relation of the daily march of transpiration to variations in the water content of foilage leaves. *Bot. Gaz.* **53**:309–330.

——— and E. E. Free, 1917. The effect of deficient soil oxygen on the roots of higher plants. *Johns Hopkins Univ. Circ. N.S.* **3**:182–185.

Livingston, L. G., 1964. The nature of plasmodesmata in normal (living) plant tissue. *Amer. J. Bot.* **51**:950–957.

Lloyd, F. E., 1908. The physiology of stomata. *Carnegie Inst. Washington Publ.* 82.

Loeffler, J. E., and J. van Overbeek, 1964. In Regulateurs Naturels de la Croissance Vegetale, pp. 77–82. CNRS, Paris.

Loehwing, W. F., 1934. Physiological aspects of the effect of continuous soil aeration on plant growth. *Plant Physiol.* **9**:567–583.

———, 1937. Root interactions of plants. *Bot. Rev.* **3**:195–239.

Loftfield, J. V. G., 1921. The behavior of stomata. *Carnegie Inst. Washington Publ.* 314.

Long, E. M., 1943. The effect of salt addition to the substrate on intake of water and nutrients by roots of approach-grafted tomato plants. *Amer. J. Bot.* **30**:594–601.

Long, W. G., D. V. Sweet, and H. B. Tukey, 1956. The loss of nutrients from plant foliage by leaching as indicated by radioisotopes. *Science* **123**:1039–1040.

Loomis, W. E., 1934. Daily growth of maize. *Amer. J. Bot.* **21**:1–6.

———, 1935. The translocation of carbohydrates in maize. *Iowa State Coll. J. Sci.* **9**:509–520.

———, R. Santamaria-P., and R. S. Gage, 1960. Cohesion of water in plants. *Plant* Physiol. **35**:300–306.

Lopushinsky, W., 1964*a*. Effect of water movement on ion movement into the xylem of tomato roots. *Plant Physiol.* **39**:494–501.

———, 1964*b*. Calcium transport in tomato roots. *Nature* **201**:518–519.

Loustalot, A. J., 1945. Influence of soil-moisture conditions on apparent photosynthesis and transpiration of pecan leaves. *J. Agr. Res.* **71**:519–532.

———, H. F. Winters, and N. F. Childers, 1947. Influence of high, medium, and low soil moisture on growth and alkaloid content of *Cinchona ledgeriana*. *Plant Physiol.* **22**:613–619.

Loweneck, M., 1930. Untersuchungen über Wurzelatmung. *Planta* **10**:185–228.

Lowry, M. W., W. C. Huggins, and L. A. Forrest, 1936. The effect of soil treatment on the mineral composition of exuded maize sap at different stages of development. *Georgia Agr. Exp. Sta. Bull.* 193.

Lund, E. J., 1931. Electric correlation between living cells in cortex and wood in the Douglas fir. *Plant Physiol.* **6**:631–652.

Lundegårdh, H., 1914. Einige Bedingungen der Bildung und Auflösung der Stärke. *Jahrb. wiss. Bot.* **53**:421–463.

———, 1940. Investigations as to the absorption and accumulation of inorganic ions. *Ann. Agr. Coll. Sweden* **8**:233.

———, 1954. The transport of water in wood. *Arkiv Bot.* **3**:89–119.

———, 1955. Mechanisms of absorption, transport, accumulation, and secretion of ions. *Ann. Rev. Plant Physiol.* **6**:1–24.

Luthin, J. N., 1957. "Drainage of Agricultural Lands." *Amer. Soc. Agron.*, Madison, Wis.

Luttge, U., and G. G. Laties, 1966. Dual mechanisms of ion absorption in relation to long distance transport in plants. *Plant Physiol.* **41**:1531–1539.

——— and J. Weigl, 1962. Mikroautoradiographische untersuchungen der aufnahme und des transportes von $^{35}SO_4^{--}$ und $^{45}Ca^{++}$ in Keimwurzeln von *Zea mays* L. und *Pisum sativum* L. *Planta* **58**:113–126.

Lutz, H. J., 1944. Determinations of certain physical properties of forest soils: I. Methods utilizing samples collected in metal cylinders. *Soil Sci.* **57**:475–487.

Lutz, J. F., 1952. Mechanical impedance and plant growth. In B. T. Shaw (ed.), "Soil Physical Conditions and Plant Growth," pp. 43–71. Academic Press Inc., New York.

Lyford, W. H., and B. F. Wilson, 1964. Development of the root system of *Acer rubrum* L. *Harvard Forest Paper* no. 10.

Lyon, T. L., and H. O. Buckman, 1943. "The Nature and Properties of Soils," 4th ed. The Macmillan Company, New York.

Lyr, H., and G. Hoffman, 1967. Growth rates and growth periodicity of tree roots. *Int. Rev. Forest. Res.* **2**:181–236.

MacCallum, W. B., 1908. The flowering stalk of the century plant. *Plant World* **11**:141–147.

McComb, A. L., and J. E. Griffith, 1946. Growth stimulation and phosphorus absorption of mycorrhizal and nonmycorrhizal northern white pine and Douglas fir seedlings in relation to fertilizer treatment. *Plant Physiol.* **21**:11–17.

——— and W. E. Loomis, 1944. Subclimax prairie. *Torrey Bot. Club Bull.* **71**:46–76.

McDermott, J. J., 1945. The effect of the moisture content of the soil upon the rate of exudation. *Amer. J. Bot.* **32**:570–574.

MacDougal, D. T., 1925. Absorption and exudation pressures of sap in plants. *Amer. Phil. Soc. Proc.* **64**:102–130.

———, 1926. The hydrostatic system of trees. *Carnegie Inst. Washington Publ.* 373.

——— and J. Dufrenoy, 1944. Mycorrhizal symbiosis in *Aplectrum, Corallorhiza* and *Pinus. Plant Physiol.* **19**:440–465.

———, 1946. Criteria of nutritive relations of fungi and seed-plants in mycorrhizae. *Plant Physiol.* **21**:1–10.

MacDowell, F. D. H., 1964. Reversible effects of chemical treatments on the rhythmic exudation of sap by tobacco plants. *Can. J. Bot.* **42**:115–122.

McIlrath, W. J., Y. P. Abrol, and F. Heiligman, 1963. Dehydration of seeds in intact tomato fruits. *Science* **142**:1681–1682.

McIntyre, D. S., 1958. Permeability measurements of soil crusts formed by raindrop impact. *Soil Sci.* **85**:185–189.

McLaren, A. D., W. A. Jensen, and L. Jacobsen, 1960. Absorption of enzymes and other proteins by barley roots. *Plant Physiol.* **35**:549–556.

McPherson, D. C., 1939. Cortical air spaces in the roots of *Zea mays. New Phytol.* **38**:190–202.

McQuilkin, W. E., 1936. Root development of pitch pine, with some comparative observations on shortleaf pine. *J. Agr. Res.* **51**:983–1016.

MacRobbie, E. A. C., 1962. Ionic relations of *Nitella translucens. J. Gen. Physiol.* **45**:861–878.

———, 1964. Factors affecting the fluxes of potassium and chloride ions in *Nitella translucens. J. Gen. Physiol.* **47**:859–877.

———, 1965. The nature of the coupling between light energy and active ion transport in *Nitella translucens. Biochim. biophys. Acta* **94**:64–73.

——— and J. Dainty, 1958. Ion transport in *Nitellopsis obtusa. J. Gen. Physiol.* **42**:335–353.

McWilliam, J. R., 1968. The nature of the perennial response in Mediterranean grasses.

II. Senescence, summer dormancy, and survival in *Phalaris. Aust. J. Agr. Res.* **19**:397–409.

——— and P. J. Kramer, 1968. The nature of the perennial response in Mediterranean grasses. I. Water relations and summer survival in *Phalaris. Aust. J. Agr. Res.* **19**:381–395.

Machlis, L., 1944. The respiratory gradient in barley roots. *Amer. J. Bot.* **31**:281–282.

Macklon, A. E. S., and P. E. Weatherley, 1965. Controlled environment studies of the nature and origins of water deficits in plants. *New Phytol.* **64**:414–427.

Magee, A. I., and W. Kalbfleisch, 1952. Moisture determinations of silage, hay and grain. *Sci. Agr.* **32**:117–126.

Maggs, D. H., 1964. Growth rates in relation to assimilate supply and demand. *J. Exp. Bot.* **15**:574–583.

Magistad, O. C., A. D. Ayers, C. H. Wadleigh, and H. G. Gauch, 1943. Effect of salt concentration, kind of salt, and climate on plant growth in sand cultures. *Plant Physiol.* **18**:151–166.

——— and R. F. Reitemeier, 1943. Soil solution concentrations at the wilting point and their correlation with plant growth. *Soil Sci.* **55**:351–360.

Magness, J. R., E. S. Degman, and J. R. Furr, 1935. Soil moisture and irrigation investigations in eastern apple orchards. *U.S. Dept. Agr. Tech. Bull.* 491.

Mansfield, T. A., 1965. Responses of stomata to short duration increases in carbon dioxide concentration. *Physiol. Plant.* **18**:79–84.

Marinos, N. G., 1963. Vacuolation in plant cells. *J. Ultrastruct. Res.* **9**:177–185.

Marshall, D. C., 1958. Measurement of sap flow in conifers by heat transport. *Plant Physiol.* **21**:95–101.

Marshall, T. J., 1959. Relations between water and soil. Tech. Comm. no. 50, Commonwealth Bureau of Soils, Harpenden, England.

Martin, E. V., 1943. Studies of evaporation and transpiration under controlled conditions. *Carnegie Inst. Washington Publ.* 550.

Martin, J. P., 1950. Effects of various leaching treatments on growth of orange seedlings in old citrus soil. *Soil Sci.* **69**:107–122.

Marvin, J. W., 1958. The physiology of maple sap flow. In K. V. Thimann (ed.), "The Physiology of Forest Trees." The Ronald Press Company, New York.

Marx, D. H., and C. B. Davey, 1967. Ectotrophic mycorrhizae as deterrents to pathogenic root infections. *Nature* **213**:1139.

Mason, A. C., 1958. The effect of soil moisture on the mineral composition of apple plants grown in pots. *J. Hort. Sci.* **33**:202–217.

Mason, T. G., and E. J. Maskell, 1928. Studies on the transport of carbohydrates in the cotton plant. I. A study of diurnal variation in the carbohydrates of leaf, bark, and wood, and of the effects of ringing. *Ann. Bot.* **42**:189–253.

Mattas, R. E., and A. W. Pauli, 1955. Trends in nitrate reduction and nitrogen fractions in young corn (*Zea mays*) plants during heat and moisture stress. *Crop Sci.* **5**:181–184.

Mauro, A., 1960. Some properties of ionic and nonionic semipermeable membranes. *Circulation* **21**:845–854.

Maximov, N. A., 1929. The plant in relation to water. George Allen & Unwin, Ltd., London.

May, L. H., E. J. Milthorpe, and F. L. Milthorpe, 1962. Pre-sowing hardening of plants to drought. *Field Crop Abstr.* **15**:93–98.

Mecklenburg, R. A., H. B. Tukey, Jr., and J. V. Morgan, 1966. A mechanism for the leaching of calcium from foliage. *Plant Physiol.* **41**:610–613.

Mederski, H. J., 1961. Determination of internal water status of plants by beta ray gauging. *Soil Sci.* **92**:143–146.

——— and W. Alles, 1968. Beta gauging leaf water status: influence of changing leaf characteristics. *Plant Physiol.* **43:**470–472.

——— and J. H. Wilson, 1960. Relation of soil moisture to ion absorption by corn plants. *Proc. Soil Sci. Soc. Amer.* **24:**149–152.

Mees, G. C., and P. E. Weatherley, 1957*a*. The mechanism of water absorption by roots. I. Preliminary studies on the effects of hydrostatic pressure gradients. *Proc. Roy. Soc.* (London) **B147:**367–380.

———, 1957*b*. The mechanism of water absorption by roots. II. The role of hydrostatic pressure gradients across the cortex. *Proc. Roy. Soc.* (London) **B147:**381–391.

Meeuse, B. J. D., 1952. The amylolytic action of maple and birch sap. *Acta Bot. Neer.* **1:**216–221.

Meidner, H., and T. A. Mansfield, 1965. Stomatal responses to illumination. *Biol. Rev.* **40:**483–509.

Melin, E., 1953. Physiology of mycorrhizal relations in plants. *Ann. Rev. Plant Physiol.* **4:**325–346.

——— and H. Nilsson, 1950. Transfer of radioactive phosphorus to pine seedlings by means of mycorrhizal hyphae. *Physiol. Plant.* **3:**88–92.

———, 1955. Ca^{45} used as an indicator of transport of cations to pine seedlings by means of mycorrhizal mycelium. *Svensk. Bot. Tidskr.* **49:**119–122.

———, H. Nilsson, and E. Hacskaylo, 1958. Translocation of cations to seedlings of *Pinus virginiana* through mycorrhizal mycelium. *Bot. Gaz.* **119:**243–246.

Mer, C. L., 1940. The factors determining the resistance to the movement of water in the leaf. *Ann. Bot.* **4:**397–401.

Mercer, F. V., 1955. The water relations of plant cells. *Proc. Linn. Soc. N. S. Wales* **80:**6–29.

Merwin, H. E., and H. Lyon, 1909. Sap pressure in the birch stem. *Bot. Gaz.* **48:**442–458.

Meyer, B. S., 1931. Effects of mineral salts upon the transpiration and water requirement of the cotton plant. *Amer. J. Bot.* **18:**79–93.

———, 1938. The water relations of plant cells. *Bot Rev.* **4:**531–547.

———, 1945. A critical evaluation of the terminology of diffusion phenomena. *Plant Physiol.* **20:**142–164.

——— and D. B. Anderson, 1952. "Plant Physiology," 2d ed. D. Van Nostrand Company, Inc., Princeton, N.J.

Meyer, R. E., and J. R. Gingrich, 1964. Osmotic stress: effects of its application to a portion of wheat root systems. *Science* **144:**1463–1464.

Michaelis, P., 1934. Ökologische Studien an der alpinen Baumgrenze. IV. Zur Kenntnis des winterlichen Wasserhaushaltes. *Jahrb. wiss. Bot.* **80:**169–247.

Mikola, P., 1965. Studies on the ectendotrophic mycorrhiza of pine. *Acta Forest. Fenn.* **79:**1–56.

Millar, B. D., 1966. Relative turgidity of leaves: temperature effects in measurement. *Science* **154:**512–513.

Milburn, J. A., and R. P. C. Johnson, 1966. The conduction of sap. II. Detection of vibrations produced by sap cavitation in *Ricinus* xylem. *Planta* **69:**43–52.

Miller, E. C., 1916. Comparative study of the root systems and leaf areas of corn and the sorghums. *J. Agr. Res.* **6:**311–332.

———, 1917. Daily variation of water and dry matter in the leaves of corn and the sorghums. *J. Agr. Res.* **10:**11–47.

———, 1938. "Plant Physiology," 2d ed. McGraw-Hill Book Company, New York.

Miller, E. E., and A. Klute, 1967. The dynamics of soil water. Part I—Mechanical forces. In R. M. Hagan et al. (eds.), "Irrigation of Agricultural Lands," pp. 209–244. *Amer. Soc. Agron.*, Madison, Wis.

Miller, L. N., 1965. Changes in radiosensitivity of pine seedlings subjected to water stress during chronic gamma irradiation. *Health Phys.* **11**:1653–1662.

Miller, R. J., and B. H. Beard, 1967. Effects of irrigation management on chemical composition of soybeans in the San Joaquin Valley. *Calif. Agr.* **21**(10):8–10.

Miller, S. A., and A. P. Mazurak, 1958. Relationships of particle and pore sizes to the growth of sunflowers. *Proc. Soil Sci. Soc. Amer.* **22**:275–278.

Milthorpe, F. L., 1955. The significance of the measurement made by the cobalt paper method. *J. Exp. Bot.* **6**:17–19.

———, 1959. Transpiration from crop plants. *Field Crops Abstr.* **12**:1–9.

———, 1960. The income and loss of water in arid and semiarid zones. *Arid Zone Res.* **15**:9–36. UNESCO, Paris.

——— and E. Spencer, 1957. Experimental studies of the factors controlling transpiration. *J. Exp. Bot.* **8**:413–437.

Minshall, W. H., 1964. Effect of nitrogen-containing nutrients on the exudation from detopped tomato plants. *Nature* **202**:925–926.

———, 1968. Effects of nitrogenous materials on translocation and stump exudation in root systems of tomato. *Can. J. Bot.* **46**:363–376.

Mitchell, H. L., R. F. Finn, and R. O. Rosendahl, 1937. The relation between mycorrhizae and the growth and nutrient absorption of coniferous seedlings in nursery beds. *Black Rock Forest Papers* **1**:57–73.

Moinat, A. D., 1943. An auto-irrigator for growing plants in the laboratory. *Plant Physiol.* **18**:280–287.

Molisch, H., 1902. Ueber localen Blutungsdruck und seine Ursachen. *Bot. Ztg.* **60**:45–63.

———, 1912. Das Offen- und Geschlossensein der Spaltöffnungen, Veranschaulicht durch eine neue Methode (Infiltrationsmethode) *Z. Bot.* **4**:106–122.

———, 1921. Über den Einfluss der Transpiration auf das Verschwinden der Stärke in den Blättern. *Ber. deut. bot. Ges.* **39**:339–344.

Mollenhauer, H. H., and D. J. Morre, 1966. Golgi apparatus and plant secretion. *Ann. Rev. Plant Physiol.* **17**:27–46.

Monteith, J. L., 1963. Gas exchange in plant communities. In L. T. Evans (ed.), "Environmental Control of Plant Growth," pp. 95–112. Academic Press Inc., New York.

———, 1965. Evaporation and environment. *Symp. Soc. Exp. Biol.* **19**:205–234.

——— and P. C. Owen, 1958. A thermocouple method for measuring relative humidity in the range 95–100%. *J. Sci. Inst.* **35**:443–446.

Montermoso, J. C., and A. R. Davis, 1942. Preliminary investigation of the rhythmic fluctuations in transpiration under constant environmental conditions. *Plant Physiol.* **17**:473–480.

Montfort, C., and H. Hahn, 1950. Atmung und assimilation als dynamisches Kennzeichen abgestufter Trockenresistenz bei Farnen und höheren Pflanzen. *Planta* **38**:503–515.

Moreland, D. E., 1950. A study of translocation of radioactive phosphorous in loblolly pine (*Pinus taeda* L.) *J. Elisha Mitchell Sci. Soc.* **66**:175–181.

Moss, D. N., 1963. The effect of environment on gas exchange of leaves. In I. Zelitch (ed.), "Stomata and Water Relations in Plants." *Conn. Agr. Exp. Sta. Bull.* 664.

———, R. B. Musgrave, and E. R. Lemon, 1961. Photosynthesis under field conditions. III. Some effects of light, carbon dioxide, temperature, and soil moisture on photosynthesis, respiration, and transpiration of corn. *Crop Sci.* **1**:83–87.

von Mothes, K., 1932. Ernährung, Struktur, und Transpiration. Ein Beitrag zur Kausalen Analyze der Xeromorphosen. *Biol. Zentr.* **52**:193–223.

———, 1956. Der Einfluss des Wasserzustandes auf Fermentprozesse und Stoffumsatz.

"Encyclopedia of Plant Physiology," vol. 3, pp. 656–664. Springer-Verlag OHG, Berlin.

Muller, C. H., 1966. The role of chemical inhibition (allelopathy) in vegetational competition. *Torrey Bot. Club Bull.* **93**:332–351.

Müller-Stoll, W. R., 1947. Der Einfluss der Ernährung auf die Xeromorphie der Hochmoorpflanzen. *Planta* **35**:225–251.

———, 1965. The problem of water outflow from roots. In B. Slavik (ed.), "Water Stress in Plants," pp. 21–29. Czech. Acad. Sci., Prague.

Münch, E., 1930. Die Stoffbewegungen in der Pflanze. Gustav Fischer Verlag KG, Jena.

Murphy, W. H., Jr., and J. T. Syverton, 1958. Absorption and translocation of mammalian viruses by plants. *Virology* **6**:623–636.

Musgrave, G. W., 1955. How much of the rain enters the soil? In A. Stefferud (ed.), "Water," pp. 151–159. *U.S. Dept. Agr. Yearb.*

Myers, G. M. P., 1951. The water permeability of unplasmolyzed tissues. *J. Exp. Bot.* **2**:129–144.

Myers, H. E., and A. L. Hallsted, 1942. The comparative effect of corn and sorghums on the yield of succeeding crops. *Proc. Soil Sci. Soc. Amer.* **7**:316–321.

Nakayama, F. S., and W. L. Ehrler, 1964. Beta ray gauging technique for measuring leaf water content changes and moisture status of plants. *Plant Physiol.* **39**:95–98.

Newman, E. I., 1966. Relationship between root growth of flax (*Linum unitatissimum*) and soil water potential. *New Phytol.* **65**:273–283.

——— and P. J. Kramer, 1966. Effects of decenylsuccinic acid on the permeability and growth of bean roots. *Plant Physiol.* **41**:606–609.

Nicholas, D. J. D., 1965. Influence of the rhizosphere on the mineral nutrition of the plant. In K. F. Baker, W. C. Snyder, et al. (eds.), "Ecology of Soil-borne Plant Pathogens," pp. 154–169. Univ. of California Press, Berkeley.

Nieman, R. H., 1962. Some effects of sodium chloride on growth, photosynthesis, and respiration of twelve crop plants. *Bot. Gaz.* **123**:279–285.

Nightingale, G. T., 1935. Effects of temperature on growth, anatomy, and metabolism of apple and peach roots. *Bot. Gaz.* **96**:581–639.

North, C. P., and A. Wallace, 1955. Soil temperature and citrus. *Calif. Agr.* **9**(11):13.

Nutman, F. J., 1933. The root-system of *Coffea arabica*. II. The effect of some soil conditions in modifying the 'normal' root-system. *Empire J. Exp. Agr.* **1**:285–296.

———, 1934. The root-system of *Coffea arabica*. III. The spatial distribution of the absorbing area of the root. *Empire J. Exp. Agr.* **2**:293–302.

Oertli, J. J., 1966. Active water transport in plants. *Physiol. Plant.* **19**:809–817.

———, 1967. The salt absorption isotherm. *Physiol. Plant.* **20**:1014–1026.

Ogata, G., L. A. Richards, and W. R. Gardner, 1960. Transpiration of alfalfa determined from soil water content changes. *Soil Sci.* **89**:179–182.

O'Leary, J. W., 1965. Root-pressure exudation in woody plants. *Bot. Gaz.* **126**:108–115.

——— and P. J. Kramer, 1964. Root pressure in conifers. *Science* **145**:284–285.

Olsen, S. R., W. D. Kemper, and R. D. Jackson, 1962. Phosphate diffusion to plant roots. *Proc. Soil Sci. Soc. Amer.* **26**:222–227.

Oosting, H. J., 1954. Ecological processes and vegetation of the maritime strand in the southeastern United States. *Bot. Rev.* **20**:226–262.

Oppenheimer, H. R., and D. L. Elze, 1941. Irrigation of citrus trees according to physiological indicators. *Palestine J. Bot. Rehovot Ser.* **4**:20–46.

——— and N. Engelberg, 1965. Mesure du degré d'ouverture des stomates de conifères méthodes anciennes et modernes. *Arid Zone Res.* **25**:317–323. UNESCO, Paris.

Orchard, B., 1967. Water deficit and the growth of crop seedlings. I. The effect of rate of change of soil water content on kale seedlings. *J. Exp. Bot.* **18**:308–320.

Ordin, L., 1958. The effect of water stress on cell wall metabolism of plant tissue. In "Radioisotopes in Scientific Research," vol. 4, pp. 553–564. Pergamon Press, New York.

———, 1960. Effect of water stress on cell wall metabolism of Avena coleoptile tissue. *Plant Physiol.* **35**:443–450.

———, T. H. Applewhite, and J. Bonner, 1956. Auxin-induced water uptake by Avena coleoptile sections. *Plant Physiol.* **31**:44–53.

——— and P. J. Kramer, 1956. Permeability of *Vicia faba* root segments to water as measured by diffusion of deuterium hydroxide. *Plant Physiol.* **31**:468-471.

Osterhout, W. J. V., 1947. Some aspects of secretion. 1. Secretion of water. *J. Gen. Physiol.* **30**:439–447.

Van Overbeek, J., 1942. Water uptake by excised root systems of the tomato due to non-osmotic forces. *Amer. J. Bot.* **29**:677–683.

———, 1956. Absorption and translocation of plant regulators. *Ann. Rev. Plant Physiol.* **7**:355–372.

Overland, L., 1966. The role of allelopathic substances in the "smother crop" barley. *Amer. J. Bot.* **53**:423–432.

Owen, P. C., 1952. The relation of germination of wheat to water potential. *J. Exp. Bot.* **3**:188–203.

——— and D. J. Watson, 1956. Effect on crop growth of rain after prolonged drought. *Nature* **177**:847.

Painter, L. I., 1966. Method of subjecting growing plants to a continuous soil moisture stress. *Agr. J.* **58**:459–460.

Pallas, J. E., and G. G. Williams, 1962. Foliar absorption and translocation of P^{32} and 2,4-dichlorophenoxyacetic acid as affected by soil-moisture tension. *Bot. Gaz.* **123**:175–180.

Parker, A. F., 1961. Bark moisture relations in disease development: present status and future needs. *Recent Advan. Bot.* II:1535–1537. Univ. of Toronto Press, Toronto.

Parker, J., 1949. Effects of variations in the root-leaf ratio on transpiration rate. *Plant Physiol.* **24**:739–743.

———, 1950. The effects of flooding on the transpiration and survival of some southeastern forest tree species. *Plant Physiol.* **25**:453–460.

———, 1952. Desiccation in conifer leaves: anatomical changes and determination of the lethal level. *Bot. Gaz.* **114**:189–198.

———, 1965. Physiological diseases of trees and shrubs. *Advan. Frontiers Plant Sci.* **12**:97–248.

———, 1968. Drought-resistance mechanisms. In T. T. Kozlowski (ed.), "Water Deficits and Plant Growth," vol. 1, pp. 195–234. Academic Press Inc., New York.

Pasquill, F., 1949. Eddy diffusion of water and heat near the ground. *Proc. Roy. Soc.* (London) **A198**:116–140.

Pate, J. S., 1965. Roots as organs of assimilation of sulfate. *Science* **149**:547–548.

Patric, J. H., J. E. Douglass, and J. D. Hewlett, 1965. Soil water absorption by mountain and Piedmont forests. *Proc. Soil Sci. Soc. Amer.* **29**:303–308.

Patrick, Z. A., and T. A. Toussoun, 1965. Plant residues and organic amendments in relation to biological control. In K. F. Baker, W. C. Snyder, et al. (eds.), "Ecology of Soil-borne Plant Pathogens," pp. 440–459. Univ. of California Press, Berkeley.

———, ———, and W. L. Koch, 1964. Effect of crop-residue decomposition on plant roots. *Ann. Rev. Phytopathol.* **2**:267–272.

Pauling, L., 1960. "The Nature of the Chemical Bond," 3d ed. Cornell Univ. Press, Ithaca, N.Y.

Pavlychenko, T. K., 1937. The soil-block washing method in quantitative root study. *Can. J. Res.* **15C**:33–57.

Pearson, G. A., 1931. Forest types in the southwest as determined by climate and soil. *U.S. Dept. Agr. Tech. Bull.* 247.

Peck, A. J., and R. M. Rabbidge, 1966. Note on an instrument for measuring water potentials particularly in soils. *C.S.I.R.O. Aust., Conf. Instrum. Plant Environ. Meas.,* p. 20. Soc. Instrum. Tech., Melbourne.

Pelishek, R. E., J. F. Osborn, and J. Letey, 1964. Carriers for air application of granulated wetting agents. *Calif. Agr.* **18**:11.

Penman, H. L., 1948. Natural evaporation from open water, bare soil and grass. *Proc. Roy. Soc.* (London) **A193**:120–145.

———, 1949. The dependence of transpiration on weather and soil conditions. *J. Soil Sci.* **1**:74–89.

———, 1956. Estimating evaporation. *Trans. Amer. Geophys. Union* **37**:43–46.

———, 1963. Vegetation and hydrology. Commonwealth Agricultural Bureaux, Farnham Royal, England.

Penman, H. L., and I. Long, 1960. Weather in wheat—an essay in micro-meteorology. *Quart. J. Roy. Meteorol. Soc.* **86**:16–50.

——— and R. K. Schofield, 1951. Some physical aspects of assimilation and transpiration. *Symp. Soc. Exp. Biol.* **5**:115–129.

Peters, D. B., 1957. Water uptake of corn roots as influenced by soil moisture content and soil moisture tension. *Proc. Soil Sci. Soc. Amer.* **21**:481–484.

——— and M. B. Russell, 1959. Relative water losses by evaporation and transpiration in field corn. *Proc. Soil Sci. Soc. Amer.* **23**:170–173.

Petrie, A. H. K., and J. I. Arthur, 1943. Physiological ontogeny in the tobacco plant. *Aust. J. Exp. Biol. Med. Sci.* **21**:191–200.

——— and J. S. Wood, 1938. Studies on the nitrogen metabolism of plants. I. The relation between the content of proteins, amino acids and water in the leaves. *Ann. Bot.* **2**:33–60.

Petritschek, K., 1953. Über die Beziehungen zwischen Geschwindigkeit und Elektrolytgehalt des aufsteigenden Saftströmes. *Flora* **140**:345–385.

Petterson, S., 1966. Artificially induced water and sulfate transport through sunflower roots. *Physiol. Plant.* **19**:581–601.

Pfeffer, W., 1877. Osmotische Untersuchungen. W. Engelmann, Leipzig.

Pharis, R. P., 1967. Seasonal fluctuations in the foliage-moisture content of well-watered conifers. *Bot. Gaz.* **128**:179–185.

Philip, J. R., 1957*a*. The physical principles of soil water movement during the irrigation cycle. *Proc. Int. Congr. Irrig. Drain.* **8**:124–154.

———, 1957*b*. Evaporation and moisture and heat fields in the soil. *J. Meteorol.* **14**:354–366.

———, 1957*c*. The theory of infiltration: 5. The influence of the initial moisture content. *Soil Sci.* **84**:329–339.

———, 1957*d*. The theory of infiltration: 1. The infiltration equation and its solution. *Soil Sci.* **83**:345–357.

———, 1958*a*. The theory of infiltration: 6. Effect of water depth over soil. *Soil Sci.* **85**:278–286.

———, 1958*b*. The osmotic cell, solute diffusibility, and the plant water economy. *Plant Physiol.* **33**:264–271.

———, 1966. Plant water relations: some physical aspects. *Ann. Rev. Plant Physiol.* **17**:245–268.

——— and D. A. de Vries, 1957. Moisture movement in porous materials under temperature gradients. *Trans. Amer. Geophys. Union* **38:**222–232.

Phillips, I. D. J., 1964. I. The importance of an aerated root system in the regulation of growth levels in the shoot of *Helianthus annuus. Ann. Bot.* **28:**17–36.

Phillis, E., and T. G. Mason, 1940. The effect of ringing on the upward movement of solutes from the roots. *Ann. Bot.* **4:**635–644.

Philpott, Jane, 1953. A blade tissue study of leaves of forty-seven species of Ficus. *Bot. Gaz.* **115:**15–35.

———, 1956. Blade tissue organization of foliage leaves of some Carolina shrub-bog species as compared with their Appalachian Mountain affinities. *Bot. Gaz.* **118:**88–105.

Pierre, W. H., and G. G. Pohlman, 1934. The phosphorous concentration of the exuded sap of corn as a measure of the available phosphorus in the soil. *J. Amer. Soc. Agron.* **25:**160–171.

Pillsbury, A. F., R. E. Pelishek, J. F. Osborn, and T. E. Szuszkiewics, 1961. Chaparral to grass conversion doubles watershed runoff. *Calif. Agr.* **15**(11):12–13.

——— and S. J. Richards, 1954. Some factors affecting rates of irrigation water entry into Ramona sandy loam soil. *Soil Sci.* **78:**211–217.

Pisek, A., and E. Cartellieri, 1932. Zur Kenntnis des Wasserhaushaltes der Pflanzen. I. Sonnenpflanzen. *Jahrb. wiss. Bot.* **75:**195–251.

Pitman, M. G., and H. D. Saddler, 1966. Active sodium and potassium transport in cells of barley roots. *Proc. Nat. Acad. Sci.* **57:**44–49.

Plymale, E. L., and R. B. Wylie, 1944. The major veins of mesomorphic leaves. *Amer. J. Bot.* **31:**99–106.

Pohlman, G. G., 1946. Effect of liming different soil layers on yield of alfalfa and on root development and nodulation. *Soil Sci.* **62:**255–266.

Pollard, J. K., and T. Sproston, 1954. Nitrogenous constituents of sap exuded from the sapwood of *Acer saccharum. Plant Physiol.* **29:**360–364.

Post, K., and J. G. Seeley, 1943. Automatic watering of greenhouse crops. *Cornell Univ. Agr. Exp. Sta. Bull.* 793.

Postlethwait, S. N., and B. Rogers, 1958. Tracing the path of the transpiration stream in trees by the use of radioactive isotopes. *Amer. J. Bot.* **45:**753–757.

Pramer, D., 1954. The movement of chloramphenicol and streptomycin in broad bean and tomato plants. *Ann. Bot.* **18:**463–470.

Preston, R. D., 1952. Movement of water in higher plants. In A. Frey-Wyssling (ed.), "Deformation and Flow in Biological Systems," pp. 257–321. North Holland Publishing Company, Amsterdam.

———, 1961. Theoretical and practical implications of the stresses in the water-conducting system. *Recent Advan. Bot. II* 1144–1149. Univ. of Toronto Press, Toronto.

Prevot, P., and F. C. Steward, 1936. Salient features of the root system relative to the problem of salt absorption. *Plant Physiol.* **11:**509–534.

Priestley, J. H., 1922. Further observations upon the mechanism of root pressure. *New Phytol.* **21:**41–48.

——— and A. Wormall, 1925. On the solutes exuded by root pressure from vines. *New Phytol.* **24:**24–38.

———, 1935. Radial growth and extension growth in the tree. *Forestry* **9:**84–95.

Probine, M. C., and R. D. Preston, 1962. Cell growth and the structure and mechanical properties of the wall in internodal cells of *Nitella opaca.* II. Mechanical properties of the walls. *J. Exp. Bot.* **13:**111–127.

Proebsting, E. L., 1943. Root distribution of some deciduous fruit trees in a California orchard. *Proc. Amer. Soc. Hort. Sci.* **43:**1–4.

——— and A. E. Gilmore, 1941. The relation of peach root toxicity to the re-establishing of peach orchards. *Proc. Amer. Soc. Hort. Sci.* **38:**21–26.

Queen, W. H., 1967. Radial movement of water and ^{32}P through suberized and unsuberized roots of grape. Ph.D. dissertation, Duke University, N.C.

van Raalte, M. H., 1940. On the oxygen supply of rice roots. *Ann. Jard. Bot. Buitenzorg* **50:**99–113.

———, 1944. On the oxidation of the environment by the roots of rice (*Oryza sativa* L.) *Ann. Bot. Gard. Buitenzorg Hors Série* **15:**34.

Raber, O., 1937. Water utilization by trees, with special reference to the economic forest species of the north temperate zone. *U.S. Dept. Agr. Misc. Publ.* 257.

Rabideau, G. S., W. G. Whaley, and C. Heimsch, 1950. Vascular development and differentiation in two maize inbreds and their hybrid. *Amer. J. Bot.* **37:**93–99.

Rains, D. W., W. E. Schmid, and E. Epstein, 1964. Absorption of cations by roots. Effects of hydrogen ions and essential role of calcium. *Plant Physiol.* **39:**274–278.

Raleigh, G. J., 1941. The effect of culture solution temperature on water intake and wilting of the muskmelon. *Proc. Amer. Soc. Hort. Sci.* **38:**487–488.

———, 1946. The effect of various ions on guttation of the tomato. *Plant Physiol.* **21:**194–200.

Raney, F. C., and Y. Mihara, 1967. Water and soil temperature. In R. M. Hagan et al. (eds.), "Irrigation of Agricultural Lands," pp. 1024–1036. *Amer. Soc. Agron.*, Madison, Wis.

———, and Y. Vaadia, 1965. Movement of tritiated water in the root system of *Helianthus annuus* in the presence and absence of transpiration. *Plant Physiol.* **40:**378–382.

Raschke, K., 1956. Über die physikalischen Beziehungen zwischen Wärmëubergangzahl, Strahlungsaustasch, Temperatur und Transpiration eines Blattes. *Planta* **48:**200–238.

———, 1958. Über den Einfluss der Diffusionswiderstände auf die Transpiration und die Temperatur eines Blattes. *Flora* **146:**546–578.

———, 1960. Heat transfer between the plant and the environment. *Ann. Rev. Plant Physiol.* **11:**111–126.

———, 1965. Die Stomata als glieder eines schwingungsfähigen CO_2 Regelsystems Experimentelles Nachweis an *Zea mays* L. *Z. Naturforsch.* **20:**1261–1270.

Ray, P. M., 1960. On the theory of osmotic water movement. *Plant Physiol.* **35:**783–795.

Rawlins, S. L., 1963. Resistance to water flow in the transpiration stream. *Conn. Agr. Exp. Sta. Bull.* 664, pp. 69–85.

———, 1964. Systematic error in leaf water potential measurements with a thermocouple psychrometer. *Science* **146:**644–646.

——— and F. N. Dalton, 1967. Psychrometric measurement of soil water potential without precise temperature control. *Proc. Soil Sci. Soc. Amer.* **31:**297–301.

Rayner, M. C., and W. Neilson-Jones, 1944. Problems in tree nutrition. Faber & Faber, Ltd., London.

Read, D. W. L., S. V. Fleck, and W. L. Pelton, 1962. Self-irrigating greenhouse pots. *Agron. J.* **54:**467–468.

Rediske, J. H., and O. Biddulph, 1953. The absorption and translocation of iron. *Plant Physiol.* **28:**594–605.

Reed, E. L., 1924. Anatomy, embryology, and ecology of *Arachis hypogea. Bot. Gaz.* **78:**289–310.

Reed, H. S., and D. T. MacDougal, 1937. Periodicity in the growth of the orange tree. *Growth* **1:**371–373.

Reed, J. F., 1939. Root and shoot growth of shortleaf and loblolly pines in relation to certain environmental conditions. *Duke Univ. Sch. Forest. Bull.* 4.

Reeves, R. C., and M. Fireman, 1967. Salt problems in relation to irrigation. In R. M. Hagan et al. (eds.), "Irrigation of Agricultural Lands," pp. 988–1008. *Amer. Soc. Agron.*, Madison, Wis.

Rehder, H., and K. Kreeb, 1961. Vergleichende Untersuchungen zur Bestimmung der Blattsaugspannung mit der gravimetrischen Methode und der Schardakow-Methode. *Ber. deut. bot. Ges.* **74**:95–98.

Reimann, E. G., C. A. Van Doren, and R. S. Stauffer, 1946. Soil moisture relationships during crop production. *Proc. Soil Sci. Soc. Amer.* **10**:41–46.

Reinders, D. E., 1938. The process of water-intake by discs of potato tuber tissue. *Proc. Kon. Akad. Neth.* **C41**:820–831.

———, 1942. Intake of water by parenchymatic tissue. *Rec. Trav. Bot. Neer.* **39**:1–140.

Reinhart, K. G., and R. E. Taylor, 1954. Infiltration and available water storage capacity in the soil. *Trans. Amer. Geophys. Union* **35**:791–795.

Renner, O., 1912. Versuche zur Mechanik der Wasserversorgung. *Ber. deut. bot. Ges.* **30**:576–580, 642–648.

———, 1915. Die Wasserversorgung der Pflanzen. *Handwörterbuch Naturwiss.* **10**:538–557.

———, 1929. Versuche zur Bestimmung des Filtrationswiderstandes der Wurzeln. *Jahrb. wiss. Bot.* **70**:805–838.

———, 1933. Wasserzustand und Saugkraft, Erwiderung an H. Walter. *Planta* **19**:644–647.

Repp, G. I., D. R. McAllister, and H. H. Wiebe, 1959. Salt resistance of protoplasm as a test for salt tolerance of agricultural plants. *Agron. J.* **51**:311–314.

Rhoades, E. D., 1964. Inundation tolerance of grasses in flooded areas. *Trans. Amer. Soc. Agr. Eng.* **7**:164–166, 169.

Richards, B. G., 1965. Thermistor hygrometer for determining the free energy of moisture in unsaturated soil. *Nature* **208**:608–609.

Richards, L. A., and H. L. Blood, 1934. Some improvements in auto-irrigator apparatus. *J. Agr. Res.* **49**:115–121.

——— and W. E. Loomis, 1942. Limitations of auto-irrigators for controlling soil moisture under growing plants. *Plant Physiol.* **17**:223–235.

——— and M. Fireman, 1943. Pressure-plate apparatus for measuring moisture sorption and transmission by soils. *Soil Sci.* **56**:395–404.

——— and L. R. Weaver, 1944. Moisture retention by some irrigated soils as related to soil moisture tension. *J. Agr. Res.* **69**:215–235.

———, 1949. Methods of measuring soil moisture tension. *Soil Sci.* **68**:95–112.

——— and R. B. Campbell, 1950. The effect of salinity on the electrical resistance of gypsum, nylon, and fiberglass soil moisture measuring units. *U.S. Reg. Salinity Lab. Res. Rep.* **42**:1–8.

——— and C. H. Wadleigh, 1952. Soil water and plant growth. In B. T. Shaw (ed.), "Soil Physical Conditions and Plant Growth," pp. 73–251. Academic Press Inc., New York.

——— (ed.), 1954. Diagnosis and improvement of saline and alkaline soils. *U.S. Dept. Agr. Handb.* 60.

——— and G. Ogata, 1958. Thermocouple for vapor pressure measurement in biological and soil systems at high humidity. *Science* **128**:1089–1090.

——— and P. L. Richards, 1962. Radial-flow cell for soil water measurements. *Proc. Soil Sci. Soc. Amer.* **26**:515–518.

Richards, S. J., R. M. Hagan, and T. M. McCalla, 1952. Soil temperature and plant growth. In B. T. Shaw (ed.), "Soil Physical Conditions and Plant Growth," pp. 303–480. Academic Press Inc., New York.

——— and A. W. Marsh, 1961. Irrigation based on soil suction measurements. *Proc. Soil Sci. Soc. Amer.* **25**:65–69.

———, L. V. Weeks, and J. C. Johnston, 1958. Effects of irrigation treatments and rates of nitrogen fertilization on young Hass avocado trees. *Proc. Amer. Soc. Hort. Sci.* **71**:292–297.

Richardson, S. D., 1953. A note on some differences in root hair formation between seedlings of sycamore and American oak. *New Phytol.* **52**:80–82.

Rickman, R. W., J. Letey, and L. H. Stolzy, 1966. Plant responses to oxygen supply and physical resistance in the root environment. *Proc. Soil Sci. Soc. Amer.* **30**:304–307.

Rider, N. E., 1957. Water losses from various land surfaces. *Quart. J. Roy. Meteorol. Soc.* **83**:181–193.

Rijtema, P. E., 1965. An analysis of actual evapotranspiration. *Wageningen Agr. Res. Rep.* 659.

Ringoet, A., 1952. Recherches sur la transpiration et le bilan d'eau de quelques plantes tropicales. *Inst. nat. étud. agron. Congo Belg. Publ. ser. sci.* 56.

Roberts, B. R., 1964. Effects of water stress on the translocation of photosynthetically assimilated carbon-14 in yellow poplar. In M. H. Zimmerman (ed.), "The Formation of Wood in Forest Trees," pp. 273–288. Academic Press Inc., New York.

Roberts, F. L., 1948. A study of the absorbing surfaces of the roots of loblolly pine. M.A. thesis, Duke Univ., N.C.

Roberts, O., and S. A. Styles, 1939. An apparent connection between the pressure of colloids and the osmotic pressure of conifer leaves. *Sci. Proc. Roy. Dublin Soc.* **22**:119–125.

Robertson, J. D., 1959. The ultrastructure of cell membranes and their derivatives. *Symp. Biochem. Soc.* **16**:3–43.

Robins, J. S., and C. E. Domingo, 1953. Some effects of severe soil moisture deficits at specific growth stages of corn. *Agron. J.* **41**:618–621.

———, 1956. Moisture deficits in relation to the growth and development of dry beans. *Agron. J.* **48**:67–70.

Robinson, F. E., 1964. Required per cent air space for normal growth of sugar cane. *Soil Sci.* **98**:206–207.

———, R. B. Campbell, and Jen-hu Chang, 1963. Assessing the utility of pan evaporation for controlling irrigation of sugar cane in Hawaii. *Proc. Soil Sci. Soc. Amer.* **55**:444–446.

Robinson, J. R., 1965. Water regulation in mammalian cells. *Soc. Exp. Biol. Symp.* **19**:237–258.

Rogers, C. F., 1929. Winter activity of the roots of perennial plants. *Science* **69**:299–300.

Rogers, W. S., 1939. Apple root growth in relation to rootstock, soil, seasonal and climatic factors. *J. Pomol. Hort. Sci.* **17**:99–130.

——— and A. B. Beakbane, 1957. Stock and scion relations. *Ann. Rev. Plant Physiol.* **8**:217–236.

Rokach, A., 1953. Water transfer from fruits to leaves in the Shamouti orange tree and related topics. *Palestine J. Bot.* **8**:146–151.

Romberger, J. A., 1963. Meristems, growth, and development in woody plants. *U.S. Dept. Agr. Tech. Bull.* 1293.

Rosa, J. T., 1921. Investigations on the hardening process in vegetable plants. *Mo. Agr. Exp. Sta. Bull.* 48.

Rose, C. W., 1966. "Agricultural Physics." Pergamon Press, New York.

——— and W. R. Stern, 1965. The drainage component of the water balance equation. *Aust. J. Soil Res.* **3**:95–100.

———, W. R. Stern, and J. E. Drummond, 1965. Determination of hydraulic conductivity as a function of depth and water content for soil *in situ*. *Aust. J. Soil Res.* **3**:1–9.

Rosenberg, N. J., 1964. Response of plants to the physical effects of soil compaction. *Advan. Agron.* **16**:181–196.

Rosene, H. F., 1937. Distribution of the velocities of absorption of water in the onion root. *Plant Physiol.* **12**:1–19.

———, 1941. Control of water transport in local root regions of attached and isolated roots by means of the osmotic pressure of the external solution. *Amer. J. Bot.* **28**:402–410.

———, 1944. Effect of cyanide on rate of exudation in excised onion roots. *Amer. J. Bot.* **31**:172–174.

———, 1947. Reversible azide inhibition of oxygen consumption and water transfer in root tissue. *J. Cell. Comp. Physiol.* **30**:15–30.

Rouschal, E., 1935. Untersuchungen über die temperatur abhangigkeit der wasseraufnahme ganzer pflanzen mit besondern berucksichtigung der übergangsreaktionen, *Sitzungsb. Akad. Wiss. Wien. Math. Naturw. Kl. Abt.* 1 **144**:313–348.

Rovira, A. D., 1965. Plant root exudates and their influence on soil organisms. In K. F. Baker, W. C. Snyder, et al. (eds.), "Ecology of Soil-borne Plant Pathogens," pp. 154–169. Univ. of California Press, Berkeley.

Ruben, S., M. Randall, and J. L. Hyde, 1941. Heavy oxygen (O^{18}) as a tracer in the study of photosynthesis. *J. Amer. Chem. Soc.* **63**:877–879.

Rudinsky, J. A., and J. P. Vité, 1959. Certain ecological and phylogenetic aspects of the pattern of water conduction in conifers. *For. Sci.* **5**:259–266.

Rufelt, H., 1956. Influence of the root pressure on the transpiration of wheat plants. *Physiol. Plant.* **9**:154–164.

Ruhland, W., 1915. Untersuchungen über die Hautdrüsen der Plumbaginaceen. Ein Beitrag zur Biologie der Halophyten. *Jahrb. wiss. Bot.* **55**:409–498.

——— (ed.), 1955. "Encyclopedia of Plant Physiology." Springer-Verlag. Berlin-Gottingen-Heidelberg.

Russell, E. W., 1961. "Soil Conditions and Plant Growth." Longmans, Green and Company. London.

Russell, M. B., 1952. Soil aeration and plant growth. In B. T. Shaw (ed.), "Soil Physical Conditions and Plant Growth," pp. 253–301. Academic Press Inc., New York.

———, F. E. Davis, and R. A. Bair, 1940. The use of tensiometers for following soil moisture conditions. *J. Am. Soc. Agron.* **32**:922–930.

——— and J. T. Woolley, 1961. Transport processes in the soil-plant system. In M. X. Zarrow (ed.), "Growth in Living Systems," pp. 695–722. Basic Books, Inc., New York.

Russell, R. S., and V. M. Shorrocks, 1959. The relationship between transpiration and the absorption of inorganic ions by intact plants. *J. Exp. Bot.* **10**:301–316.

——— and D. A. Barber, 1960. The relationship between salt uptake and the absorption of water by intact plants. *Ann. Rev. Plant Physiol.* **11**:127–140.

——— and J. Sanderson, 1967. Nutrient uptake by different parts of the intact roots of plants. *J. Exp. Bot.* **18**:491–508.

Rutter, A. J., and K. Sands, 1958. The relation of leaf water to soil moisture tension in *Pinus sylvestris*. I. The effect of soil moisture on diurnal changes in water balance. *New Phytol.* **57**:50–65.

——— and F. H. Whitehead (eds.), 1963. "The Water Relations of Plants." British Ecol. Soc. Symp. 3, Blackwell Scientific Publications, London.

Sadasivan, T. S., 1961. Physiology of wilt disease. *Ann. Rev. Plant Physiol.* **12:**449–468.

Sandstrom, B., 1950. The ion absorption in roots lacking epidermis. *Physiol. Plant.* **3:**496–505.

Sayre, J. D., 1920. The relation of hairy leaf coverings to the resistance of leaves to transpiration. *Ohio J. Sci.* **20:**55–86.

Schieferstein, R. H., and W. E. Loomis, 1959. Development of cuticular layers in Angiosperm leaves. *Amer. J. Bot.* **46:**625–635.

van Schilfgaarde, J. (ed.), 1965. Proceedings, drainage for efficient crop production conference. Amer. Soc. Agr. Eng., St. Joseph, Mich.

Schmidt, H., 1930. Zur Funktion der Hydathoden von Saxifraga. *Planta* **10:**314–344.

Scholander, P. F., L. Van Dam, and Ṣ. I. Scholander, 1955. Gas exchange in the roots of mangroves. *Amer. J. Bot.* **42:** 92–98.

———, B. Ruud, and H. Leivestad, 1957. The rise of sap in a tropical liana. *Plant Physiol.* **32:**1–6.

———, H. T. Hammel, E. A. Hemmingsen, and E. D. Bradstreet, 1964. Hydrostatic pressure and osmotic potential in leaves of mangroves and some other plants. *Proc. Nat. Acad. Sci.* **52:**119–125.

———, ———, E. D. Bradstreet, and E. A. Hemmingsen, 1965. Sap pressure in vascular plants. *Science* **148:**339–346.

———, E. D. Bradstreet, H. T. Hammel, and E. A. Hemmingsen, 1966. Sap concentration in halophytes and some other plants. *Plant Physiol.* **41:**529–532.

Schramm, R. J., 1960. Anatomical and physiological development of roots in relation to aeration of the substrate. *Diss. Abstr.* **21:**2089.

Schroeder, C. A., and P. A. Wieland, 1956. Diurnal fluctuations in size in various parts of the avocado tree and fruit. *Proc. Amer. Soc. Hort. Sci.* **68:**253–258.

Schroeder, G. L., H. W. Kramer, and R. D. Evans, 1965. Diffusion of radon in several naturally occurring soil types. *J. Geophys. Res.* **70:**471–474.

Schroeder, R. A., 1939. The effect of root temperature upon the absorption of water by the cucumber. *Univ. Mo. Res. Bull.* **309:**1–27.

Scofield, C. S., 1945. The water requirement of alfalfa. *U.S. Dept. Agr. Circ.* 735.

Scott, F. M., 1949. Plasmodesmata in xylem vessels. *Bot. Gaz.* **110:**492–495.

———, 1950. Internal suberization of tissues. *Bot. Gaz.* **111:**378–394.

———, 1963. Root hair zone of soil-grown plants. *Nature* **199:**1009–1010.

———, 1964. Lipid deposition in intercellular space. *Nature* **203:**164–168.

———, 1965. The anatomy of plant roots. In K. F. Baker et al. (eds.), "International Symposium on Factors Determining the Behavior of Plant Pathogens in Soil," pp. 145–153. Univ. of California Press, Berkeley.

———, B. G. Bystrom, and E. Bowler, 1963. Root hairs, cuticle, and pits. *Science* **140:**63–64.

Scott, L. I., and J. H. Priestley, 1928. The root as an absorbing organ. I. A reconsideration of the entry of water and salts into the absorbing region. *New Phytol.* **27:**125–141.

Seifriz, W., 1942. "The Structure of Protoplasm." Iowa State College Press, Ames, Iowa.

Setterfield, G., and S. T. Bayley, 1961. Structure and physiology of cell walls. *Ann. Rev. Plant Physiol.* **12:**35–62.

Shah, C. B., and R. S. Loomis, 1965. Ribonucleic acid and protein metabolism in sugar beet during drought. *Physiol. Plant.* **18:**240–254.

Shantz, H. L., 1925. Soil moisture in relation to the growth of plants. *J. Amer. Soc. Agron.* **17:**705–711.

——— and L. N. Piemeisel, 1927. The water requirement of plants at Akron, Colorado. *J. Agr. Res.* **34:**1093–1190.

Shardakov, V. S., 1948. New field method for the determination of the suction pressure of plants. *Dokl. Akad. Nauk SSSR* **60**:169–172. (In Russian)

Shaw, B., and L. D. Baver, 1939. Heat conductivity as an index of soil moisture. *J. Amer. Soc. Agron.* **31**:886–891.

Shaw, B. T. (ed.), 1952. "Soil Physical Conditions and Plant Growth." Academic Press Inc., New York.

Shaw, M. D., and W. C. Arble, 1959. Bibliography on methods for determining soil moisture. *Eng. Res. Bull.* B–78, Coll. of Engineering and Architecture, University Park, Pennsylvania.

Shimshi, D., 1963*a*. Effect of chemical closure of stomata on transpiration in varied soil and atmospheric environments. *Plant Physiol.* **38**:709–712.

———, 1963*b*. Effect of soil moisture and phenylmercuric acetate upon stomatal aperture, transpiration, and photosynthesis. *Plant Physiol.* **38**:713–721.

Shirk, H. G., 1942. Freezable water content and the oxygen respiration in wheat and rye grain at different stages of ripening. *Amer. J. Bot.* **29**:105–109.

Shirley, H. L., 1936. Lethal high temperatures for conifers, and the cooling effect of transpiration. *J. Agr. Res.* **53**:239–258.

Shmueli, E., 1953. Irrigation studies in the Jordan Valley. I: Physiological activity of the banana in relation to soil moisture. *Bull. Res. Counc. Israel* **3**:228–247.

——— and O. P. Cohen, 1964. A critique of Walter's hydrature concept and of his evaluation of water status measurements. *Israel J. Bot.* **13**:199–207.

Shull, C. A., 1916. Measurement of the surface forces in soils. *Bot. Gaz.* **62**:1–31.

———, 1930. Absorption of water and the forces involved. *J. Amer. Soc. Agron.* **22**:459–471.

Sibirsky, W., 1935. Die Bestimmung der Bodenfeuchtigkeit nach der Carbidmethode. 3d *Int. Congr. Soil Sci. Trans.* **1**:10–13.

Sierp, H., and A. Brewig, 1935. Quantitative Untersuchungen über die Wasserabsorptionzone der Wurzeln. *Jahrb. wiss. Bot.* **82**:99–122.

Sinclair, W. B., and E. T. Bartholomew, 1944. Effects of rootstock and environment on the composition of oranges and grapefruit. *Hilgardia* **16**:125–176.

Sinnott, E. W., 1939. Growth and differentiation in living plant meristems. *Proc. Nat. Acad. Sci.* **25**:55–58.

——— and R. Bloch, 1939. Cell polarity and the differentiation of root hairs. *Proc. Nat. Acad. Sci.* **25**:248–252.

Skau, C. M., and R. H. Swanson, 1963. An improved heat pulse velocity meter as an indicator of sap speed and transpiration. *J. Geophys. Res.* **68**:4743–4749.

Skelding, A. D., and W. J. Rees, 1952. An inhibitor of salt absorption in the root tissue of red beet. *Ann. Bot.* **16**:513–529.

Skene, K. G. M., 1967. Gibberellin-like substances in root exudate of *Vitis vinifera*. *Planta* **74**:250–262.

——— and G. H. Kerridge, 1967. Effect of root temperature on cytokinin activity in root exudate of *Vitis vinifera* L. *Plant Physiol.* **42**:1131–1139.

Skidmore, E. L., and J. F. Stone, 1964. Physiological role in transpiration rate of the cotton plant. *Agron. J.* **56**:405–410.

Skoog, F., T. C. Broyer, and K. A. Grossenbacher, 1938. Effects of auxin on rates, periodicity, and osmotic relations in exudation. *Amer. J. Bot.* **25**:749–759.

Slankis, V., 1958. The role of auxin and other exudates in mycorrhizal symbiosis of forest trees. In K. V. Thimann (ed.), "The Physiology of Forest Trees," chap. 21. The Ronald Press Company, New York.

Slatyer, R. O., 1955. Studies of the water relations of crop plants grown under natural rainfall in northern Australia. *Aust. J. Agr. Res.* **6**:365–377.

———, 1956. Absorption of water from atmospheres of different humidity and its transport through plants. *Aust. J. Biol. Sci.* **9**:552–558.

———, 1957*a*. The influence of progressive increases in total soil moisture stress on transpiration, growth, and internal water relationships of plants. *Aust. J. Biol. Sci.* **10**:320–336.

———, 1957*b*. The significance of the permanent wilting percentage in studies of plant and soil water relations. *Bot. Rev.* **23**:585–636.

———, 1958. The measurement of diffusion pressure deficit in plants by a method of vapour equilibration. *Aust. J. Biol. Sci.* **11**:349–365.

———, 1960*a*. Aspects of the tissue water relationships of an important arid zone species (*Acacia aneura* F. Muell.) in comparison with two mesophytes. *Bull. Res. Counc. Israel* **8D**:159–168.

———, 1960*b*. Absorption of water by plants. *Bot. Rev.* **26**:331–392.

———, 1961. Effects of several osmotic substrates on water relations of tomato. *Aust. J. Biol. Sci.* **14**:519–540.

———, 1962. Methodology of a water balance study conducted on a desert woodland (*Acacia aneura* F. Muell.) community in central Australia. *Arid Zone Res.* **16**:15–26. UNESCO, Paris.

———, 1964. Efficiency of water utilization by arid zone vegetation. *Ann. Arid Zone* **3**:1–12.

———, 1966. Some physical aspects of non-stomatal control of leaf transpiration. *Agr. Meteorol.* **3**:281–292.

———, 1967. "Plant-water Relationships." Academic Press Inc., New York.

——— and J. F. Bierhuizen, 1964*a*. Transpiration from cotton leaves under a range of environmental conditions in relation to internal and external diffusive resistances. *Aust. J. Biol. Sci.* **17**:115–130.

——— and ———, 1964*b*. The influence of several transpiration suppressants on transpiration, photosynthesis, and water-use efficiency of cotton leaves. *Aust. J. Biol. Sci.* **17**:131–146.

——— and ———, 1964*c*. A differential psychrometer for continuous measurements of transpiration. *Plant Physiol.* **39**:1051–1056.

——— and S. A. Taylor, 1960. Terminology in plant and soil-water relations. *Nature* **187**: 922.

——— and I. C. McIlroy, 1961. Practical micro-climatology. UNESCO, Paris.

——— and P. G. Jarvis, 1966. A gaseous-diffusion porometer for continuous measurement of diffusive resistance of leaves. *Science* **151**:574–576.

——— and J. V. Lake, 1966. Resistance to water transport in plants—whose misconception? *Nature* **212**:1585–1586.

——— and E. Shmueli, 1967. Measurements of internal water status and transpiration. In R. M. Hagan et al. (eds.), "Irrigation of Agricultural Lands," pp. 337–353. *Amer. Soc. Agron.*, Madison, Wis.

Slavik, B., 1963*a*. Relationship between the osmotic potential of the cell sap and the water saturation deficit during the wilting of leaf tissue. *Biol. Plant.* (Praha) **5**:258–264.

———, 1963*b*. On the problem of the relationship between hydration of leaf tissue and intensity of photosynthesis and respiration. In A. J. Rutter and F. H. Whitehead (eds.), "The Water Relations of Plants," pp. 225–234. John Wiley & Sons, Inc., New York.

——— (ed.), 1965. "Water Stress in Plants." Czech. Acad. Sci., Prague.

Slavikova, J., 1964. Horizontales gradient der Saugkraft eines Wurzelastes und seine Zusammenhang mit dem Wassertransport in der Wurzel. *Acta Hort. bot. prag.* 73–79.

———, 1967. Compensation of root suction force within a single root system. *Biol. Plant.* (*Praha*) **9**:20–27.

Smith, H. and P. F. Wareing, 1966. Apical dominance and the effect of gravity on nutrient distribution. *Planta* **70**:87–94.

Smith, R. C., 1960. Influence of upward translocation on uptake of ions in corn plants. *Amer. J. Bot.* **47**:724–729.

——— and E. Epstein, 1964. Ion absorption by shoot tissue: kinetics of potassium and rubidium absorption by corn leaf tissue. *Plant Physiol.* **39**:992–996.

Smith, R. M., and D. R. Browning, 1946. Some suggested laboratory standards of subsoil permeability. *Proc. Soil Sci. Soc. Amer.* **11**:21–26.

Smith, S. N., 1934. Response of inbred lines and crosses in maize to variations of nitrogen and phosphorus supplied as nutrients. *Amer. Soc. Agron. J.* **26**:785–804.

Smith, W. O., 1943. Thermal transfer of moisture in soils. *Trans. Amer. Geophys. Union* **24**:511–523.

Smyth, D. H., 1965. Water movement across the mammalian gut. *Soc. Exp. Biol. Symp.* **19**:307–328.

Sopper, W. E., and H. W. Lull (eds.), 1967. "Forest Hydrology." Pergamon Press, New York.

Soran, V., and D. Cosma, 1962. Effectul transpiratiei asupra activatatii absorbante a diferitelor regiuni ale sistemului radical. *Studi. Univ. Barbes-Bolyai Ser. Biol. Fasc.* **1**:75–87. (French summary)

Spanner, D. C., 1951. The Peltier effect and its use in the measurement of suction pressure. *J. Exp. Bot.* **2**:145–168.

———, 1956. Energetics and mathematical treatment of diffusion. "Encyclopedia of Plant Physiology," vol. 2, pp. 125–138. Springer-Verlag OHG, Berlin.

———, 1958. The translocation of sugar in sieve tubes. *J. Exp. Bot.* **9**:332–342.

———, 1964. "Introduction to Thermodynamics." Academic Press Inc., New York.

Sperlich, A., and A. Hampel, 1936. Über das Bluten der Wurzel im Entwicklungsgange einer einjährigen Pflanzen (*Helianthus annuus*), *Jahrb. wiss. Bot.* **83**:406–422.

Spoehr, H. A., 1919. The carbohydrate economy of cacti. *Carnegie Inst. Washington Publ.* 287.

——— and H. W. Milner, 1939. Starch dissolution and amylolytic activity of leaves. *Proc. Amer. Phil. Soc.* **81**:37–78.

Sponsler, O. L., J. D. Bath, and J. W. Ellis, 1940. Water bound to gelatin as shown by structural studies. *J. Phys. Chem.* **44**:996–1006.

Stadelmann, E., 1963. Zur Vergleich und Umrechnung von Permeabilitätskonstanten für Wasser. *Protoplasma* **57**:660–718.

Stålfelt, M. G., 1935. Die Spaltöffnungsweite als Assimilationsfaktor. *Planta* **23**:715–759.

———, 1956*a*. Morphologie und Anatomie des Blattes als Transpirationsorganen. "Encyclopedia of Plant Physiology," vol. 3, pp. 324–341. Springer-Verlag OHG, Berlin.

———, 1956*b*. Die cuticuläre Transpiration. "Encyclopedia of Plant Physiology," vol. 3, pp. 342–350. Springer-Verlag OHG, Berlin.

———, 1956*c*. Die stomatäre Transpiration und die Physiologie der Spaltöffnungen. "Encyclopedia of Plant Physiology," vol. 3, pp. 351–426. Springer-Verlag OHG, Berlin.

———, 1966. The role of the epidermal cells in the stomatal movements. *Physiol. Plant.* **19**:241–256.

Stanhill, G., 1957. The effect of differences in soil-moisture status on plant growth: a review and evaluation. *Soil Sci.* **84**:205–214.

———, 1961. A comparison of methods of calculating potential evapotranspiration from climatic data. *Israel J. Agr. Res.* **11**:159–171.

Staple, W. J., and J. J. Lehane, 1962. Variability in soil moisture sampling. *Can. J. Soil Sci.* **42:**157–164.

Staverman, A. J., 1951. The theory of measurement of osmotic pressure. *Rev. Trav. Chim.* **70:**344–352.

Stebbins, G. L., and S. S. Shah, 1960. Developmental studies of cell differentiation in the epidermis of monocotyledons. II. Cytological features of stomatal development in the Gramineae. *Develop. Biol.* **2:**477–500.

Stenlid, G., 1958. Salt losses and redistribution of salts in higher plants. "Encyclopedia of Plant Physiology," vol. 4, pp. 615–637. Springer-Verlag OHG, Berlin.

Stevens, C. L., 1931. Root growth of white pine (*Pinus strobus* L.). *Yale Univ. Sch. Forest. Bull.* **32:**1–62.

——— and R. L. Eggert, 1945. Observations on the causes of flow of sap in red maple. *Plant Physiol.* **20:**636–648.

Steward, F. C., 1933. Observations upon the effects of time, oxygen and salt concentration upon absorption and respiration by storage tissue. *Protoplasma* **18:**208–242.

———, W. E. Berry, and T. C. Broyer, 1936. The absorption and accumulation of solutes by living plant cells. VIII. The effect of oxygen upon respiration and salt accumulation. *Ann. Bot.* **50:**345–366.

———, P. Prevot, and J. A. Harrison, 1942. Absorption and accumulation of rubidium bromide by barley plants. Localization in the root of cation accumulation and of transfer to the shoot. *Plant Physiol.* **17:**411–421.

——— and J. F. Sutcliffe, 1959. Plants in relation to inorganic salts. In F. C. Steward (ed.), "Plant Physiology," vol. 2, pp. 253–478. Academic Press Inc., New York.

Stewart, C. M., 1967. Moisture content of living trees. *Nature* **214:**138–140.

Stice, N. W., and L. J. Booher, 1965. Plastic tube irrigators with electric control. *Calif. Agr.* **19:**4–5.

Stocker, O., 1929. Das Wasserdefizit von Gefässpflanzen in verschiedenen Klimazonen. *Planta* **7:**382–387.

———, 1956. Messmethoden der Transpiration. "Encyclopedia of Plant Physiology," vol. 3, pp. 293–311. Springer-Verlag OHG, Berlin.

———, 1960. Physiological and morphological changes in plants due to water deficiency. *Arid Zone Res.* **15:**63–104. UNESCO, Paris.

——— and W. Holdheide, 1937. Die Assimilation Helgoländer Gezeitenalgen während der Ebbzeit. *Z. Bot.* **32:**1–59.

——— and H. Ross, 1956. Reaktions- und Restitutionsphase der plasmaviskosität bei Dürre- und Schüttelreizen. *Naturwiss.* **33:**283–284.

Stocking, C. R., 1945. The calculation of tensions in *Cucurbita pepo*. *Amer. J. Bot.* **32:**126–134.

———, 1956. Excretion by glandular organs. "Encyclopedia of Plant Physiology," vol. 3, pp. 503–510. Springer-Verlag OHG, Berlin.

Stolwijk, J. A. J., and K. V. Thimann, 1957. On the uptake of carbon dioxide and bicarbonate by roots and its influence on growth. *Plant Physiol.* **32:**513–520.

Stolzy, L. H., O. C. Taylor, W. M. Dugger, Jr., and J. D. Mesereau, 1964. Physiological changes in and ozone susceptibility of the tomato plant after short periods of inadequate oxygen diffusion to the roots. *Proc. Soil Sci. Soc. Amer.* **28:**305–308.

Stone, E. C., 1957. Dew as an ecological factor. I. A review of the literature. *Ecology* **38:**407–413.

———, 1957. Dew as an ecological factor. II. The effect of artificial dew on the survival of *Pinus ponderosa* and associated species. *Ecology* **38:**414–422.

——— and G. H. Schubert, 1959. Root regeneration by ponderosa pine seedlings lifted at different times of the year. *Forest Sci.* **5:**322–332.

Stone, J. F., D. Kirknam, and A. A. Read, 1955. Soil moisture determination by a portable neutron scatter meter. *Proc. Soil Sci. Soc. Amer.* **19:**418–423.

Stout, P. R., and D. R. Hoagland, 1939. Upward and lateral movement of salt in certain plants as indicated by radioactive isotopes of potassium, sodium, and phosphorus absorbed by roots. *Amer. J. Bot.* **26**:320–324.

Street, H. E., 1966. The physiology of root growth. *Ann. Rev. Plant Physiol.* **17**:315–344.

———, and J. S. Lowe, 1950. The carbohydrate nutrition of tomato roots. II. The mechanism of sucrose absorption by excised roots. *Ann. Bot.* **14**:307–329.

Strogonov, B. P., 1964. "Physiological Basis of Salt Tolerance of Plants," Eng. trans. by Podjakoff-Mayber and Mayer. Oldbourne Press, London.

Strugger, S., 1943. Der Aufsteigende Saftstrom in der Pflanze. *Naturwiss.* **31**:181–194.

———, 1949. "Praktikum der Zell- und Gewebephysiologie der Pflanzen," 2d ed. Springer-Verlag OHG, Berlin.

Stuckey, I. H., 1941. Seasonal growth of grass roots. *Amer. J. Bot.* **28**:486–491.

Sutcliffe, J. F., 1952. The influence of internal ion concentration on potassium accumulation and salt respiration in cells of red beet root tissue. *J. Exp. Bot.* **3**:59–76.

———, 1954. The exchangeability of potassium and bromide ions in cells of red beet root tissue. *J. Exp. Bot.* **5**:313–326.

———, 1962. "Mineral Salts Absorption in Plants." Pergamon Press, New York.

Swanson, C. A., 1943. Transpiration in American holly in relation to leaf structure. *Ohio J. Sci.* **43**:43–46.

Swanson, R. H., 1966. Seasonal course of transpiration of lodgepole pine and Engelmann spruce. In W. E. Sopper and H. W. Lull (eds.), "Forest Hydrology." Pergamon Press, New York.

——— and R. Lee, 1966. Measurement of water movement from and through shrubs and trees. *J. Forest.* **64**:187–190.

Swinbank, W. C., 1951. The measurement of vertical transfer of heat and water vapour by eddies in the lower atmosphere. *J. Meteorol.* **8**:135–145.

Tackett, J. L., and R. W. Pearson, 1964. Oxygen requirements of cotton seedling roots for penetration of compacted soil cores. *Proc. Soil Sci. Soc. Amer.* **28**:600–605.

Tagawa, T., 1934. The relation between the absorption of water by plant root and the concentration and nature of the surrounding solution. *Jap. J. Bot.* **7**:33–60.

Tal, M., 1966. Abnormal stomatal behavior in wilty mutants of tomato. *Plant Physiol.* **41**:1387–1391.

Tamm, C. O., 1958. The atmosphere. "Encyclopedia of Plant Physiology," vol. 4, pp. 233–242. Springer-Verlag OHG, Berlin.

Tanford, C., 1963. The structure of water and of aqueous solutions. In "Temperature—Its Measurement and Control in Science and Industry," vol. 3, pp. 123–129. Reinhold Publishing Corporation, New York.

Tanner, C. B., 1960. Energy balance approach to evapotranspiration from crops. *Proc. Soil Sci. Soc. Amer.* **24**:1–9.

———, 1967. Measurement of evapotranspiration. In R. M. Hagan et al. (eds.), "Irrigation of Agricultural Lands," pp. 534–574. *Amer. Soc. Agron.*, Madison, Wis.

——— and W. L. Pelton, 1960. Potential evapotranspiration estimates by the approximate energy balance method of Penman. *J. Geophys. Res.* **65**:3391–3413.

Taylor, C. A., H. F. Blaney, and W. W. McLaughlin, 1934. The wilting-range in certain soils and the ultimate wilting-point. *Amer. Geophys. Union Trans.* **15**:436–444.

Taylor, F. H., 1956. Variation in sugar content of maple sap. *Vt. Agr. Exp. Sta. Bull.* **587**:1–39.

Taylor, S. A., 1965. Managing irrigation water on the farm. *Amer. Soc. Agr. Eng. Trans.* **8**:433–436.

——— and N. C. Heuser, 1953. Water entry and downward movement in undisturbed soil cores. *Proc. Soil Sci. Soc. Amer.* **17**:195–201.

Thoday, D., 1918. On turgescence and the absorption of water by the cells of plants. *New Phytol.* **17:**108–113.

Thomas, W. A., 1967. Dye and calcium ascent in dogwood trees. *Plant Physiol.* **42:** 1800–1802.

Thompson, E. W., and R. D. Preston, 1967. Proteins in the cell walls of some green algae. *Nature* **213:**684–685.

Thompson, G. A., 1965. Cellular membranes. In J. Bonner and J. E. Varner (eds.), "Plant Biochemistry." Academic Press Inc., New York.

Thompson, L. M., 1952. "Soils and Soil Fertility." McGraw-Hill Book Company, New York.

Thornthwaite, C. W., 1948. An approach toward a rational classification of climate. *Geogr. Rev.* **38:**55–94.

———, 1954. A re-examination of the concept and measurement of potential evapotranspiration. *Johns Hopkins Univ. Climatol. Publ. no. 7.*

Thut, H. F., 1932. The movement of water through some submerged plants. *Amer. J. Bot.* **19:**693–709.

——— and W. E. Loomis, 1944. Relation of light to growth of plants. *Plant Physiol.* **19:** 117–130.

Tibbals, E. C., E. K. Carr, D. M. Gates, and F. Kreith, 1964. Radiation and convection in conifers. *Amer. J. Bot.* **51:**529–538.

Ting, I. P., and W. E. Loomis, 1963. Diffusion through stomates. *Amer. J. Bot.* 866–872.

Tisdall, A. L., 1951. Antecedent soil moisture and its relation to infiltration. *Aust. J. Agr. Res.* **2:**342–354.

Todd, G. W., and B. Y. Yoo, 1964. Enzymatic changes in detached wheat leaves as affected by water stress. *Phyton* **21:**61–68.

Tomlinson, P. B., 1963. Stem structure in arborescent monocotyledons. In M. H. Zimmerman (ed.), "The Formation of Wood in Trees," pp. 65–86. Academic Press Inc., New York.

Torii, K., and G. G. Laties, 1966. Mechanisms of ion uptake in relation to vacuolation of corn roots. *Plant Physiol.* **41:**863–870.

Torrey, J. G., 1965. Physiological bases of organization and development in the root. "Encyclopedia of Plant Physiology," vol. 15, pp. 1256–1327. Springer-Verlag OHG, Berlin.

Toumey, J. W., 1929. Initial root habit in American trees and its bearing on regeneration. *Proc. 4th Int. Bot. Congr.* **1:**713–728.

Transeau, E. N., 1905. Forest centers of eastern North America. *Amer. Nat.* **39:**875–889.

———, 1925. "General Botany." Harcourt, Brace & World, Inc., New York.

Traube, M., 1867. Experimente zur Theorie der Zellbildung und Endosmose. *Archiv. Anat. Physiol. wiss. Med.* 87–165.

Trousdell, K. B., and M. D. Hoover, 1955. A change in groundwater after clearcutting of loblolly pine in the coastal plain. *J. Forest.* **53:**493–498.

Tukey, H. B., Jr., and H. J. Amling, 1958. Leaching of foliage by rain and dew as an explanation of differences in the nutrient composition of greenhouse- and field-grown plants. *Mich. Agr. Exp. Sta. Quart. Bull.* **40:**876–881.

———, H. B. Tukey, and S. H. Wittwer, 1958. Loss of nutrients by foliar leaching as determined by radioisotopes. *Proc. Amer. Soc. Hort. Sci.* **71:**496–506.

Tukey, L. D., 1964. A linear electronic device for continuous measurement and recording of fruit enlargement and contraction. *Proc. Amer. Soc. Hort. Sci.* **84:**653–660.

Turner, L. M., 1936. Root growth of seedlings of *Pinus echinata* and *Pinus taeda*. *J. Agr. Res.* **53:**145–149.

Turrell, F. M., 1936. The area of the internal exposed surface of dicotyledon leaves. *Amer. J. Bot.* **23:**255–264.

———, 1944. Correlation between internal surface and transpiration rate in mesomorphic and xeromorphic leaves grown under artificial light. *Bot. Gaz.* **105:**413–425.

———, M. Chervenak, and A. P. Vanselow, 1952. Inorganic constituents of exudate from Eureka lemons treated with elemental sulfur dust. *Amer. J. Bot.* **39:**693–699.

Uhvits, R., 1946. Effect of osmotic pressure on water absorption and germination of alfalfa seeds. *Amer. J. Bot.* **33:**278–285.

Úlehla, J., 1963. Changes in sap exudation of maize and occurrence of lags in exudation during the growing season. *Biol. Plant.* (*Praha*) **5:**190–197.

Ulrich, J. M., and A. D. McLaren, 1965. The absorption and translocation of C^{14}-labeled proteins in young tomato plants. *Amer. J. Bot.* **52:**120–126.

Ursprung, A., 1929. The osmotic quantities of the plant cell. *Proc. Int. Congr. Plant Sci.* **2:**1081–1094.

——— and G. Blum, 1916. Über den Einfluss der Aussenbedingungen auf den osmotischen Wert. *Ber. deut. bot. Ges.* **34:**123–142.

Ussing, H. H., 1953. Transport through biological membranes. *Ann. Rev. Physiol.* **15:** 1–20.

———, 1954. Ion transport across biological membranes. In H. T. Clark (ed.), "Ion Transport across Membranes," pp. 3–22. Academic Press Inc., New York.

Vaadia, Y., 1960. Autonomic diurnal fluctuations in rate of exudation and root pressure of decapitated sunflower plants. *Physiol. Plant.* **13:**701–717.

———, F. C. Raney, and R. M. Hagan, 1961. Plant water deficits and physiological processes. *Ann. Rev. Plant Physiol.* **12:**265–292.

——— and Y. Waisel, 1963. Water absorption by the aerial organs of plants. *Physiol. Plant.* **16:**44–51.

Vaclavik, J., 1966. The maintaining of constant soil moisture levels (lower than maximum capillary capacity) in pot experiments. *Biol. Plant.* (*Praha*) **8:**80–85.

Valoras, N., and J. Letey, 1966. Soil oxygen and water relationships to rice growth. *Soil Sci.* **101:**210–215.

Van Fleet, D. S., 1961. Histochemistry and function of the endodermis. *Bot. Rev.* **27:** 166–220.

Veihmeyer, F. J., 1927. Some factors affecting the irrigation requirements of deciduous orchards. *Hilgardia* **2:**125–284.

———, 1956. Soil moisture. "Encyclopedia of Plant Physiology," vol. 3, pp. 64–123. Springer-Verlag OHG, Berlin.

——— and A. H. Hendrickson, 1927. Soil moisture conditions in relation to plant growth. *Plant Physiol.* **2:**71–82.

——— and ———, 1938. Soil moisture as an indication of root distribution in deciduous orchards. *Plant Physiol.* **13:**169–177.

——— and ———, 1949. Methods of measuring field capacity and permanent wilting percentage of soils. *Soil Sci.* **68:**75–94.

——— and ———, 1950. Soil moisture in relation to plant growth. *Ann. Rev. Plant Physiol.* **1:**285–304.

Vesque, J., 1878. L'absorption comparée directement à la transpiration. *Ann. Sci. Nat. Bot. ser. 6,* **6:**201–222.

Veto, F., 1963. Mobilization of fluids in biological objects by means of temperature gradient. *Acta Physiol. Acad. Sci. Hung.* **24:**119–128.

Viets, F. G., Jr., 1944. Calcium and other polyvalent cations as accelerators of ion accumulation by excised barley roots. *Plant Physiol.* **19:**466–480.

———, 1962. Fertilizers and the efficient use of water. *Advan. Agron.* **14:**223–264.

Visser, W. C., 1964. Moisture requirements of crops and rate of moisture depletion of the soil. *Inst. Land Water Manage. Res. Tech. Bull.* 32.

Vité, P. J., 1961. The influence of water supply on oleoresin exudation pressure and resistance to bark beetle attack in *Pinus ponderosa*. *Contrib. Boyce Thompson Inst.* **21**:37–66.

Vlamis, J., and A. R. Davis, 1944. Effects of oxygen tension on certain physiological responses of rice, barley and tomato. *Plant Physiol.* **19**:33–51.

Voigt, G. K., B. N. Richards, and E. C. Mannion, 1964. Nutrient utilization by young pitch pine. *Proc. Soil Sci. Soc. Amer.* **28**:707–709.

Vomocil, J. A., 1954. In situ measurement of soil bulk density. *Agr. Eng.* **35**:651–654.

——— and W. J. Flocker, 1961. Effect of soil compaction on storage and movement of soil air and water. *Trans. Amer. Soc. Agr. Eng.* **4**:242–245.

de Vries, D. A., and A. J. Peck, 1958. On the cylindrical probe method of measuring thermal conductivity with special reference to soils. I. Extension of theory and discussion of probe characteristics. *Aust. J. Phys.* **11**:255–271.

de Vries, H., 1884. Eine Methode zur Analyse der Turgorkraft. *Jahrb. wiss Bot.* **14**: 427–601.

Vyvyan, M. C., 1955. Interrelation of scion and rootstock in fruit trees. *Ann. Bot. N.S.* 401–423.

Wadleigh, C. H., 1946. The integrated soil moisture stress upon a root system in a large container of saline soil. *Soil Sci.* **61**:225–238.

——— and A. D. Ayers, 1945. Growth and biochemical composition of bean plants as conditioned by soil moisture tension and salt concentration. *Plant Physiol.* **20**:106–132.

———, H. G. Gauch, and O. C. Magistad, 1946. Growth and rubber accumulation in guayule as conditioned by soil salinity and irrigation regime. *U.S. Dept. Agr. Tech. Bull.* 925.

———, ———, and D. G. Strong, 1947. Root penetration and moisture extraction in saline soil by crop plants. *Soil Sci.* **63**:341–349.

——— and H. G. Gauch, 1948. Rate of leaf elongation as affected by the intensity of the total soil moisture stress. *Plant Physiol.* **23**:485–495.

Waggoner, P. E., and B. A. Bravdo, 1967. Stomata and the hydrologic cycle. *Proc. Nat. Acad. Sci.* **57**:1096–1102.

——— and N. W. Simmonds, 1966. Stomata and transpiration of droopy potatoes. *Plant Physiol.* **41**:1268–1271.

——— and W. E. Reifsnyder, 1968. Simulation of the temperature, humidity and evaporation profiles in a leaf canopy. *J. Appl. Meteorol.* **7**:400–409.

Walker, D. A., and I. Zelitch, 1963. Some affects of metabolic inhibitors, temperature, and anaerobic conditions on stomatal movement. *Plant Physiol.* **38**:390–396.

Wallace, A. (ed.), 1963. "Solute Uptake by Intact Plants." Published by editor, Los Angeles.

———, S. M. Soufi, and N. Hemaidan, 1966. Day-night differences in accumulation and translocation of ions by tobacco plants. *Plant Physiol.* **41**:102–104.

———, R. T. Ashcraft, and O. R. Lunt, 1967. Day-night periodicity of exudation in detopped tobacco. *Plant Physiol.* **42**:238–242.

Wallihan, E. F., 1946. Studies of the dielectric method of measuring soil moisture. *Proc. Soil Sci. Soc. Amer.* **10**:39–40.

———, 1964. Modification and use of an electric hygrometer for estimating relative stomatal apertures. *Plant Physiol.* **39**:86–90.

Walter, H., 1931. "Die Hydratur der Pflanze." Gustav Fischer Verlag KG, Stuttgart.

———, 1955. The water economy and the hydrature of plants. *Ann. Rev. Plant Physiol.* **6**:239–252.

———, 1963. Zur Klärung des spezifischen Wasserzustandes in Plasma und in der Zellwand bei höheren Pflanze und seine Bestimmung. *Ber. deut. bot. Ges.* **76**:40–71.

———, 1965. Zur Klärung des spezifischen. Wasserzustandes im Plasma. *Ber. deut. bot. Ges.* **78**:104–114.

Wander, I. W., 1949. An interpretation of the cause of water-repellent sandy soils found in the citrus groves of central Florida. *Science* **110**:299–300.

Waring, R. H., and B. D. Cleary, 1967. Plant moisture stress: Evaluation by pressure bomb. *Science* **155**:1248, 1253–1254.

Weatherley, P. E., 1950. Studies in the water relations of the cotton plant. I. The field measurement of water deficits in leaves. *New Phytol.* **40**:81–97.

———, 1951. Studies in the water relations of the cotton plant. I. The field measurements of water deficits in leaves. *New Phytol.* **50**:36–51.

———, 1963. The pathway of water movement across the root cortex and leaf mesophyll of transpiring plants. In A. J. Rutter and F. H. Whitehead (eds.), "The Water Relations of Plants," pp. 85–100. John Wiley & Sons, Inc., New York.

———, 1965. The state and movement of water in the leaf. *Symp. Soc. Exp. Biol.* **19**:157–184.

——— and R. O. Slatyer, 1957. Relationship between relative turgidity and diffusion pressure deficit in leaves. *Nature* **179**:1085–1086.

Weatherspoon, C. P., 1968. The significance of the mesophyll resistance in transpiration. Ph.D. dissertation, Duke Univ., N.C.

Weaver, H. A., and V. C. Jamison, 1951. Limitations in the use of electrical resistance soil moisture units. *Agron. J.* **43**:602–605.

Weaver, J. E., 1919. The ecological relations of roots. *Carnegie Inst. Washington Publ.* 286.

———, 1920. Root development in the grassland formation. *Carnegie Inst. Washington Publ.* 292.

———, F. C. Jean, and J. W. Crist, 1922. Development and activities of roots of crop plants. *Carnegie Inst. Washington Publ.* 316.

———, 1925. Investigations on the root habits of plants. *Amer. J. Bot.* **12**:502–509.

——— and E. Zink, 1946. Length of life of roots of ten species of perennial range and pasture grasses. *Plant Physiol.* **21**:201–217.

———, 1926. "Root Development of Field Crops." McGraw-Hill Book Company, New York.

——— and W. E. Bruner, 1927. "Root Development of Vegetable Crops." McGraw-Hill Book Company, New York.

——— and W. J. Himmel, 1930. Relation of increased water content and decreased aeration to root development in hydrophytes. *Plant Physiol.* **5**:69–92.

——— and F. E. Clements, 1938. "Plant Ecology," 2d ed. McGraw-Hill Book Company, New York.

——— and J. W. Crist, 1922. Relation of hardpan to root penetration in the Great Plains. *Ecology* **3**:237–249.

——— and R. W. Darland, 1947. A method of measuring vigor of range grasses. *Ecology* **28**:146–162.

Webb, E. K., 1960. An investigation of the evaporation from Lake Eucumbene. *C.S.I.R.O., Div. Meteorol. Phys. Tech. Paper* 10.

Weigl, J., and U. Lüttge, 1962. Mikroautoradiographische Untersuchungen über die Aufnahme von $^{35}SO_4^{2-}$ durch Wurzeln von *Zea mays* L. Die Funktion der primären Endodermis. *Planta* **59**:15–28.

Weiling, F., 1962. Über Pinocytose-Mechanismen im Verlauf der Meiose bei *Lycoper-*

scium und *Cucurbita* unter Berücksichtigung ihrer Bedeutung sowie der Literatur uber Pinocytose bei Tier und Mensch: II. Kritische Besprechung und Ausdeutung der Beobachtungen. *Protoplasma* **55:**452–496.

Weinmann, H., and M. Le Roux, 1946. A critical study of the torsion balance method of measuring transpiration. *S. Afr. J. Sci.* **42:**147–153. *Biol. Abstr.* **21:**7176.

Weisz, P. B., and M. S. Fuller, 1962. "The Science of Botany." McGraw-Hill Book Company, New York.

Welbank, P. J., 1963. Toxin production during decay of *Agropyron repens* (couch grass) and other species. *Weed Res.* **3:**205–214.

Weller, D. M., 1931. Root pressure and root pressure liquids of the sugar cane plant. *Hawaii. Plant. Rec.* **35:**349–382.

Wenger, K. F., 1955. Light and mycorrhiza development. *Ecology* **36:**518–520.

Went, F. W., 1938. Specific factors other than auxin affecting growth and root formation. *Plant Physiol.* **13:**55–80.

———, 1943. Effect of the root system on tomato stem growth. *Plant Physiol.* **18:**51–65.

West, S. H., 1962. Protein, nucleotide and ribonucleic acid metabolism in corn during germination under water stress. *Plant Physiol.* **37:**565–571.

White, L. M., and W. H. Ross, 1939. Effect of various grades of fertilizers on the salt content of the soil solution. *J. Agr. Res.* **59:**81–100.

White, P. R., 1938. Root-pressure—an unappreciated force in sap movement. *Amer. J. Bot.* **25:**223–227.

———, E. Schuker, J. R. Kern, and F. H. Fuller, 1958. Root-pressure in gymnosperms. *Science* **128:**308–309.

Whiteman, P. C., and D. Koller, 1964. Saturation deficit of the mesophyll evaporating surfaces in a desert halophyte. *Science* **146:**1320–1321.

Whitfield, C. J., 1932. Ecological aspects of transpiration. II. Pikes Peak and Santa Barbara regions: edaphic and climatic aspects. *Bot. Gaz.* **94:**183–196.

Whitmore, F. W., and R. Zahner, 1967. Evidence for a direct effect of water stress on tracheid wall metabolism in pine. *Forest Sci.* **13:**397–400.

Wiebe, H. H., and P. J. Kramer, 1954. Translocation of radioactive isotopes from various regions of roots of barley seedlings. *Plant Physiol.* **29:**342–348.

——— and S. E. Wihrheim, 1962. The influence of internal moisture stress on translocation. In "Radioisotopes in Soil-plant Nutrition Studies," pp. 279–288. Int. At. Energy Agency, Vienna.

Wiegert, R. G., 1964. The ingestion of xylem sap by meadow spittlebugs, *Philaenus spumarius* L. *Amer. Midland Nat.* **71:**422–428.

Wieler, A., 1893. Dus Bluten der Pflanzen. *Beitr. Biol. Pflanz.* **6:**1–211.

Wiersum, L. K., 1948. Transfer of solutes across the young root. *Rec. Trav. Bot. Neer.* **41:**1–79.

———, 1961. Utilization of soil by the plant root system. *Plant and Soil* **15:**189–192.

———, 1962. Uptake of nitrogen and phosphorus in relation to soil structure and nutrient mobility. *Plant and Soil* **16:**62–70.

Wiggans, C. C., 1936. The effect of orchard plants on subsoil moisture. *Proc. Amer. Soc. Hort. Sci.* **33:**103–107.

———, 1937. Some further observations on the depletion of subsoil moisture by apple trees. *Proc. Amer. Soc. Hort. Sci.* **34:**160–163.

———, 1938. Some results from orchard irrigation in eastern Nebraska. *Proc. Amer. Soc. Hort. Sci.* **36:**74–76.

Wilcox, H., 1954. Primary organization of active and dormant roots of noble fir, *Abies procera*. *Amer. J. Bot.* **41:**812–821.

———, 1962. Growth studies of the root of incense cedar *Libocedrus decurrens*. II.

Morphological features of the root system and growth behavior. *Amer. J. Bot.* **49:**237–245.

Wilde, S. A., E. C. Steinbrenner, R. S. Pierce, R. C. Dozen, and D. T. Pronin, 1953. Influence of forest cover on the state of the ground water table. *Proc. Soil Sci. Soc. Amer.* **17:**65–67.

Will, G. M., 1966. Root growth and dry-matter production in a high-producing stand of *Pinus radiata. N.Z. Forest Serv. Res. Note* 44.

——— and E. L. Stone, 1967. Pumice soils as a medium for tree growth. 1. Moisture storage capacity. *N.Z. J. Forest.* **12:**189–199.

Williams, H. F., 1933. Absorption of water by the leaves of common mesophytes. *Elisha Mitchell Sci. Soc. J.* **48:**83–99.

Williams, R. F., 1955. Redistribution of mineral elements during development. *Ann. Rev. Plant Physiol.* **6:**25–40.

Williams, W. T., 1950. Studies in stomatal behaviour. IV. The water-relations of the epidermis. *J. Exp. Bot.* **1:**114–131.

——— and F. A. Amer, 1957. Transpiration from wilting leaves. *J. Exp. Bot.* **8:**1–19.

Williamson, C. E., 1950. Ethylene, a metabolic product of diseased or injured plants. *Phytopathol.* **40:**205–208.

Williamson, R. E., 1964. The effect of root aeration on plant growth. *Proc. Soil Sci. Soc. Amer.* **28:**86–90.

Wilm, H. G., and E. G. Dunford, 1948. Effect of timber cutting on water available for stream flow from a lodgepole pine forest. *U.S. Dept. Agr. Tech. Bull.* 968.

Wilson, B. F., 1967. Root growth around barriers. *Bot. Gaz.* **128:**79–82.

Wilson, C. C., 1947. The porometer method for the continuous estimation of dimensions of stomates. *Plant Physiol.* **22:**582–589.

———, 1948. Diurnal fluctuations in growth in length of tomato stems. *Plant Physiol.* **23:**156–157.

———, W. R. Boggess, and P. J. Kramer, 1953. Diurnal fluctuations in the moisture content of some herbaceous plants. *Amer. J. Bot.* **40:**97–100.

———, and P. J. Kramer, 1949. Relation between root respiration and absorption. *Plant Physiol.* **24:**55–59.

Wilson, J. D., 1929. A double-walled pot for the auto-irrigation of plants. *Bull. Torrey Bot. Club* **56:**139–153.

——— and B. E. Livingston, 1937. Lag in water absorption by plants in water culture with respect to changes in wind. *Plant Physiol.* **12:**135–150.

Wilson, J. W., 1967. The components of leaf water potential. I. Osmotic and matric potentials. *Aust. J. Biol. Sci.* **20:**329–347.

Wind, G. P., 1955*a*. A field experiment concerning capillary rise of moisture in a heavy clay soil. *Neth. J. Agr. Sci.* **3:**60–69.

———, 1955*b*. Flow of water through plant roots. *Neth. J. Agr. Sci.* **3:**259–264.

———, 1960. Capillary rise and some applications of the theory of moisture movement in unsaturated soils. *Versl. Meded.* **5:**1–15.

Wittwer, S. H., 1964. Foliar absorption of plant nutrients. *Advan. Frontiers Plant Sci.* **8:**161–182.

——— and F. G. Teubner, 1959. Foliar absorption of mineral nutrients. *Ann. Rev. Plant Physiol.* **10:**13–32.

Wolf, F. A., 1962. Aromatic or oriental tobaccos. Duke Univ. Press, Durham, N.C.

Wong, C. L., and W. R. Blevin, 1967. Infrared reflectances of plant leaves. *Aust. J. Biol. Sci.* **20:**501–508.

Woo, K. B., L. Boersma, and L. N. Stone, 1966. Dynamic simulation of the transpiration process. *Water Resources Res.* **2(1):**85–97.

Woodhams, D. H., and T. T. Kozlowski, 1954. Effects of soil moisture stress on carbohydrate development and growth in plants. *Amer. J. Bot.* **41**:316–320.

Woodroof, J. G., and N. C. Woodroof, 1934. Pecan root growth and development. *J. Agr. Res.* **49**:511–530.

Woods, F. W., 1957. Factors limiting root penetration in deep sands of the southeastern Coastal Plain. *Ecology* **38**:357–359.

———, 1960. Biological antagonisms due to phytotoxic root exudates. *Bot. Rev.* **26**:546–569.

Woolley, J. T., 1961. Mechanisms by which wind influences transpiration. *Plant Physiol.* **36**:112–114.

———, 1965. Radial exchange of labeled water in intact maize roots. *Plant Physiol.* **40**:711–717.

———, 1966. Drainage requirements of plants. Proc. conf. on drainage for efficient crop production, pp. 2–5. Amer. Soc. Agr. Eng., St. Joseph, Mich.

———, 1967. Relative permeabilities of plastic films to water and carbon dioxide. *Plant Physiol.* **42**:641–643.

Wylie, R. B., 1938. Concerning the conductive capacity of the minor veins of foliage leaves. *Amer. J. Bot.* **25**:567–572.

———, 1939. Relations between tissue organization and vein distribution in dicotyledon leaves. *Amer. J. Bot.* **26**:219–225.

———, 1943. The role of the epidermis in foliar organization and its relations to the minor venation. *Amer. J. Bot.* **30**:273–280.

———, 1952. The bundle sheath extension in leaves of dicotyledons. *Amer. J. Bot.* **39**:645–651.

Yapp, R. H., 1912. *Spiraea ulmaria* and its bearing on the problem of xeromorphy in marsh plants. *Ann. Bot.* **46**:159–181.

Yelenosky, G., 1964. Tolerance of trees to deficiencies of soil aeration. *Proc. Int. Shade Tree Conf.* **40**:127–147.

Young, K. K., and J. D. Dixon, 1966. Overestimation of water content at field capacity by use of sieved sample data. *Soil Sci.* **101**:104–107.

Yu, G. H., and P. J. Kramer, 1967. Radial salt transport in corn roots. *Plant Physiol.* **42**:985–990.

Zahner, R., 1968. Water deficits and growth of trees. In T. T. Kozlowski (ed.), "Water Deficits and Plant Growth," vol. 2, pp. 191–254. Academic Press Inc., New York.

Zak, B., 1964. Role of mycorrhizae in root disease. *Ann. Rev. Phytopathol.* **2**:377–392.

Zelitch, I., 1961. Biochemical control of stomatal opening in leaves. *Proc. Nat. Acad. Sci.* **47**:1423–1433.

———, 1963. The control and mechanisms of stomatal movement. In I. Zelitch (ed.), Stomata and water relations in plants, pp. 18–42. *Conn. Agr. Exp. Sta. Bull.* 664. New Haven, Conn.

———, 1965. Environmental and biochemical control of stomatal movement in leaves. *Biol. Rev.* **40**:463–482.

———, and P. E. Waggoner, 1962. Effect of chemical control of stomata on transpiration and photosynthesis. *Proc. Nat. Acad. Sci.* **48**:1101–1108.

Zentmyer, G. A., 1966. Soil aeration and plant disease. Proc. conf. on drainage for efficient crop production, pp. 15–16. Amer. Soc. Agr. Eng., St Joseph, Mich.

Ziegenspeck, H., 1945. Fluoroskopische Versuche an Blättern, über Leitung, Transpiration und Abscheidung von Wasser. *Biol. Gen.* (*Vienna*) **18**:254–326.

Zimmerman, M. H., 1964. Sap movement in trees. *Biorheol.* **2**:15–27.

Zimmerman, P. W., and M. H. Connard, 1934. Reversal of direction of translocation of solutes in stems. *Contrib. Boyce Thompson Inst.* **6**:297–301.

Zur, B., 1967. Osmotic control of the matric soil-water potential. II. Soil-plant system. *Soil Sci.* **103**:30–38.

Name Index

Name Index

Subject Index

Subject Index

This book was set in Helvetica by Brown Bros. Linotypers, Inc., and printed on permanent paper by Halliday Lithograph Corporation, and bound by The Maple Press Company. The designer was Marsha Cohen; the drawings were done by J. & R. Technical Services, Inc. The editors were James R. Young and Judy Reed. Peter D. Guilmette supervised the production.